Ferdinand Freiherr von Mueller

Systematic Census of Australian Plants

With chronologic, literary and geographic annotations. Vol. 1

Ferdinand Freiherr von Mueller

Systematic Census of Australian Plants

With chronologic, literary and geographic annotations. Vol. 1

ISBN/EAN: 9783337319601

Printed in Europe, USA, Canada, Australia, Japan

Cover: Foto ©berggeist007 / pixelio.de

More available books at **www.hansebooks.com**

SYSTEMATIC CENSUS

OF

AUSTRALIAN PLANTS,

WITH

𝔠𝔥𝔯𝔬𝔫𝔬𝔩𝔬𝔤𝔦𝔠, 𝔏𝔦𝔱𝔢𝔯𝔞𝔯𝔶 𝔞𝔫𝔡 𝔊𝔢𝔬𝔤𝔯𝔞𝔭𝔥𝔦𝔠 𝔄𝔫𝔫𝔬𝔱𝔞𝔱𝔦𝔬𝔫𝔰;

BY

BARON FERDINAND VON MUELLER,

K.C.M.G., M.D., Ph.D., F.R.S.,

F.L.S., F.G.S., F.R.G.S., C.M.Z.S., H.F.R.H.S. &c.

GOVERNMENT BOTANIST FOR THE COLONY OF VICTORIA.

PART I.–VASCULARES.

"LEVATE OCULOS IN EXCELSUM ET VIDETE, QUIS CREAVIT HAEC OMNIA, QUIS EDUCIT MULTITUDINEM HORUM."—*Isaiah* xl. 26.

MELBOURNE:

PRINTED FOR THE VICTORIAN GOVERNMENT

By M'CARRON, BIRD & Co., 37 FLINDERS LANE WEST.

1882.

THIS work needs hardly to be introduced by prefatory observations to the notice of professed naturalists, who will recognise its scope and object at a glance. To those however, who may be less conversant in phytographic science, and who may use these pages, some explanatory preface would be needful, in order that they may fully understand the views, which guided the author through the details of this enumerative essay on Australian indigenous plants. Again, to all recipients it may be of some interest, to be informed of the causes, which gave rise to the present literary issue. Of the seven volumes of the "Flora Australiensis," elaborated by Mr. Bentham, mainly through the aid of the Botanic Departments of Kew and Melbourne, the earlier ones were written at a time, when a large portion of Australia remained yet untraversed by any explorers. Since then the territorial occupation as a whole has become nearly doubled, while new lines of geographic research have opened up vast additional tracts of lands for future settlements. Thus, the Flora of our island-continent became also more revealed, with the result of about 850 species of vascular plants having been added already to those described in the "Flora Australiensis"; but, irrespective of this, we have learned in later years much more concerning the regional distribution of the species formerly recorded. All these additional data, accumulated gradually during the last two decennia, though mainly kept together in the volumes of the "Fragmenta Phytographiae Australiae," became necessarily much disconnected and dispersed during the progress of discoveries and the early records connected therewith, so much so, that much difficulty was experienced of late, to command a clear and easy view over these scattered literary fragments. To collect and rearrange them into one or more volumes, conformous with those of the "Flora Australiensis," and supplementary to that work, proved beyond the range of operations of the venerable savant, who spent much of his precious time during fully sixteen years for grandly systematising throughout the native vegetation of these great parts of the British Dominions,—because his unrivalled experience and scientific valour were called more pressingly into channels of research, by which not only the realm of plants of Australia, but indeed that of the whole globe, became drawn both comprehensively and connectedly into the cycle of his observations. Perhaps it may fall yet to the share of the writer of the present pages, to furnish supplements to the "Flora Australiensis"; indeed, he has it under contemplation, to add one volume, solely to be devoted to a descriptive enumeration of all the additional species, and to be arranged in consonance with Bentham's views, whereas another volume would be rendered exclusively the vehicle of record of all new localities, from whence any species of the "Flora Australiensis" were obtained, when perhaps also such alterations, as largely augmented material may suggest, could be made to the descriptive text. But to complete the "Flora" in this manner, so far as our present state of knowledge will admit of it, would involve the thorough revision of the whole vast collections now stored in our local Botanical Museum; therefore, the connected re-examination of at least 80,000 specimens, from which the pages of the "Flora" largely, and those of the "Fragmenta" solely, have emanated. Such a task, to be performed with conscientious care, would need continued and close application throughout several years. Under these circumstances it seemed best, to pave the way for subsequent purposes through the issue of a preliminary publication, by which anyhow the use of both works, above referred to, would be facilitated. A postponement of the work for supplementary volumes, thus arising, would bring with it one advantage also, that of gaining time for carrying the augmentation of the Flora still further, as we would meanwhile profit from phytologic searches of travellers and settlers in wide regions of Australia, as yet in reference to vegetation but scantily investigated or altogether unexplored.

By deferring therefore the elaboration of supplemental volumes, we could aim more at future exhaustiveness, whereas the work as left by its illustrious author will meet mainly all present wants, and must remain the foundation for all future systematic study of the vegetation peculiar to this part of the globe. In offering now a separate statistic volume, the writer has availed himself of the opportunity, to place on record independently his views on preferable systematic sequences of orders, wherever occasionally his opinions differed from those enunciated by other phytographers. He furthermore could draw the limits of some genera and many species in accordance with original observations of his own; besides he was able to apply more rigorously the rules for systematic naming in strict application of chronologic data: But to avoid misapprehensions, the author at the very outset endeavours to anticipate objections, which may perhaps be raised against some portion of the nomenclature adopted in these pages. If, for instance, Haller is regarded as the earliest indicator of Cyperaceae, it must be conceded, that his wording of the order was "Cyperi" in 1742, as well shown in Pfeiffer's extensive work. If however in accordance with the more exact limitation of this ordinal group of plants we ascribe it under the name of

Cyperaceae to De Candolle, as defined in 1805, or to St. Hilaire as circumscribed in the same year, or to A. L. De Jussieu, who already in 1789 adopted the same order under the appellation Cyperoideae, then we evidently act unjustly in not recognising prior claims; indeed, so far as wording is concerned, we lose the substance for the shadow, especially as no necessity exists, to strive for etymologic uniformity at the sacrifice of fair priority. To overcome difficulties or ambiguities of these kinds, we could use the preposition "from" (to be rendered *nach* in German and *d'après* in French) for indicating the originator of an order, genus or species, after the quotation of any later accepted authority. Indeed this principle was adopted already for a list of Australian plant-genera, published by the Royal Society of New South Wales in 1881; and had the scope assigned to the present pages admitted of it, the same rule would have been extended to the species so far as applicable.

To accomplish the object of rendering this publication, as a commentary index, one for easy reference, the schematic form appeared the most eligible. But to bring it within the means available for issue, all specific synonymy and any secondary notations had to be abandoned. Indeed for these, in quoting the "Flora" and the "Fragmenta" throughout, all needful indications are afforded. The geographic notes had to be limited to the initial letters, indicative of our main colonial territories, in which respect a plan was followed, precisely the same as that adopted in 1866 for a geographic list of nearly one thousand species of trees, then already known to constitute the very varied arboreous vegetation of Australia (see volume of the Intercolonial Exhibition of 1866, and of the second Paris "Exposition," 1867.) On that occasion also the terms Choripetaleae and Synpetaleae were already used, to distinguish the chief divisions of the Dicotyledoneae, while the sequence of orders, effected then, was also mainly the same as that followed in the present pages, the Apetaleae of Jussieu or Monochlamydeae of De Candolle having been merged into the petaliferous divisions. But the system, built up by these two great masters in botanic science, is in its main features so genuinely natural, that no subsequent research could bring about very material changes, except in the one particular direction above indicated, a design more or less happily carried out from the time of the earlier writings of Brongniart to that of the latest essays of Jean Mueller, because it was felt, that so long as the Monochlamydeae remained isolated and associated with the Gymnospermeae, so long would we have an imperfect natural system. To proceed in chief alterations further than this, would give a less acceptable system. Even among the Thalamiflorae and Calyciflorae, represented in Australia, as placed together by Bentham, we had already not less than 58 genera, which are entirely apetalous or contain species, in which the corolla remains undeveloped, leaving out of consideration numerous absolutely extra-Australian genera wholly or partly apetalous. Again in Haloragis, Myriophyllum, Cotula, Soliva and some other genera the corolla is absent in the non-antheriferous flowers, while even the very commencement of the Candollean arrangement is made with the apetalous genera Clematis, Thalictrum and Anemone, soon followed by the equally well-known Caltha. Furthermore the Euphorbiaceae, which in the train of their alliances must carry with them always the Urticeae, count among their numerous genera more than one third provided with petals. Besides, in Proteaceae the floral envelope may be regarded as homologous to that of the closely allied Loranthaceae, with an absence of a calyx, comparable to the suppression of that organ in Diplolaena, Asterolasia and few other thus far exceptional genera. Again, a firm thread runs uninterruptedly through all the orders of Curvembryonatae or Amyliferae, whether placed in Thalamiflorae or Calyciflorae or Monochlamydeae or even as regards Plumbagineae into Corolliflorae; nor is there any real difficulty of finding for the rest of the Monochlamydeae proper places among naturally allied orders of supposed higher organisation. Nevertheless affinity is variously radial, not altogether uniserial, as beautifully demonstrated already by Linné in his map of ordinal alliances of plants, published by Fabricius and Giseke, or as lucidly exhibited as long ago as the middle of the last century by Bernard de Jussieu in the Royal Garden of Trianon, through arranging the plants in a class-ground, a method adopted in our Botanic Garden here also already in 1857. In using the term Calyceae for the three main-divisions of the Monocotyledoneae, the author wishes it to be understood, that it remains an open question, whether the flowers of two of these divisons are to be regarded as mono- or di-chlamydeous. But if in the latter case a petaleous development is attributed to the flowers, we are then bound by homology to carry the term of petals also to Junceae even and Restiaceae, and a change in the naming of the three latter sections would be needful accordingly. In suppressing throughout the epigynous divisions for unison with the perigynous, into which they often pass so gradually, simplification is aimed at, without causing any disarrangement, or without attempting alterations in the great received systems except nominally. Furthermore we should always regard the apocarpic orders as those of the highest development, because the fruit—the final effort of all growth—when multiplying into separate and complete carpic elements, exhibits the most advanced vegetable organisation.

Much could it have been wished, to allot in this systematic census also columns of citation to those of the larger works, which are exclusively devoted to the Australian Flora. But Robert Brown's ever-memorable "Prodromus" records only vascular Acotyledoneae, Monocotyledoneae, some Apetaleae and a portion of the Synpetaleae, as far as known in 1810. The two important volumes, issued by Lehmann and his coadjutors as "Plantae Preissianae," refer solely to South-West Australian species. Sir Joseph Hooker's grand "Flora Tasmaniae" became necessarily limited to the plants of that island, although prefaced by an extensive general essay.

A future re-edition of this census from augmented material, in ampler paginal form, might afford space also for noting these works. The "Flora Tasmaniae" will replace however the "Flora Australiensis" in the quotation-column of the second part of this census, in which all the "Evasculares" of Australia are to be enumerated, the total of them having risen already to about 3750 recorded species. For, be it understood, that the genial and learned Director of the Royal Botanic Garden of Kew has laid a firm foundation for the system of evascular Acotyledoneae of All-Australia in his Tasmanian Flora, that work being the first, in which any extensive and connectedly elaborated account of Mosses, Lichens, Fungs and Algs of Australia has been presented.

It was beyond the scope of these pages, to extend the statistics here given; otherwise the writer would have gladly assigned geographic columns also to Europe, Asia, Africa, North and South America for indicating the respective range of many Australian species of plants into other regions of the globe. Whoever may be interested in such phytogeographic subjects is referred to an able article, written by Dr. Engler, Director of the Botanic Garden of Kiel, for the volume of 1881 of his "Jahrbücher," or to the print of a discourse, delivered a few months ago before the School of Mines at Ballarat. To sum up some of the results of this statistic volume, tables are appended, setting forth the number of the species in each order as well as the grand totals arrived at, clearly immigrated plants being excluded from these pages throughout. Future phytological explorations are not likely, to add beyond several hundred specific forms, considered in conservative limitations, to those now adduced; but extensive additions are sure to be made yet to the records of regional distribution. Indeed even while this first part of the work went through the press, so many new localities became known, as to render an appendix now already necessary. For these addenda I am largely indebted either in notations or material to the Rev. Dr. Woolls, Professor Tate, the Rev. B. Scortechini and Mr. F. M. Bailey, while the supplemental notes concerning West Australian plants emanated chiefly from gatherings, which we owe to the Hon. J. Forrest, C.M.G. It should however be observed, that the whole of the vast additional collections, which accumulated in our Botanic Museum since the volumes of the "Flora Australiensis" successively appeared, could not receive in all instances timely close and critical attention for the elaboration of the present work, though its author had the advantage of departmental help from Mr. G. Luehmann in preliminary sorting and comparing of much of the supplemental material; and here it is an apt place, to recognise likewise the aid, afforded by Mr. Léon Henry, in writing out primarily the requisite notes from the "Flora" and the "Fragmenta." Nearly one hundred species more might have been taken into account already, for estimating the number of species in various orders; but the samples hitherto secured did not admit of obtaining accurate specific data, though the ordinal position could be recognised. Thus we know, that already some access to Anonaceae, Menispermaceae, Lauraceae, Euphorbiaceae, Rutaceae and a few other orders could be recorded.

The chronologic references for species, here more extensively carried out, than ever before, amount to nearly 10,000, including those for genera, in this first part of the present publication. The bibliographic works of Pritzel, Pfeiffer and Jackson have done excellent services in these particular inquiries. The position, given to several of the orders on this occasion, is to some extent assailable; thus—as well-known—reasons exist for moving Piperaceae to Nymphaeaceae, Aristolochieae and Nepenthaceae to Sarraceniaceae, Viniferae to Araliaceae, Thymeleae to Rhamnaceae, Balanophoreae and Podostemoneae to Halorageae, Droseraceae to Saxifrageae, Rubiaceae to Loganiaceae, Fluviales to Hydrocharideae, Casuarineae to Coniferae, although the two latter are neither anatomically nor morphologically approaching each other, notwithstanding some deceptive external resemblance, in which however Exocarpus also shares, not to speak of a few other genera. But in considerations like these we should not insist in a dogmatic spirit on the full acceptance of hitherto recognised systematic arrangements, particularly as not yet all the forms of the world's vegetation are known, and as at any moment the discovery of a new plant may turn the scale in weighing the affinity of its allies.

In incurring the responsibility of restoring some generic names, which had sunk almost into oblivion, the writer has been guided by the impartial rules of strict priority, feeling assured, that phytographers will become quite as quickly accustomed to changes from Stylidium to Candollea, or Ionidium to Hybanthus, or Chenolea to Bassia, as they become reconciled to the long forgotten appellations of Trema, Hovea, Centipeda, Galeola, Floriscopa, Nunnezharia and others, thereby discarding names, sanctioned by lengthened use and familiar to all of us, with a readiness as surprising as universal. To alterations or restitutions of specific names shall only be alluded here with a few words. Why Lamarck for an Eleusine selected designedly the appellation E. cruciata, or why Schleiden called a Wolffia purposely W. Michelii, can be as easily perceived as the advisability of withdrawing the specific name antarctica from a Vitis, Cymodocea or Dicksonia. Where authors bestowed several names at the same time on a species, it should be free to those, who effect the reductions, to choose a collective designation for the consolidated species; cases in point are afforded by Euphrasia Brownii, Fimbristylis communis, Pappophorum commune, Danthonia penicillata.

As to the circumscription of species themselves, it will require yet much of assiduous research of future generations for arriving at a full understanding of the diagnostic value, which has to be attached respectively to the thousands of specific and therefore original forms in the Australian and any other vegetation. Of the plants, enumerated as species in the present pages, a considerable number may prove, that they owe their particularity to hybridism. In such cases they should occupy a subordinate position, and should also here, as elsewhere, receive their double parental name;— thus Lasiopetalum Tepperi would require to have its appellation changed into Lasiopetalum Baueri × discolor; but as the instances of ascertained hybridism among our native plants are as yet so very few, it was deemed best to admit at present the very limited number of known bastards under ordinary specific rank.

The geographical limitations in this work coincide with the political boundaries of the colonial territories, except that the tropic of Capricorn eastward to the 138th degree separates what is here called Northern Australia (N.A.), from the South- and West-Australian extratropic possessions. Such geographic segregations are necessarily quite arbitrary, though they serve our present purpose of assigning to each of the colonial divisions of Australia its number of specified plants; the limitation is the same as that adopted in the "Flora Australiensis," and as regards abbreviations also identical with the method of indications, chosen for the list of Australian trees in 1866. Here may it further be remarked, that the plants of the small and isolated insular spot, called Norfolk-Island, have been counted with those of N.S.W., to whose dominions this isle as well as Lord Howe's Island politically belongs. To draw the species into physiographic and regional complexes must be the work of future periods, when climatic and geologic circumstances throughout Australia shall have become more extensively known. The geographic columns in these pages indicate simply the occurrence of plants within any of the colonial areas, but have been extended even to such species, which merely may pass boundary-lines. In this way are noted for W.A. and S.A. such tropical plants, as reach not beyond Shark-Bay or the vicinity of the MacDonnell-Range; in the same manner Victoria is credited with those plants of New South Wales, which barely advance into East-Gippsland, while South-Australia obtains credit for several Tasmanian plants, which are confined to very limited stretches of country, chiefly in the extreme south-east of that province; and again a few of the New England plants are recorded for Queensland, although they pass merely on to the nearest adjacent ranges across the boundary. At the whole however it can be foreseen, that many a species, still accepted as genuine for these lists, will in the course of further and more facilitated research be eliminated from any future editions of this work, though still larger arrays, containing many a novelty, will have to be mustered for systematic rolls. The lines of demarkation between truly indigenous and more recently immigrated plants can no longer in all cases be drawn with precision; but whereas Alchemilla vulgaris and Veronica serpillifolia were found along with several European Carices in untrodden parts of the Australian Alps during the author's earliest explorations, Alchemilla arvensis and Veronica peregrina were at first only noticed near settlements. The occurrence of Arabis glabra, Geum urbanum, Agrimonia Eupatoria, Eupatorium cannabinum, Carpesium cernuum and some others may therefore readily be disputed as indigenous, and some questions concerning the nativity of various of our plants will probably remain for ever involved in doubts. While concluding these introductory remarks, it is incumbent on the writer, to acknowledge in grateful terms the consideration, shown him by the Hon. James Macpherson Grant, M.L.A., the Ministerial Chief of the Department, in allowing this work to be issued under the auspices of the Victorian Government and at departmental expenditure. Encouraged by generous support, the author will cheerfully continue through such time, as may still be allotted for his worldly career, to devote his strength and resources also in future mainly for the furtherance of Australian phytography, on which much of his efforts were concentrated through more than a third of a century; and as these enquiries are largely carried on without worldly gain, he feels free, to claim the co-operation of any educated and high-minded colonists in these great dominions of the British Crown for promoting studies, which cannot fail to be of great industrial advantage at all future time, which are calculated to advance continuously the cause of education in this part of the world also, which must far and wide contribute to intellectual enjoyments, and above all should lead to religious reverence of that Supreme Godly power, so gloriously revealed in nature's wondrous works.

DICOTYLEDONEAE.

Ray, Method. Plant. nova 2 (1682).

CHORIPETALEAE HYPOGYNAE.

F. v. Mueller, native plants of Victoria I, 1 (1879).

RANUNCULACEAE.

A. L. de Jussieu, Rec. de l'Acad. des sc. (1773) from B. de Jussien (1759).

CLEMATIS, Linné, gen. pl. 163 (1737) from l'Ecluse (1576).
C. aristata, R. Brown in De Candolle, syst. veg. I, 147 (1818) ... — S.A. T. V. N.S.W. Q. — B.fl.I,6 M.fr.X,2.
C. pubescens, Huegel, enum. pl. Nov. Holl. austr. occ. I (1837) W.A. — — — — — — B.fl.I,6
C. glycinoides, De Candolle, syst. veg. I, 145 (1818) — — — — N.S.W. Q. N.A. B.fl.I,7 M.fr.X,2.
C. microphylla, De Candolle, syst. veg. I, 147 (1818) ... W.A. S.A. T. V. N.S.W. Q. — B.fl.I,7 M.fr.X,2; XI,20,27.
C. Fawcettii, F. v. Mueller, fragm. phytogr. Austr. X, 1 (1876) — — — — N.S.W. — — M.fr.X,1.

ANEMONE, Tournefort, inst. rei herb. 275, t. 147 (1700) from Hippocrates, Theophrastos and Dioscorides.
A. crassifolia, Hooker, icones pl. t. 257 (1840) ... — — — T. — — — B.fl.I,8 M.fr.X,1.

MYOSURUS, Dillenius, catal. pl. Giess. 100, t. 4 (1719) from Tabernaemontanus (1588). (Myosuros).
M. minimus, Linné, sp. pl. 284 (1753) ... ... — S.A. — V. N.S.W. — B.fl.I,8 M.fr.X,2; XI,27.

RANUNCULUS, Tournefort, inst. 285, t. 140 (1700) from Bock (1552).
R. aquatilis, Dodonæus stirp. hist. pempt. 387 (1583) ... ... — S.A. T. V. — — — B.fl.I,10
R. Millani, F. v. Mueller in Hook. Kew misc. VII, 358 (1855)... — — — V. N.S.W. — B.fl.I,10 M.fr.X,2.
R. anemoneus, F. v. Mueller in Trans. Phil. Soc. Vict. I, 97 (1855) — — — V. N.S.W. — B.fl.I,11
R. Gunnianus, Hooker, Journ. of Bot. I, 245, t. 233 (1834) ... — — T. V. N.S.W. — B.fl.I,11
R. lappaceus, Smith in Rees's Cyclopædia, XXIX (1815) ... W.A. S.A. T. V. N.S.W. Q. — B.fl.I,12 M.fr.X,2.
R. Muelleri, Bentham, Fl. Austr. I, 13 (1863) ... ... — — — V. N.S.W. — B.fl.I,13
R. rivularis, Bks. & Sol. in G. Forst. fl. ins. austr. prodr. 90 (1786) — — — S.A. T. V. N.S.W. Q. — B.fl.I,13 M.fr.X,2.
R. hirtus, Bks. & Sol. in G. Forst. fl. ins. austr. prodr. 90 (1786) W.A. — — T. V. N.S.W. Q. — B.fl.I,13 M.fr.X,2.
R. parviflorus, Linné, sp. pl. sec. ed. 780 (1762) ... ... — S.A. T. V. N.S.W. Q. — B.fl.I,14 M.fr.X,2; XI,27.

CALTHA, Ruppius, flora Jenensis 119 (1718) from Bock (1552). (Trollius).
C. introloba, F. v. Mueller in Trans. Phil. Soc. Vict. I, 98 (1855) — — — T. V. N.S.W. — — B.fl.I,15 M.fr.X,2; XI,136.

NYMPHAEACEAE.

Salisbury in Koenig & Sims's Ann. of Bot. II. 70 (1805).

CABOMBA, Aublet, Hist. pl. Guian. I, 321, t. 124 (1775). (Brasenia, Hydropeltis).
C. peltata, F. v. Mueller, native pl. of Vict. 15 (1874) ... ... — — — V. N.S.W. Q. — B.fl.I,60 M.fr.X,77.

NYMPHAEA, Tournefort, inst. 260, t. 137 8 (1700) from Theophrastos and Dioscorides.
N. stellata, Willdenow, spec. pl. II, 1153 (1799) ... ... — — — — — N.A. — M.fr.II,142; X,77.
N. gigantea, Hooker, Bot. Mag. t. 4647 (1852) ... ... — — — — — N.S.W. Q. N.A. B.fl.I,61 M.fr.I,141; X,78.

NELUMBIO, Tournefort, inst. 261 (1700). (Nelumbium).
N. nucifera, Gaertner, de fruct. I, 73 (1788) ... ... — — — — — Q. N.A. B.fl.I,62 M.fr.X,77.

DILLENIACEAE.

Salisbury, Parad. Lond. I, 73 (1801).

WORMIA, Rottboell, Nye Saml. Vid. Selsk. Skrivt. II, 532 (1783).
W. alata, R. Brown in De Cand. syst. veg. I, 434 (1818) ... — — — — Q. — B.fl.I,16 M.fr.VIII,124.

HIBBERTIA, Andrews, Bot. Reposit. t. 126 (1800). (Candollea, Pleurandra, Hemistemma, Adrastaea, Ochrolasia, Hemistephus, Huttia.)
H. Bankeii, F. v. Mueller, pl. of Vict. I, 14, implied (1860) ... — — — — Q. — B.fl.I,21 M.fr.I,1; XI,92.
H. Brownii, Bentham, Fl. Austr. I, 21 (1863) ... ... — — — — Q. — B.fl.I,21 M.fr.II,1.
H. dealbata, F. v. Mueller, pl. of Vict. I, 14, implied (1860) ... — — — — N.A. B.fl.I,21 M.fr.II,1.
H. candicans, Bentham, Fl. Austr. I, 21 (1863) ... ... — — — — Q. N.A. B.fl.I,21
H. Benthami, H. angustifolia, Benth. fl. Austr. I, 21 (1863) ... — — — — N.A. B.fl.I,21 M.fr.II,1.
H. Muelleri, Bentham, Fl. Austr. I, 21 (1863) ... ... — — — — N.A. B.fl.I,21
H. ledifolia, Bentham, Fl. Austr. I, 22 (1863) ... ... — — — — N.A. B.fl.I,22
H. verrucosa, F. v. Mueller, pl. of Vict. I, 14, implied (1860) ... — — — — — B.fl.I,22 M.fr.X,02.
H. spicata, F. v. Mueller, fragm. II, 1 (1860) ... ... W.A. — — — B.fl.I,22 M.fr.II,1; XI,20,92.
H. polystachya, Bentham, Fl. Austr. I, 22 (1863)... ... W.A. — — — B.fl.I,22 M.fr.XI,92.
H. rhadinopoda, F. v. Mueller, fragm. XI, 91 (1880) W.A. — — — — M.fr.XI,91.
H. furfuracea, Bentham, Fl. Austr. I, 23 (1863) ... ... W.A. — — — B.fl.I,23 M.fr.I.161; XI,92.
H. hypericoides, F. v. Mueller, pl. of Vict. I, 13, implied (1860) W.A. — — — B.fl.I,23 M.fr.X,02.
H. microphylla, Steudel in Lehmann pl. Preiss. I, 273 (1844) ... W.A. — — — B.fl.I,23 M.fr.I,217.
H. recurvifolia, Bentham, Fl. Austr. I, 24 (1863) ... W.A. — — — B.fl.I,24
H. lineata, Steudel in Lehmann pl. Preiss. I, 272 (1844)... W.A. — — — B.fl.I,24 M.fr.X,02.
H. acerosa, Bentham, Fl. Austr. I, 24 (1863) ... ... W.A. — — — B.fl.I,24 M.fr.XI,92.
H. aurea, Steudel in Lehmann pl. Preiss. I, 272 (1844) ... W.A. — — — B.fl.I,25 M.fr.XI,02.
H. crassifolia, Bentham, Fl. Austr. I, 25 (1863) ... W.A. — — — B.fl.I,25
H. nitida, F. v. Mueller, pl. of Vict. I, 15 (1860)... ... — — — N.S.W. — B.fl.I,25
H. bracteata, Bentham, Fl. Austr. I, 26 (1863) ... ... — — — N.S.W. — B.fl.I,25
H. densiflora, F. v. Mueller, pl. of Vict. I, 15 (1860) ... S.A. T. V. N.S.W. — B.fl.I,26 M.fr.VII,125; XI,02.
H. hirsuta, Bentham, Fl. Austr. 26 (1863) ... ... — — T. — — B.fl.I,26
H. riparia, F. v. Mueller, fragm. IV, 151 (1864) ... — — — — Q. — M.fr.IV,151,
H. stricta, R. Brown in De Candolle, syst. veg. I, 422 (1818) ... W.A. S.A. T. V. N.S.W. Q. — B.fl.I,27
H. humifusa, F. v. Mueller, pl. of Vict. I, 16, t. Suppl. I (1800) — — — V. — — B.fl.I,27

B

	W.A.	S.A.	T.	V.	N.S.W.	Q.	N.A.		
H. Billardieri, F. v. Mueller, pl. of Vict. I, 14 (1800)	—	S.A.	T.	V.	N.S.W.	Q.	—	B.fl.I,28	M.fr.IV,116.
H. gracilipes, Bentham, Fl. Austr. I, 28 (1863)	W.A.	—	—	—	—	—	—	B.fl.I,28	M.fr.XI,92.
H. acicularis, F. v. Mueller, pl. of Vict. I, 17 (1860) ...	—	S.A.	T.	V.	N.S.W.	Q.	—	B.fl.I,29	M.fr.XI,93.
H. mucronata, F. v. Mueller, pl. of Vict. I, 17 (1860)	W.A.	—	—	—	—	—	—	B.fl.I,29	M.fr.XI,92.
H. hermannifolia, De Candolle, syst. veg. I, 431 (1818) ...	—	—	—	—	N.S.W.	—	—	B.fl.I,30	
H. velutina, R. Brown in Bentham, Fl. Austr. I, 30 (1863)	—	—	—	—	—	Q.	—	B.fl.I.30	
H. oblongata, R. Brown in De Candolle, syst. veg. I, 431 (1818)	—	—	—	—	—	—	N.A.	B.fl.I,30	M.fr.XI,94.
H. melanoides, F. v. Mueller, fragm. IV, 116 (1864) ...	—	—	—	—	—	Q.	—	—	M.fr.IV,110.
H. tomentosa, R. Brown in De Candolle, syst. veg. I, 432 (1818)	—	—	—	—	—	—	N.A.	B.fl.I,30	
H. cistifolia, R. Brown in De Candolle, syst. veg. I, 431 (1818)	—	—	—	—	—	—	N.A.	B.fl.I,30	
H. echiifolia, R. Brown in Bentham, Fl. Austr. I. 31 (1863) ...	—	—	—	—	—	—	N.A.	B.fl.I,31	
H. scabra, R. Brown in Bentham, Fl. Austr. I, 31 (1863) ...	—	—	—	—	—	—	N.A.	B.fl.I,31	
H. lepidota, R. Brown in De Candolle, syst. veg. I, 432 (1818)...	—	—	—	—	N.S.W.	Q.	—	B.fl.I,31	
H. vestita, Cunningham in Bentham, Fl. Austr. I, 31 (1863) ...	—	—	—	—	N.S.W.	—	—	B.fl.I,31	
H. serpyllifolia, R. Brown in De Cand. syst. veg. I, 430 (1818)	—	—	T.	V.	N.S.W.	—	—	D.fl.I,32	
H. pedunculata, R. Brown in De Cand. syst. veg. I, 430 (1818)	—	—	—	V.	N.S.W.	—	—	B.fl.I,32	
H. ochrolasia, Bentham, Fl. Austr. I, 32 (1863)	W.A.	—	—	—	—	—	—	B.fl.I,32	M.fr.VII,125;XI,03.
H. angustifolia, Salisbury, parad. Lond. t. 73 (1807) ...	—	—	T.	V.	—	—	—	B.fl.I,33	
H. fasciculata, R. Brown in De Candolle, syst. veg. I, 428 (1818)	—	S.A.	T.	V.	N.S.W.	—	—	B.fl.I,33	
H. virgata, R. Brown in De Candolle, syst. veg. I, 428 (1818) ...	—	S.A.	T.	V.	N.S.W.	—	—	B.fl.I,34	M.fr.XI,04.
H. inclusa, Bentham, Fl. Austr. I, 34 (1863)	W.A.	—	—	—	—	—	—	B.fl.I,34	
H. rostellata, Turcz. in Bull. Soc. Mosc. XXII, part III, 8 (1849)	W.A.	—	—	—	—	—	—	B.fl.I,34	M.fr.XI,93.
H. glomerata, Bentham, Fl. Austr. I, 34 (1863)	W.A.	—	—	—	—	—	—	B.fl.I,34	
H. argentea, Steudel in Lehmann, pl. Preiss. I, 268 (1844) ...	W.A.	—	—	—	—	—	—	B.fl.I,35	
H. pilosa, Steudel in Lehmann, pl. Preiss. I, 272 (1844)... ...	W.A.	—	—	—	—	—	—	B.fl.I,35	M.fr.XI,04.
H. montana, Steudel in Lehmann, pl. Preiss. I, 270 (1844) ...	W.A.	—	—	—	—	—	—	B.fl.I,34	
H. linearis, R. Brown in De Candolle, syst. veg. I, 428 (1818) ...	—	—	—	V.	N.S.W.	Q.	—	B.fl.I,36	
H. diffusa, R. Brown in De Candolle, syst. veg. I, 429 (1818) ...	—	—	—	V.	N.S.W.	—	—	B.fl.I,36	
H. saligna, R. Brown in De Candolle, syst. veg. I, 427 (1818) ...	—	—	—	—	N.S.W.	—	—	B.fl.I,37	M.fr.XI,94.
H. volubilis, Andrews in Bot. Reposit. t. 126 (1600) ...	—	—	—	—	N.S.W.	Q.	—	B.fl.I,37	M.fr.IV,110;VII,125; XI,04.
H. grossularifolia, Salisbury, parad. Lond. t. 73 (1806) ...	W.A.	—	—	—	—	—	—	B.fl.I,37	M.fr.XI,94.
H. dentata, R. Brown in De Candolle, syst. veg. I, 426 (1818) ...	—	—	—	V.	N.S.W.	Q.	—	B.fl.I,38	M.fr.IV,125;XI,94.
H. Cunninghamii, Aiton in Hook. Bot. Mag. t. 318 (1832) ...	W.A.	—	—	—	—	—	—	B.fl.I,39	M.fr.XI,93.
H. perfoliata, Huegel, enum. pl. Nov. Holl. austr. occ. 3 (1837)	W.A.	—	—	—	—	—	—	B.fl.I,38	M.fr.XI,93.
H. bracteosa, Turczaninow, in Bull. Soc. Mosc. XXV, 140 (1852)	W.A.	—	—	—	—	—	—	B.fl.I,38	M.fr.XI,94.
H. longifolia, F. v. Mueller, fragm. IV, 115 (1864)	—	—	—	—	—	Q.	—	—	M.fr.VII,37.
H. oenotheroides, F. v. Mueller, fragm. VII, 37 (1870) ...	—	—	—	—	—	Q.	—	—	M.fr.VII,37.
H. glaberrima, F. v. Mueller, fragm. III, 1 (1862) ...	—	S.A.	—	—	—	—	—	B.fl.I,39	M.fr.III,11;XI,94.
H. Mylnoi, Bentham, Fl. Austr. I, 39 (1863)	W.A.	—	—	—	—	—	—	B.fl.I,39	
H. lasiopus, Bentham, Fl. Austr. I, 40 (1863)	W.A.	—	—	—	—	—	—	B.fl.I,40	
H. potentilliflora, F. v. Mueller in Bentham, Fl. Austr. I, 40 (1863)	W.A.	—	—	—	—	—	—	B.fl.I,40	M.fr.XI,94.
H. pungens, Bentham, Fl. Austr. I, 40 (1863)	W.A.	—	—	—	—	—	—	B.fl.I,40	
H. nutans, Bentham, Fl. Austr. I, 40 (1863)	W.A.	—	—	—	—	—	—	B.fl.I,41	
H. leptopus, Bentham, Fl. Austr. I, 41 (1863)	W.A.	—	—	—	—	—	—	B.fl.I,41	
H. stellaris, Endlicher in Hueg. enum. 3 (1837)	W.A.	—	—	—	—	—	—	B.fl.I,41	M.fr.XI,93.
H. obcuneata, Salisbury, parad. Lond. t. 73 (1807) ...	W.A.	—	—	—	—	—	—	B.fl.I,42	M.fr.XI,94.
H. glomerosa,F.v.M.;Candollea glomerosa,Benth.Fl.Aus.I,43(1863)	W.A.	—	—	—	—	—	—	B.fl.I,43	
H. teretifolia, F. v. Mueller, fragm. IV, 117 (1864)	W.A.	—	—	—	—	—	—	B.fl.I,43	M.fr.IV,117;XI,95.
H. desmophylla, F. v. Mueller, fragm. XI, 95 (1880) ...	W.A.	—	—	—	—	—	—	B.fl.I,43	M.fr.XI,95.
H. helianthemoides, F. v. M.; Candollea helianthemoides, Turcz. in Bull. Mosc. XXII, part III, 8 (1849)	W.A.	—	—	—	—	—	—	B.fl.I,43	
H. depressa, Steudel in Lehm. pl. Preiss. I, 268 (1844) ...	W.A.	—	—	—	—	—	—	B.fl.I,44	M.fr.XI,95.
H. Huegelii, F. v. Mueller, fragm. XI, 95 (1880) ...	W.A.	—	—	—	—	—	—	B.fl.I,44	M.fr.XI,95.
H. subvaginata, F. v. Mueller, fragm. XI, 95 (1880) ...	W.A.	—	—	—	—	—	—	B.fl.I,45	M.fr.XI,95.
H. vaginata, F. v. Mueller, fragm. XI, 96 (1880) ...	W.A.	—	—	—	—	—	—	B.fl.I,45	M.fr.XI,95.
H. subexcisa, Steudel in Lehmann, pl. Preiss. I, 269 (1844) ...	W.A.	—	—	—	—	—	—	B.fl.I,46	M.fr.XI,95.
H. aquamosa, Turczaninow in Bull. Soc. Mosc. XXII, part III, 9 (1840)	W.A.	—	—	—	—	—	—	B.fl.I,46	
H. uncinata, F. v. M.; Candollea uncinata, Benth. Fl. Aus. I, 46 (1863)	W.A.	—	—	—	—	—	—	B.fl.I,46	
H. salicifolia, F.v. Mueller, fragm. V, 191 (1866) ...	—	—	—	—	N.S.W.	Q.	—	B.fl.I,46	M.fr.I,161;VII,124;XI,94.
H. Huttii, F. v. Mueller, fragm. VII, 123 (1870)...	W.A.	—	—	—	—	—	—	B.fl.I,47	M.fr.VII,123;XI,20.
H. Goyderi, F. v. Mueller, fragm. XI, 123 (1870) ...	—	—	—	—	—	—	N.A.	—	M.fr.VII,123.

TETRACERA, Linné, gen. pl. 345 (1737).

						Q.			
T. Nordtiana, F. v. Mueller, fragm. V, 1 (1865) ...	—	—	—	—	—	Q.	—	—	M.fr.V,1.
T. Daemeliana, F. v. Mueller, fragm. V, 191 (1866) ...	—	—	—	—	—	Q.	—	—	M.fr.V,101.
T. Wuthiana, F. v. Mueller, fragm. X, 49 (1876) ...	—	—	—	—	—	Q.	—	—	M.fr.X,49.

DILLENIA, Linné, gen. pl. 162 (1737).

						Q.			
D. Andreana, F. v. Mueller, fragm. V, 175 (1866) ...	—	—	—	—	—	Q.	—	—	M.fr.V,175.

PACHYNEMA, R. Brown in De Candolle, syst. veg. I, 412 (1818).

							N.A.		
P. junceum, Bentham, Fl. Austr. I, 47 (1803)	—	—	—	—	—	—	N.A.	B.fl.I,47	M.fr.X,124.
P. complanatum, R. Brown in De Cand. syst. veg. I, 412 (1818)	—	—	—	—	—	—	N.A.	B.fl.I,48	M.fr.X,124.
P. dilatatum, Bentham, Fl. Austr. I, 48 (1863) ...	—	—	—	—	—	—	N.A.	B.fl.I,48	M.fr.X,124.

MAGNOLIACEAE.

J. de St. Hilaire, Expos. fam. nat. II, 74, t. 83-84 (1805).

DRIMYS, R. & G. Forster, char. gen. 83, t. 42 (1776). (Tasmania).

D. aromatica, F. v. Mueller, pl. of Vict. I, 20 (1860) ...	—	—	T.	V.	N.S.W.	—	—	B.fl.I,49	M.fr.VII,18.
D. membranea, F. v. Mueller, fragm. V, 175 (1866) ...	—	—	—	—	—	Q.	—	—	M.fr.V,175;VII 18.
D. dipetala, F. v. Mueller, pl. of Vict. I, 21 (1860) ...	—	—	—	—	N.S.W.	Q.	—	B.fl.I,49	M.fr.VII,18.
D. Howeana, F. v. Mueller, fragm. VII, 17 (1870) ...	—	—	—	—	N.S.W.	—	—	—	M.fr.VII,17.

ANONACEAE.

A. L. de Jussieu, Gen. 283 (1789) from B. de Jussieu (1759).

UVARIA, Linné, Amoen. acad. 404 (1747).
U. membranacea, Bentham, Fl. Austr. I, 51 (1863) ... — — — — — Q. — B.fl.I,51
U. Goezeana, F. v. Mueller, fragm. VII, 125 (1871) ... — — — — — Q. — — M.fr.VII,125.

FITZALANIA, F. v. Mueller, fragm. IV, 33 (1863).
F. heteropetala, F. v. Mueller, fragm. IV, 33 (1863) ... ... — — — — — Q. — B.fl.I,51 M.fr.III,1;VII,126.

CANANGA, Rumphius, herb. Amboin. II, 195, t. 65 (1741).
C. odorata, J. Hooker and Thomson, fl. Indic. 130 (1855) ... — — — — — Q. — — M.fr.VI,1.

ANCANA, F. v. Mueller, fragm. V, 27 (1865).
A. stenopetala, F. v. Mueller, fragm. V, 27 (1865) — — — — N.S.W. Q. — — M.fr.V,27,189;VII,121.

POLYALTHIA, Blume, Fl. Jav. Anon. 68 (1829).
P. nitidissima, Bentham, Fl. Austr. I, 51 (1863) ... ... — — — — N.S.W. Q. — B.fl.I,51 M.fr.III,2;VII,126.

POPOWIA, Endlicher, Gen. 831 (1839).
P. australis, Bentham, Fl. Austr. I, 52 (1863) ... ... — — — — — — N.A. B.fl.I,52

MELODORUM, Loureiro, Fl. Cochinch. I, 351 (1790).
M. Leichhardtii, Bentham, Fl. Austr. I, 52 (1863) ... — — — — N.S.W. Q. — B.fl.I,52 M.fr.III,41,VI,2;VII,126.
M. Uhrii, F. v. Mueller, fragm. VI, 2 (1867) ... ... — — — — — Q. — M.fr.VI,2.
M. Maccreai, F. v. Mueller, fragm. VI, 176 (1868) ... — — — — — Q. — M.fr.VI,176,250.

MILIUSIA, Leschenault in Mém. Genèv. V, 213 (1832).
Species undetermined ... ... — — — — — Q. — —

SACCOPETALUM, Bennett in Horsf. pl. Jav. rar. 115, t. 35 (1838).
S. Didwilli, Bentham, Fl. Austr. I, 53 (1863) ... — — — — — Q. — B.fl.I,53
S. Brahei, F. v. Mueller, fragm. VIII, 159 (1874) ... — — — — — Q. — — M.fr.VIII,159.

EUPOMATIA, R. Brown in App. Flind. voy. II, 597, t. 2 (1814).
E. Bennettii, F. v. Mueller, fragm. I, 45 (1859) ... ... — — — N.S.W. Q. — B.fl.I,54 M.fr.I,45;VI,177.
E. laurina, R. Brown in App. Flind. voy. II, 597, t. 2 (1814) ... — — V. N.S.W. Q. — B.fl.I,54 M.fr.I,45;VI,2;VII,126.

MONIMIEAE.

A. L. de Jussieu in Ann. du Mus. XIV, 30 (1809).

DORYPHORA, Endlicher, Gen. 3015 (1837) (Learoea).
D. sassafras, Endlicher, iconogr. gen. pl., t. 10 (1838) ... — — — N.S.W. — — D.fl.V,283

ATHEROSPERMA, Labillardière, Nov. Holl. pl. spec. II, 74, t. 224 (1806).
A. moschatum, Labillardière, Nov. Holl. pl. spec. II, 74, t. 224(1806) — — T. V. N.S.W. — — B.fl.V,284 M.fr.VIII,142;X,106.

DAPHNANDRA, Bentham, Fl. Austr. V, 285 (1870).
D. micrantha, Bentham, Fl. Austr. V, 285 (1870)... ... ... — — — N.S.W. Q. — B.fl.V,285
D. repandula, F. v. Mueller, fragm. X, 105 (1877) ... ... — — — — Q. — M.fr.X,105.

MOLLINEDIA, Ruiz & Pavon, prodr. 83, t. 15 (1794). (Kibara, Wilkiea).
M. Huegeliana, Tulasne in Ann. sc. nat. sér. quatr. III, 45 (1855) — — — N.S.W. Q. — D.fl.V,286
M. Wardellii, F. v. Mueller, fragm. V, 155 (1866) ... — — — — — Q. — B.fl.V,287 M.fr.V,155;VI,252.
M. loxocarya, Bentham, Fl. Austr. V, 287 (1870)... ... — — — — — Q. — B.fl.V,287
M. acuminata, F. v. Mueller, fragm. V, 155 (1866) ... — — — — — Q. — B.fl.V,287 M.fr.V,155.
M. macrophylla, Tulasne in Ann. sc. nat. sér. quatr. III, 45 (1855) — — — N.S.W. Q. — B.fl.V,288
M. longipes, Bentham, Fl. Austr. V, 289 (1870) ... ... — — — — — Q. — B.fl.V,289
M. laxiflora, Bentham, Fl. Austr. V, 289 (1870) ... ... — — — — — Q. — B.fl.V,289
M. pubescens, Bentham, Fl. Austr. V, 290 (1870) ... ... — — — N.S.W. Q. — B.fl.V,290

HEDYCARYA, R. & G. Forster, char. gen. 127, t. 64 (1776).
H. Cunninghami, Tulasne in Archiv. du Mus. hist. nat. VIII, 408 (1855) — — — V. N.S.W. Q. — B.fl.V,291

PALMERIA, F. v. Mueller, fragm. IV, 151 (1864).
P. scandens, F. v. Mueller, fragm. IV, 152 (1864) — — — — N.S.W. Q. — B.fl.V,291 M.fr.IV,152.

PIPTOCALYX, Oliver in Bentham's Fl. Austr. V, 292 (1870).
P. Moorei, Oliver in Bentham's Fl. Austr. V, 292 (1870)... — — — N.S.W. — — B.fl.V,292 M.fr.X,106.

MYRISTICEAE.

R. Brown, prodr. 399 (1810) from A. L. de Jussieu (1800).

MYRISTICA, Linné, gen. pl. ed. II, 524 (1742).
M. insipida, R. Brown, prodr. 400 (1810) ... ... — Q. N.A. B.fl.V,281

LAURACEAE.

Ventenat, Tabl. I, 245 (1799).

CRYPTOCARYA, R. Brown, prodr. 402 (1810). (Caryodaphne).
C. Murrayi, F. v. Mueller, fragm. V, 170 (1866) ... — — — — — Q. — B.fl.V,295 M.fr.V,170;X,253.
C. Mackinnoniana, F. v. Mueller, fragm. V, 109 (1866) ... — — — — — Q. — B.fl.V,290 M.fr.V,169.
C. patentinervis, F. v. Mueller in De Candolle, prodr. XV, 508(1864) — — — N.S.W. Q. — B.fl.V,296 M.fr.V,166.
C. obovata, R. Brown, prodr. 402 (1810) ... ... — — — N.S.W. Q. — B.fl.V,296 M.fr.V,170.
C. glaucescens, R. Brown, prodr. 402 (1810) ... ... — — — N.S.W. Q. N.A. B.fl.V,297 M.fr.V,165,166.
C. triplinervis, R. Brown, prodr. 402 (1810) ... ... — — — N.S.W. Q. — D.fl.V,297 M.fr.V,166.
C. cinnamomifolia, Bentham, Fl. Austr. V, 298 (1870) ... ... — — — — — Q. — B.fl.V,298
C. Melssneri, F. v. Mueller, fragm. V, 170 (1866)... ... — — — N.S.W. Q. — B.fl.V,298 M.fr.V,170.
C. australis, Bentham, Fl. Austr. V, 299 (1870) ... ... — — — N.S.W. Q. — B.fl.V,299 M.fr.V,167.

BEILSCHMIEDIA, Nees in Wallich, pl. Asiat. rarior II, 69 (1831). (Nesodaphne).
B. obtusifolia, Bentham, in B. & H. gen. III, 1521 (1880) ... — — — N.S.W. Q. — B.fl.V,299 M.fr.V,166.

4

ENDIANDRA, R. Brown, prodr. 402 (1810)
E. glauca, R. Brown, prodr. 402 (1810) ... ... — — — — — Q. — B.fl.V,300
E. hypotephra, F. v. Mueller, fragm. V, 166 (1866) ... ... — — — — — Q. — B.fl.V,301 M.fr.V,166.
E. discolor, Bentham, Fl. Austr. V, 301 (1870) ... ... — — — — N.S.W. Q. — B.fl.V,301
E. Sieberi, Nees, Syst. Laurin. 194 (1836) ... ... — — — — N.S.W. — — B.fl.V,301 M.fr.V,166.
E. virens, F. v. Mueller, fragm. II, 90 (1860) ... ... — — — — N.S.W. Q. — B.fl.V,302 M.fr.V,166.
E. Muelleri, Meissner in De Candolle, prodr. XV, 509 (1864) ... — — — — N.S.W. Q. — B.fl.V,302
E. pubens, Meissner in De Candolle, prodr. XV, 509 (1864) ... — — — — N.S.W. Q. — B.fl.V,302

CINNAMOMUM, Burman, fl. zeil. 62 (1737).
C. Tamala, C. G. & T. F. Nees, Syst. Laurin. 56 (1836)... ... — — — — — Q. — B.fl.V,303 M.fr.V,165.

LITSEA, Lamarck, Diction. III, 574 (1780). (Glabraria 1771, Tetranthera, Cylicodaphne, Litsaea).
L. chinensis, Lamarck, Diction. III, 574 (1780) ... ... — — — — — Q. N.A. B.fl.V,305
L. Bindoniana, F. v. Mueller, Tetranthera, fragm. V, 167 (1866) — — — — — Q. — B.fl.V,305 M.fr.V,167.
L. Hexanthus, A. L. de Jussieu in Annal. du Mus. VI, 212 (1805) — — — — — Q. — B.fl.V,305
L. reticulata, Bentham in B. & J. H. gen. pl. III, part I, 161 (1880) — — — — — Q. — B.fl.V,306
L. Zeylanica, C. G. & T. F. Nees, in Amoen. Bonn. I, 58t. 5 (1823) — — — — — Q. — B.fl.V,307
L. dealbata, Nees, Syst. Laurin. 630 (1836) ... ... — — — — N.S.W. Q. — B.fl.V,307

CASSYTHA, Osbeck in Linné, sp. pl. 35 (1753).
C. nodiflora, Meissner in De Candolle, prodr. XV, 252 (1864) ... W.A. — — — — — — B.fl.V,309
C. glabella, R. Brown, prodr. 404 (1810) ... ... W.A. S.A. T. V. N.S.W. Q. N.A. B.fl.V,309 M.fr.V,167.
C. flava, Nees in Lehmann, pl. Preiss. I, 620 (1845) ... ... W.A. — — — — — — B.fl.V,310
C. pubescens, R. Brown, prodr. 404 (1810)... ... ... W.A. S.A. T. V. N.S.W. Q. — B.fl.V,310 M.fr.V,167.
C. phaeolasia, F. v. Mueller, fragm. V, 167 (1866) ... ... — — — V. N.S.W. — — B.fl.V,310
C. paniculata, R. Brown, prodr. 404 (1810) ... ... — — — — N.S.W. Q. — B.fl.V,311
C. filiformis, Linné, sp. pl. 35 (1753) ... ... — — — — — Q. N.A. B.fl.V,311 M.fr.V,167.
C. melantha, R. Brown, prodr. 404 (1810) ... ... W.A. S.A. T. V. N.S.W. — — B.fl.V,311 M.fr.V,167.
C. micrantha, Meissner in De Candolle, prodr. XV, 256 (1864)... W.A. — — — — — — B.fl.V,312
C. racemosa, Nees in Lehmann, pl. Preiss. I, 621 (1845)... ... W.A. — — — — Q. — B.fl.V,312
C. pomiformis, Nees in Lehmann, pl. Preiss. I, 620 (1845) ... W.A. — — — — — — B.fl.V,313

HERNANDIA, Plumier, Gen. 6, t. 40 (1703).
H. peltata, Meissner in De Candolle, prodr. XV, 263 (1864) ... — — — — — Q. — B.fl.V,314
H. bivalvis, Bentham, Fl. Austr. V, 314 (1870) ... ... — — — — — Q. — B.fl.V,314

MENISPERMEAE.
A. L. de Jussieu, Gen. 284 (1789).
TINOSPORA, Miers in Ann. and Mag. of nat. hist. sec. ser. VII, 35 (1851).
T. smilacina, Bentham in Journ. Linn. Soc. V, suppl. II, 2 (1861) — — — — — Q. N.A. B.fl.I,55 M.fr.IX,82.
T. Walcottii, F. v. Mueller in Bentham's Fl. Austr. I, 56 (1863) — — — — — — N.A. B.fl.I,56

FAWCETTIA, F. v. Mueller, fragm. X, 93 (1877).
F. tinosporoides, F. v. Mueller, fragm. X, 93 (1877) ... ... — — — — N.S.W. — — M.fr.X,93;XI,136.

COCCULUS, De Candolle, sys. veg. I, 515 (1818) from C. Baubin (1623). (Pericampylus, Legnephora).
C. Moorei, F. v. Mueller, fragm. I, 162 (1859) ... ... — — — — N.S.W. Q. — B.fl.I,56 M.fr.I,162;IX,81,107.

HYPSERPA, Miers in Ann. and Mag. of nat. hist. sec. ser. VII, 36 (1851). (Selwynia).
H. Selwyni, F. v. Mueller, fragm. IX, 82 (1875) ... ... — — — — — Q. — M.fr.IV,153.

TRISTICHOCALYX, F. v. Mueller, fragm. IV, 27 (1863).
T. pubescens, F. v. Mueller, fragm. IV, 27 (1863) ... ... — — — — — Q. — B.fl.I,58 M.fr.IV,27.
T. diffusus, Miers, contrib. to Bot. III, 286 (1871) ... ... — — — — — Q. — M.fr.

SARCOPETALUM, F. v. Mueller, pl. of Vict. I, 26 (1860).
S. Harveyanum, F. v. Mueller, pl. of Vict. 27 and 221 (1860) ... — — V. N.S.W. Q. — B.fl.I,57 M.fr.IX,83.

LEICHHARDTIA, F. v. Mueller, fragm. X, 67 (1876).
L. clambcides, F. v. Mueller, fragm. X, 68 (1876) ... ... — — — — — Q. — M.fr.X,68.

STEPHANIA, Loureiro, Fl. Cochinch. II, 608 (1790).
S. hernandifolia, Walpers, report bot. syst. I, 96 (1842)... ... — — V. N.S.W. Q. N.A. B.fl.I,57 M.fr.X,94.

PACHYGONE, Miers in Ann. and Mag. nat. hist. sec. ser. VII, 37 (1851).
P. Hullsii, F. v. Mueller, fragm. IX, 81 (1875) ... ... — — — — — Q. N.A. M.fr.V,147.

PLEOGYNE, Miers in Taylor's Ann. and Mag. of nat. hist. sec. ser. VII, 37 (1851). (Microclisia).
P. Cunninghami,Miers in Ann.&Mag.of nat.hist.sec.ser.VII,37(1851) — — — — — Q. — B.fl.I,59 M.fr.IX,82.

CARBONIA, F. v. Mueller, fragm. IX, 84 (1875).
C. multisepalea, F. v. Mueller, fragm. IX, 171 (1875) ... ... — — — — N.S.W. — — M.fr.IX,84.

ADELIOPSIS, Bentham & J. Hooker, Gen. pl. I, 436 (1862).
A. decumbens, Bentham, Fl. Austr. I, 59 (1863) ... ... — — — — — Q. — B.fl.I,59 M.fr.IX,83.

PAPAVERACEAE.
A. L. de Jussieu, Gen. 235 (1789) from B. de Jussieu (1759).
PAPAVER, Tournefort, inst. 237, t. 119 (1700).
P. aculeatum, Thunberg, Fl. Capensis 431 (1813)... ... ... — — S.A. T. V. N.S.W. Q. — B.fl.I,63

CAPPARIDEAE.
Ventenat, Tabl. III, 118 (1799).
CLEOME, Linné, syst. nat. 9 (1735); Linné, gen. 200 (1737). (Polanisia).
C. oxalidea, F. v. Mueller, fragm. I, 69 (1858) ... ... — — — — — Q. N.A. B.fl.I,90 M.fr.I,69.
C. tetrandra, Banks in De Candolle prodr. I, 240 (1824)... ... — — — — — Q. N.A. B.fl.I,90 M.fr.VII,40;IX,174.
C. viscosa, Linné, sp. pl. 672 (1753) ... ... — S.A. — N.S.W. Q. N.A. B.fl.I,90 M.fr.IX,174.

ROEPERIA, F. v. Mueller in Hooker's Kew Misc. IX, 15 (1857).   (Tetratelia).
R. clemoides, F. v. Mueller in Hooker's Kew Misc. IX, 15 (1857) — — — — — Q. N.A. B.fl.I,91 M.fr.V,105;IX,174.
EMBLINGIA, F. v. Mueller, fragm. II, 2 (1860).
E. calceoliflora, F. v. Mueller, fragm. II, 3 (1860)   ... ... W.A. — — — — — — B.fl.I,92 M.fr.II,3;XI,20.
CADABA, Forskael, Fl. Aeg. Arab. 67 (1775).
C. capparoides, De Candolle, prodr. I, 244 (1824)... ... ... — — — — — — N.A. B.fl.I,92
CAPPARIS, Tournefort, inst. 261, t. 139 (1700) from Theophrastos, Dioscorides and Plinius. (Busbeckea, Busbecquea).
C. umbellata, R. Brown in De Candolle, prodr. I, 247 (1824) ... — — — — — Q. N.A. B.fl.I,93 M.fr.V,104;IX,172.
C. lasiantha, R. Brown in De Candolle, prodr. I, 247 (1824) ... — S.A. — — N.S.W. Q. N.A. B.fl.I,94 M.fr.IX,173.
C. quiniflora, De Candolle, prodr. I, 247 (1824)   ... ... — — — — — Q. N.A. B.fl.I,94 M.fr.IX,173.
C. spinosa, Linné, sp. pl. 503 (1753) ... ... W.A. — — — — — N.A. B.fl.I,94 M.fr.I,143;V,104;IX,173.
C. sarmentosa, Cunningham in Bentham's Fl. Austr. I, 95 (1863) — — — — — Q. — B.fl.I,95 M.fr.V,104;IX,173.
C. ornans, F. v. Mueller in Bentham's Fl. Austr. I, 95 (1863) ... — — — — — Q. — B.fl.I,95 M.fr.IX,173.
C. nobilis, F. v. Mueller in Bentham's Fl. Austr. I, 95 (1863) ... — — — — N.S.W. Q. — B.fl.I,95 M.fr.I,163;V,104.
C. canescens, Banks in De Candolle, prodr. I, 246 (1824)   ... — — — — — Q. — B.fl.I,96 M.fr.IX.173.
C. lucida, R. Brown in Bentham's Fl. Austr. I, 96 (1863) ... — — — — — Q. N.A. B.fl.I,96 M.fr.I,163;V,104;IX,173.
C. Mitchelli, Lindley in Mitchell's Three Exped. I, 315 (1838)... — S.A. — V. N.S.W. Q. N.A. B.fl.I,96 M.fr.IX.173.
C. loranthifolia, Lindley in Mitchell's Trop. Austr. 220 (1848) ... — — — — N.S.W. Q. — B.fl.I,97
C. umbonata, Lindley in Mitchell's Trop. Austr. 257 (1848) ... — — — — — Q. N.A. B.fl.I,97 M.fr.IX,173.
C. uberiflora, F. v. Mueller, fragm. IX, 172 (1875) ... ... — — — — — Q. — — M.fr.IX,172.
C. Shanesiana, F. v. Mueller, fragm. X, 94 (1877)   ... ... — — — — — Q. — — M.fr.X,94.
C. humistrata, F. v. Mueller, fragm. V, 156 (1866)   ... ... — — — — — Q. — — M.fr.V,156.
C. Thozetiana, F. v. Mueller, fragm. V. 104 (1866)   ... ... — — — — — Q. — — M.fr.V,156.
APOPHYLLUM, F. v. Mueller in Hooker's Kew Misc. IX, 307 (1857).
A. anomalum, F. v. Mueller in Hooker's Kew Misc. IX, 307 (1857) — — — — N.S.W. Q. N.A. B.fl.I,97 M.fr.IX,173;XI,27.

CRUCIFERAE.
A. L. de Jussieu, Gen. 237 (1789) from B. de Jussieu (1759).

NASTURTIUM, R. Brown in W. T. Aiton's hortus Kewensis IV, 110 (1812) from Linné, syst. nat. 9 (1735).
N. terrestre, R. Brown in Aiton's hort. Kewens. IV, 110 (1812) — S.A. T. V. N.S.W. Q. — B.fl.I,66 M.fr.XI,60.
BARBARAEA, Beckmann, Lex. botan. 33 (1801).   (Barbarea).
B. vulgaris, R. Brown in De Candolle, prodr. I, 140 (1824) ... — T. V. N.S.W. — — B.fl.I,66 M.fr.VIII,142;XI,6,60.
ARABIS, Linné, gen. pl. 198 (1737) from Dalechamps (1554).   (Turritis).
A. glabra, Crantz, stirp. Austriac. 36 (1767)   ... ... — — V. N.S.W. — B.fl.I,67
CARDAMINE, Tournefort, inst. 224, t. 109 (1700) from l'Ecluse (1576).
C. stylosa, De Candolle, regn. veg. syst. nat. II, 24 (1821)   ... — — T. V. N.S.W. Q. — B.fl.I,68 M.fr.VII,19;XI,60.
C. dictyosperma, Hooker, Journ. of Bot. I, 246 (1834) ... ... — — T. V. N.S.W. — — B.fl.I,68 M.fr.VII,19;X,6,60.
C. radicata, J. Hooker in icones pl. t. 882 (1852) ... ... — — — T. — — Q. B.fl.I,60 M.fr.XI,60.
C. laciniata, F. v. Mueller in Transac. Phil. Soc. Vict. I, 34 (1854) — S.A. — V. N.S.W. — — B.fl.I,60 M.fr.XI,27,60.
C. hirsuta, Linné, sp. pl. 655 (1753)... ... ... W.A. S.A. T. V. N.S.W. Q. — B.fl.I,70 M.fr.VII,19;XI,60.
C. eustyla, F. v. Mueller in Transact. Vict. Inst. I, 114 (1854) — S.A. — V. N.S.W. Q. N.A. B.fl.I,71 M.fr.VII,19;XI,60.
ALYSSUM, Tournefort, inst. 217 (1700) from l'Ecluse (1576).   (Meniocus, Alyssum).
A. minimum, Pallas, Reise Prov. Russ. Reich. III, 746 (1776) W.A. S.A. — V. N.S.W. Q. N.A. B.fl.I,71 M.fr.XI,27,60.
WILKIA, Scopoli, introd. hist. nat. 317 (1777).   (Malcolmia).
W. Africana, F. v. Mueller, native plants of Vict. I, 33 (1879)... — — — V. N.S.W. — — M.fr.VII,18.
SISYMBRIUM, Tournefort, inst. 225, t. 109 (1700).   (Blennodia).
S. filifolium, F. v. Mueller in Transact. Phil. Soc. Vict. I, 34 (1854) — S.A. — — N.S.W. — B.fl.I,73
S. trisectum, F. v. Mueller in Transact. Vict. Inst. I, 114 (1854) W.A. S.A. — V. N.S.W. — B.fl.I,74 M.fr.VII,20;XI,60.
S. nasturtioides, F. v. Mueller in Transact. Vict. Inst. I, 115 (1854) — — V. N.S.W. — — B.fl.I,74
S. Richardsii, F. v. Mueller, fragm. X, 105 (1877)   ... W.A. — — — — — — M.fr.X,105.
S. eremigenum, F. v. Mueller, fragm. II, 143 (1861) ... — — — N.S.W. Q. — B.fl.I,74 M.fr.II,143.
S. Lucae, F. v. Mueller, fragm. XI, 59 (1879)   ... ... — — — V. N.S.W. — — B.fl.I,74 M.fr.XI,59.
S. cardaminoides, F. v. Mueller in Bentham, Fl. Austr. I, 75 (1863) — S.A. — V. N.S.W. — — B.fl.I,75 M.fr.XI,27,59.
ERYSIMUM, Linné, gen. pl. 198 (1737).   (Blennodia).
E. curvipes, F. v. Mueller in Linnaea XXV, 368 (1852) ... ... — S.A. — N.S.W. — — B.fl.I,75 M.fr.VII,20.
E. brevipes, F. v. Mueller in Linnaea XXV, 367 (1852) ... W.A. S.A. — V. N.S.W. Q. — B.fl.I,75 M.VII,20;XI,6.
E. blennodioides, F. v. Mueller in Linnaea XXV, 367 (1852) ... — S.A. — V. N.S.W. Q. — B.fl.I,76 M.fr.VII,20;27,60.
E. Cunninghamii, Bentham, Fl. Austr. I, 76 (1863)   ... ... — S.A. — V. N.S.W. — — B.fl.I,76
E. Blennodia, F. v. Mueller, fragm. X, 78 (1877)... ... — S.A. — — N.S.W. — — B.fl.I,76 M.fr.VII,20;X,78.
E. capsellinum, F. v. Mueller, native pl. of Vict. I, 35 (1879) ... — — V. N.S.W. — — B.fl.I,77
STENOPETALUM, R. Brown in De Candolle, Mém. du Mus. VII, 239 (1821).
S. velutinum, F. v. Mueller, pl. Vict. I, 49 (1860)   ... ... — — — — — Q. — B.fl.I,78 M.fr.VII,20;XI,27.
S. lineare, R. Brown in De Candolle, syst. veg. I, 513 (1821) ... W.A. S.A. — T. V. N.S.W. Q. — B.fl.I,78 M.fr.VII,20;XI,6.
S. sphaerocarpum, F. v. Mueller in Trans. Ph. Soc. Vict. I, 35 (1854) W.A. S.A. — — — — — B.fl.I,78 M.fr.VII,20.
S. nutans, F. v. Mueller, fragm. III, 27 (1862)   ... ... W.A. — — — — — — B.fl.I,79 M.fr.III,27;VII,20;XI,6.
S. robustum, Endlicher in Huegel enum. 4 (1837)   ... ... W.A. — — — — — — B.fl.I,79 M.fr.VII,20;V,5,60.
S. pedicellare, F. v. Mueller in Benth. Fl. Austr. I, 79 (1863) ... W.A. — — — — — — B.fl.I,79 M.fr.XI,6.
S. croceum, Bunge in Lehmann, pl. Preiss. I, 258 (1844) ... W.A. — — — — — — — M.fr.XI,6.
GEOCOCCUS, Drummond and Harvey in Hook. Kew Misc. VII, 52 (1855).
G. pusillus, Drumm. & Harv. in Hook. Kew Misc. VII, 52 (1855) W.A. — — V. N.S.W. — — B.fl.I,80 M.fr.VII,19;XI,6,26.
MENKEA, Lehmann, Ind. sem. hort. bot. Hamb. 8 (1843).
M. australis, Lehmann in Ind. sem. hort. bot. Hamb. 8 (1843)... W.A. S.A. — V. N.S.W. — — B.fl.I,80 M.fr.II,142.
M. draboides, J. Hooker in Benth. Fl. Austr. I, 80 (1863) ... W.A. — — — — — — B.fl.I,80 M.fr.II,142;XII,19.
M. sphaerocarpa, F. v. Mueller, fragm. VIII, 223 (1874) ... — S.A. — — — — — — M.fr.VIII,223.

CAPSELLA, Medicus, Pflanzen-Gattungen 85 (1792). (Hutchinsia partly, Thlaspi partly, Microlepidium).
C. elliptica, C. A. Meyer, Verz. der Pfl. v. Caucas. 194 (1831) ... W.A. S.A. T. V. N.S.W. — — B.fl.I,81 M.fr.VII,19;XI,60.
C. antipoda, F. v. Mueller. pl. of Vict. I, 44 (1860) ... ... — T. V. — — B.fl.I,81
C. pilosula, F. v. Mueller, pl. of Vict. I, 44 (1860) ... ... — S.A. — V. N.S.W. — — B.fl.I,82 M.fr.XI,60.
C. Tasmanica, F. v. Mueller, fragm. XI, 26 (1878) ... ... — T. — — — — B.fl.I,87
C. cochlearina, F. v. Mueller, pl. of Vict. I, 51 (1860) ... — S.A. — — N.S.W. - - — B.fl.I,88 M.fr.XI,26.
C. Drummondii, F. v. Mueller, fragm. XI, 26 (1878) ... W.A. S.A. — — — — B.fl.I,88 M.fr.VII,90;XI,26.
C. humistrata, F. v. Mueller, fragm. XI, 25 (1878) ... ... — S.A. — — N.S.W. — - — M.fr.XI,25.

SENEBIERA, De Candolle in Mém. soc. hist. nat. Par. I, 140 (1799).
S. integrifolia, De Candolle in Mém. soc. his. nat. Par. I, 144 (1799) — — — — — Q. — D.fl.I,82

LEPIDIUM, Tournefort, inst. 215, t. 103 (1700) from Diosc. and Plinius. (Iberis partly, Lepia, Monoploca).
L. strongylophyllum, F. v. Mueller in Benth. Fl. Austr. I, 84 (1863) — S.A. — — — Q. N.A. B.fl.I,84 M.fr.XI,6.
L. linifolium, Bentham, Fl. Austr. I, 84 (1863) ... ... W.A. — — — — — B.fl.I,84 M.fr.VII,20;XI,6.
L. leptopetalum, F. v. Mueller, pl. of Vict. I, 48 (1860) ... — S.A. — V. N.S.W. — — B.fl.I,84
L. rotundum, De Candolle, syst. veg. II, 537 (1818) ... W.A. — — — — — B.fl.I,85 M.fr.VII,20;XI,6.
L. phlebopetalum, F. v. Mueller, pl. of Vict. I, 47 (1860) — S.A. — V. N.S.W. — — B.fl.I,85 M.fr.VII,19;XI,60.
L. podicellosum, F. v. Mueller, fragm. XI, 27 (1878) ... — — — — — — N.A. — M.fr.XI,27.
L. monoplocoides, F. v. Mueller in Trans. Phil. Soc. Vict. 35 (1854) — S.A. — V. N.S.W. — — B.fl.I,85 M.fr.XI,27.
L. papillosum, F. v. Mueller in Linnaea XXV, 370 (1852) ... W.A. S.A. — V. N.S.W. — — B.fl.I,86 M.fr.VII,19;XI,6,28.
L. foliosum, Desvaux, Journ. de Bot. III, 164 et 180 (1813) ... W.A. S.A. T. V. N.S.W. — — D.fl.I,86 M.fr.VII,19.
L. ruderale, Linné, sp. pl. 645 (1753) ... ... W.A. S.A. T. V. N.S.W. Q. N.A. B.fl.I,86 M.fr.VII,19;XI,27,61.

CAKILE, Tournefort, coroll. 49, t. 483 (1700).
C. maritima, Scopoli, Fl. Carniolica, 344 (1760) ... ... S.A. T. V. N.S.W. — — — M.fr.VII,19;XI,6,28,60.

VIOLACEAE.
De Candolle, Fl. Franc. IV, 801 (1805).

VIOLA, Tournefort, inst. 419, t. 236 (1700) from Plinius.
V. betonicifolia, Smith in Rees's Cyclopedia XXXVII (1817) ... — S.A. T. V. N.S.W. Q. — B.fl.I,99 M.fr.X,82.
V. hederacea, Labillardière, Nov. Holl. pl. spec. I, 66, t. 91 (1804) — S.A. T. V. N.S.W. Q. — B.fl.I,90 M.fr.X,82.
V. Cunninghamii, J. Hooker, Fl. N. Zel. I, 16 (1853) ... ... — T. — — — — B.fl.I,100
V. Caleyana, G. Don, gen. syst. dichlam. pl. I, 329 (1831) ... — — T. V. N.S.W. — — B.fl.I,100 M.fr.X,82.

HYBANTHUS, Jacquin, Stirp. Amer. hist. 77, t. 75 (1763). (Ionidium, Pigea, Vlamingia).
H. floribundus, F. v. Mueller, nat. pl. of Vict. I, 45 (1870) W.A. S.A. — V. N.S.W. — — B.fl.I,112 M.fr.X,81.
H. Vernonii, F. v. Mueller, nat. pl. of Vict. I, 45 (1879) ... — — — V. N.S.W. — — B.fl.I,103 M.fr.X,81.
H. filiformis, F. v. Mueller, native pl. of Vict. I, 44 (1879) ... — — — V. N.S.W. Q. — B.fl.I,103 M.fr.X,81.
H. Tatei, F. v. Mueller in Trans. R. Socy. of S. Aust. IV, 102 (1882) — S.A. — — — — — —
H. debilissimus, F. v. Mueller, fragm. XI, 4 (1878) ... ... W.A. — — — — — — M.fr.XI,4.
H. calycinus, F. v. Mueller, fragm. XI, 5 (1877) ... ... W.A. — — — — — B.fl.I,104 M.fr.XI,4.
H. enneaspermus, F. v. Mueller, fragm. X, 81 (1877) ... ... — S.A. — — N.S.W. Q. N.A. B.fl.I,101 M.fr.X,81.

HYMENANTHERA, R. Brown in Tuck. Cong. 442 (1818).
H. Banksii, F. v. Mueller, first. gen. report 9 (1853) ... ... — — T. V. N.S.W. — — B.fl.I,104 M.fr.X,82.

FLACOURTIEAE.
Cl. Richard in Mém. du Mus. I, 366 (1815).

COCHLOSPERMUM, Kunth, Malvac. 6 (1822).
C. Fraseri, Planchon in Hook. Lond. Journ. VI, 307 (1847) ... — — — — — N.A. B.fl.I,106
C. heteronemum, F. v. Mueller in Hook. Kew. Misc. IX, 15 (1857) — — — — — N.A. B.fl.I,106 M.fr.IX,61.
C. Gillivraei, Bentham, Fl. Austr. I, 106 (1863) ... ... — — — — — Q. B.fl.I,106
C. Gregorii, F. v. Mueller, fragm. I, 71 (1858) ... ... — — — — — Q. N.A. B.fl.I,106 M.fr.IX,61.

SCOLOPIA, Schreber, Gen. 335 (1789). (Phoberos).
S. Brownii, F. v. Mueller, fragm. III, 11 (1862) ... ... — — — N.S.W. Q. — B.fl.I,107 M.fr.III,11;IX,61.

XYLOSMA, G. Forster, prodr. 380 (1786).
X. ovata, Bentham, Fl. Austr. I, 108 (1863) ... ... — — — N.S.W. Q. — B.fl.I,108 M.fr.IX,60.

STREPTOTHAMNUS, F. v. Mueller, fragm. III, 28 (1862).
S. Moorei, F. v. Mueller, fragm. III, 28 (1862) ... ... — — — N.S.W. — — B.fl.I,108 M.fr.III,28.
S. Beckleri, F. v. Mueller, fragm. III, 28 (1862) ... ... — — — N.S.W. — — B.fl.I,108 M.fr.IX,61.

SAMYDACEAE.
J. Gaertner, de Fruct. III, 328 (1805).

CASEARIA, Jacquin, Stirp. Amer. hist. 132, t. 85 (1763).
C. esculenta, Roxburgh, Fl. Ind. II, 422 (1832) ... ... — — — — — Q. — B.fl.VI,309
C. Dallachii, F. v. Mueller, fragm. V, 107 (1866)... ... ... — — — — — Q. — B.fl.III,309 M.fr.VI,232.

HOMALIUM, Jacquin, Stirp. Amer. hist. 170, t. 183 (1763). (Blackwellia).
H. Vitiense, Bentham in Journ. Linn. Soc. IV, 36 (1859) ... — — — — — Q. — B.fl.III,310
H. brachybotrys, F. v. Mueller, fragm. II, 127 (1861) ... ... — — — — — Q. — B.fl.III,310 M.fr.II,127.

PITTOSPOREAE.
R. Brown in Flinders' voy. 542 (1814).

PITTOSPORUM, Banks in Gaertner, de Fruct. I, 286, t. 59 (1788).
P. rhombifolium, Cunningham in Hook. icon. pl. t. 621 (1844)... — — — N.S.W. Q. — B.fl.I,110 M.fr.XII,4.
P. melanospermum, F. v. Mueller, fragm. I, 70 (1858) ... ... — — — — — N.A. B.fl.I,111 M.fr.VI,167;XII,4.
P. undulatum, Andrews, bot. reposit. 383 (1800)... ... ... — — T. V. N.S.W. Q. — B.fl.I,111 M.fr.VI,167;XII,4.
P. revolutum, Aiton, Hort. Kew, sec. ed. II, 27 (1811) ... ... — — — V. N.S.W. Q. — B.fl.I,111 M.fr.VI,167&186;VII,140.
P. venulosum, F. v. Mueller, fragm. VI, 186 (1868) ... ... — — — — — Q. — M.fr.VII,140;XII,4.
P. ferrugineum, Aiton, Hort. Kew, sec. ed. II, 27 (1811) ... ... — — — — — Q. B.fl.I,112 M.fr.II,78;III,165;VI,167

P. rubiginosum, Cunningham in Ann. of Nat. Hist. IV, 106 (1842) — — — — — Q. — B.fi.I,112 M.fr.VI,167;VII,140;IX,
P. phillyroides, De Candolle, prodr. I, 347 (1824) ... ... W.A. S.A. — V. N.S.W. Q. N.A. B.fi.I,112 M.fr.VI,167. [190.
P. bicolor, Hooker, Journ. Bot. I, 249 (1834) ... ... — — T. V. N.S.W. — — B.fi.I,113 M.fr.VIII,142;XII,4.
P. erioloma, C. Moore et F. v. Mueller, fragm. VII, 139 (1871) — — — — N.S.W. — — — M.fr.VII,130.

HYMENOSPORUM, R. Brown in F. v. Mueller, fragm. II, 77 (1860).]140.
H. flavum, F. v. Mueller, fragm. II, 77 (1860) ... ... — — — — N.S.W. Q. — B.fi.I,114 M.fr.II,77;VI,168;VII,

BURSARIA, Cavanilles, icon. et descr. pl. IV, 30, t. 350 (1797).
B. spinosa, Cavanilles, icon. pl. IV, 30, t. 350 (1797) ... ... W.A. S.A. T. V. N.S.W. Q. N.A. B.fi.I,115 M.fr,XII,3.

MARIANTHUS, Huegel, enum. pl. Nov. Holl. 8 (1837). (Oncosporum, Rhytidosporum, Calopetalon).
M. procumbens, Bentham, Fl. Austr. I, 117 (1863) ... ... — — T. V. N.S.W. Q. — B.fi.I,117 M.fr.XII,2.
M. microphyllus, Bentham, Fl. Austr. I, 117 (1863) ... ... W.A. — — — — — — B.fi.I,117 M.fr.II,145.
M. villosus, Bentham, Fl. Austr. I, 117 (1863) ... ... W.A. — — — — — — B.fi.I,117
M. parviflorus, F. v. Mueller, fragm. II, 144 (1861) ... ... W.A. — — — — — — B.fi.I,118 M.fr.II,144;XII,2.
M. bignoniaceus, F. v. Mueller in Trans. Phil. Soc. Vict. 6 (1854) — S.A. — V. — — — B.fi.I,118 M.fr.XII,2.
M. Drummondianus, Bentham, Fl. Austr. I, 119 (1863)... ... W.A. — — — — — — B.fi.I,119 M.fr.XII,2.
M. coeruleo-punctatus, Klotzsch in Link and Otto, icon. pl. 28,
t. 12 (1841) ... ... W.A. — — — — — — B.fi.I,119 M.fr.XII,2.
M. candidus, Huegel, enum. pl. Nov. Holl. 8 (1837) ... W.A. — — — — — — B.fi.I,120 M.fr.XII,2.
M. erubescens, Putterlick in Endl. nov. stirp. dec. 60 (1839) ... W.A. — — — — — — B.fi.I,120 M.fr.XII,2.
M. ringens, F. v. Mueller, fragm. I, 218 (1859) ... ... W.A. — — — — — — B.fi.I,120 M.fr.I,218;XII,2.
M. pictus, Lindley in Bot. Regist. XXV, App. XXV (1839) — W.A. — — — — — — B.fi.I,121 M.fr.I,217;II,182;XII,2.

CITRIOBATUS, Cunningham, in Loudon's hort. Brit. supp. I, 585 (1832). (Ixiosporum).
C. multiflorus, Cunningham in Lond. hort. Brit. I, 585 (1832) ... — — — N.S.W. Q. — B.fi.I,121 M.fr.II,77;VI,168;XII,3.
C. pauciflorus, Cunningham in Lond. hort. Brit. I, 585 (1832) ... — — — — — Q. N.A. B.fi.I,122 M.fr.II,76;VI,168;VII,180,

BILLARDIERA, Smith, Sp. Bot. New Holl. I, t. 1 (1793). (Pronaya, Campylanthera).
B. mollis, Labillardière, Nov. Holl. plant. spec. I, 64, t. 89 (1804) — — T. V. N.S.W. — — B.fi.I,123 .
B. scandens, Smith, Spec. bot. New Holl. I, t. 1 (1793)... — — S.A. T. V. N.S.W. Q. — B.fi.I,123 M.fr.XI,123.
B. coriacea, Bentham, Fl. Austr. I, 124 (1863) ... ... W.A. — — — — — — B.fi.I,124 M.fr.II,1.
B. cymosa, F. v. Mueller in Transact. Vict. Inst. I, 29 (1855) ... — S.A. — V. N.S.W, — — B.fi.I,124 M.fr.VI,168;VII,140.
B. elegans, F. v. Mueller, pl. of Vict. I, 78 (1860) ... W.A. — — — — — — B.fi.I,125 M.fr.XII,2.
B. variifolia, De Candolle, prodr. I, 346 (1824) ... ... W.A. — — — — — — B.fi.I,125 M.fr.
B. Lehmanniana, F. v. Mueller, pl. of Vict. I, 78 (1860) ... W.A. — — — — — — B.fi.I,125 M.fr.XII,2.
B. floribunda, F. v. Mueller in South Sc. Record II, 1 (1882) ... W.A. — — — — — — B.fi.I,126 M.fr.XII,1.

SOLLYA, Lindley, Bot. Regist. XVII, t. 1400 (1831). (Xerosollya.)
S. heterophylla, Lindley, Bot. Regist. XVII, t. 1406 (1831) ... W.A. — — — — — — B.fi.I,126 M.fr.XII,3.
S. parviflora, Turcz. in Bull. Soc. Mosc. XXVII, 361 (1854) ... W.A. — — — — — — B.fi.I,126 M.fr.XII,3.

CHEIRANTHERA, Cunningham, in Brongn. Bot. voy. Coquille, t. 77 (1829).
C. linearis, Cunningham in Bot. Regist. 1719 (1834) ... — S.A. — V. N.S.W. Q. — B.fi.I,127 M.fr.I,97;II,180;IX,168.
C. filifolia, Turczaninow in Bull. Soc. Mosc. XXVII, 364 (1854) W.A. — — — — — — B.fi.I,127 M.fr.I,97;II,79;IX,168.
C. parviflora, Bentham, Fl. Austr. I, 128 (1863) ... ... W.A. — — — — — — B.fi.I,128 M.fr.XII,3.

DROSERACEAE.
Salisbury, Parad. Lond. 95 (1809).
ALDROVANDA, Monti in Act. Bonon. II, 404, t. 12 (1747).
A. vesiculosa, Linné, sp. pl. 281 (1753) ... ... — — — — — — Q. — — M.fr.VI,104;X,79.

DROSERA, Linné, gen. pl. 89 (1737). (Londera).
D. Indica, Linné, sp. pl. 282 (1753) ... ... W.A. S.A. — V. N.S.W. Q. N.A. B.fi.II,456 M.fr.X,79.
D. Adelae, F. v. Mueller, fragm. IV, 154 (1864) ... ... — — — — — — Q. — — M.fr.X,80.
D. Arcturi, Hooker, Journ. of Bot. I, 247 (1834) ... ... — — T. V. N.S.W. — — B.fi.II,456
D. glanduligera, Lehmann, pugill. VIII, 37 (1844) ... ... W.A. S.A. — V. N.S.W. — — B.fi.II,457 M.fr.X,80.
D. pygmaea, De Candolle, prodr. I, 317 (1824) ... ... W.A. — — — — — — B.fi.II,457 M.fr.X,80.
D. platystigma, Lehmann, pugill. VIII, 37 (1844)... ... W.A. — — — — — — B.fi.II,457 M.fr.X,80.
D. pulchella, Lehmann, pugill. VIII, 38 (1844) ... ... W.A. — — — — — — B.fi.II,458 M.fr.X,80.
D. leucoblasta, Bentham, Fl. Austr. II, 458 (1864) ... ... W.A. — — — — — — B.fi.II,458 M.fr.X,80.
D. nitidula, Planchon in Ann. des sc. nat. IX, 285 (1848) ... W.A. — — — — — — B.fi.II,458
D. paleacea, R. Brown in De Candolle, prodr. I, 318 (1824) ... W.A. — — — — — — B.fi.II,458
D. parvula, Planchon in Ann. des sc. nat. IX, 287 (1848) ... W.A. — — — — — — B.fi.II,458
D. Burmanni, Vahl, Symbol. III, 50 (1794) ... ... — — S.A. — — — Q. N.A. B.fi.II,459 M.fr.X,80.
D. spathulata, Labillardière, Nov. Holl. pl. spec. I, 79, t. 106 (1804) — S.A. — T. V. N.S.W. — — B.fi.II,459
D. Drummondii, Lehmann, pl. Preiss. II, 235 (1847) ... W.A. — — — — — — B.fi.II,460
D. scorpioides, Planchon in Ann. des sc. nat. IX, 288 (1848) ... W.A. — — — — — — B.fi.II,460 M.fr.VIII,185;X,85.
D. petiolaris, R. Brown in De Candolle, prodr. I, 318 (1824) ... — — — — — — Q. N.A. B.fi.II,460 M.fr.X,80.
D. binata, Labillardière, Nov. Holl. pl. spec. I, 78, t. 105 (1804) — S.A. — T. V. N.S.W. Q. — B.fi.II,461 M.fr.X,80.
D. zonaria, Planchon in Ann. des sc. nat. IX, 303 (1848) ... W.A. — — — — — — B.fi.II,462
D. bulbosa, Hooker, icon. pl. t. 375 (1841)... ... ... W.A. — — — — — — B.fi.II,462 M.fr.X,80.
D. rosulata, Lehmann, pugill. VIII, 36 (1844) ... ... W.A. — — — — — — B.fi.II,462
D. Whittakerii, Planchon in Ann. des sc. nat. IX, 302 (1848) ... — S.A. — V. — — — B.fi.II,463
D. macrophylla, Lindley, Bot. Regist. XXV, App. XX, (1839)... W.A. — — — — — — B.fi.II,463
D. squamosa, Bentham, Fl. Austr. II, 463 (1864) ... ... W.A. — — — — — — B.fi.II,463
D. erythrorrhiza, Lindley, Bot. Regist. XXV, App. XX (1830)... W.A. — — — — — — B.fi.II,463
D. stolonifera, Endlicher in Hueg. enum. 5 (1837) ... ... W.A. — — — — — — B.fi.II,464
D. humilis, Planchon in Ann. des sc. nat. IX, 300 (1848) ... W.A. — — — — — — B.fi.II,464
D. ramellosa, Lehmann, pugill. VIII, 40 (1844) ... ... W.A. — — — — — — B.fi.II,464
D. platypoda, Turczaninow in Bull. del'Ac. deSt. Petersb. 343 (1852) W.A. — — — — — — B.fi.II,464 M.fr.VIII,185;X,80.
D. auriculata, Backhouse in Ann. des sc. nat. IX, 295 (1848) ... — S.A. T. V. N.S.W. — — B.fi.II,465 M.fr.X,80.
D. lunata, Buchanan in De Cand. prodr. I, 319 (1824) ... — — — — — — Q. — —
D. peltata, Smith in Willd. spec. pl. I, 1546 (1797) ... ... — S.A. T. V. N.S.W. Q. — B.fi.II,465 M.fr.X,80.

D. Neesii, Lehmann, pugill. VIII, 42 (1844) ... ... W.A. — — — — — — R.fl.II,466
D. gigantea, Lindley, Bot. Regist. XXV, App. XX (1839) ... W.A. — — — — — — B.fl.II,466
D. myriantha, Planchon in Ann. des sc. nat. IX, 291 (1848) ... W.A. — — — — — — B.fl.II,466
D. pallida, Lindley, Bot. Regist. XXV, App. XX (1839) ... W.A. — — — — — — B.fl.II,467
D. penicillaris, Bentham, Fl. Austr. II, 467 (1864) ... ... W.A. — — — — — — B.fl.II,467
D. filicaulis, Endlicher in Hueg. enum. 6 (1837) ... ... W.A. — — — — — — B.fl.II,467 M.fr.VIII,185;X,51.
D. Huegelii, Endlicher in Hueg. enum. 6 (1837) ... ... W.A. — — — — — — B.fl.II,468
D. macrantha, Endlicher in Hueg. enum. 6 (1837) ... ... W.A. — — — — — — B.fl.II,468
D. Menziesii, R. Brown in De Candolle, prodr. I, 319 (1824) ... W.A. S.A. T. V. N.S.W. — B.fl.II,468
D. calycina, Planchon in Ann. des sc. nat. IX, 290 (1848) ... W.A. — — — — — — B.fl.II,468
D. heterophylla, Lindley, Bot. Regist. XXV, App. XX (1839)... W.A. — — — — — — B.fl.II,469
D. Banksii, R. Brown in De Candolle, prodr. I, 319 (1824) ... — — — — — Q. B.fl.II,460

BYBLIS, Salisbury, Parad. Lond. t. 95 (1808).
B. liniflora, Salisbury, Parad. Lond. t. 95 (1808) ... ... — — — — — Q. N.A. B.fl.I,470 M.fr.X,81.
B. gigantea, Lindley Bot. Regist. XXV, App. XXI (1839) ... W.A. — — — — — — B.fl.II,470

ELATINEAE.
Cambessèdes in Mém. du Mus. XVIII, 225 (1829).

ELATINE, Linné, gen. pl. 118 (1737).
E. Americana, Arnott in Edinb. journ. nat. sc. I, 431 (1830) ... W.A. S.A. T. V. N.S.W. Q. — B.fl.I,178 M.fr.XI,27.

BERGIA, Linné, mantiss. II, 152 (1771).
B. ammannioides, Roth, nov. pl. spec. 219 (1821) ... ... — — — V. N.S.W. Q. N.A. B.fl.I,180 M.fr.II,147.
B. pedicellaris, F. v. Mueller in Benth. Fl. Austr. I, 180 (1863) — — — — — N.A. B.fl.I,180 M.fr.II,145.
B. perennis, F. v. Mueller in Benth. Fl. Austr. I, 180 (1863) ... — — — — — N.A. B.fl.I,181 M.fr.II,146.

HYPERICINAE.
J. de St. Hilaire, Expos. fam. II, 23, t. 75 (1805).

HYPERICUM, Tournefort, inst. 254, t. 131 (1700), from Dioscorides and Plinius.
H. Japonicum, Thunberg, Fl. Jap. 295, t. 31 (1784) ... ... W.A. S.A. T. V. N.S.W. Q. N.A. B.fl.I,182

GUTTIFERAE.
A. L. de Jussieu, Gen. 243 (1769).

CALOPHYLLUM, Linné, gen. 154 (1737).
C. inophyllum, Linné, sp. pl. 513 (1753) ... ... — — — — — Q. — B.fl.I,183 M.fr.IX,175.
C. tomentosum, Wight, Illustrat. Pl. Ind. Orient. I, 128 (1840) - — — — — Q. — — M.fr.IX,174.

POLYGALEAE.
A. L. de Jussieu, in Ann. du Mus. XIV, 386 (1809).

SALOMONIA, Loureiro, Fl. Cochinch. I, 14 (1790).
S. oblongifolia, De Candolle, prodr. I, 334 (1824)... ... — — — — N.S.W. Q. — B.fl.I,138 M.fr.XI,4.

POLYGALA, Tournefort, inst. 174, t. 79 (1700).
P. Sibirica, Linné, sp. pl. 702 (1753) ... ... — — — V. N.S.W. Q. — B.fl.I,139 M.fr.XI,4.
P. leptalea, De Candolle, prodr. I, 325 (1824) ... ... - — — — — Q. N.A. B.fl.I,139 M.fr.XI,4.
P. eriocephala, F. v. Mueller in Benth. Fl. Austr. I, 139 (1863) — — — — — — — B.fl.I,139
P. persicarifolia, De Candolle, prodr. I, 320 (1818) ... ... — — — — — — N.A. — M.fr.XI,4.
P. Chinensis, Linné, sp. pl. 704 (1753) ... ... — — — — — Q. N.A. B fl.I,140 M.fr.XI,4.
P. stenoclada, Bentham, Fl. Austr. I, 141 (1863)... ... ... — — — — — — N.A. B.fl.I,141

COMESPERMA, Labillardière, Nov. Holl. pl. spec. II, 21 (1806).
C. sphaerocarpum, Steetz in Lehm. pl. Preiss. II, 314 (1847) ... — — — — N.S.W. Q. — B.fl.I,143 M.fr.XI,3.
C. scoparium, Steetz in Lehm. pl. Preiss. II, 300 (1847)... ... W.A. S.A. — V. N.S.W. — — B.fl.I,143
C. aphyllum, R. Brown in Benth. Fl. Austr. I, 143 (1863) ... — — — — — — N.A. B.fl.I,143
C. spinosum, F. v. Mueller, fragm. I, 144 (1859) ... ... W.A. — — — — — — B.fl.I,144 M.fr.I,144.
C. volubile, Labillardière, Nov. Holl. pl. spec. II, 24, t. 163 (1806) W.A. S.A. T. V. N.S.W. Q. — B.fl.I,144 M.fr.X,3.
C. ciliatum, Steetz in Lehm. pl. Preiss. II, 304 (1847) ... W.A. — — — — — — B.fl.I,144 M.fr.XI,3.
C. integerrimum, Endlicher in Hueg. enum. (1837) ... ... — — — — — Q. N.A. B.fl.I,145 M.fr.XI,3.
C. secundum, Banks in De Candolle, prodr. I, 334 (1924) ... — — — — — Q. N.A. B.fl.I,145
C. Drummondii, Steetz in Lehm. pl. Preiss. II, 301 (1847) ... W.A. — — — — — — B.fl.I,145 M.fr.X,2.
C. praecelsum, F. v. Mueller, fragm. XI, 2 (1878) ... ... — — — — — — — — M.fr.XI,2.
C. retusum, Labillardière, Nov. Holl. pl. spec. II, 22, t.160 (1806) — — — T. V. N.S.W. Q. — B.fl.I,145 M.fr.XI,2.
C. sylvestre, Lindley in Mitch. Trop. Austr. 342 (1848) ... — — — T. V. N.S.W. — — B.fl.I,146 M.fr.I,49;XI,2.
C. acerosum, Steetz in Lehm. pl. Preiss. II, 299 (1847) ... W.A. — — — — — — B.fl.I,146 M.fr.I,48;XI,2.
C. ericinum, De Candolle, prodr. I, 334 (1824) ... ... — — — T. V. N.S.W. Q. — B.fl.I,146 M.fr.I,48;XI,2.
C. viscidulum, F. v. Mueller, fragm. X, 4 (1876) ... ... — — — — — — — — M.fr.X,4;XI,2.
C. confertum, Labillardière, Nov. Holl. pl. spec. II, 23, t. 161 (1806) W.A. — — — — — — B.fl.I,147 M.fr.XI,3.
C. flavum, De Candolle, prodr. I, 334 (1824) ... ... — — — — — -- B.fl.I,147 M.fr.XI,3.
C. calymega, Labillardière, Nov. Holl. pl. spec. II, 23, t. 162 (1806) W.A. S.A. T. V. N.S.W. — — B.fl.I,147 M.fr.XI,3.
C. rhadinocarpum, F. v. Mueller, fragm. XI, 1 (1878) ... ... — — — — — — — — M.fr.XI,1.
C. lanceolatum, R. Brown in Benth. Fl. Austr. I, 148 (1863) ... W.A. — — T. V. N.S.W. — — B.fl.I,148 M.fr.XI,1.
C. defoliatum, F. v. Mueller, fragm. XI, 189 (1862) ... ... — — — — — — — — M.fr.XI,4.
C. nudiusculum, De Candolle, prodr. I, 334 (1824) ... ... W.A. — — — — — — B.fl.I,148 M.fr.XI,2.
C. virgatum, Labillardière, Nov. Holl. pl. spec. II, 21, 159 (1806) W.A. — — — — — — B.fl.I,149 M.fr.XI,4.
C. polygaloides, F. v. Mueller in Trans. Phil. Soc. Vict. I, 7 (1854) W.A. S.A. — V. — — — B.fl.I,149

XANTHOPHYLLUM, Roxburgh, Pl. Corom. III, 81, t. 284 (1819).
X. Macintyrii, F. v. Mueller, fragm. V, 57 (1865) ... ... — — — — — Q. — — M.fr.V,57;XI,4.

9

TREMANDREAE.
R. Brown in Flind. voy. II, 554 App. (1814).

PLATYTHECA, Steetz in Lehmann, pl. Preiss. I, 220 (1845).
P. galioides, Steetz in Lehm. pl. Preiss. I, 220 (1845) ... ... W.A. — — — — — ... B.fl.I,136 M.fr.XII.

TETRATHECA, Smith, Spec. Bot. New Holl. I, t. 2 (1793).
T. Harperi, F. v. Mueller, fragm. V, 49 (1865) ... — — — — — — M.fr.V,189.
T. ciliata, Lindley in Mitch. three Exped. II, 206 (1838) ... — S.A. T. V. — — — B.fl.I,130 M.fr.XII.
T. efoliata, F. v. Mueller, fragm. X, 3 (1870) ... ... W.A. — — — — — M.fr.X,3.
T. aphylla, F. v. M. in South. Science Record, vol. II, May(1882) W.A. — — — — — M.fr.XII.
T. ericifolia, Smith, Exot. Bot. I, 37, t. 20 (1804) ... — S.A. T. V. N.S.W. Q. — B.fl.I,131 M.fr.XII.
T. juncea, Smith, specim. of Bot. of New Holl. 5, t. 2 (1793) ... — — — N.S.W. — — B.fl.I,132 M fr.XII.
T. affinis, Endlicher in Hueg. enum. 7 (1837) ... ... W.A. — — — — — B.fl.I,132 M.fr.XII.
T. nuda, Lindley, Bot. Regist. XXV, App. XXXVIII (1839) ... W.A. — — — — — B.fl.I,133 M.fr.XII.
T. virgata, Steetz in Lehm. pl. Preiss. I, 212 (1845) ... W.A. — — — — — B.fl.I,133 M.fr.XII.
T. confertifolia, Steetz in Lehm. pl. Preiss. I, 214 (1845) ... W.A. — — — — — B.fl.I,133 M.fr.XII.
T. setigera, Endlicher in Hueg. enum. 8 (1837) ... ... W.A. — — — — — B.fl.I,133 M.fr.XII.
T. hirsuta, Lindley, Bot. Regist. XXV, App. XXXVIII (1839) W.A. — — — — — B.fl.I,134 M.fr.XII.
T. pilifera, Lindley, Bot. Regist. XXV, App. XXXVIII (1839) W.A. — — — — — B.fl.I,135 M.fr.XII.
T. filiformis, Bentham, Fl. Austr. I, 135 (1863) ... ... W.A. — — — — — B.fl.I,135 M.fr.XII.

TREMANDRA, R. Brown in App. Flind. voy. II, 544 (1814).
T. stelligera, R. Brown in De Candolle, prodr. I, 344 (1824) ... W.A. — — — — — B.fl.I,136 M.fr.XII.
T. diffusa, R. Brown in De Candolle, prodr. I, 344 (1824) ... W.A. — — — — — B.fl.I,137 M.fr.XII.

MELIACEAE.
Ventenat, Tabl. III, 159 (1799).

HEDRAIANTHERA, F. v. Mueller, fragm. V, 58 (1865).
H. porphyropetala, F. v. Mueller, fragm. V, 58 (1865) ... — — — — Q. — — M.fr.V,58.

TURRAEA, Linné, mantiss. II, 150 (1771).
T. pubescens, Hellenius in K. Sv. Vent. Acad. Handl. 26, t. 10 (1788)— — — N.S.W. Q. N.A. B.fl.I,379 M.fr.V,144.

MELIA, Linné, gen. pl. 127 (1737).
M. Azedarach, Linné, spec. pl. 384 (1753) ... ... — — — N.S.W. Q. N.A. B.fl.I,380 M.fr.V,144;IX,134;XI,135.

DYSOXYLUM, Blume, Bijdrag. 172 (1825). (Dysoxylon, Epicharis, Hartigheca).
D. oppositifolium, F. v. Mueller, fragm. V, 144 (1866) ... — — — — Q. — M.fr.V,144;IX,61.
D. latifolium, Bentham, Fl. Austr. I, 381 (1863) ... — — — — Q. — B.fl.I,381 M.fr.IX,61.
D. Klanderi, F. v. Mueller, fragm. V, 176 (1866) ... — — — — Q. — M.fr.V,176;IX,61,134.
D. Fraserianum, Bentham, Fl. Austr. I, 381 (1863) ... — — — — Q. — B.fl.I,381 M.fr.IX,61.
D. arborescens, Miquel in Annal. Mus. Bot. Lugd. IV, 24 (1868) ... — — — N.S.W. Q. — B.fl.I,381 M.fr.IX,144.
D. Muelleri, Bentham, Fl. Austr. I, 381 (1863) ... — — — N.S.W. Q. — B.fl.I,381 M.fr.V,145.
D. Schiffneri, F. v. Mueller in Melb. Chemist, 53 (1881) ... — — — N.S.W. Q. — M.fr.XI,133.
D. Becklerianum, C. de Candolle in Monogr. phan. I, 509 (1878) — — — N.S.W. Q. — M.fr.V,133.
D. Lessertianum, Bentham, Fl. Austr. I, 382 (1863) ... — — — N.S.W. Q. — B.fl.I,382 M.fr.V,145;IX,61.
D. Schultzii, C. de Candolle in Monogr. phanerogr. I, 509 (1878) — — — — — N.A. M.fr.XI,133.
D. rufum, Bentham. Fl. Austr. I, 382 (1863) ... — — — N.S.W. Q. — B.fl.I,382 M.fr.V,145.
D. Nernstii, F. v. Mueller, fragm. V, 176 (1866) ... ... — — — — — — M.fr.V,176.

AMOORA, Roxburgh, pl. Corom. III, 54, t. 258 (1819).
A. nitidula, Bentham, Fl. Austr. I, 383 (1863) ... — — — N.S.W. Q. — B.fl.I,383

SYNOUM, Adr. de Jussieu in Mém. du Mus. XIX, 226, t. 15 (1830).
A. glandulosum, de Jussieu in Mém. du Mus. XIX, 227, t. 15 (1830) ... — — — N.S.W. Q. — B.fl.I,384 M.fr.V,145;IX,134.
A. Muelleri, C. de Candolle in Monogr. phanerogr. I, 593 (1878) — — — — Q. — B.fl.I,384 M.fr.V,145;XI,138.

OWENIA, F. v. Mueller in Hook. Kew Misc. IX, 303 (1857).
O. acidula, F. v. Mueller in Hook. Kew Misc. IX, 304 (1857) ... — S.A. — N.S.W. Q. — B.fl.I,385 M.fr.III,14;IX,134.
O. vernicosa, F. v. Mueller, fragm. III, 15 (1862) ... — — — — — N.A. B.fl.I,385 M.fr.III,15;IX,134.
O. venosa, F. v. Mueller in Hook. Kew Misc. IX, 304 (1857) ... — — — — Q. — B.fl.I,386 M.fr.III,14;IX,134.
O. cepiodora, F. v. Mueller, fragm. XI, 82 (1880) ... — — — N.S.W. — — M.fr.XI,82.
O. reticulata, F. v. Mueller in Hook. Kew Misc. IX, 305 (1857) W.A. — — — — N.A. B.fl.I,386 M.fr.III,13;IX,134.

AGLAIA, Loureiro, Fl. Cochinch. I, 173 (1790), non Allemand (1770). (Milnea, Nemedra).
A. elaeagnoidea, Bentham, Fl. Austr. I, 383 (1863) ... — — — — — N.A. B.fl.I,383 M.fr.V,145;IX,134.

HEARNIA, F. v. Mueller, fragm. V, 55 (1865). (Aglaiopsis).
H. sapindina, F. v. Mueller, fragm. V, 55 (1865) ... — — — — — — M.fr.V,55.

CARAPA, Aublet, Hist. pl. Guian. II, Suppl. 32, t. 387 (1775). (Xylocarpus).
C. Moluccensis, Lamarck, Encycl. meth. Bot. I, 621 (1783) ... — — — — Q. N.A. B.fl.I,387 M.fr.V,145;IX,134.

CEDRELA, P. Browne, Nat. hist. of Jamaica 158 (1756).
C. australis, F. v. Mueller, fragm. I, 4 (1858) ... ... — — — N.S.W. Q. — B.fl.I,387 M.fr.IX,133;XI,82.

FLINDERSIA, R. Brown in Flind. voy. II, 595 (1814). (Oxleya, Strzeleckia).
F. australis, R. Brown in Flind. voy. II, 595, t. 1 (1814) ... ... — — — N.S.W. Q. — B.fl.I,388 M.fr.I,65;II,174;IX,133.
F. Bourjotiana, F. v. Mueller, fragm. IX, 133 (1875) ... ... — — — — Q. — M.fr.IX,133.
F. Schottiana, F. v. Mueller, fragm. III, 25 (1862) ... ... — — — N.S.W. Q. — B.fl.I,388 M.fr.II,65;VI,143;IX,133.
F. Oxleyana, F. v. Mueller, fragm. I, 65 (1858) ... ... — — — N.S.W. Q. — B.fl.I,389 M.fr.I,65;III,25;IX,133.
F. Pimenteliana, F. v. Mueller, fragm. IX, 132 (1875) ... — — — — Q. — M.fr.IX,132.
F. Bennettiana, F. v. Mueller in Benth. Fl. Austr. I, 389 (1863) ... — — — N.S.W. Q. — B.fl.I,389 M.fr.IX,131;XI,135.
F. Ifflaiana, F. v. Mueller, fragm. X, 94 (1877) ... ... — — — — Q. — M.fr.X,94.
F. Brayleyana, F. v. Mueller, fragm. V, 143 (1866) ... ... — — — — Q. — M.fr.I,143;VI,232.
F. Strzeleckiana, F. v. Mueller, fragm. I, 65 (1858) ... ... — — — N.S.W. Q. — B.fl.I,389 M.fr.IX,133.

c

OCHNACEAE.
De Candolle in Ann. du Mus. XVII, 308 (1811).

BRACKENRIDGEA, Asa Gray. Bot. Wilk. Expl. Exped. 361, t. 42 (1854).
B. Australiana, F. v. Mueller, fragm. V, 29 (1865) ... ... — — — — — Q. — — M.fr.V,29.

RUTACEAE.
A. L. de Jussieu, Gen. 296 (1789).

ZIERIA, Smith in Transact. Linn. Soc. IV, 216 (1798).
Z. laevigata, Smith in Rees's Cyclopaedia, XXXIX (1819) ... — — V. N.S.W. Q. — B.fl.I,304 M.fr.IX,116.
Z. pilosa, Rudge in Transact. Linn. Soc. X, 293, t. 17 (1809) ... — — — N.S.W. — — B.fl.I,305 M.fr.I,100;IX,110.
Z. obcordata, Cunningham in Field. N.S.Wales, 330 (1825) ... — — — N.S.W. Q. — B.fl.I,305 M.fr.I,101;IX,116.
Z. cytisoides, Smith in Rees's Cyclopaedia, XXXIX (1819) ... — — T. V. N.S.W. Q. — B.fl.I,306 M.fr.IX,116.
Z. Smithii, Andrews' Bot. Rep. t. 606 (1810) ... ... — — T. V. N.S.W. Q. — B.fl.I,306 M.fr.I,100;IX,116.
Z. granulata, C. Moore in Benth. Fl. Austr. I, 307 (1863) ... — — — — N.S.W. Q. — B.fl.I,307 M.fr.IX,116.
Z. veronicea, F. v. Mueller in Trans. phil. Soc. Vict. I, 11 (1854) — S.A. T. V. — — — B.fl.I,305 M.fr.IX,116.

BORONIA, Smith, Tracts relat. nat. hist. 285 (1798). (Cyanothamnus).
B. grandisepala, F. v. Mueller, fragm. I, 66 (1858) ... ... — — — N.A. B.fl.I,311 M.fr.I,66;II,179.
B. artemisifolia, F. v. Mueller, fragm. I. 66 (1858) ... ... — — N.A. B.fl.I,311 M.fr.I,66;IX,111.
B. affinis, R. Brown in Benth. Fl. Austr. I, 311 (1863) ... — — — N.A. B.fl.I,311
B. filicifolia, Cunningham in Benth. Fl. Austr. I, 311 (1863) ... — — — — N.A. B.fl.I,311
B. alata, Smith in Transact. Linn. Soc. VIII, 283 (1807) ... W.A. — — — — B.fl.I,312 M.fr.IX,111.
B. algida, F. v. Mueller in Transact. phil. Soc. Vict. I, 100 (1855) — — V. N.S.W. — B.fl.I,312 M.fr.IX,111.
B. Edwardsii, Bentham, Fl. Austr. I, 312 (1863) ... ... — S.A. — — — B.fl.I,312
B. ternata, Endlicher, nov. stirp. dec. 6 (1839) ... ... W.A. — — — — B.fl.I,312 M.fr.IX,111.
B. ericifolia, Bentham, Fl. Austr. I, 313 (1863) ... ... W.A. — — — — B.fl.I,313 M.fr.IX,111.
B. inconspicua, Bentham, Fl. Austr. I, 313 (1863) ... W.A. — — — — B.fl.I,313
B. eriantha, Lindley in Mitch. Trop. Austr. 208 (1848) ... — — — Q. — B.fl.I,313
B. alulata, Solander in Benth. Fl. Austr. I, 313 (1863) ... — — — Q. — B.fl.I,313
B. ledifolia, J. Gay, monogr. des Lasiopetalées 29 (1821) ... — — N.S.W. Q. — B.fl.I,314 M.fr.IX,111.
B. lanceolata, F. v. Mueller, fragm. I, 66 (1858) ... ... — — — Q. N.A. B.fl.I,314 M.fr.I,66;VII,38;IX,112;N
B. Bowmanii, F. v. Mueller, fragm. IV, 135 (1864) ... — — — Q. — M.fr.VII,38. [I
B. Fraseri, Hooker, Bot. Mag. t. 4052 (1844) ... ... — — N.S.W. — B.fl.I,315 M.fr.IX,111.
B. mollis, Cunningham in Lindl. Bot. Reg. t. 1597 (1841) ... — — N.S.W. — B.fl.I,315 M.fr.IX,111.
B. megastigma, Nees in Lehm. pl. Preiss. II, 227 (1847) ... W.A. — — — — B.fl.I,315 M.fr.II,97;IV,112;IX,112.
B. heterophylla, F. v. Mueller, fragm. II, 98 (1861) ... W.A. — — — — B.fl.I,315 M.fr.II,98;IX,112.
B. elatior, Bartling in Lehm. pl. Preiss. I, 170 (1844) ... W.A. — — — — B.fl.I,316 M.fr.II,98;IX,122;XI,97.
B. tetrandra, Labillardière, Nov. Holl. pl. spec. I, 98, t. 125 (1804) W.A. — — — — B.fl.I,316
B. crassifolia, Bartling in Lehm. pl. Preiss. I, 169 (1844) ... W.A. — — — — B.fl.I,316 M.fr.IX,112.
B. albiflora, R. Brown in Benth. Fl. Austr. I, 317 (1863) ... W.A. — — — — B.fl.I,317 M.fr.IX,114.
B. lanuginosa, Endlicher in Hueg. enum. 16 (1837) ... W.A. — — — — B.fl.I,317 M.fr.IX,114.
B. pulchella, Turczaninow in Bull. Soc. Mosc. XXV, 162 (1852) W.A. — — — — B.fl.I,318 M.fr.IX,114.
B. gracilipes, F. v. Mueller, fragm. IX, 99 (1861) ... W.A. — — — — B.fl.I,318 M.fr.II,99;IX,114.
B. microphylla, Sieber in Spreng. syst. cur. post. 148 (1827) ... — — N.S.W. Q. — B.fl.I,318
B. pinnata, Smith, Tracts relat. nat. hist. 290, t. 4 (1798) ... — S.A. T. V. N.S.W. Q. — B.fl.I,318 M.fr.IX,114.
B. pilosa, Labillardière, Nov. Holl. pl. spec. I, 97, t. 124 (1804) ... — T. V. — — — B.fl.I,319
B. spinescens, Bentham, Fl. Austr. I, 319 (1863) ... ... W.A. — — — — B.fl.I,319
B. bæckeacea, F. v. Mueller, fragm. IX, 112 (1875) ... W.A. — — — — M.fr.IX,112.
B. coerulescens, F. v. Mueller in Trans. phil. Soc. Vict. II (1854) W.A. S.A. — V. N.S.W. — B.fl.I,320 M.fr.IX,112.
B. defoliata, F. v. Mueller, fragm. IX, 113 (1875) ... W.A. — — — — M.fr.IX,113.
B. Busselliana, F. v. Mueller, fragm. II, 113 (1875) ... W.A. — — — — M.fr.IX,113.
B. subcoerulea, F. v. Mueller, fragm. II, 100 (1860) ... W.A. — — — — B.fl.I,320 M.fr.II,100;IX,112&198.
B. pulcifolia, Smith, Tracts relat. nat. hist. 297, t. 7 (1798) — S.A. T. V. N.S.W. Q. — B.fl.I,320 M.fr.IX,114.
B. falcifolia, Cunningham in Lindl. Bot. Reg. 47 (1841)... — — N.S.W. Q. — B.fl.I,322 M.fr.IX,114.
B. penicillata, Bentham, Fl. Austr. I, 322 (1863)... ... W.A. — — — — B.fl.I,322
B. crassipes, Bartling in Lehm. pl. Preiss. I, 168 (1844) ... W.A. — — — — B.fl.I,322 M.fr.IX,116.
B. subcssilis, Bentham, Fl. Austr. I, 322 (1863) ... ... W.A. — — — — B.fl.I,322
B. capitata, Bentham, Fl. Austr. I, 323 (1863) ... ... W.A. — — — — B.fl.I,323 M.fr.IX,116.
B. nematophylla, F. v. Mueller, fragm. II, 100 (1860) ... W.A. — — — — B.fl.I,323 M.fr.IX,116.
B. crenulata, Smith in Transact. Linn. Soc. VIII, 284 (1807) ... W.A. — — — — B.fl.I,324 M.fr.XI,115&198.
B. serrulata, Smith, Tracts relat. nat. hist. 292 (1798) ... — — N.S.W. — B.fl.I,324 M.fr.IX,115.
B. rhomboidea, Hooker icon. pl. t. 722 (1848) ... — — T. — B.fl.I,324
B. parviflora, Smith, Tracts relat. nat. hist. 293, t. 5 (1798) ... — S.A. T. V. N.S.W. Q. — B.fl.I,324 M.fr.IX,115.
B. viminea, Lindley, Bot. Regist. XXV, App. XVII (1839) ... W.A. — — — — B.fl.I,324
B. filifolia, F. v. Mueller, fragm. II, 101 (1861) ... ... W.A. — — — — B.fl.I,325 M.fr.I,31;IX,114.
B. clavellifolia, F. v. Mueller in Trans. phil. Soc. Vict. I. 12 (1854) W.A. S.A. — V. N.S.W. — B.fl.I,325 M.fr.I,90;IX,114&198.
B. scabra, Lindley, Bot. Regist. XXV, App. XVII (1839) ... W.A. — — — — B.fl.I,326
B. thymifolia, Turcz. in Bull. Soc. Mosc. XXV, part II, 165 (1852) W.A. — — — — B.fl.I,326 M.fr.I,99;II,99,IX,116.
B. ovata, Lindley, Bot. Regist. 47 (1841) ... ... W.A. — — — — B.fl.I,327 M.fr.IX,116.
B. Barkeriana, F. v. Mueller, fragm. XI, 96 (1880) ... — — T. — M.fr.XI,96.
B. denticulata, Smith in Transact. Linn. Soc. VIII, 284 (1807) ... W.A. — — — N.S.W. — B.fl.I,327 M.fr.IX,115&198.
B. Machardiana, F. v. Mueller, fragm. IX, 115 (1875) ... W.A. — — — — B.fl.I,327 M.fr.IX,115.
B. spathulata, Lindley, Bot. Regist. XXV, App. XVII (1839)... W.A. — — — — B.fl.I,327 M.fr.IX,115.
B. juncea, Bartling in Lehm. pl. Preiss. I, 166 (1844) ... W.A. — — — — B.fl.I,327 M.fr.IX,115.
B. cymosa, Endlicher in Hueg. enum. 16 (1837) ... ... W.A. — — — — B.fl.I,328 M.fr.II,101;IX,116.

ERIOSTEMON, Smith in Transact. Linn. Soc. IX, 221 (1798). (Crowea, Phebalium, Asterolasia, Microcybe, Pleurandrop
Geleznowia, Urocarpus, Sandfordia, Actinostigma).
E. virgatus, Cunningham in Hooker's Journ. of Bot. II, 417 (1840) — — T. — — — B.fl.I,332 M.fr.IX,111.
E. nodiflorus, Lindley, Bot. Regist. XXV, App. XVII (1839) ... W.A. — — — — B.fl.I,336 M.fr.IX,110.

E. aploatus, A. Richard, sert. Astrolab. 76, t. 27 (1833) ... W.A. — — — — — — B.fl.I,336 M.fr.IX,110.
E. pungens, Lindley in Mitch. Three Exped. II, 156 (1838) ... — S.A. — V. N.S.W. — B.fl.I,338 M.fr.IX,110.
E. montanus, F. v. M., pl. of Vict. I, 120 (1860) ... ... — — T. — — — B.fl.I,338 M.fr.IX,109.
E. phylicifolius, F. v. M., fragm. I, 105 (1859) ... ... — — — V. — — — B.fl.I,339 M.fr.I,105.
E. umbellatus, Turcz. in Bull. Soc. Mosc. XXII, part III, 15 (1849) — — — V. N.S.W. — B.fl.I,339 M.fr.I,104;IX,109.
E. Ralstoni, F. v. M., fragm. II, 101, t. 14 (1861) ... ... — — — V. N.S.W. — B.fl.I,339 M.fr.III,165;IX,109.
E. Hillebrandi, F. v. M. in Transact. phil. Soc. Vict. I, 10 (1854) ... — S.A. T. V. — — — B.fl.I,340 M.fr.I,4;IX,109.
E. lamprophyllus, F. v. M. in Journ. pharm. Soc. Vict. II, 43 (1858) — — — V. — — B.fl.I,340 M.fr.IX,109.
E. elatior, F. v. M., fragm. I, 181 (1859) ... ... — — — N.S.W. — B.fl.I,340 M.fr.IX,109.
E. amblens, F. v. M., fragm. VI, 166 (1868) ... ... — — — N.S.W. — — M.fr.VI,166;IX,110.
E. Oldfieldii, F. v. M., fragm. I, 3 (1858) ... ... — — T. — — — B.fl.I,340 M.fr.IX,109.
E. rotundifolius, Cunningham in Hueg. enum. 15 (1837)... ... — — — N.S.W. Q. — B.fl.I,341
E. microphyllus, F. v. M. in Transact. phil. Soc. Vict. I, 99 (1834) — S.A. — — — — — B.fl.I,341 M.fr.IX,109.
E. phylicoides, F. v. M., fragm. I, 107 (1859) ... ... — — — V. N.S.W. — B.fl.I,341 M.fr.IX,107.
E. ozothamnoides, F. v. M., fragm. I, 103 (1859) ... ... — — — V. N.S.W. — B.fl.I,342 M.fr.IX,108.
E. Mortoni, F. v. M., fragm. IX, 108 (1875) ... ... — — — — N.S.W. — B.fl.I,342 M.fr.IX,108.
E. sediflorus, F. v. M., fragm. I, 102 (1859) ... ... — S.A. T. V. N.S.W. — B.fl.I,342 M.fr.I,102.
E. Nottii, F. v. M., fragm. VI, 22 (1867) ... ... — — — — — Q. — M.fr.VI,22.
E. leptolotus, Sprengel, syst. II, 322 (1825) ... ... — S.A. — V. N.S.W. Q. — B.fl.I,343 M.fr.I,104;IX,107.
E. alpinus, F. v. M., fragm. I, 103 (1859) ... ... — — — V. N.S.W. — — M.fr.I,103.
E. tuberculosus, F. v. M., pl. of Vict. I, 130 (1862) ... W.A. — — — — — B.fl.I,343 M.fr.IX,108.
E. Benthami, F. v. M., fragm. IX, 108 (1875) ... ... W.A. — — — — — B.fl.I,343 M.fr.IX,108.
E. Maxwelli, F. v. M., fragm. IX, 108 (1875) ... ... W.A. — — — — — — M.fr.IX,108.
E. filifolius, F. v. M., fragm. IX, 108 (1875) ... ... W.A. — — — — — B.fl.I,344 M.fr.IX,108.
E. squameus, Labillardière, Nov. Holl. pl. spec. I, 111, t. 141 (1804) — — T. V. N.S.W. Q. — B.fl.I,344 M.fr.I,104;IX,107.
E. anceps, Sprengel, syst. II, 322 (1825) ... ... W.A. — — — — — B.fl.I,345 M.fr.I,103;IX,108.
E. ovatifolius, F. v. M., fragm. I, 103 (1859) ... ... — — — V. N.S.W. — B.fl.I,345 M.fr.I,103.
E. bilobus, F. v. M., fragm. I, 102 (1859) ... ... W.A. — — — — — B.fl.I,345 M.fr.IX,108.
E. Becklevi, F. v. M., fragm. IX, 109 (1875) ... ... — — — — N.S.W. — — M.fr.IX,108.
E. capitatus, F. v. M., fragm. I, 116 (1859) ... ... W.A. S.A. — — — — B.fl.I,346 M.fr.I,106;IX,107.
E. Geleznowii, F. v. M., fragm. I, 107 (1859) ... ... W.A. — — — — — B.fl.I,347 M.fr.IX,107.
E. correlfolius, F. v. M., fragm. I, 105 (1859) ... ... — — — V. N.S.W. — B.fl.I,350 M.fr.IX,107.
E. Cunninghami, F. v. M., fragm. IX, 107 (1875) ... ... — — — — N.S.W. — B.fl.I,351 M.fr.IX,107.
E. mollis, Asterolasia mollis, Benth. Fl. Austr. I, 351 (1863) ... — — — N.S.W. — B.fl.I,340
E. plouramtroides, F. v. M., fragm. I, 106 (1859) ... ... — — — V. — — B.fl.I,349 M.fr.IX,107.
E. trymalioides, F. v. M., fragm. I, 106 (1859) ... ... — — — V. N.S.W. — B.fl.I,351 M.fr.IX,107.
E. Hookeri, F. v. M., fragm. I, 104 (1859) ... ... W.A. — — — — — B.fl.I,352 M.fr.VII,22;IX,107.
E. pallidus, Asterolasia pallida, Benth. Fl. Austr. I, 352 (1863) W.A. — — — — — B.fl.I,352
E. Drummondii, F. v. M., fragm. I, 105 (1859) ... ... W.A. — — — — — B.fl.I,352 M.fr.VII,22.
E. grandiflorus, F. v. M., fragm. I, 105 (1859) ... ... W.A. — — — — — B.fl.I,353 M.fr.VI,21.
E. Crowei, F. v. Mueller, pl. of Vict. I, 119 (1860) ... ... W.A. — — V. N.S.W. Q. — B.fl.I,320 M.fr.IX,106.
E. Turczaninowii, F. v. Mueller, pl. of Vict. I, 120 (1860) ... W.A. — — — — — B.fl.I,330 M.fr.IX,106.
E. lanceolatus, K. F. Gaertner, de fruct. III, 154, t. 210 (1807) — — — V. N.S.W. — — B.fl.I,331 M.fr.IX,110.
E. Banksii, Cunningham in Hueg. enum. 15 (1837) ... ... — — — — — Q. — B.fl.I,332
E. trachyphyllus, F. v. M. in Transact. phil. Soc. Vict. I, 99 (1854) — — — V. N.S.W. — B.fl.I,333 M.fr.IX,110.
E. myoporoides, De Candolle, prodr. I, 720 (1824) ... ... — — — V. N.S.W. Q. — B.fl.I,333 M.fr.IX,110.
E. hispidulus, Sieber in Spreng. syst. cur. post. 164 (1827) ... — — — V. N.S.W. — B.fl.I,333 M.fr.IX,111.
E. buxifolius, Smith in Rees's Cyclop. XIII, 2 (1809) ... ... — — — N.S.W. — B.fl.I,333 M.fr.IX,111.
E. obovalis, Cunningham in Field's N. S. Wales, 331 (1825) ... — S.A. T. V. N.S.W. — B.fl.I,334 M.fr.IX,111.
E. scaber, Paxton, Mag. of Bot. XIII, 127 (1846)... ... — — — V. N.S.W. — B.fl.I,334 M.fr.IX,111.
E. Brucei, F. v. M., fragm. VII, 38 (1869) ... ... W.A. — — — — — — M.fr.IX,110.
E. linearis, Cunningham in Hueg. enum. 16 (1837) ... ... — — — N.S.W. — B.fl.I,334
E. difformis, Cunningham in Hueg. enum. 15 (1837) ... ... W.A. S.A. — V. N.S.W. — B.fl.I,335 M.fr.IX,110.
E. parvifolius, R. Brown in Benth. Fl. Austr. I, 335 (1863) ... — — — — — Q. — B.fl.I,335
E. ericifolius, Cunningham in Benth. Fl. Austr. I, 335 (1863) ... — — — N.S.W. — B.fl.I,335

PHILOTHECA, Rudge in Transact. Linn. Soc. XI, 298, t. 21 (1815). (Drummondita).
P. australis, Rudge in Transact. Linn. Soc. XI, 298, t. 21 (1815) ... — — — N.S.W. Q. — B.fl.I,348 M.fr.VII,20,141.
P. ericoides, F. v. M., fragm. I, 107 (1859) ... ... W.A. — — — — — B.fl.I,349 M.fr.VII,21.
P. calida, F. v. M., fragm. VII, 21 (1869)... ... ... — — — — — Q. — M.fr.VII,38.

CORREA, Smith in Transact. Linn. Soc. IV, 219 (1798). (Didymeria).
C. aemula, F. v. M., fragm. I, 3 (1859) ... ... — S.A. — V. — — — B.fl.I,353 M.fr.IX,117.
C. alba, Andrews, Bot. Reposit. t. 18 (1797) ... ... — S.A. T. V. N.S.W. — B.fl.I,354 M.fr.IX,117.
C. speciosa, Andrews, Bot. Reposit. t. 653 (1811)... ... ... — S.A. T. V. N.S.W. Q. — B.fl.I,354 M.fr.IX,117.
C. Lawrenciana, Hooker, Journ. of Bot. I, 254 (1834) ... ... — — T. V. N.S.W. — B.fl.I,355 M.fr.VIII,142;IX,117.
C. decumbens, F. v. M. in Transact. phil. Soc. Vict. I, 30 (1854) ... — S.A. — V. — — — B.fl.I,355 M.fr.IX,117.

NEMATOLEPIS, Turczaninow in Bull. Soc. Mosc. XXV, part III, 158 (1852). (Symphyopetalum).
N. phlebaloides, Turczaninow in Bull. Soc. Mosc. part III, 158 (1852) W.A. — — — — — B.fl.I,356 M.fr.IX,106.
N. Euphemiae, F. v. M., fragm. III, 140 (1863) ... ... W.A. — — — — — B.fl.I,345 M.fr.IX,106.

CHORILAENA, Endlicher in Hueg. enum. 17 (1837).
C. quercifolia, Endlicher in Hueg. enum. 17 (1837) ... ... W.A. — — — — — B.fl.I,357 M.fr.IX,106.
C. hirsuta, Bentham, Fl. Austr. I, 357 (1863) ... ... W.A. — — — — — B.fl.I,357 M.fr.IX,106.

DIPLOLAENA, R. Brown in Append. Flind. voy. II, 546 (1814).
D. Dampieri, Desfontaines in Mém. du Mus. 452, t. 20 (1817) ... W.A. — — — — — B.fl.I,358 M.fr.IX,106.

BOSISTOA, F. v. M. in Benth. Fl. Austr. I, 359 (1863).
B. sapindiformis, F. v. M. in Benth. Fl. Austr. I, 359 (1863) ... — — — N.S.W. Q. — B.fl.I,359 M.fr.IX,103.
B. euodiformis, F. v. M., fragm. IX, 174 (1875) ... ... — — — N.S.W. — — M.fr.IX,174.

ACRADENIA, Kippist in Transact. Linn. Soc. XXI, 207 (1855).
A. Frankliniae, Kippist in Transact Linn. Soc. XXI, 207 (1855) — — T. — — — — B.fl.I,323 M.fr.VI,167;IX,103.

EUODIA, R. et G. Forster, char. gen. 13, t. 7 (1776). (Melicope, Medicosma).
E. pentacocca, F. v. M., fragm. III, 41 (1862) ... ... — — — N.S.W. Q. — — M.fr.III,168;IV,175.
E. Cunninghami, F. v. M., fragm. III, 2 (1862) ... ... — — — N.S.W. Q. — B.fl.I,362 M.fr.I,27;IX,102.
E. Fareana, F. v. M., fragm. IX, 101 (1875) ... ... — — — — Q. — — M.fr.IX,101.
E. erythrococca, F. v. M., fragm. I, 26 (1858) ... ... — — — N.S.W. Q. — B.fl.I,360 M.fr.II,103,178;IX,102.
E. contermina, Moore & F. v. M., fragm. VII, 144 (1871) ... — — — N.S.W. — — — M.fr.VII,144.
E. octandra, F. v. M., fragm. II, 102 (1861) ... ... — — — N.S.W. — — B.fl.I,360 M.fr.II,102;IX,102.
E. micrococca, F. v. M., fragm. I, 144 (1859) ... ... — — — N.S.W. Q. — B.fl.I,361 M.fr.II,103,180;IX,102.
E. xanthoxyloides, F. v. M., fragm. IV, 155 (1864) ... ... — — — — Q. — — M.fr.IV,155.
E. Bonwickii, F. v. M., fragm. V, 56 (1865) ... ... — — — — Q. — — M.fr.V,179.
E. alata, F. v. M., fragm. VII, 142 (1871) ... ... — — — — Q. ... — M.fr.VII,142.
E. accedens, Blume, Bijdr. 246 (1825) ... ... — — — N.S.W. Q. ... — M.fr.V,4,56;VII,22;IX,102.
E. littoralis, Endlicher, prodr. Fl. Norfolk. 86 (1833) ... — — — N.S.W. — — — M.fr.IX,102.
E. polybotrya, Moore & F. v. M., fragm. VII, 143 (1871) ... — — — N.S.W. — — — M.fr.IX,193.
E. vitiflora, F. v. M., fragm. VII, 144 (1871) ... ... — — — — Q. — .. M.fr.VII,144.

BROMBYA, F. v. M., fragm. V, 4 (1865).
B. platynema, F. v. M., fragm. V, 4 (1865) ... — — — — — Q. — — M.fr.V,56;IX,102.

PAGETIA, F. v. M., fragm. V, 178 (1866).
P. medicinalis, F. v. M., fragm. V, 178 (1866) ... ... — — — — — Q. — — M.fr.V,178;IX,103.

BOUCHARDATIA, Baillon in Adans. VII, 350, t. 10 (1867).
B. neurococca, Baillon in Adans. IX, 110 (1860) ... ... — — — N.S.W. Q. — B.fl.I,360 M.fr.I,28;II,103,178;IX,103.

XANTHOXYLUM, Catesby in Linné, hort. Cliff. 487 (1737). (Zanthoxylum, Blackburnia).
X. brachyacanthum, F. v. M., pl. of Vict. I, 108 (1862) — — — N.S.W. Q. — B.fl.I,363 M.fr.VII,141;IX,104.
X. Blackburnia, Bentham, Fl. Austr. I, 363 (1863) ... — — — N.S.W. — — B.fl.I,363
X. parviflorum, Bentham, Fl. Austr. I, 363 (1863) ... — — — — N.A. B.fl.I,363 M.fr.VII,141;IX,104.
X. torvum, F. v. M., fragm. VII, 140 (1871) ... ... — — — — Q. — M.fr.VII,140;IX,104.

GEIJERA, Schott, Rutac. 4, t. 7 (1834). (Coatesia).
G. Muelleri, Bentham, Fl. Austr. I, 364 (1863) ... ... — — — — — Q. — B.fl.I,364 M.fr.I,26;IX,104.
G. salicifolia, Schott, Rutac. t. 4 (1834) ... ... — — — N.S.W. Q. — B.fl.I,364 M.fr.VII,22;IX,105;XI,134.
G. parviflora, Lindley in Mitch. Trop. Austr. 102 (1848) ... W.A. S.A. — V. N.S.W. Q. — B.fl.I,364 M.fr.VII,22;XI,105;XI,130.

PLEIOCOCCA, F. v. Mueller, fragm. IX, 117 (1875).
P. Wilcoxiana, F. v. M., fragm. IX, 117 (1875) ... ... — — — N.S.W. — — — M.fr.IX,117.

ACRONYCHIA, R. & G. Forster, char. gen. 53, t. 27 (1776). (Cyminosma).
A. Baueri, Schott, Rutac. t. 3 (1834) ... ... — — — N.S.W. Q. — B.fl.I,366 M.fr.I,26;IX,103.
A. laevis, R. & G. Forster, char. gen. 53, t. 27 (1776) ... — — — V. N.S.W. Q. — B.fl.I,366 M.fr.I,27;IX,104.
A. imperforata, F. v. M., fragm. I, 26 (1858) ... ... — — — N.S.W. Q. — B.fl.I,367 M.fr.I,26;IX,104.
A. melicopoides, F. v. M., fragm. V, 3 (1865) ... ... — — — N.S.W. Q. — M.fr.IV,117;VII,145.
A. acidula, F. v. M., fragm. IV, 154 (1864) ... ... — — — — — Q. — M.fr.IV,154.
A. vestita, F. v. M., fragm. IV, 155 (1864) ... ... — — — — — Q. — M.fr.IV,155;IX,104.
A. Endlicheri, Schott, fragm. bot. III, t. 2 ... ... — — — N.S.W. — — M.fr.IX,104.
A. tetrandra, F. v. M., fragm. IX, 104 (1875) ... ... — — — — — Q. — M.fr.V,179;IX,104.

HALFORDIA, F. v. Mueller, fragm. V, 43 (1865).
H. drupifera, F. v. M., fragm. V, 43 (1865) ... ... — — — — Q. — M.fr.V,5,180;IX,103.
H. scleroxyla, F. v. M., fragm. VII, 142 (1871) ... ... — — — — Q. — M.fr.VII,142;IX,103.

GLYCOSMIS, Correa in Ann. du Mus. VI, 384 (1805).
G. pentaphylla, Correa in Annal. du Mus. VI, 384 (1805) ... ... — — — — — Q. — B.fl.I,367 M.fr.IX,105.

MICROMELUM, Blume, Bijdrag. 137 (1825).
M. pubescens, Blume, Bijdrag. 137 (1825) ... ... — — — — — Q. N.A. B.fl.I,366 M.fr.I,25;IX,105.

MURRAYA, Koenig in Linné, mant. alt. 563 (1771). (Chalcas, 1767).
M. exotica, Koenig in Linné, mant. alt. 563 (1771) ... — — — — Q. — B.fl.I,369 M.fr.IX,106.
M. crenulata, Oliver in Journ. Linn. Soc. V, suppl. II, 29 (1861) — — — — Q. — B.fl.I,369

CLAUSENA, Burman, Fl. Indica index & t. 29 (1768).
C. brevistyla, Oliver in Journ. Linn. Soc. V, suppl. II. 31 (1861) — — — — Q. — B.fl.I,369 M.fr.IX,106.

ATALANTIA, Correa in Ann. du Mus. VI, 383 (1805).
A. glauca, J. Hooker in B. & H. Gen. pl. I, 305 (1862) ... — — — N.S.W. Q. — B.fl.I,370 M.fr.XI,105.
A. recurva, Bentham, Fl. Austr. I, 370 (1863) ... ... — — — N.A. B.fl.I,370 M.fr.VII,142.

CITRUS, Linné, gen. pl. 230 (1737), from Plinius.
C. Planchonii, F. v. M., Report Intercol. Exhib. 23 (1867) ... — — — N S W. Q. — B.fl.I,371 M.fr.IX,105. [IX,105.
C. Australasica, F. v. M., fragm. I, 26 (1858) ... ... — — — N.S.W. Q. — B.fl.I,371 M.fr.I,26;II,178;III,167;

PENTACERAS, J. Hooker in B. & H. Gen. pl. I, 298 (1862).
P. australis, J. Hooker in B. & H. Gen. pl. I, 298 (1862) ... — — — N.S.W. Q. — B.fl.I,365 M.fr.IX,104.

SIMARUBEAE.
De Candolle in Ann. du Mus. XVII, 323 (1811).

AILANTUS, Desfontaines, in Act. Acad. Paris, 265 (1786).
A. imberbiflora, F. v. M., fragm. III, 42 (1862) ... ... — — — N.S.W. Q. — B.fl.I,373 M.fr.III,42;VI,166.

BRUCEA, J. S. Miller, icon. plantar. t. 25 (1780).
B. Sumatrana, Roxburgh, Fl. Indic. I, 469 (1820) ... ... — — — N.A. B.fl.I,373 M.fr.IX,166.

HYPTIANDRA, J. Hooker in B. & H. Gen. pl. I, 294 (1862).
H. Bidwilli, J. Hooker in B. & H. Gen. pl. 294 (1862) ... — — — N.S W. Q. — B.fl.I,374 M.fr.VI,165.

CADELLIA, F. v. Mueller, fragm. II, 25 (1860).
C. pentastylis, F. v. M., fragm. II, 26, t. 12 (1860) ... ... — — — — N.S.W. — — B.fl.I,374 M.fr.II,23&171.
C. monostylis, Bentham, Fl. Austr. I, 375 (1863)... ... — — — — N.S.W. Q. — B.fl.I,375 M.fr.IV,166;VIII,34.
SURIANA, Plumier, nov. plant. Amer. gen. 37, t. 40 (1703).
S. maritima, Linné, sp. pl. 284 (1753) ... ... — — — — — Q. — B.fl.I,375
HARRISONIA, R. Brown in Mém. du Mus. XII, 517 (1825). (Ebelingia).
H. Brownii, A. L. de Juss. in Mém. du Mus. Par. XII, 530, t. 28 (1825) — — — — — — N.A. B.fl.I,376

ZYGOPHYLLEAE.
R. Brown in App. Flind. voy. II, 545 (1814).
NITRARIA, Linné, syst. nat. X, 1044 (1759).
N. Schoberi, Linné, sp. pl. ed. sec. 638 (1762) ... ... W.A. S.A. — V. N.S.W. — — B.fl.I,291 M.fr.XI,30.
ZYGOPHYLLUM, Linné, syst. nat. 8 (1735), Linné, gen. 126 (1737). (Sarcozygium).
Z. apiculatum, F. v. M. in Linnaea XXV, 373 (1852) ... ... — S.A. T. V. N.S.W. Q. — B.fl.I,292 M.fr.XI,29.
Z. glaucescens, F. v. M. pl. of Vict. I, 228 (1862) ... — S.A. — V. N.S.W. — — B.fl.I,293 M.fr.XI,29.
Z. iodocarpum, F. v. M. in Linnaea XXV, 372 (1852) ... W.A. S.A. — V. N.S.W. — — B.fl.I,293 M.fr.XI,29.
Z. prismatothecum, F. v. M. in Linnaea XXV, 373 (1852) ... — S.A. — — — — B.fl.I,293
Z. ammophilum, F. v. M. in Linnaea XXV, 376 (1852) ... W.A. S.A. — V. N.S.W. — — M.fr.XI,28.
Z. Billardieri, de Candolle, prodr. I, 705 (1824) ... W.A. S.A. T. V. N.S.W. — — B.fl.I,293 M.fr.XI,29.
Z. fruticulosum, de Candolle, prodr. I, 705 (1824) ... W.A. S.A. — V. N.S.W. Q. — B.fl.I,294 M.fr.XI,29.
Z. Howittii, F. v. M., fragm. III, 150 (1863) ... ... — S.A. — — — — — M.fr.III,150;XI,29.
TRIBULUS, L'Obel, Icon. II, 84 (1581), from Theophr. Diosc. & Plinius. (Tribulopsis).
T. terrestris, L'Obel, stirp. icones II, 84 (1581) ... ... W.A. S.A. — V. N.S.W. Q. N.A. B.fl.I,288 M.fr.XI,30.
T. ranunculiflorus, F. v. M., fragm. I, 48 (1858) ... ... — — — — — N.A. B.fl.I,288 M.fr.I,48.
T. hystrix, R. Brown in App. Sturt. Exped. 6 (1849) ... — S.A. — — — — N.A. B.fl.I,289 M.fr.XI,30.
T. macrocarpus, F. v. M. in Benth. Fl. Austr. I, 289 (1863) ... — — — — — N.A. B.fl.I,289
T. platypterus, Bentham, Fl. Austr. I, 289 (1863) ... — — — — — N.A. B.fl.I,289
T. hirsutus, Bentham, Fl. Austr. I, 289 (1863) ... ... — — — — — N.A. B.fl.I,289
T. minutus, Leichhardt in Benth. Fl. Austr. I, 291 (1863) ... — — — — Q. — B.fl.I,291
T. Brownii, F. v. M., pl. of Vict. I, 90 (1860) ... ... — — — — — Q. N.A. B.fl.I,290 M.fr.I,48.
T. bicolor, F. v. M., pl. of Vict. I, 90 (1860) ... ... — — — — — N.A. B.fl.I,290 M.fr.I,47.
T. Solandri, F. v. M., pl. of Vict. I, 90 (1860) ... ... — — — — Q. N.A. B.fl.I,290 M.fr.I,47; XI,30.

LINEAE.
De Candolle. Théor. élém. bot. 214 (1813).
HUGONIA, Linné, gen. pl. 134 (1737).
H. Jenkinsii, F. v. M., fragm. V, 7 (1865) ... ... — — — — — — Q. — — M.fr.IX,190.
ERYTHROXYLUM, P. Brown, Hist. pl. of Jamaica 278 (1756), from Theophr. Diosc. & Plinius. (Erythroxylon).
E. australe, F. v. M. in Transact. Vict. Inst. III, 22 (1855) ... — — — — — Q. — B.fl.I,284
E. ellipticum, R. Brown in Benth. Fl. Austr. I, 284 (1863) ... — — — — — N.A. B.fl.I,284
LINUM, Tournefort, inst. 339, t. 176, (1700), from Theophr. Diosc. & Plinius.
L. marginale, Cunningh. in Hook. Lond. Jour. Bot. VII, 169 (1848) W.A. S.A. T. V. N.S.W. Q. — B.fl.I,283

GERANIACEAE.
A. L. de Jussieu, Gen. 268 (1789), from B. de Jussieu (1759).
GERANIUM, Tournefort, inst. 266, t. 142 (1700), from Diosc. & Plinius.
G. Carolinianum, Linné, sp. pl. 682 (1753)... ... ... W.A. S.A. T. V. N.S.W. Q. — B.fl.I,296
G. sessiliflorum, Cavanilles, Diss. 106, t. 77 (1790) ... ... — T. V. N.S.W. — B.fl.I,297
ERODIUM, L. Héritier, Geraniologia, t. 1 (1787).
E. cygnorum, Nees in Lehm. pl. Preiss. I, 162 (1844) ... ... W.A. S.A. — V. N.S.W. Q. N.A. B.fl.I,297 M.fr.XI,20.
PELARGONIUM, L'Héritier, Geraniologia, t. 7 (1787).
P. australe, Willdenow, sp. pl. III, 675 (1800) ... ... W.A. S.A. T. V. N.S.W. Q. — B.fl.I,298
P. Rodneyanum, Mitchell, three Exped. II, 144 (1838) ... W.A. S.A. — V. N.S.W. — — B.fl.I,299
OXALIS, Linné, gen. pl. 134 (1737), from Plinius.
O. Magellanica, G. Forster in Comm. Goetting. IX, 33 (1789) ... — — T. V. — — B.fl.I,300
O. corniculata, Linné, sp. pl. 435 (1753) ... ... W.A. S.A. T. V. N.S.W. Q. — B.fl.I,301

MALVACEAE.
Adanson in Mém. Acad. fr. 224 (1761).
LAVATERA, Tournefort in Act. Acad. Par. 86, t. 3 (1706).
L. plebeja, Sims in Bot. Mag. t. 2269 (1821) ... ... W.A. S.A. T. V. N.S.W. — — B.fl.I,185 M.fr.IX,130.
MALVASTRUM, Asa Gray in Mém. Amer. Acad. IV, 21 (1849). (Malva partly).
M. spicatum, A. Gray in Mém. Amer. Acad. IV, 21 (1849) — S.A. — N.S.W. Q. N.A. B.fl.I,187 M.fr.VI,170;IX,130.
M. tricuspidatum, A. Gray in Mém. Amer. Acad. IV, 21 (1849) — — — — N.S.W. Q. — B.fl.I,187
PLAGIANTHUS, R. & G. Forster, char. gen. 85, t. 43 (1776). (Asterotrichon, Blepharanthemum, Lawrencia, Halothamnus).
P. sidoides, Hooker, Bot. Mag. t. 3396 (1835) ... ... — — — — — — B.fl.I,185 M.fr.VI,169.
P. pulchellus, A. Gray, Bot. Amer. Expl. Exped. I, 181 (1854) — — T. V. N.S.W. — — B.fl.I,180 M.fr.IX,130.
P. spicatus, Bentham in proceed. Linn. Soc. VI, 103 (1862) ... W.A. S.A. T. V. N.S.W. — — B.fl.I,180 M.fr.VI,170;IX,130.
P. glomeratus, Bentham in proceed. Linn. Soc. VI, 103 (1862) ... W.A. S.A. — N.S.W. Q. — B.fl.I,180 M.fr.IX,130.
P. diffusus, Bentham, Fl. Austr. I, 190 (1863) ... ... W.A. — — — — N.A. B.fl.I,190
P. microphyllus, F. v. M., fragm. I, 29 (1858) ... ... W.A. S.A. — V. N.S.W. — B.fl.I,190 M.fr.VII,170;IX,13.
P. Berthae, F. v. M., fragm. V, 103 (1866) ... ... W.A. S.A. — — — — — M.fr.V,103.

14

SIDA, Linné, gen. pl. 205 (1737), from Theophrastos.
S. corrugata, Lindley in Mitch. Three Exped. II, 13 (1838) ... W.A. S.A. — V. N.S.W. Q. N.A. B.fl.I,192
S. intricata, F. v. M. in Transact. phil. Soc. Vict. I, 19 (1860)... W.A. S.A. — — N.S.W. — N.A. B.fl.I,193
S. macropoda, F. v. M. in Benth. Fl. Austr. I, 193 (1863) ... — — — — — N.A. B.fl.I,193
S. virgata, Hooker in Mitch. Trop. Austr. 301 (1848) ... ... — S.A. — — N.S.W. Q. N.A. B.fl.I,194
S. echinocarpa, F. v. M., fragm. XI, 62 (1879) ... ... — — — — — N.A. — M.fr.XI,62.
S. cryphiopetala, F. v. M., fragm. III, 4 (1862) ... ... W.A. S.A — — — — Q. N.A. B.fl.I,194 M.fr.III,4.
S. petrophila, F. v. M. in Linnaea XXV, 381 (1852) ... ... — S.A. — — N.S.W. Q. N.A. B.fl.I,194
S. calychymenia, J. Gay in de Candolle, prodr. I, 402 (1824) .. W.A. S.A. — — — — N.A. B.fl.I,194
S. physocalyx, F. v. M., fragm. III, 3 (1862) ... ... — — — — — N.A. B.fl.I,195 M.fr.III,3.
S. subspicata, F. v. M. in Benth. Fl. Austr. I, 195 (1863) ... — — — — N.S.W. Q. N.A. B.fl.I,195
S. pleianthus, F. v. M. in Benth Fl. Austr. I, 195 (1863) ... — — — — — Q. — B.fl.I,195 M.fr.VI,109.
S. spinosa, Linné, sp. pl. 683 (1753)... ... ... — — — — — Q. N.A. B.fl.I,190
S. rhombifolia, Linné, sp. pl. 684 (1753) ... ... — — — — N.S.W. Q. N.A. B.fl.I,196
S. cordifolia, Linné, sp. pl. 684 (1753) ... ... — — — — — Q. N.A. B fl.I,196
S. platycalyx, F. v. M. in Benth. Fl. Austr. I, 197 (1863) ... ... — — — — — — N.A. B.fl.I,197
S. cleisocalyx, F. v. M., fragm. X, 73 (1876) ... ... — — — — — — N.A. — M.fr.X,73.
S. inclusa, Bentham, Fl. Austr. I, 197 (1863) ... ... — S.A. — — — N.A. B.fl.I,197 M.fr.IX,181;XI,32.
S. cardiophylla, F. v. M., fragm. VIII, 242 (1874) ... W.A. S.A. — — — — N.A. — M.fr.VIII,242.
S. Hookeriana, Miquel in Lehm. pl. Preiss. I, 242 (1844) W.A. — — — — — — B.fl.I,197
S. lepida, F. v. M., fragm. VI, 168 (1868) ... ... — — — — — — N.A. — M.fr.XI,32.

ABUTILON, Tournefort, inst. 99, t. 25 (1700), from Camerarius (1586).
A. tubulosum, Hooker in Mitch. Trop. Austr. 390 (1848) ... — S.A. — — N.S.W. Q. — B.fl.I,200 M.fr.IX,131.
A. amplum, Bentham, Fl. Austr. I, 200 (1863) ... ... — — — — — — N.A. B.fl.I,200
A. leucopetalum, F. v. M. in Benth. Fl. Austr. I, 200 (1863) .. — S.A. — — N.S.W. — N.A. B.fl.I,200 M.fr.II,12.
A. Mitchelli, Bentham, Fl. Austr. I, 201 (1863) ... ... — S.A. — — — Q. — B.fl.I,201
A. micropetalum, F. v. M. in Benth. Fl. Austr. I, 201 (1863) ... — — — — N.S.W. Q. — B.fl.I,201 M.fr.VI,109 S.;IX.131;XI,
A. cryptopetalum, F. v. M. in Benth. Fl. Austr. I, 201 (1863) .. W.A. S.A. — — N.S.W. — N.A. B.fl.I,201 M.fr.II,11;IX,13. [32.
A. geranioides, Bentham, Fl. Austr. I, 202 (1863) ... ... — — — — — N.A. B.fl.I,202
A. otocarpum, F. v. M. in Transact. phil. Soc. Vict. I, 13 (1854) — S.A. — V. N.S.W. Q. N.A. B.fl.I,202 M.fr.XI,32.
A. longilobum, F. v. M., fragm. IX, 130 (1875) ... W.A. — — — — — — M.fr.IX,130.
A. subviscosum, Bentham, Fl. Austr. I, 202 (1863) — ... ... — — — — — Q. — B.fl.I,202
A. Indicum, G. Don, gen. syst. I, 504 (1831) — ... ... — — — — — Q. N.A. B.fl.I,202
A. auritum, G. Don, gen. syst. I, 500 (1831) — ... ... — — — — — — B.fl.I,203 M.fr.XI,32.
A. Avicennae, Gaertner, de fruct. II, 251, t. 135 (1791) — S.A. — V. N.S.W. — Q. — B.fl.I,203
A. graveolens, Wight & Arnott, Prodr. Fl. Pen. Ind. or. I, 56 (1840) — ... — — — — Q. — B.fl.I,204
A. oxycarpum, F. v. M. in Benth. Fl. Austr. I, 204 (1863) ... W.A. S.A. — — N.S.W. Q. N.A. B.fl.I,204 M.fr.II,12;IX,131;XI,32.
A. muticum, G. Don, gen. syst. I, 502 (1831) ... ... — — — — — Q. N.A. B.fl.I,204
A. Cunninghamii, Bentham, Fl. Austr. I, 205 (1863) ... ... — — — — — Q. N.A. B.fl.I,205
A. exonemum, F. v. M., fragm. XI, 63 (1879) ... ... — — — — — — N.A. — M.fr.XI,63.
A. Fraseri, Hooker in Walpers, annal. II, 158 (1851) ... ... — S.A. — — N.S.W. Q. N.A. B.fl.I,205
A. halophilum, F. v. M. in Linnaea XXV, 381 (1852) ·... ... W.A. S.A. — — — — B.fl.I,206 M.fr.XI,27.
A. macrum, F. v. M., fragm. IX, 59 (1875) ... ... — S.A. — — — — — M.fr.IX,59.
A. crispum, G. Don, gen. syst. I, 502 (1831) ... ... — — — — — N.A. B.fl.I,206 M.fr.VI,169.

URENA, Dillenius, Hort. Eltham. 430, t. 319 (1732).
U. lobata, Linné, sp. pl. 692 (1753) ... ... — — — — — Q. — B.fl.I,206 M.fr.VI,160;IX,130.
U. Armitiana, F. v. M., fragm. X, 78 (1877) ... ... — — — — — Q. — M.fr.X,78.

PAVONIA, Cavanilles, monadelph. dissert. II, App. II (1786). (Greevesia).
P. hastata, Cavanilles, dissert. III, 138, t. 47 (1786) ... ... — N.S.W. Q. — B.fl.I,207 M.fr.IX,130.

HOWITTIA, F. v. Mueller in Transact. Vict. Inst. I, 116 (1855).
H. trilocularis, F. v. M. in Transact. Vict. Inst. I, 116 (1855) ... — S.A. — V. N.S.W. — — B.fl.I,198 M.fr.VI,109.

HIBISCUS, Linné, syst. nat. (1735); Linné, gen. pl. 207 (1737). (Fugosia partly, Abelmoschus, Paritium).
H. ficulneus, Linné, sp. pl. 695 (1753) ... ... — — — — — Q. N.A. B.fl.I,209 M.fr.I,67.
H. Abelmoschus, Linné, sp. pl. 696 (1753)... ... ... — — — — — Q. — M.fr.VI,109.
H. rhodopetalus, F. v. M. in Benth. Fl. Austr. I, 209 (1863) ... — — — — — Q. N.A. B.fl.I,209 M.fr.IX,229;XI,31.
H. Manihot, Linné, sp. pl. 696 (1753) ... ... — — — — — Q. — B.fl.I,210
H. Notho-Manihot, F. v. M., fragm. V, 57 (1865) ... ... — — — — — — — M.fr.IX,130.
H. Trionum, Linné, sp. pl. 697 (1753) ... ... — S.A. — N.S.W. Q. N.A. B.fl.I,210 M.fr.II,115;IX,129;XI,31.
H. brachysiphonius, F. v. M., fragm. I, 67 (1858) ... ... — S.A. — N.S.W. Q. — B.fl.I,210 M.fr.I,67;VI,109.
H. Drummondii, Turczan. in Bull. Soc. Mosc. XXXI, 195 (1858) W.A. — — — — — — B.fl.I,211 M.fr.III,5;IX,129;XI,31.
H. microlaenus, F. v. M., fragm. II, 116 (1861) ... ... W.A. S.A. — — — N.A. B.fl.I,211 M.fr.III,5;IX,130.
H. Pinonianus, Gaudichaud in Freyc. voy. Bot. 476, t. 100 (1826) W.A. S.A. — — — — N.A. B.fl.I,212 M.fr.II,117;III,166;VI,209;
H. cannabinus, Linné, sp. pl. ed. sec., 979 (1763)... ... — — — — — Q. N.A. B.fl.I,212 M.fr.II,118;IX,129;XI,31. [IX,129;XI,31.
H. divaricatus, Graham in Edinb. phil. Journ. (1830) ... ... — N.S.W. Q. — B.fl.I,212 M.fr.VI,169.
H. Fitzgeraldi, F. v. M., fragm. VIII, 242 (1874) ... ... — — — — — — — M.fr.VIII,242.
H. Elsworthii, F. v. M., fragm. VIII, 241 (1874)... ... ... — — — — — — — M.fr.VIII,241.
H. heterophyllus, Ventenat, jard. de la Malm., t. 103 (1804) ... — N.S.W. Q. — B.fl.I,212 M.fr.VI,170;XI,31.
H. diversifolius, N. Jacquin, icon. pl. rar. 551 (1789) ... ... — N.S.W. Q. — B.fl.I,213 M.fr.II,117;VI,169.
H. splendens, Fraser in Edinb. phil. journ. (1830) ... ... — N.S.W. Q. — B.fl.I,213 M.fr.VI,170;XI,31&123.
H. zonatus, F. v. M., fragm. I, 221 (1859)... ... ... — — — — — Q. N.A. B.fl.I,213 M.fr.VI,170.
H. Cotesii, F. v. M., fragm. III, 5 (1862)... ... ... — — — — — Q. — B.fl.I,214 M.fr.III,5.
H. leptocladus, Bentham, Fl. Austr. I, 214 (1863) ... ... — — — — — Q. N.A. B.fl.I,214 M.fr.VI,169;IX,129.
H. setulosus, F. v. M., fragm. I, 221 (1859) ... ... — — — — — Q. N.A. B.fl.I,214 M.fr.I,221;XI,31.
H. Goldsworthii, F. v. M., fragm. XI, 30 (1878) ... ... — — — — — — N.A. — M.fr.XI,30.
H. pentaphyllus, F. v. M., fragm. III, 13 (1860) ... ... — — — — — — N.A. — M.fr.VI,169;XI,32.
H. geranioides, Cunningham in Benth. Fl. Austr. I, 215 (1863) ... — — — — — N.A. B.fl.I,215
H. vitifolius, Linné, sp. pl. 696 (1753) ... ... — — — — — Q. — B.fl.I,215 M.fr.II,114;VI,169.
H. panduriformis, N. Burman, Fl. Ind. 151, t. 47 (1768) ... ... — — — — — N.A. B.fl.I,215 M.fr.II,115;VI,109;IX,129.
H. Normani, F. v. M., fragm. III, 4 (1862) .. .. — — — — — Q. — B.fl.I,216 M.fr.III,4;VI,169.

H. Krichauffii, F. v. M., Rep. Babb. Exped. 7 (1858) ... ... — S.A. — — N.S.W. Q. — B.fl.I,216 M.fr.VI,169;IX,129.
H. Farragei, F. v. M., fragm. VIII, 241 (1874) ... ... W.A. S.A. — — N.S.W. — — — M.fr.IX,129&196;XI,32.
H. Sturtii, Hooker in Mitch. trop. Austr. 363 (1848) ... — S.A. — — N.S.W. Q. N.A. B.fl.I,216 M.fr.II,13.
H. phyllochlaenus, F. v. M., fragm. IX, 128 (1875) ... ... — — — — — Q. — — M.fr.IX,128.
H. lluegelii, Endlicher in Huog. enum. 10 (1837)... ... W.A. S.A. — — — — — B.fl.I,217 M.fr.IX,128.
H. Wrayac, Lindley, bot. Regist. t. 69 (1840) ... ... W.A. S.A. — — — — — — M.fr.IX,128.
H. tiliaceus, Linné, sp. pl. 694 (1753) ... ... — — — — N.S.W. Q. N.A. B.fl.I,218 M.fr.VI,170;IX,129;XI,32.
H. cuneiformis, de Candolle, prodr. I, 454 (1824)... ... ... W.A. — — — — — N.A. B.fl.I,219 M.fr.XI,32.
H. hakeaefolius, Giordano, Mém. di nuov. sp. d'Ibisco (1833) ... W.A. S.A. — — — — — B.fl.I,220 M.fr.IX,128.

LAGUNARIA, G. Don, gen. syst. dichlam. pl. I, 485 (1831). ·
L. Patersoni, G. Don, gen. syst. I, 485 (1831) ... ... — — — — N.S.W. Q. — B.fl.I,218 M.fr.IX,130;XI,32.

THESPESIA, Solander in Ann. du Mus. IX, 290 (1807).
T. populnea, Solander in Ann. du Mus. IX, 290, t. 8 (1807) — — — — — Q. N.A. B.fl.I,221 M.fr.VII,170;IX,130;XI,32.

GOSSYPIUM, Linné, gen. pl. 200 (1737), from Camerarius (1586). (Fugosia partly, Sturtia). [IX,127;XI,63.
G. Sturtii, F. v. M., fragm. III, 6 (1862) ... ... — S.A. — — N.S.W. — — B.fl.I,222 M.fr.III,174; VI,169 & 251;
G. australe, F. v. M., fragm. I, 46 (1858) ... ... — S.A. — — — Q. N.A. B.fl.I,220 M.fr III,6;IX,127,XI,32&63
G. thespesioides, F. v. M., fragm. IX, 127 (1875) — — — — — — N.A. B.fl.I,220 M.fr.V.177;VI,109;XI,32&63
G. Cunninghamii, Todaro, Monogr. Gen. Gossyp. 7 (1878) — — — — — — N.A. B.fl.I,220 M.fr.VI,169;IX,129;XI,32&
G. costulatum, Todaro, Monogr. Gen. Gossyp. 7 (1878) ... — — — — — — N.A. B.fl.I,221 M.fr.IX,127;XI,32&63. [63.
G. flaviflorum, F. v. M., fragm. IX, 127 (1875) ... — — — — — — N.A. — M.fr.XI,63.
G. populifolium, F. v. M., fragm. IX, 127 (1875)... — — — — — — N.A. B.fl.I,221 M.fr.XI,32&63.
G. Robinsoni, F. v. M., fragm. IX, 126 (1875) ... W.A. — — — — — N.A. — M.fr.XI,63.

ADANSONIA, Linné, sp. pl. I, 1190 (1753).
A. Gregorii, F. v. M. in Hook. Kew misc. IX, 14 (1857) — — — — — N.A. B.fl.I,223 M.fr.VII,40;XI,32.

CAMPTOSTEMON, Masters in J. Hook. Icon. pl. XII, 18, t. 1119 (1876).
C. Schultzii, Masters in J. Hook. icon. pl. XII, 18, t. 1119 (1876) — — — — — — N.A. — M.fr.XI,32.

BOMBAX, Linné, sp. pl. I, 511 (1753).
B. Malabaricus, de Candolle, prodr. I, 479 (1824) ... — — — — — Q. N.A. B.fl.I,223 M.fr.XI,32.

STERCULIACEAE.
Ventenat, Jard. de la Malm. II, 91 (1804).

STERCULIA, Linné, Fl. Zeyl. 166 (1747).
S. quadrifida, R. Brown in Benn. pl. Jav. rar. 233 (1852) ... — — — — N.S.W. Q. N.A. B.fl.I,227 M.fr.VI,173.
S. laurifolia, F. v. M. fragm. VI, 172 (1868) ... — — — — — Q. — — M.fr.VI,172.

BRACHYCHITON, Schott & Endlicher, Melat. bot. 34 (1832). (Trichosiphon, Poecilodermis, Delabechea).
B. paradoxum, Schott, Melatem. 34 (1832) ... ... — — — — — — N.A. B.fl.I,228 —
B. Bidwilli, Hooker, Bot. Mag. t. 5133 (1859) ... ... — — — — — Q. — B.fl.I,228 —
B. discolor, F. v. M., fragm. I, 1 (1858) ... ... — — N.S.W. Q. N.A. B.fl.I,228 M.fr.I,242;II,177.
B. incanum, R. Brown in Benn. pl. Jav. rar. 234 (1852)... ... — — — — — Q. N.A. B.fl.I,229 M.fr.VI,173.
B. luridum, C. Moore in F. v. M., fragm. I, 1 (1858) ... — — N.S.W. Q. — B.fl.I,228 M.fr.I,1;II,177.
B. platanoides, R. Brown in Benn. pl. Jav. rar. 234 (1852) — — — — — Q. N.A. B.fl.I,229 M.fr.VI,173.
B. acerifolium, F. v. M., fragm. I, 1 (1858) ... ... — — N.S.W. Q. — B.fl.I,229 M.fr.I,1;VI,173.
B. diversifolium, R. Brown in Benn. pl. Jav. rar. 234 (1852) W.A. — — — — — N.A. B.fl.I,230 —
B. populneum, R. Brown in Benn. pl. Jav. rar. 234 (1852) — — — V. N.S.W. Q. — B.fl.I,230 —
B. Delabechii, F. v. M., pl. of Vict. I, 157 (1860)... ... — — — — — Q. — B.fl.I,230 —
B. Gregorii, F. v. M. in Hook. Kew misc. IX, 199 (1857) W.A. S.A. — — — — — — M.fr.IX,137.

TARRIETIA, Blume, Bijdr. 227 (1825). (Argyrodendron).
T. argyrodendron, Bentham, Fl. Austr. I, 230 (1863) ... — — — — N.S.W. Q. — B.fl.I,231 M.fr.VI,173;IX,42.
T. trifoliolata, F. v. M., fragm. IX, 43 (1875) ... — — — — N.S.W. Q. — — M.fr.IX,43.

HERITIERA, Dryander in Aiton, Hort. Kew III, 546 (1789).
H. littoralis, Dryander in Aiton, Hort. Kew III, 546 (1789) ... — — — — — Q. N.A. B.fl.I,231 M.fr.VI,173.

UNGERIA, Schott & Endlicher, Melat. 27—31 (1832).
U. floribunda, Schott & Endlicher, Melat. 27, t. 4 (1832) ... — — — — N.S.W. — — B.fl.I,232 M.fr.IX,169.

HELICTERES, Plukenet Phytogr. 181, t. 245 (1692).
H. Isora, Linné, sp. pl. 963 (1753) ... ... — — — — — — N.A. B.fl.I,232 M.fr.VI,173.

METHORIUM, Schott & Endlicher, Melat. 29—30 (1832).
M. spicatum, Helicteres, Colebrooke in G. Don, syst. I, 507 (1831) — — — — — Q. — — M.fr.VII,39.
M. canum, Schott, Melatem. 29, t. 5 (1832) ... ... — — — — — — N.A. B.fl.I,232 —
M. dentatum, F. v. M. in Benth. Fl. Austr. I, 232 (1863) ... — — — — — — N.A. B.fl.I,232 —
M. semiglabrum, F. v. M., fragm. V, 43 (1865) ... ... — — — — — Q. — — M.fr.V,43.

MELHANIA, Forskael, Fl. Aeg. Arab. 64 (1775).
M. incana, Heyne in Wight & Arn. prodr. 68 (1834) ... ... — S.A. — — N.S.W. Q. N.A. B.fl.I,234 M.fr.I,69;VI,174.

MELOCHIA, Dillenius, Hort. Eltham. 221, t. 176 (1732). (Riedleya).
M. pyramidata, Linné, sp. pl. 674 (1753) ... ... — — — — — Q. N.A. B.fl.I,234 M.fr.VI,173.
M. corchorifolia, Linné, sp. pl. 675 (1753) ... ... — — — — — Q. N.A. B.fl.I,235 M.fr.VI,173.

HERMANNIA, Tournefort, inst. 656, t. 432 (1700). (Gilesia, Mahernia partly).
H. Gilesii, F. v. M., fragm. IX, 42 (1875) ... ... — S.A. — — — — — — M.fr.IX,42.

DICARPIDIUM, F. v. Mueller in Hook. Kew misc. IX, 302 (1857).
D. moncioum, F. v. M., in Hook. Kew misc. IX, 302 (1857) ... — — — — — — N.A. B.fl.I,235 —

WALTHERIA, Linné, gen. pl. 203 (1737).
W. Indica, Linné, sp. pl. 637 (1753) ... ... — S.A. — — — Q. N.A. B.fl.I,236 M.fr.VI,173.

ABROMA, N. Jacquin, Hort. Vindob. III, t. 1 (1776).
A. fastuosa, Salisbury, Parad. Lond. 102 (1807) ... ... — — — — — Q. — B.fl.I,236 M.fr.VI,173.

COMMERÇONIA, R. & G. Forster, char. gen. 43, t. 22 (1776). (Ruelingia, Achilleopsis).
C. salvifolia, Bentham, Fl. Austr. I, 238 (1863) ... ... — — — — — Q. — B.fl.I,238
C. dasyphylla, Andrews, Bot. Reposit. t. 603 (1804) ... ... — — — V. N.S.W. Q. — B.fl.I,238 M.fr.X,22.
C. rugosa, Rulingia, Steetz in pl. Preiss. II, 352 (1847) ... ... — — — N.S.W. — — B.fl.I,238
C. Preissti, Steudel in Lehm. pl. Preiss. I, 237 (1844) ... W.A. — — — — — — R.fl.I,239
C. cinerea, Steudel in Lehm. pl. Preiss. I, 238 (1844) ... W.A. — — — — — — B.fl.I,239
C. cygnorum, Steudel in Lehm. pl. Preiss. I, 237 (1844)... ... W.A. — — — — — — B.fl.I,239
C. platycalyx, Rulingia, Benth. fl. Austr. I, 240 (1863) ... W.A. — — — — — B.fl.I,240
C. parviflora, Rulingia, Endlicher in Hueg. enum. 12 (1837) ... W.A. — — — — — — B.fl.I,240
C. hermanninefolia, Rulingia, Steetz in pl. Preiss. II, 353 (1847) — — — N.S.W. — — B.fl.I,240
C. Kempeana, F. v. M., fragm. XI, 113 (1881) ... ... — S.A. — — — — — M.fr.XI,113.
C. laxophylla, F. v. M., fragm. X, 22 (1876) ... ... — S.A. — — — — N.A. B.fl.I,240 M.fr.I,68;X,22.
C. magniflora, F. v. M., fragm. VIII, 223 (1874) ... ... — S.A. — — — — — M.fr.X,22;XI,114.
C. cuneata, Rulingia, Turczan. in Bull. Mosc. XXV, 151 (1852) W.A. — — — — — B.fl.I,241
C. rotundifolia, Rulingia, Turcz. in Bull. de Mosc. XXV, 151 (1852) W.A. — — — — — B.fl.I,241
C. densiflora, F. v. M., fragm. X, 21 (1876) ... ... W.A. — — — — — B.fl.I,241 M.fr.VI,175;XI,114.
C. Fraseri, J. Gay in Mém. du Mus. Par. X, 215, t. 15 (1823) ... — — — V. N.S.W. — — B.fl.I,242
C. Leichhardtii, Bentham, Fl. Austr. I, 242 (1863) ... ... — — — — Q. — R.fl.I,242 M.fr.VI,175;X,22;XI,114.
C. echinata, R. & G. Forster, char. gen. 43, t. 22 (1776)... ... — — — N.S.W. Q. — R fl.I,243 M.fr.VI,175;X,22.
C. Gaudichaudi, J. Gay in Mém. du Mus. Par. X, 213, t. 14 (1823) W.A. — — — — — B.fl.I,243
C. crispa, Turczaninow in Bull. Soc. Mosc. XIX, 501 (1846) ... W.A. — — — — — B.fl.I,243
C. melanopetala, F. v. M., fragm. X, 21 (1876) ... ... W.A. — — — — — M.fr.X,21.
C. pulchella, Turczaninow in Bull. Soc. Mosc. XIX, 502 (1846) W.A. — — — — — R.fl.I,244 M.fr.X,22.
C. microphylla, Bentham, Fl. Austr. I, 244 (1863) ... ... W.A. — — — — — B.fl.I,244
C. craurophylla, F. v. M., fragm. IX, 59 (1875) ... ... W.A. — — — — — — M.fr.IX,59.

HANNAFORDIA, F. v. Mueller, fragm. II, 9 (1860).
H. quadrivalvis, F. v. M., fragm. II, 9 (1860) ... ... W.A. — — — — — B.fl.I,249 M.fr.XI,20&115.
H. Shanesii, F. v. M., fragm. VI, 175 (1868) ... ... — — — — Q. — M.fr.VI,175.
H. Bissillii, F. v. M., fragm. X, 95 (1877) ... ... W.A. S.A. — — N.S.W. — — M.fr.XI,115.

SERINGEA, Sprengel, Anleit. II, 649 (1818). (Keraudronia).
S. platyphylla, J. Gay in Mém. du Mus. Par. VII, 443 (1821) ... — — — — N.S.W. — — B.fl.I,244 M.fr.II,5;VI,175;X,97
S. lanceolata, Steetz in Lehm. pl. Preiss. II, 349 (1847) ... ... — — — — Q. — B.fl.I,245 M.fr.VI,174;X,97.
S. corollata, Steetz in Lehm. pl. Preiss. II, 330 (1847) ... ... — S.A. — — — N.A. B.fl.I,246 M.fr.II,5;VI,174;X,96.
S. Hillii, Keraudrenia, F. v. M. in Benth. Fl. Austr. I, 246 (1863) — — — N.S.W. Q. — B.fl.I,216 M.fr.VI,174;X,97.
S. nephrosperma, F. v. M. in Hook. Kew misc. IX, 15 (1857) ... — S.A. — — — N.A. B.fl.I,246 M.fr.II,5;X,97.
S. microphylla, F. v. M., fragm. II, 5 (1860) ... ... W.A. — — — — — R.fl.I,247 M.fr.II,5.
S. adenolasia, F. v. M., fragm. X, 96 (1877) ... ... — — — — Q. N.A. — M.fr.X,96.
S. integrifolia, F. v. M., fragm. II, 5 (1860) ... ... W.A. S.A. — — — — B.fl.I,247 M.fr.II,5;VI,175;X,97

THOMASIA, J. Gay in Mém. du Mus. VII, t. 6—7 (1821). (Leucothamnus, Rhynchostemon).
T. macrocarpa, Huegel in Endlicher, nov. stirp. dec. 32 (1839)... W.A. — — — — — B.fl.I,250
T. rugosa, Turczaninow in Bull. Soc. Mosc. XIX, 501 (1846) ... W.A. — — — — — B.fl.I,250
T. montana, Steudel in Lehm. pl. Preiss. I, 230 (1844) ... ... W.A. — — — — — B.fl.I,250
T. tenuivestita, F. v. M., fragm. II, 7 (1860) ... ... W.A. — — — — — B.fl.I,251 M.fr.II,7.
T. solanacea, J. Gay in Mém. du Mus. VII, 456, t. 21 (1821) ... W.A. — — — — — B.fl.I,251
T. brachystachys, Turczan. in Bull. Soc. Mosc. XXV, 143 (1852) W.A. — — — — — B.fl.I,251
T. discolor, Steudel in Lehm. pl. Preiss. I, 233 (1844) ... ... W.A. — — — — — B.fl.I,251
T. quercifolia, J. Gay in Mém. du Mus. VII, 459, t. 21 (1821) ... W.A. — — — — — B.fl.I,252
T. foliosa, J. Gay in Mém. du Mus. VII, 454, t. 22 (1821) ... W.A. — — — — — B.fl.I,252
T. triloba, Turczaninow in Bull. Soc. Mosc. XIX, 500 (1846) ... W.A. — — — — — B.fl.I,252
T. triphylla, J. Gay in Mém. du Mus. VII, 453 (1821) ... ... W.A. — — — — — B.fl.I,252
T. purpurea, J. Gay in Mém. du Mus. VII, 452, t. 21 (1821) ... W.A. — — — — — B.fl.I,253
T. macrocalyx, Steudel in Lehm. pl. Preiss. I, 230 (1844) ... W.A. — — — — — R.fl.I,253 M.fr.VI,175.
T. pauciflora, Lindley, Bot. Regist. XXV, App. XVIII (1839)... W.A. — — — — — B.fl.I,254
T. rhynchocarpa, Turcz. in Bull. Soc. Mosc. XXV, 142 (1852) ... W.A. — — — — — B.fl.I,254 M.fr.II,8;VI,175.
T. grandiflora, Lindley, Bot. Regist. XXV, App. XVIII (1839) W.A. — — — — — R.fl.I,254
T. cognata, Steudel in Lehm. pl. Preiss. I, 232 (1844) ... W.A. — — — — — B.fl.I,254
T. rulingioides, Steudel in Lehm. pl. Preiss. I, 232 (1844) ... W.A. — — — — — B.fl.I,255
T. angustifolia, Steudel in Lehm. pl. Preiss. I, 232 (1844) ... W.A. — — — — — B.fl.I,255
T. petalocalyx, F. v. M. in Transact. phil. Soc. Vict. I, 35 (1854) W.A. S.A. — V. — — B.fl.I,255
T. Sarotes, Turczaninow in Bull. Soc. Mosc. XXV, 145 (1852)... W.A. — — — — — B.fl.I,255

GUICHENOTIA, J. Gay in Mém. du Mus. VII, 448, t. 5 (1821). (Sarotes).
G. ledifolia, J. Gay in Mém. du Mus. VII, 449, t. 5 (1821) ... W.A. — — — — — B.fl.I,258 M.fr.VII,174;XI,114.
G. macrantha, Turczaninow in Bull. Soc. Mosc. XIX, 500 (1846) W.A. — — — — — R.fl.I,258
G. semilasiata, Bentham, Fl. Austr. I, 258 (1863) ... ... W.A. — — — — — B.fl.I,258 M.fr.II,4.
G. Sarotes, Bentham, Fl. Austr. I, 258 (1863) ... ... W.A. — — — — — B.fl.I,258
G. micrantha, Bentham, Fl. Austr. I, 258 (1863) ... ... W.A. — — — — — B.fl.I,258 M.fr.II,7.

H. ElswoSIOPETALUM, Smith in Transact. Linn. Soc. IV, 216 (1798). (Corethrostylis, Asterochiton).
H. heterophyllum, F. v. M., fragm. XI, 112 (1881) ... ... — — — — — — R.fl.I,259 M.fr.XI,112.
H. diversifolium, Bentham, Austr. I, 256 (1863) ... ... — — — — — — M.fr.XI,113.
H. splendens, Fraenkham, Austr. I, 256 (1863) ... ... — — — — — — M.fr.XI,113.
H. zeastus, F. v. M., fragm. XI, 113 (1881) ... ... — — — — — — M.fr.XI,113.
H. Coatesii, F. v. M., fragm. Bot. Mag. I, 276 (1835) ... W.A. S.A. T. — — — B.fl.I,260 M.fr.XI,110.
H. leptocladus, Bentham, Fl. Austr. I, 414 (1810) ... ... — T. V. N.S.W. — — B.fl.I,201 M.fr.VI,174;XI,110.
H. setulosum, F. v. M., fragm. I, 225 (1844) ... ... W.A. — — — — — B.fl.I,261 M.fr.XI,112.
H. Goldsworthii, F. v. M., fragm. XI, 50 (81) ... ... W.A. — — — — — M.fr.XI,107.
H. pentaphyllus, F. v. M., fragm. XI, 108 (81) ... ... W.A. — — — — — M.fr.XI,107.
H. geranioides, Cunningham in Benth. Fl. Austr. I, 36 (1854) ... — S.A. — V. N.S.W. — B.fl.I,261
H. vitifolius, Linné, sp. pl. 690 (1753) ... 97, t. 10 (1811) ... W.A. — — — — — B.fl.I,262 M.fr.VI,174.
H. panduriformis, N. Burman, Fl. Ind. 151, t. 47 (1' ... W.A. — — — — — M.fr.VI,174.
H. Normani, F. v. M., fragm. III, 4 (1862) ... or, (1882)... W.A. — — — — — M.fr.XII.

L. oppositifolium, F. v. M., fragm. II, 5 (1800) ... ... W.A — — — — — — B.fl.I,262 M.fr.XI,112.
L. micranthum, J. Hooker, Fl. Tasman. I, 51 (1860) ... — — T. — — — — B.fl.I,262
L. macrophyllum, Graham in Bot. Mag. t. 3098 (1844) ... .. — — — N.S.W. — — B.fl.I,262
L. Baueri, Steetz in Lehm. pl. Preiss. II, 339 (1847) ... .. — S.A. — V. N.S.W. — — B fl.I,263 M.fr.VI,174;XI,110.
L. Tepperi, F. v. M., fragm. XI, 100 (1881) ... ... — S.A. — — — — — — M.fr.XI,109.
L. rufum, R. Brown in Benth. Fl. Austr. I, 263 (1863) ... ... —.' — — N.S.W. — — B.fl.I,263
L. ferrugineum, Smith in Bot. Reposit. t. 308 (1799) ... — — — V. N.S.W. — — B.fl.I,263
L. acutiflorum, Turczaninow in Bull. Soc. Mosc. XXV, 145 (1852) W.A. — — — — — — B.fl.I,264 M.fr.XI,109.
L. Oldfieldii, F. v. M., fragm. II, 6 (1860)... ... ... W.A. — — — — — — M.fr.II,6.
L. quinquenervium, Turcz. in Bull. Soc. Mosc. XXV, 146 (1852) W.A. — — — — — — M.fr.XI,109.
L. Drummondii, Bentham, Fl. Austr. I, 264 (1863) ... ... W.A. — — — — — — B.fl.I,264 M.fr.VI,174;XI,112.
L. rosmarinifolium, Bentham, Fl. Austr. I, 264 (1863) ... ... W.A. — — — — — — B.fl.I,264 M.fr.XI,112.
L. cordifolium, Endlicher in Hueg. enum. 10 (1837) ... ... W.A. — — — — — — B.fl.I,265 M.fr.VI,174;XI,112.
L. Schulzenii, F. v. M., pl. of Vict. I, 145 implied (1861) ... — S.A. — V. — — — B.fl.I,265
L. floribundum, Bentham, Fl. Austr. I, 265 (1863) ... ... W.A. — — — — — — B.fl.I,265 M.fr.XI,112.
L. molle, Bentham, Fl. Austr. I, 265 (1863) ... ... W.A. — — — — — — B.fl.I,265 M.fr.XI,111.
L. membranaceum, Bentham, Fl. Austr. I, 266 (1863) ... ... W.A. — — — — — — B.fl.I,266 M.fr.XI,112.
L. bracteatum, Bentham, Fl. Austr. I, 266 (1863) ... ... W.A. — — — — — — B.fl.I,266

LYSIOSEPALUM, F. v. Mueller, fragm. I, 143 (1859).
L. Barryanum, F. v. M., fragm. I, 143 (1859) ... ... W.A. — — — — — — B.fl.I,267 M.fr.VI,174.
L. rugosum, Bentham, Fl. Austr. I, 267 (1863) ... ... W.A. — — — — — — B.fl.I,267

TILIACEAE.
A. L. de Jussieu, Gen. 289 (1789).

BERRYA, Roxburgh, Pl. Coromand. III, 60, t. 264 (1819).
B. ammonilla, Roxburgh, pl. Coromand. III, 60, t. 264 (1819) ... — — — — — Q. — B.fl.I,268
GREWIA, Linné, syst. nat. 9 (1735); Linné, gen. pl. 276 (1737).
G. orientalis, Linné, sp. pl. 964 (1753) ... ... — — — — — Q. N.A. B.fl.I,270
G. multiflora, A. L. de Jussieu in Ann. du Mus. IV, 89, t. 47 (1804) — — — — — Q. N.A. B.fl.I,270 M.fr.VI,172.
G. breviflora, Bentham, Fl. Austr. I, 270 (1863) ... ... — — — — — — N.A. B.fl.I,270
G. latifolia, F. v. M. in Benth. Fl. Austr. I, 271 (1863) ... ... — — — — N.S.W. Q. B.fl.I,271 M.fr.VI,172;VIII,4.
G. polygama, Roxburgh, Fl. Ind. II, 588 (1832) ... ... — — — — — Q. N.A. B.fl.I,271
G. xanthopetala, F. v. M. in Benth. Fl. Austr. I, 271 (1863) ... — — — — — — N.A. B.fl.I,271
G. scabrella, Bentham, Fl. Austr. I, 272 (1863) ... ... — — — — — Q. N.A. B.fl.I,272
G. orbifolia, F. v. M. in Benth. Fl. Austr. I, 272 (1863)... ... — — — — — — N.A. B.fl.I,272
G. pleiostigma, F. v. M., fragm. VIII, 4 (1872) ... ... — — — — — Q. — — M.fr.VIII,4.
TRIUMFETTA, Plumier, nov. gen. 40, t. 8 (1703).
T. procumbens, G. Forster, prodr. fl. insul. Austr. 35 (1786) ... — — — — — Q. B.fl.I,273 M.fr.III,8;VI,172.
T. appendiculata, F. v. M., fragm. III, 7 (1862) ... ... — — — — — — N.A. B.fl.I,273 M.fr.VIII,5.
T. chactocarpa, F. v. M., fragm. XI, 61 (1879) ... ... — — — — — — N.A. — M.fr.XI,61.
T. glaucescens, R. Brown in Benth. Fl. Austr. I, 273 (1863) ... — — — — — — N.A. B.fl.I,273
T. denticulata, R. Brown in Benth. Fl. Austr. I, 273 (1863) ... — — — — — — N.A. B.fl.I,274
T. micrantha, F. v. M., fragm. III, 7 (1862) ... ... — — — — — — N.A. B.fl.I,274 M.fr.VIII,5;XI,62.
T. leptacantha, F. v. M., fragm. XI, 62 (1879) ... ... — — — — — — N.A. — M.fr.XI,62.
T. plumigera, F. v. M., fragm. I, 69 (1858) ... ... — — — — — — N.A. B.fl.I,274 M.fr.III,8;VI,172.
T. pilosa, Roth, nov. pl. spec. 223 (1821) ... ... — — — — — Q. — M.fr.IV,28;VIII,5.
T. parviflora, Bentham, Fl. Austr. I, 274 (1863) ... ... — — — — — — N.A. B.fl.I,274
NETTOA, Baillon in Adans. VI, 238, t. 7 (1866).
N. chrozophorifolia, Baillon in Adans. VI, 238, 242(1866) ... — — — — — — N.A. —
CORCHORUS, Tournefort, inst. 259, t. 135 (1700).
C. echinatus, Bentham, Fl. Austr. I, 276 (1863) ... ... — — — — — — N.A. B.fl.I,276 M.fr.III,8;VIII,5.
C. hygrophilus, Cunningham in Benth. Fl. Austr. I, 276 (1863) — — — — — Q. — B.fl.I,276 M.fr.VIII,5.
C. Cunninghamii, F. v. M., fragm. III, 8 (1862) ... ... — — — N.S.W. Q. — B.fl.I,276 M.fr.167;VI,172.
C. olitorius, Linné, sp. pl. 529 (1753) ... ... — — — — — — N.A. B.fl.I,276 M.fr.VIII,5.
C. tridens, Linné, mantiss. alter. 566 (1771) ... ... — — — — — — N.A. B.fl.I,276
C. acutangulus, Lamarck, encycl. meth. II, 104 (1786) ... — — — — — Q. N.A. B.fl.I,277 M.fr.VIII,6.
C. fascicularis, Lamarck, encycl. meth. II, 104 (1786) ... — — — — — — N.A. B.fl.I,277
C. pumilio, Bentham, Fl. Austr. I, 277 (1863) ... ... — — — — — — N.A. B.fl.I,277
C. vermicularis, F. v. M., fragm. III, 10 (1862) ... ... — — — — — — N.A. B.fl.I,278 M.fr.VIII,5.
C. tomentellus, F. v. M., fragm. III, 10 (1862) ... ... — — — — — — N.A. B.fl.I,278 M.fr.III,10.
C. sidoides, F. v. M., fragm. III, 9 (1862) ... ... — — — — — Q. N.A. B.fl.I,278 M.fr.III,8;XI,62.
C. leptocarpus, Cunningham in Benth. Fl. Austr. I, 278 (1863) ... W A. S.A. — — — — N.A. B.fl.I,278
C. elachocarpus, F. v. M., fragm. VIII, 6 (1872) ... ... — — — — — — N.A. — M.fr.VIII,6. [102
C. Walcottii, F. v. M., fragm. III, 9 (1862) ... ... — — — — — — N.A. B.fl.I,278 M.fr.VI,172;VIII,6; XI.
C. trilocularis, Linné, mantiss. 77 (1771) ... ... — — — — — — N.A. — M.fr.VIII,5.
SLOANEA, Linné, hort. Cliffort. 210 (1737). (Echinocarpus.)
S. Australis, F. v. M., fragm. V, 91 (1864) ... ... — — — — N.S.W. Q. — B.fl.I,279 M.fr.VI,172;VIII,2.
S. Langii, F. v. M., fragm. V, 28 (1865) ... ... — — — — — — Q. — M.fr.VIII,2.
S. Macbrydei, F. v. M., fragm. VI, 170 (1868) ... ... — — — — — — Q. — M.fr.VIII,2.
S. Woollsii, F. v. M., fragm. VI, 171 (1868) ... ... — — — — N.S.W. Q. — M.fr.VIII,2.
ARISTOTELIA, L'Héritier, stirp. II, 21, t. 16 (1784). (Friesia.)
A. peduncularis, J. Hooker, Fl. Nov. Fel. I, 33 (1853) ... ... — — T. — — — — B.fl.I,280 M.fr.VIII,2.
A. Australasica, F. v. M., fragm. II, 79 (1860) ... ... — — — N.S.W. — — B.fl.I,280 M.fr.VIII,2.
A. megalosperma, F. v. M., fragm. IX, 84 (1875)... ... ... — — — — — — — — M.fr.IX,84.
ELAEOCARPUS, Burman, thesaur. Zeil. 39, t. 40 (1737). (Elaiocarpus, Dicera.) [VIII.
E. holopetalus, F. v. M., fragm. II, 143 (1861) ... ... — — — — V. N.S.W. — — B.fl.I,281 M.fr.IV,173;VI,172;
E. Arnhemicus, F. v. M., Docum. intercol. Exhib. 24 (1867) ... — — — — — — — N.A. — M.fr.X,4.

D

E. obovatus, G. Don, gen. syst. dichl. pl. I, 599 (1831) .. ... — — — — N.S.W. Q. N.A. B.fl.I,281 M.fr.II,80;VI,172.
E. cyaneus, Aiton, Epit. hort. Kew, addend. 367 (1814) .. — — T. V. N.S.W. Q. — B.fl.I,281 M.fr.VI,172;VIII,2. [12.
E. grandis, F. v. M., fragm. II, 81 (1860) ... ... — — — — N.S.W. Q. — B.fl.I,281 M.fr.II.175;VI.172;VIII,
E. foveolatus, F. v. M., fragm. V, 157 (1866) ... ... — — — — — — Q. — M.fr.VI,172;VIII,2.
E. ruminatus, F. v. M., fragm. VIII, 1 (1872) ... ... — — — — — Q. — — M.fr.X,4.
E. Grahami, F. v. M., fragm. X, 3 (1876) ... ... — — — — — Q. — — M.fr.X,3.
E. sericopetalus, F. v. M., fragm. VI. 171 (1908)... ... ... — — — — — Q. — — M.fr.VI,171.

EUPHORBIACEAE.

A. L. de Jussieu, Gen. 384 (1789) from B. de Jussieu (1759).

CALYCOPEPLUS, Planchon in Bull. de la Soc. Bot. de Fr. VIII, 31 (1861).
C. ephedroides, Planchon in Bull. Soc. Bot. Fr. VIII, 31 (1861) W.A. — — — — — — B.fl.VI,53
C. marginatus, Bentham, Fl. Austr. VI, 53 (1873) ... ... W.A. — — — — — — B.fl.VI,53

EUPHORBIA, Linné, gen. pl. 152 (1737) from Plinius.
E. Atoto, G. Forster, florul. ins. Austr. prodr. 36 (1786) ... .— — — — — Q. N.A. B.fl.VI,46
E. Sparmanni, Boissier, Cent. Euphorb. 5 (1860) ... ... — — — — N.S.W. — — B.fl.VI,46
E. Mitchelliana, Boissier in De Candolle, prodr. XV, 25 (1866) W.A. — — — — Q. N.A. B.fl.VI,47 M.fr.VI,182.
E. schizolepis, F. v. M. in Benth. Fl. Austr. VI, 47 (1873) ... .— — — — — N.A. B.fl.VI,47
E. Schultzii, Bentham in Fl. Austr. VI, 47 (1873) ... .— — — — — N.A. B.fl.VI,47
E. Armstrongiana, Boissier in De Candolle, prodr. XV, 47 (1866) — — — — — N.A. B.fl.VI,48
E. erythrantha, F. v. M., fragm. 11, 152 (1861) ... ... — — 8.A. — — N.S.W. Q. N.A. B.fl.VI,48 M.fr.XI,64.
E. Muelleri, Boissier in De Candolle, prodr. XV, 27 (1866) ... — — — — — N.A. B.fl.VI,48
E. Drummondii, Boissier, Cent. Euphorb. 14 (1860) ... W.A. S.A. T. V. N.S.W. Q. N.A. B.fl.VI,49 M.fr.VI,182.
E. alsinflora, Baillon, Adans. VI, 288 (1866) ... ... — — — — — N.A. B.fl.VI,49
E. Wheeleri, Baillon, Adans. VI, 286 (1866) ... ... — — 8.A. — — — N.A. B.fl.VI,49
E. myrtoides, Boissier in De Candolle, prodr. XV, 15 (1866) W.A. — — — — — N.A. B.fl.VI,50
E. micradenia, Boissier in De Candolle, prodr. XV, 27 (1866) ... — — — — — Q. N.A. B.fl.VI,50
E. Macgillivrayi, Boissier in De Candolle, prodr. XV, 26 (1866) — — — — N.S.W. — — B.fl.VI,50
E. serrulata, Reinwardt in Blum. Bijdrag. 635 (1826) ... ... — — — — — Q. N.A. B.fl.VI,51
E. filipes, Bentham, Fl. Austr. VI, 51 (1873) ... ... — — — — — — N.A. B.fl.VI,51
E. Careyi, F. v. M., fragm. XI, 64 (1879) ... ... — — — — — Q. — N.A. — M.fr.XI,64.
E. pilulifera, Linné, sp. pl. 454 (1753) ... ... — — — — — Q. — B.fl.VI,51
E. Norfolkiana, Boissier in De Cand., prodr. XV, 206 (1866) ... — — — — N.S.W. — — M.fr.IX,169.
E. eremophila, Cunningham in Mitch. Trop. Austr. 348 (1848)... W.A. S.A. — V. N.S.W. Q. N.A. B.fl.VI,52

MONOTAXIS, Brongniart in Duperr. Voy. Bot. 223, t. 49 (1829). (Hippocrepandra.)
M. macrophylla, Bentham, Fl. Austr. VI, 79 (1873) ... ... — — — — — N.S.W. Q. — B.fl.VI,79
M. linifolia, Brongniart in Duperr. Voy. Bot. 223, t. 49 (1829)... — — — — N.S.W. — — B.fl.VI,79
M. occidentalis, Endlicher in Hueg. enum. 19 (1837) ... W.A. — — — — — — B.fl.VI,79
M. lurida, Bentham, Fl. Austr. VI, 80 (1873) ... ... W.A. — — — — — — B.fl.VI,80
M. megacarpa, F. v. M., fragm. IV, 143 (1865) ... ... W.A. — — — — — — B.fl.VI,80 M.fr.IV,143.
M. gracilis, Baillon, Adans. VI, 293 (1866) ... ... W.A. — — — — — — B.fl.VI,81
M. grandiflora, Endlicher in Hueg. enum. 19 (1837) ... ... W.A. — — — — — — B.fl.VI,81
M. luteiflora, F. v. M., fragm. X, 51 (1877) ... ... W.A. — — — — — — — M.fr.X,51.

PORANTHERA, Rudge in Transact. Linn. Soc. X. 302, t. 22 (1811).
P. ericifolia, Rudge in transact. Linn. Soc. X, 302, t. 22 (1811) — — — — N.S.W. — — B.fl.VI,55
P. ericoides, Klotzsch in Lehm. pl. Preiss. II, 239 (1847) ... W.A. S.A. — V. — — — B.fl.VI,55
P. Huegelii, Klotzsch in Lehm. pl. Preiss. II, 231 (1847) W.A. — — — — — — B.fl.VI,55
P. corymbosa, Brongniart in Duperr. Voy.Cog.Bot.219, t. 56(1826) — — — V. N.S.W. — — B.fl.VI,56 M.fr.VI,183.
P. microphylla, Brongniart in Duperr.Voy.Cog.Bot.218,t.50(1826) W.A. S.A. T. V. N.S.W. Q. N.A. B.fl.VI,56 M.fr.VI,20.

MICRANTHEUM, Desfontaines in Mém. du Mus. hist. nat. IV, 253 (1818). (Caletia.)
M. ericoides, Desfontaines in Mém. du Mus. IV, 253, t. 14 (1818) — — — — — N.S.W. Q. — B.fl.VI,57
M. hexandrum, J. Hooker in Lond. Journ. VI, 283 (1847) — — — S.A. T. V. N.S.W. — — B.fl.VI,57

PSEUDANTHUS, Sieber in Sprengel, syst. veg. cur. post. 25 (1827). (Stachystemon.)
P. pimeloides, Sieber in Sprengel, syst. veg. cur. post. 25 (1827) — — — — N.S.W. Q. — B.fl.VI,59
P. ovalifolius, F. v. M. in Transact. phil. Inst. Vict. II, 66 (1857) — — T. V. N.S.W. — — B.fl.VI,59
P. micranthus, Bentham, Fl. Austr. VI, 59 (1873) ... ... — S.A. — — — — — B.fl.VI,59
P. divaricatissimus, Bentham, Fl. Austr. VI, 60 (1873) ... ... — — — — N.S.W. — — B.fl.VI,60
P. orientalis, F. v. M., fragm. II, 14 (1860) ... ... — — — — N.S.W. Q. — B.fl.VI,60 M.fr.II,14.
P. occidentalis, F. v. M., fragm. I, 107 (1859) ... ... W.A. — — — — — — B.fl.VI,60 M.fr.I,107.
P. nematophorus, F. v. M., fragm. II, 14 (1860) ... ... W.A. — — — — — — B.fl.VI,60 M.11,14.
P. polyandrus, F. v. M., fragm. II, 153 (1861) ... ... W.A. — — — — — — B.fl.VI,62 M.fr.II,153;IV,173
P. brachyphyllus, F. v. M. inTransact.R.Soc.N.S.Wales 11 (1881) W.A. — — — — — — B.fl.VI,62
P. vermicularis, F. v. M. in Transact. R. Soc. N.S.W. 11 (1881) W.A. — — — — — — B.fl.VI,62

BEYERIA, Miquel in Ann. sc. nat. sér. trois. I, 350 (1844). (Calyptrostigma, Beyeriopsis.)
B. viscosa, Miquel in Ann. des. sc. nat. sér. trois. I, 350, t. 15 (1844) W.A. S.A. T. V. N.S.W. Q. — B.fl.VI,64 M.fr.I,230.
B. lasiocarpa, F. v. M. in Benth. Fl. Austr. VI, 65 (1873) ... — — — — N.S.W. — — B.fl.VI,65 M.fr.VI,182.
B. opaca, F. v. M. in Transact. phil. Soc. Vict. I, 16 (1854) ... — S.A. T. V. N.S.W. — — B.fl.VI,65 M.fr.VI,182.
B. uncinata, F. v. M. in Benth. Fl. Austr. VI, 65 (1873) ... S.A. — — — — — — B.fl.VI,65
B. latifolia, Baillon, Adans. VI, 304 (1866) ... ... W.A. — — — — — — B.fl.VI,66
B. cygnorum, Baillon, Adans. VI, 309 (1866) ... ... W.A. — — — — — — B.fl.VI,66
B. cinerea, Baillon, Adans. VI, 309 (1866) ... ... W.A. — — — — — — B.fl.VI,66
B. cyanescens, Bentham, Fl. Austr. VI, 66 (1873) ... ... W.A. — — — — — — B.fl.VI,66
B. lepidopetala, F. v. M., fragm. I, 230 (1859) ... ... W.A. — — — — — — B.fl.VI,67
B. similis, Baillon, Adans. VI, 309 (1866) ... ... W.A. — — — — — — B.fl.VI,67
B. brevifolia, Baillon, Adans. VI, 309 (1866) ... ... W.A. — — — — — — B.fl.VI,67
B. Drummondii, J. Mueller in Linnaea XXXIV, 58 (1865) ... W.A. — — — — — — B.fl.VI,68
B. tristigma, F. v. M., fragm. VI, 181 (1868) ... ... — — — — — Q. — B.fl.VI,68 M.fr.VI,181.

RICINOCARPUS, Desfontaines in Mém. du Mus. III, 459, t. 22 (1817). (Roeperia.)
R. pinifolius, Desfontaines in Mém. du Mus. III, 459, t. 22 (1817) — — T. V. N.S.W. Q. — B.fl.VI,70 M.fr.VI,182.
R. tuberculatus, J. Mueller in Linnaea XXXIV, 60 (1865) ... W.A. — — — — — — B.fl.VI,70
R. cyaneoscus, J. Mueller in Linnaea XXXIV, 60 (1865) ... W.A. — — — — — — B.fl.VI,70
R. psilocladus, Bentham, Fl. Austr. VI. 71 (1873) ... W.A. — — — — — — B.fl.VI,70
R. glaucus, Endlicher in Hueg. enum. 18 (1837) ... ... W.A. — — — — — — B.fl.VI,71 M.fr.VI,182.
R. major, J. Mueller in Linnaea XXXIV, 59 (1865) ... ... — T. — — — — B.fl.VI,72
R. Bowmanii, F. v. M., fragm. I, 161 (1859) ... ... — — — — N.S.W. Q. — B.fl.VI,72 M.fr.I,161.
R. ledifolius, F. v. M., fragm. I, 76 (1858)... ... ... — — — — — Q. — B.fl.VI,72 M.fr.I,76.
R. rosmarinifolius, Bentham, Fl. Austr. VI, 72 (1873) ... ... — — — — — — N.A. B.fl.VI,72
R. marginatus, Bentham, Fl. Austr. VI, 73 (1873) ... ... — — — — — — N.A. B.fl.VI,73
R. speciosus, J. Mueller in De Candolle, prodr. XV, 204 (1863)... — — — — N.S.W. — B.fl.VI,73
R. trichophorus, J. Mueller in Linnaea XXXIV, 60 (1865) W.A. — — — — — — B.fl.VI,73
R. muricatus, J. Mueller in Linnaea XXXIV, 61 (1865)... ... W.A. — — — — — — B.fl.VI,73
R. velutinus, F. v. M., fragm. IX, 2 (1875) ... ... W.A. — — — — — — — M.fr.IX,2.

BERTYA, Planchon in Hook. Lond. Journ. Bot. IV, 472, t. 16 (1845).
B. gummifera, Planchon in Hook. Lond. Journ. IV, 473 (1845) — — — N.S.W. — — B.fl.VI,75
B. pinifolia, Planchon in Hook. Lond. Journ. IV, 473 (1845) ... — — — N.S.W. Q. — B.fl.VI,75
B. quadrisepala, F. v. M., fragm. X, 52 (1876) ... W.A. — — — — — — M.fr.X,52.
B. Cunninghamii, Planchon in Hook. Lond. Journ. IV, 473 (1845) — — — V. N.S.W. — B.fl.VI,75 M.fr.X,52.
B. dimerostigma, F. v. M. in Wing's S. Science Journ. II (1882) W.A. — — — — — — — M.fr.XII.
B. rosmarinifolia, Planchon in Hook. Lond. Journ. IV, 473 (1845) — — T. — N.S.W. — B.fl.VI,76
B. Mitchelli, J. Mueller in Linnaea XXXIV, 63 (1865) ... ... — S.A. — V. N.S.W. — B.fl.VI,76
B. oppositifolia, F. v. M. & O'Shanesy in S. Sc. Rec. II, May (1882) — — — — — Q. — B.fl.VI,76 M.fr.XII.
B. oleaefolia, Planchon in Hook. Lond. Journ. IV, 473 (1845)... — — — — N.S.W. Q. — B.fl.VI,76
B. rotundifolia, F. v. M., fragm. IV, 34 (1864) ... ... — S.A. — — — — — B.fl.VI,77 M.fr.IV,34.
B. pedicellata, F. v. M., fragm. IV, 143 (1865) ... ... — — — — — Q. — B.fl.VI,77 M.fr.IV,143.
B. pomadorrioides, F. v. M., fragm. IV, 34 (1864) ... ... — — — — — Q. — B.fl.VI,77 M.fr.IV,34.
B. Findlayi, F. v. M., fragm. VIII, 141 (1874) ... ... — — — V. N.S.W. — — M.fr.VIII,141.

AMPEREA, Adr. de Jussieu, Euphorb. gen. 35, t. 10 (1824).
A. protensa, Nees in Lehm. pl. Preiss. II, 229 (1847) ... W.A. — — — — — — B.fl.VI,82
A. volubilis, F. v. M. in Benth. Fl. Austr. VI, 82 (1873) ... W.A. — — — — — — B.fl.VI,82 M.fr.VI,182.
A. micrantha, Bentham, Fl. Austr. VI, 83 (1873) ... W.A. — — — — — — B.fl.VI,83
A. conferta, Bentham, Fl. Austr. VI, 83 (1873) ... ... W.A. — — — — — — B.fl.VI,83
A. ericoides, Adr. de Jussieu, Tent. Euphorb. 112, t. 10 (1824) W.A. — — — — — — B.fl.VI,83
A. spartioides, Bronguiart in Duperr. Voy. Cog. 221, t. 49 (1826) — S.A. T. V. N.S.W. Q. — B.fl.VI,84

ANTIDESMA, Linné, Fl. Zeyl. 160 (1747).
A. Ghaesembilla, Gaertner, de Fruct. I, 189, t. 39 (1788) ... — — — — — — N.A. B.fl.VI,183 M.fr.VI,183.
A. Dallachyanum, Baillon, Adans. VI, 337 (1866) ... ... — — — — — Q. N.A. B.fl.VI,85 M.fr.VI,183.
A. Bunius, Sprengel, syst. veg. I, 826 (1825) ... ... — — — — — Q. — B.fl.VI,86
A. parviflorum, Thwaites & F. v. M., fragm. IV, 86 (1864) ... — — — — — Q. — B.fl.VI,86 M.fr.IV,86;VI,183.
A. Schultzii, Bentham, Fl. Austr. VI, 86 (1873) ... ... — — — — — — N.A. B.fl.VI,86
A. erostre, F. v. M. in Benth. Fl. Austr. VI, 87 (1873) ... — — — — — Q. — B.fl.VI,87
A. sinuatum, Bentham, Fl. Austr. VI, 87 (1873) ... ... — — — — — Q. — B.fl.VI,87

ANDRACHNE, Linné, syst. nat. 9 (1735); Linné, gen. pl. 287 (1737).
A. Decaisnei, Bentham, Fl. Austr. VI, 88 (1873)... ... ... W.A. — — — — Q. N.A. B.fl.VI,88 M.fr.VI,183.

ACTEPHILA, Blume, Bijdr. XII, 581 (1825).
A. grandifolia, Baillon, Adans. VI, 330 & 360, t. 10 (1866) ... — — — N.S.W. Q. — B.fl.VI,89
A. Mooreana, Baillon, Adans. VI, 330 & 336 (1866) ... ... — — — N.S.W. Q. — B.fl.VI,89
A. latifolia, Bentham, Fl. Austr. VI, 89 (1873) ... ... — — — — — Q. — B.fl.VI,89
A. petiolaris, Bentham, Fl. Austr. VI, 89 (1873) ... ... — — — — — Q. — B.fl.VI,89
A. sessilifolia, Bentham, Fl. Austr. VI, 90 (1873) ... ... — — — — — Q. — B.fl.VI,90

DISSILIARIA, F. v. M. in Baillon, Adans. VII, 356, pl. 1 (1867). (Choriceras.)
D. baloghioides, F. v. M. in Baillon, Adans. VII, 356 (1867) ... — — — — — Q. — B.fl.VI,90
D. Muelleri, Baillon, Adans. VII, 359, t. 1 (1867) ... ... — — — — — Q. — B.fl.VI,91
D. tricornis, Bentham, Fl. Austr. VI, 91 (1873) ... ... — — — — — Q. N.A. B.fl.VI,91

PETALOSTIGMA, F. v. M. in Hook. Kew Misc. IX, 17 (1857). (Hylococcus.)
P. quadriloculare, F. v. M. in Hook. Kew Misc. IX, 17 (1857) — — — — N.S.W. Q. N.A. B.fl.VI,92 M.fr.VI,182.

PHYLLANTHUS, Linné, gen. pl. 282 (1737), from J. Commelyn. (Glochidion, Synostemon, Bradleia.)
P. Ferdinandi, J. Mueller in Regensb. Flora, 370 (1865) ... — — — — N.S.W. Q. — B.fl.VI,96 M.fr.VI,183.
P. lobocarpus, Bentham, Fl. Austr. VI, 97 (1873) ... ... — — — — — Q. — B.fl.VI,97
P. ditassoides, J. Mueller in Regensb. Flora, 487 (1864)... ... — — — — — — N.A. B.fl.VI,97
P. J. Mueller in De Candolle, prodr. XV, 327 (1866) ... — — — — — — N.A. B.fl.VI,97
P. thesioides, Bentham, Fl. Austr. VI, 98 (1873)... ... ... — — — N.S.W. Q. — B.fl.VI,98
P. Hirtellus, J. Mueller in De Candolle, prodr. XV, 326 (1866) — — — — — Q. — B.fl.VI,98 M.fr.III,89.
P. rigens, J. Mueller in Regensb. Flora, 513 (1864) ... ... — — — N.S.W. Q. — B.fl.VI,99
P. ochrophyllus, Bentham, Fl. Austr. VI, 99 (1873) ... ... — — — — — N.A. B.fl.VI,99
P. rigidulus, F. v. M. in Linnaea XXXIV, 72 (1865) ... ... — — — — — N.A. B.fl.VI,99
P. ramosissimus, J. Mueller in Linnaea XXXIV, 70 (1865) ... — — — — N.S.W. Q. — B.fl.VI,100
P. Tatei, F. v. M. in Wing's South. Science Record II, 55 (1882) — S.A. — — — — — — B.fl.VI,100
P. rhytidospermus, F. v. M. in Linnaea XXXIV, 70 (1865) ... — — — — — — N.A. B.fl.VI,100
P. albiflorus, F. v. M. in Linnaea XXXIV, 70 (1865) ... ... — — — — — Q. — B.fl.VI,100
P. crassifolius, J. Mueller in Regensb. Flora, 513 (1864) ... — — — — — Q. — B.fl.VI,101
P. elachophyllus, F. v. M. in Benth. Fl. Austr. VI, 101 (1873)... W.A. — — — — — — B.fl.VI,101
P. uberiflorus, F. v. M. in Adansonia, 341 (1866)... ... ... — — — — — Q. — B.fl.VI,101
P. reticulatus, Poiret, encycl. meth. V, 298 (1804) ... ... — — — — — — N.A. B.fl.VI,101
P. baccatus, F. v. M. in Benth. Fl. Austr. VI, 102 (1873) ... — — — — — — N.A. B.fl.VI,102

P. Urinaria, Linné, sp. pl. 982 (1753) ... ... — — — — — — N.A. B.fl.VI,102
P. trachygyne, Bentham, Fl. Austr. VI, 103 (1873) ... ... — — — — — — — N.A. B.fl.VI,103
P. Maderaspatanus, Linné, sp. pl. 982 (1753) ... ... — — — — — — Q. N.A. B.fl.VI,103
P. Mitchelli, Bentham, Fl. Austr. VI, 103 (1873) ... ... — — — — — — Q. — B.fl.VI,103
P. Gastroemii, J. Mueller in De Cand., prodr. XV, 358 (1866)... — — — — N.S.W. Q. — B.fl.VI,104
P. Dallachyanus, Bentham, Fl. Austr. VI, 104 (1873) ... ... — — — — — — Q. — B.fl.VI,104
P. subcrenulatus, F. v. M., fragm. I, 108 (1859) ... ... — — — — — N.S.W. Q. — B.fl.VI,104   M.fr.I,108.
P. calycinus, Labillardière, Nov. Holl. pl. spec. II, 75, t. 225 (1806) W.A. S.A. — — — — — B.fl.VI,105
P. flagellaris, Bentham, Fl. Austr. VI, 106 (1873) ... ... — — — — — — N.A. B.fl.VI,106
P. similis, J. Mueller in Linnaea XXXIV, 71 (1865) ... ... — — — — — Q. — B.fl.VI,106
P. microcladus, J. Mueller in Linnaea XXXIV, 71 (1865) ... — — — — — N.S.W. Q. — B.fl.VI,106
P. grandisepalus, F. v. M. in Linnaea XXXIV, 72 (1863) ... ... — — — — — — N A. D.fl.VI,106
P. Carpentariae, J. Mueller in Linnaea XXXIV, 72 (1863) ... — — — — — — N.A. B.fl.VI,107
P. Fuerurohrii, F. v. M. in Transact. phil. Soc. Vict. I, 15 (1854) W.A. S.A. — V. N.S.W. Q. — B.fl.VI,107
P. hebecarpus, Bentham, Fl. Austr. VI, 108 (1873) ... ... — — — — — Q. N.A. B.fl.VI,108
P. lacunarius, F. v. M. in Transact. phil. Soc. Vict. I, 14 (1854) — S.A. — V. N.S.W. — — B.fl.VI,108
P. trachyspermus, F. v. M. in Transact. phil. Soc. Vict. I, 14 (1854) — S.A — V. N.S.W. — — B.fl.VI,108
P. australis, J. Hooker in Lond. Journ. VI, 284 (1847) ... ... — — T. — N.S.W. — B.fl VI,108
P. thymoides, Sieber in Linnaea XXVIII, 566 (1866) ... — 8.A. — V. N.S.W. — — .B.fl.VI,109
P. scaber, Klotzsch in Lehm. pl. Preiss. I, 179 (1844) ... ... W.A. — — — — — — B.fl.VI,100
P. indigoferoides, Bentham, Fl. Austr. VI, 110 (1873) ... ... — — — — — — N.A. B.fl.VI,110
P. aridus, Bentham, Fl. Austr. VI, 110 (1873) ... ... — — — — — Q. N.A. B.fl.VI,110
P. Gunnii, J. Hooker in Lond. Journ. VI, 284 (1847) ... ... — S.A. T. V. N.S.W. — — B.fl.VI,110   M.fr.VI,183.
P. simplex, Retzius, observ. V, 29 (1789) ... ... — — — — — Q. — B.fl.VI,111
P. filicaulis, Bentham, Fl. Austr. VI, 111 (1873) ... ... — — — — — N.S.W. — B.fl.VI,111
P. minutiflorus, F. v. M. in Linnaea XXXIV, 75 (1863)... ... — — — — — Q. N.A. B.fl.VI,112
P. Armstrongii, Bentham, Fl. Austr. VI, 112 (1873) ... ... — — — — — — N.A. B.fl.VI,112
P. buxifolius, Reinwardt in Blume, catal. plant. Buitenz. 106 (1823) — — — — — Q. — M.fr.X,121.

BREYNIA, R. & G. Forster, char. gen. 145, t. 177G). (Melanthesa, Melanthesiopsis.)
B. cernua, J. Mueller in De Candolle, prodr. XV, 430 (1866) ... — — — — — Q. N.A. B.fl.VI,113
B. oblongifolia, J. Mueller in De Candolle, prodr. XV, 110 (1866) — — — — N.S.W. Q. — B.fl.VI,114
B. stipitata, J. Mueller in De Candolle, prodr. XV, 442 (1866) .. — — — — — Q. N.A. B.fl.VI,114
B. rhynchocarpa, Bentham, Fl. Austr. VI, 114 (1873) ... — — — — — — N.A. B.fl.VI,114

SECURINEGA, A. L. de Jussieu, Gen. 388 (1789). (Fluggea.)
S. Abyssinica, A. Richard, tentam. fl. Abyss. II, 256 (1851) ... — — — — — Q. N.A. B.fl.VI,116
S. Leucopyrus, J. Mueller in De Cand., prodr. XV, 451 (1866)... — — — — — — Q. — B.fl.VI,116

NEOROEPERA, F. & J. Mueller in De Candolle, prodr. XV, 489 (1866).
N. buxifolia, F. & J. Mueller in De Cand., prodr. XV, 489 (1866) — — — — — Q. — B.fl.VI,116
N. Banksii, Bentham, Fl. Austr. VI, 117 (1873) ... ... — — — — — — Q. — B.fl.VI,117

BISCHOFFIA, Blume, Bijdr. 1168 (1825).
B. Javanica, Blume, Bijdr. 1168 (1825) ... ... — — — — — Q. — — M.fr.VIII,141.

HEMICYCLIA, Wight & Arnott in Edinb. phil. Journ. XIV, 297 (1833).
H. sepiaria, Wight & Arnott in Edinb. phil. Journ. XIV, 297 (1833) ... — — — — N.A B.fl.VI,117   M.fr.IV,119;VI,182.
H. Australasica, J. Mueller in De Candolle, prodr. XV, 487 (1866) — — — — N.S.W. Q. — B.fl.VI,118
H. laslogyna, F. v. M., fragm. IV, 119 (1864) ... ... — — — — — N.A. B.fl.VI,118   M.fr.IV,119.

APOROSA, Blume, Bijdrag. 514 (1824).
A. Australiana, F. v. M., fragm. XII ined. ... ... — — — — — Q. — — M.fr.VIII,141;XII.

BRIDELIA, Willdenow, Spec. IV, 978 (1805).
B. exaltata, F. v. M., fragm. III, 32 (1862) ... ... — — — — N.S.W. — B.fl.VI,119   M.fr.III,32.
B. ovata, Decaisne in nouv. ann. du Mus. V, 484 (1835)... ... — — — — — N.A. B.fl.VI,120
B. tomentosa, Blume, Bijdrag. 597 (1825) ... ... — — — — — Q. N.A. B.fl.VI,120   M.fr.VI,182.
B. faginea, F. v. M. in Benth. Fl. Austr. VI, 120 (1873)... — — — — — — Q. — B.fl.VI,120

CLEISTANTHUS, J. Hooker, icon. pl. t. 779 (1847). (Lebedicra, Amanoa.)
C. Cunninghamii, J. Mueller in De Cand., prodr. XV, 506 (1866) — — — — N.S.W. Q. — B.fl.VI,122
C. apodus, Bentham, Fl. Austr. VI, 122 (1873) ... ... — — — — — — Q. — B.fl.VI,122
C. Dallachyanus, Baillon in Benth. Fl. Austr. VI, 122 (1873) ... — — — — — Q. — B.fl.VI,122   M.fr.VI,182.
C. semiopacus, F. v. M. in Benth. Fl. Austr. VI, 123 (1873) ... — — — — — — Q. — B.fl.VI,123

CROTON, Linné, gen. pl. 288 (1737).
C. Schultzii, Bentham, Fl. Austr. VI, 124 (1873) ... ... — — — — — N.A. B.fl.VI,124
C. insularis, Baillon, Adans. II, 217 (1862) ... ... — — — — — Q. — B.fl.VI,124
C. phlebalioides, F. v. M. in Benth. Fl. Austr. VI, 125 (1873) ... — — — — N.S.W. Q. — B.fl.VI,125   M.fr.IV,140.
C. opponens, F. v. M. in Benth. Fl. Austr. VI, 125 (1873) ... — — — — — Q. — B.fl.VI,125
C. tomentellus, F. v. M., fragm. IV, 141 (1864) ... ... — — — — — N.A. B.fl.VI,126   M.fr.IV,141.
C. Verreauxii, Baillon, Etud. Euph. 357 (1858) ... ... — — — — N.S.W. Q. N.A. B.fl.VI,126   M.fr.IV,141;VI,186.
C. acronychioides, F. v. M., fragm. IV, 142 (1864) ... ... — — — — — — Q. — B.fl.VI,127   M.fr.IV,142.
C. triacros, F. v. M., fragm. VI, 185 (1867) ... ... — — — — — Q. — B.fl.VI,127   M.fr.VI,185.
C. Arnhemicus, F. v. M. in Linnaea XXXIV, 112 (1865) ... — — — — — Q. N.A. B.fl.VI,127   M.fr.VI,186.

ALEURITES, R. & G. Forster, char. gen. 111, t. 56 (1776).
A. triloba, R. & G. Forster, char. gen. 111, t. 56 (1776)... — — — — — — Q. — B.fl.VI,128   M.fr.VI,182.

CLAOXYLON, Adr. de Jussieu, Euphorb. gen. 43, t. 14 (1824).
C. angustifolium, J. Mueller in Linnaea XXXIV, 165 (1865) ... — — — — — Q. — B.fl.VI,129   M.fr.VI,183.
C. tenerifolium, F. v. M. in Baillon, Adans. VI, 323 (1866) ... — — — — — Q. — B.fl.VI,130   M.fr.IV,142.
C. australe, Baillon, Etud. Euphorb. 493 (1858) ... ... — — — V. N.S.W. Q. — B.fl.VI,130
C. Hillii, Bentham, Fl. Austr. VI, 131 (1873) ... ... — — — — — — Q. — B.fl.VI,131

ACALYPHA, Royen in Linné, hort. Cliffort. 495 (1737).
A. nemorum, F. v. M in Linnaea XXXIV, 38 (1865) ... ... — — — — N.S.W. Q. — B.fl.VI,132

A. cremorum, F. v. M. in Regensb. Flora 440 (1864) ... ... — — — — — Q. — B.fl.VI,132
A. capillipes, F. v. M. in Linnaea XXXIV, 40 (1865) ... ... — — — — N.S.W. — — B.fl.VI,133
ADRIANA, Gaudichaud in Ann. sc. nat. V, 223 (1825). (Trachycaryon.)
A. tomentosa, Gaudichand in Ann. sc. nat. V, 223 (1825) ... W.A. S.A. — V. N.S.W. Q. N.A. B.fl.VI,134 M.fr.IX,2.
A. quadripartita, Gaudichaud in Freyc. voy. Bot. 489 (1826) ... W.A. S.A. — V. — — — B.fl.VI,135
ALCHORNEA, Solander in Swartz, prodr. 698 (1788). (Coelebogyne, Cladodes.)
A. ilicifolia, J. Mueller in Linnaea XXXIV, 170 (1865) ... ... — — — — N.S.W. Q. — B.fl.VI,136
A. Thozetiana, F. v. M. in Benth. Fl. Austr. VI, 137 (1873) ... — — — — — Q. — B.fl.VI,137
TRAGIA, Plumier, gen. 14, t. 12 (1703).
T. Novae Hollandiae, J. Mueller in Linnaea XXXIV, 180 (1865) — — — — — Q. — B.fl.VI,137 M.fr.VI,183.
MALLOTUS, Loureiro, Fl. Cochinch. II, 635 (1790). (Echinus, Rottlera, Echinocroton.)
M. ricinoides, J. Mueller in Linnaea XXXIV, 187 (1865) — — — — — — Q. — B.fl.VI,140 M.fr.IV,138;VI,184.
M. Claineneis, Loureiro, Fl. Cochinchin. II, 635 (1770) ... ... — — — — — B.fl.VI,140 M.fr.VI,184.
M. claoxyloides, J. Mueller in Linnaea XXXIV, 192 (1865) ... — — — N.S.W. Q. — B.fl.VI,140 M.fr.VI,184.
M. Philippinensis, J. Mueller in Linnaea XXXIV, 196 (1865) ... — — — N.S.W. Q. — B.fl.VI,141 M.fr.VI,183.
M. angustifolius, Bentham in Fl. Austr. VI, 141 (1873)... ... — — — — — Q. — B.fl.VI,142
M. polyadenos, F. v. M., fragm. VI, 184 (1867) ... ... — — — — — Q. — B.fl.VI,142 M.fr.VI,184.
M. repandus, J. Mueller in Linnaea XXXIV, 197 (1865)... — — — — — Q. — B.fl.VI,142 M.fr.VI,185.
M. nesophilus, F. v. M. in Linnaea XXXIV, 196 (1865)... — — — — Q. N.A. B.fl.VI,143 M.fr.VI,184.
M. discolor, F. v. M. in Benth. Fl. Austr. VI, 143 (1873) — — — N.S.W. Q. — B.fl.VI,143 M.fr.IV,139.
MACARANGA, Petit-Thouars, Gen. Madag. n. 88 (1800). (Mappa.)
M. Dallachyi, F. v. M. in Benth. Fl. Austr. VI, 144 (1873) ... — — — — Q. — B.fl.VI,144 M.fr.VI,184.
M. subdentata, Bentham, Fl. Austr. VI, 145 (1873) ... — — — — Q. — B.fl.VI,145
M. inamoena, F. v. M. in Benth. Fl. Austr. VI, 145 (1873) ... — — — — Q. — B.fl.VI,145
M. involucrata, Baillon, Etud. Euphorb. 432 (1858) ... — — — — Q. — B.fl.VI,146 M.fr.IV,139;VI,183.
M. Tanaria, J. Mueller in De Candolle, prodr. XV, 997 (1866)... — — — N.S.W. Q. N.A. B.fl.VI,146 M.fr.VI,183.
CODIAEUM, Rumphius, herbar. Amboin. IV, 65, t. 25 (1743).
C. variegatum, Blume, Bijdrag. 605 (1825)... ... — — — — — Q. — B.fl.VI,147 M.fr.VI,182.
BALOGHIA, Endlicher, prodr. fl. Norfolk. 84 (1833).
B. lucida, Endlicher, prodr. fl. Norfolk. 84 (1833) ... — — — — N.S.W. Q. — B.fl.VI,148 M.fr.VI,162.
FONTAINEA, Heckel, thèse inaugurale, Montpell. (1870).
F. Pancheri, Heckel, thèse inaugurale (1870) ... ... — — — — — Q. — B.fl.VI,149
OMALANTHUS, Adr. de Jussieu, Euphorb. gen. 50, t. 16 (1824). (Carumbium, Wartmannia.)
O. populifolius, Graham in Bot. Mag. 2780 (1827) ... — — — V. N.S.W. Q. — B.fl.VI,150 M.fr.I,32.
O. stillingifolius, F. v. M., fragm. I, 32 (1858) ... ... — — — — N.S.W. Q. — B.fl.VI,150 M.fr.I,32.
SEBASTIANIA, Sprengel, Neue Entd. II, 118 (1821). (Microstachys, Elachocroton.)
S. chamelaea, J. Mueller in De Cand., prodr. XV, 1173 (1866)... — — — — — Q. N.A. B.fl.VI,151
EXCAECARIA, Linné, syst. veg. ed. decim. 1288 (1759).
E. Agallocha, Linné, syst. ed. decim. 1288 (1759)... ... — — — — N.S.W. Q. N.A. B.fl.VI,152 M.fr.XI,04.
E. Dallachyana, Baillon, Adans. VJ, 324 (1866)... ... — — — — N.S.W. Q. — B.fl.VI,153
E. parvifolia, J. Mueller in Flora, 433 (1864) ... ... — — — — — — N.A. B.fl.VI,153

URTICACEAE.
Ventenat, Table III, 524 (1790).

CELTIS, Tournefort, inst. 612, t. 383 (1700), from Camerarius (1586). (Solenostigma.)
C. Philippinensis, Blanco, Fl. de Filipinas. 197 (1837) ... ... — — — — — Q. N.A. B.fl.VI,156
C. paniculata, Planchon in Ann. sc. nat. sér. trois. V, 305 (1848) — — — N.S.W. Q. N.A. B.fl.VI,156 M.fr.IV,88.
C. amblyphylla, F. v. M., fragm. IX, 76 (1875) ... ... — — — N.S.W. — — — M.fr.IX,70.
ULMUS, Tournefort, inst. 601, t. 372 (1700), from Dodoens (1583).
U. parviflora, N. Jacquin, pl. rar. hort. Schoenbr. descr. et icon.
III, t. 260 (1798) ... ... — — — — — — — M.fr.VI,101;XI,133.
TREMA, Loureiro, Fl. Cochinch. II, 562 (1790). (Sponia.)
T. cannabina, Loureiro, Fl. Cochinch. II, 680 (1790) ... — — — V. N.S.W. Q. N.A. B.fl.VI,158 M.fr.VIII,247.
APHANANTHE, Planchon in Ann. sc. nat. X, 265 (1848).
A. Philippinensis, Planchon in Ann. sc. nat. X, 337 (1848) ... — — — — N.S.W. Q. — B.fl.VI,160
FICUS, Tournefort, inst. 662 (1700), from Plinius. (Urostigma, Covellia.)
F. colossea, F. v. M. in Benth. Fl. Austr. VI, 163 (1873) ... — — — — — Q. — B.fl.VI,163
F. pilosa, Reinwardt in Blume, Bijdr. 446 (1825)... ... — — — — — Q. — B.fl.VI,164 M.fr.VI,195.
F. nesophila, F. v. M., Docum. Interc. Exhib. 26 (1866) — — — — — Q. N.A. B.fl.VI,165 M.fr.VI,195.
F. Cunninghamii, Miquel in Ann. Mus. Lugd.Bot. III,286 (1867)— — — — — Q. — B.fl.VI,165 M.fr.VI,195;VIII,246.
F. Henneana, Miquel in Ann. Mus. Lugd. Bot. III, 286 (1867) ... — — — — Q. N.A. B.fl.VI,165
F. validinervis, F. v. M. in Benth. Fl. Austr. VI, 166 (1873) ... — — — — — Q. — B.fl.VI,166
F. retusa, Linné, mantissa plant. 129 (1767) ... ... — — — — — Q. N.A. B.fl.VI,166 M.fr.VI,195.
F. eugenioides, F. v. M., Docum. Interc. Exhib. 26 (1866) ... — — — N.S.W. Q. — B.fl.VI,166 M.fr.VI,195.
F. Benjamina, Linné, mantissa plant. 129 (1767) ... ... — — — — — Q. — B.fl.VI,167 M.fr.VI,195.
F. Muelleri, Miquel in Ann. Mus. Lugd. Bot. III, 287 (1867) ... — — — N.S.W. Q. — B.fl.VI,167 M.fr.VI,195;VIII,246.
F. leucotricha, Miquel in Ann. Mus. Lugd. Bot. III, 285 (1867) — — — — — — N.A. B.fl.VI,167 M.fr.VI,195.
F. rubiginosa, Desfontaines, catal. pl. hort. Paris. 209 (1804) ... — — — — N.S.W. Q. — B.fl.VI,168 M.fr.VI,194;VIII,247.
F. columnaris, F. v. M. & Moore in proc. Accl. Soc. Vict. III,
71 (1874) ... ... — — — N.S.W. — — — M.fr.VIII,247.
F. puberula, Cunningham in Hook. Lond. Journ. VI, 502 (1847) — — — — — — N.A. B.fl.VI,169
F. platypoda, Cunningham in Hook. Lond. Journ. VI, 561 (1847) — S.A. — — — Q. N.A. B.fl.VI,169 M.fr.VI,194;VIII,247.
F. dictyophleba, F. v. M. in Benth. Fl. Austr. VI, 170 (1873)... — — — — — Q. — B.fl.VI,170

F. macrophylla, Desfontaines, catal. pl. hort. Paris. 209 (1804) — — — — N.S.W. Q. — B.fl.VI,170 M.fr.VI,194;VIII,247.
F. maguilolia, F. v. M., fragm. IV, 50 (1864) ... — — — — — Q. — B.fl.VI,171 M.fr.IV,50&177;VI,196.
F. ehreticides, F. v. M. in Benth. Fl. Austr. VI, 171 (1873) ... — — — — — Q. — B.fl.VI,171
F. pumila, Linné, spec. plant. 1060 (1753)... ... ... — — — — N.S.W. Q. — B.fl.VI,171 M.fr.VI,196. [246.
F. coronulata, T. v. M. in Journ. Bot. Neerland 241 (1861) ... — — — — — N.A. B.fl.VI,172 M.fr.IV,49; VI,196;VIII,
F. leptoclada, Bentham, Fl. Austr. VI, 172 (1873) ... ... — — — — — Q. — B.fl.VI,172
F. deprossa, Bentham, Fl. Austr. VI, 172 (1873)... ... ... — — — — — Q. — B.fl.VI,172
F. Philippinensis, Miquel in Hooker, Lond. Journ. VII, 435 (1848) — — — — — Q. — B.fl.VI,173
F. mollior, F. v. M. in Benth. Fl. Austr. VI, 173 (1873) ... — — — — — Q. — B.fl.VI,173
F. stenocarpa, F. v. M. in Benth. Fl. Austr. VI, 174 (1873) ... — — — — V. N.S.W. Q. — B.fl.VI,174 M.fr.IX,152. [114.
F. scabra, G. Forster, florul. ins. Austr. prodr. 76 (1786) — — — — V. N.S.W. Q. — B.fl.VI,174 M.fr.VI,196;IX,152;X,
F. orbicularis, Cunningham in Hook. Lond. Journ. VII, 426 (1848) — S.A. — — — — N.A. B.fl.VI,175 M.fr.VI,196.
F. aculeata, Cunningham in Hook. Lond. Journ. VII, 426 (1848) — — — — — N.A. B.fl.VI,175
F. opposita, Miquel in Hook. Lond. Journ. VII, 426 (1848) ... — — — — N.S.W. Q. — B.fl.VI,176 M.fr.VI,196;VIII,246.
F. scobina, Bentham, Fl. Austr. VI, 176 (1873) ... ... — — — — — N.A. B.fl.VI,176
F. hispida, Linné, fil. suppl. 442 (1781) ... ... — — — — — Q. N.A. B.fl.VI,176
F. fasciculata, F. v. M. in Benth. Fl. Austr. VI, 177 (1873) ... — — — — — Q. — B.fl.VI,177
F. cassaria, F. v. M. in Benth. Fl. Austr. VI, 177 (1873) ... — — — — — Q. — B.fl.VI,177
F. glomerata, Willdenow, spec. pl. IV, 1148 (1806) ... ... — — — — — Q. N.A. B.fl.VI,178 M.fr.VI,195.
F. pleurocarpa, F. v. M., fragm. VIII, 246 (1874) ... — — — — — Q. — M.fr.VIII,246.

CUDRANIA, Trécul in Ann. des. sc. nat. VIII, 122 (1847).
C. Javanensis, Trécul in Ann. des. sc. nat. VIII, 123 (1847) ... — — — — N.S.W. Q. — B.fl.VI,179 M.fr.VIII,247.

ANTIARIS, Leschenault in Ann. du Mus. XVI, 476 (1810).
A. macrophylla, R. Brown in Flind. voy. II, 602, t. 5 (1814) ... — — — — N.A. B.fl.VI,179 M.fr.VI,194.

MALAISIA, Blanco, Fl. Filip. 789 (1837). [247.
M. tortuosa, Blanco, Fl. Filip. 789 (1837) ... ... — — — — N.S.W. Q. N.A. B.fl.VI,180 M.fr.VI,193&255;VIII,

PSEUDOMORUS, Bureau in Ann. des sc. nat. sér. cinq. XI, 372 (1872).
P. Brunoniana, Bureau in Ann. des sc. nat. sér. cinq. XI,372 (1872) — — — — N.S.W. Q. — B.fl.VI,181

FATOUA, Gaudichaud in Freyc. voy. Bot. 509 (1826).
F. pilosa, Gaudichaud in Freyc. voy. Bot. 509 (1826) ... ... — — — — N.A. B.fl.VI,182

ELATOSTEMMA, R. & G. Forster, char. gen. 105, t. 53 (1776).
E. reticulatum, Weddell in Ann. des sc. nat. sér. 4, I, 188 (1854) — — — N.S.W. Q. — B.fl.VI,183
E. stipitatum, Weddell in Ann. des sc. nat. sér. 4, I, 190 (1854) — — — N.S.W. Q. — B.fl.VI,184

PROCRIS, Commerçon in A. L. de Jussieu, gen. 403 (1789).
P. montana, Steudel, nomencl. bot. II. 398 (1841) ... ... — — — N.S.W. — — M.fr.IX,169.

BOEHMERIA, N. J. Jacquin, stirp. Amer. hist. 246, t. 157 (1763).
B. calophleba, Moore & Mueller, fragm. VIII, 11 (1870)... ... — — — N.S.W. — — B.fl.VI,184 M.fr.VIII,11.
B. australis, Endlicher, prodr. Fl. Norfolk, 38 (1833) ... ... — — — N.S.W. — — — M.fr.IX,160.

PIPTURUS, Weddell in Ann. des sc. nat. sér. quatr. I, 196 (1854).
P. propinquus, Weddell in Archiv. du Mus. 447, t. 15 (1856) ... — — — — N.S.W. Q. — B.fl.VI,185 M.fr.VI,197.

POUZOLZIA, Gaudichaud in Freyc. voy. Bot. 503 (1826). (Memorialis, Hyrtanandra.)
P. Indica, Gaudichaud in Freyc. voy. Bot. 503 (1826) ... — — — — Q. N.A. B.fl.VI,186 M.fr.IV,87.
P. quinquenervis, Bennett, Pl. Jav. rar. 66 (1852) ... ... — — — — Q. — B.fl.VI,187

PARIETARIA, Tournefort, inst. 509, t. 289 (1700), from C. Bauhin (1623). (Frcirea.)
P. debilis, G. Forster, florul. ins. Austr. prodr. 73 (1786) — W.A. S.A. T. V. N.S.W. Q. N.A. B.fl.VI,188 M.fr.XI,20.

AUSTRALINA, Gaudichaud in Freyc. voy. Bot. 407 (1826).
A. pusilla, Gaudichaud in Freyc. voy. Bot. 505 (1826) ... — — — T. V. N.S.W. — — M.fr.VIII,247.

FLEURYA, Gaudichaud in Freyc. voy. Bot. 497 (1826).
F. interrupta, Gaudichaud in Freyc. voy. Bot. 497 (1826) ... — — — — — Q. — M.fr.X,114.

URTICA, Tournefort, inst. 534, t. 308 (1700), from Plinius.
U. incisa, Poiret, Encycl. meth. suppl. IV, 223 (1817) ... — S.A. T. V. N.S.W. Q. — B.fl.VI,190

LAPORTEA, Gaudichaud in Freyc. voy. Bot. 498 (1826).
L. gigas, Weddell in Arch. du Mus. IX, 129, t. 3 & 4 (1856) ... — — — N.S.W. Q. — B.fl.VI,192
L. photiniphylla, Weddell in Arch. du Mus. IX, 138 (1856) ... — — — N.S.W. Q. — B.fl.VI,192
L. moroides, Weddell in Arch. du Mus. IX, 142 (1856) ... — — — N.S.W. Q. — B fl.VI,192 M.fr.VIII,248.

CUPULIFERAE.
L. Cl. Richard, Anal. de fr. 32 & 92 (1808).

FAGUS, Tournefort, inst. 584, t. 351 (1700), from Camerarius (1586).
F. Gunnii, J. Hooker, icon. plant., t. 881 (1852) ... ... — — — T. — — B.fl.VI,210
F. Cunninghamii, Hooker, Journ. of Bot. II, 152, t. 7 (1840) ... — — — T. V. — — B.fl.VI,210
F. Moorei, F. v. M., fragm. V, 109 (1866)... ... ... — — — N.S.W. — — B.fl.VI,211 M.fr.V,109.

BALANOPS, Baillon, Hist. des pl. VI, 237 & 258 (1876).
B. Australiana, F. v. M., fragm. X, 114 (1877) ... ... — — — — Q. — M.fr.X,114.

CASUARINEAE.
Mirbel in Ann. du Mus. XVI, 451 (1810).

CASUARINA, Rumphius, herb. Amboin. III, 67, t. 58 (1743).
C. quadrivalvis, Labill., voy. Holl. pl. spec. II, 67, t. 218 (1806) — S.A. T. V. N.S.W. — B.fl.VI,195 M.fr.VI,18&19;X,62.
C. trichodon, Miquel in Lehm. pl. Preiss. I, 641 (1845) ... W.A. — — — — — B.fl.VI,196 M.fr.VI,17.
C. lepidophloia, F. v. M., fragm. X, 115 (1877) ... ... — — S.A. — N.S.W. — — M.fr.X,115.
C. glauca, Sieber in Sprengel, syst. III, 803 (1826) ... — — S.A. — V. N.S.W. Q. — B.fl.VI,196 M.fr.VI,18.

C. Hncgeliana, Miquel in Lehm. pl. Preiss. I, 640 (1845) ... W.A. — — — — — — B.fl.VI,196 M.fr.VI,19;X,62.
C. equisetifolia, R. & G. Forster, char. gen. 103, t. 52 (1776) ... — — — — N.S.W. Q. N.A. B.fl.VI,197 M.fr.VI,17.
C. suberosa, Otto & Dietrich, allgem. Gartenzcit. 155 (1841) ... — — T. V. N.S.W. Q. — B.fl.VI,197 M.fr.VI,203:254.
C. Cunninghamiana, Miquel, Revis. Casuarin. 56, t. 6 (1848) ... — — — — N.S.W. Q. — B.fl.VI,198
C. distyla, Ventenat, pl. dans le jard. de Cels. t. 62 (1800) ... W.A. S.A. T. V. N.S.W. — — B.fl.VI,198 M.fr.VI,20.
C. Inophloia, F. v. M. & Bailey in Melb. Chemist, April (1882) — — — — — Q. — — M.fr.XII.
C. Fraseriana, Miquel, Revis. Casuarin. 50, t. 6 (1848) ... ... W.A. — — — — — — B.fl.VI,199 M.fr.VI,19.
C. nana, Sieber in Sprengel, syst. III, 804 (1826)... ... ... — — — N.S.W. — — B.fl.VI,199
C. humilis, Otto & Dietrich, allgem. Gartenzeit. 163 (1841) ... W.A. — — — — — — B.fl.VI,200 M.fr.VI,18.
C. torulosa, Alton, hort. Kew. III, 320 (1789) ... ... — — — — N.S.W. Q. — B.fl.VI,200 M.fr.VI,17.
C. decussata, Bentham, Fl. Austr. VI, 200 (1873) ... ... W.A. — — — — — — B.fl.VI,200
C. Decaisneana, F. v. M., fragm. I, 61 (1858) ... ... — — S.A. — — — N.A. B.fl.VI,201 M.fr.I,61;X,62.
C. Drummondiana, Miquel, Revis. Casuar. 26, t. 1 (1848) ... W.A. — — — — — — B.fl.VI,201
C. microstachya, Miquel in Lehm. pl. Preiss. I, 642 (1845) ... W.A. — — — — — — B.fl.VI,201 M.fr.VI,17;X,62.
C. acutivalvis, F. v. M., fragm. X, 61 (1876) ... ... W.A. — — — — — — — M.fr.X,61.
C. bicuspidata, Bentham, Fl. Austr. VI, 202 (1873) ... W.A. S.A. T. — — — — B & .VI,202
C. thuyoides, Miquel in Lehm. pl. Preiss. I, 641 (1845) ... W.A. — — — — — — B.fl.VI,202 M.fr.VI,17.
C. corniculata, F. v. M., fragm. X, 62 (1876) ... ... W.A. — — — — — — — M.fr.X,62.
C. acuaria, F. v. M., fragm. VI, 16 (1868)... ... ... W.A. — — — — — — B.fl.VI,202 M.fr.VI,10&250.

PIPERACEAE.

L. Cl. Richard in Humb. Boupl. & Kunth, Nov. gen. I, 46 (1815).

PIPER, Linné, gen. pl. 333 (1737), from C. Bauhin (1623). (Macropiper.)
P. subpeltatum, Willdenow, spec. plant. V, 166 (1810) .. ... — — — — — Q. — B.fl.VI,204
P. excelsum, G. Forster, florul. ins. Austr. prodr. 5 (1786) ... — — — — N.S.W. Q. — B.fl.VI,204
P. Novae Hollandiae, Miquel in Medd. Akad. Vetensk. Amsterd.
    sec. ser. II, 8 (1866)... ... ... — — — — N.S.W. Q. — B.fl.VI,204
P. Banksii, Miquel in Medd. Akad. Amst. sec. ser. II, 9 (1866) ... — — — — — Q. — B.fl.VI,205
P. triandrum, F. v. M., fragm. V, 107 (1966) ... ... — — — — — Q. — B.fl.VI,205 M.fr.V,107.
P. hederaceum, Cunningham in De Cand. prodr. XVI, 306 (1869) ... — — — N.S.W. — — B.fl.VI,205

PEPEROMIA, Ruiz & Pavon, fl. Peruv. & Chil. prodr. 8, t. 2 (1794).
P. leptostachya, Hooker & Arnott, Bot. Beech. 96 (1841) ... — — — N.S.W. Q. — B.fl.VI,206 M.fr.IX,76.
P. reflexa, A. Dietrich, spec. plant. I, 180 (1831) ... ... — — — N.S.W. Q. — B.fl.VI,206
P. Urvilleana, Ach. Richard in d'Urville, voy. Bot. 356 (1832)... — — — — — — — M.fr.IX,76.

PODOSTEMONEAE.

L. Cl. Richard in Humb. Bonpl. & Kunth, Nov. gen. I, 246 (1815).

Genus and species not yet determined ... ... — Q. — —

NEPENTHACEAE.

Reichenbach, Conspectus regni vegetabilis 45 (1828).

NEPENTHES, Linné, syst. nat. 9 (1735); Linné, gen. pl. 273 (1737).
N. Kennedyana, F. v. M., fragm. V, 154 (1866) ... ... — — — — — Q. — B.fl.VI,40 M.fr.V,154.

ARISTOLOCHIEAE.

A. L. de Jussieu, in Ann. du Mus. V, 221 (1804).

ARISTOLOCHIA, Tournefort, inst. 162, t. 71 (1700), from Hippocrates, Theophrastos & Dioscorides.
A. deltantha, F. v. M., fragm. VI, 179 (1869) ... ... — — — — — Q. -- B.fl.VI,208 M.fr.VI,179&255.
A. praevenosa, F. v. M., fragm. II, 166 (1861) ... ... — — — N.S.W. — — B.fl.VI,208 M.fr.VI,180.
A. pubera, R. Brown, prodr. 349 (1810) ... ... — — — N.S.W. — — B.fl.VI,208
A. Thozetii, F. v. M., fragm. II, 167 (1861) ... ... — — — — — Q. — B.fl.VI,208 M.fr.II,167.
A. Indica, Linné, spec. plant. 960 (1753) ... ... — — — — — — — B.fl.VI,209

BALONOPHOREAE.

L. Cl. Richard in Mém. du Mus. VIII, 404 & 429 (1822).

BALONOPHORA, R. & G. Forster, char. gen. 99, t. 50 (1776).
B. fungosa, R. & G. Forster, char. gen. 99, t. 50 (1776) ... ... — — — — — Q. — B.fl.VI,232

VINIFERAE.

J. de St. Hilaire, Exp. fam. II, 48, t. 79 (1805).

VITIS, Tournefort, inst. 613, t. 384 (1700), e Latinis. (Cissus.)
V. Baudiniana, F. v. M., fragm. IV, 136 (1864) ... ... — — — V. N.S.W. Q. — B.fl.I,447 M.fr.V,212;VI,178.
V. oblonga, Bentham, Fl. Austr. I, 447 (1863) ... ... — — — — — Q. — B.fl.I,447 M.fr.VI,178;IX,125.
V. cordata, Wallich in Bentham's flora Hongk. 54 (1861) ... — — — — — Q.A. B.fl.I,447 M.fr.II,73;VI,178;IX,126.
V. adnata, Wallich in Wight & Arnott, prodr. 126 (1834) ... — — — — — — N.A. B.fl.I,448 M.fr.VI,178.
V. nitens, F. v. M., fragm. II, 73 (1860) ... ... — — — — — — — B.fl.I,448 M.fr.VI,178.
V. brachypoda, F. v. M., fr. IX, 125 (1875) ... ... — — — — — — — M.fr.IX,125.
V. saponaria, Seemann, Bonplandia, 254 (1859) ... ... — — — — — — — B.fl.I,448 M.fr.VI,178.
V. acris, F. v. M., fragm. II, 75 (1860) ... ... — — — N.S.W. Q. — B.fl.I,448 M.fr.VI,178.
V. trifolia, Linné, spec. plant. 203 (1753) ... ... — — — — — Q. N.A. B.fl.I,449 M.fr.II,75;VI,177;IX,126.
V. clematidea, F. v. M., fragm. II, 74 (1860) ... ... — — — N.S.W. Q. — B.fl.I,449 M.fr.VI,177;IX,126.
V. acetosa, F. v. M., plant. of Vict. I, 94 (1860) ... ... — — — — — Q. N.A. B.fl.I,449 M.fr.V,177;IX,126.
V. hypoglauca, F. v. M., pl. of Vict. I, 95 (1860) ... ... — — V. N.S.W. Q. — B.fl.I,450 M.fr.VI,177.
V. stenolifolia, F. v. M. in Benth. Fl. Austr. I, 450 (1863) ... — — — N.S.W. Q. — B.fl.I,450 M.fr.IX,126.
V. penninervis, F. v. M., fr. VI, 177 (1868) ... ... — — — — — — — M.fr.VI,177.

24

V. opaca, F. v. M. in Benth. Fl. Austr. I, 450 (1863) ... ... — — — — N.S.W. Q. — B.fl.I,450 M.fr.VI,177;IX,123.
V. augustissima, F. v. M., fragm. 1, 141 (1859) ... ... W.A. — — — — — — — B.fl.I,450 M.fr.I,244;II,180;IX,126.

LEEA, Linné, mantisa. 17 (1767).
L. staphylea, Roxburgh, hort. Beng. 18 (1814) ... ... — — — — — Q. — B.fl.I,415 M.fr.VI,178.
L. aculeata, Blume, Bijdrag. 107 (1825) ... ... — — — — — Q. N.A. — M.fr.VI,178.

SAPINDACEAE.

A. L. de Jussieu, Gen. 246 (1780).

CARDIOSPERMUM, Linné, syst. nat. (1735); Linné, gen. pl. 117 (1737).
C. Halicacabum, Linné, sp. pl. 366 (1753) ... ... — — — — Q. N.A. B.fl.I,453

GANOPHYLLUM, Blume, Mus. Bot. Lugd. I, 230 (1850).
G. falcatum, Blume, Mus. Bot. Lugd. I, 230 (1850) ... ... — — — — Q. — — M.fr.VII,24;IX,103.

ATALAYA, Blume, Rumphia III, 186 (1847). (Pseudatalaya, Sapindus partly.)
A. coriacea, Radlkofer in Sitzungsber. d. Akad. zu München. 326 (1878) — — — — N.S.W.
A. salicifolia, Blume, Rumphia III, 186 (1847) ... ... — — — — — N.A. B.fl.I,463
A. australis, Radlk. in Sitzungsber. der Akad. zu München. 327 (1878) — — — — — Q. — B.fl.I,464
A. multiflora, Bentham, Fl. Austr. I, 463 (1863) ... ... — — — — N.S.W. Q. — B fl.I,463 M.fr.IX,00.
A. hemiglauca, F. v. M. in Benth. Fl. Austr. I, 463 (1863) ... — S.A. — — N.S.W. Q. N.A. B.fl.I,463 M.fr.IX,90.
A. variifolia, F. v. M. in Benth. Fl. Austr. I, 463 (1863) ... — — — — — — N.A. B.fl.I,463 M.fr.I,45.

DIPLOGLOTTIS, J. Hooker in H. & B. gen. pl. 395 (1862).
D. Cunninghamii, J. Hooker in H. & B. gen. pl. 395 (1862) ... — — — N.S.W. Q. — B.fl.I,454 M.fr.V,145;IX,00.

ERIOGLOSSUM, Blume, Bijdr. 229 (1829). (Pancovia partly.)
E. edule, Blume, Bijdr. 229 (1829) ... ... — — — — N.A. B.fl.I,454 M.fr.IX,100.

CASTANOSPORA, F. v. Mueller, fragm. IX, 92 (1875).
C. Alphandi, F. v. M., fragm. IX, 92 (1875) ... ... — — — — Q. — — M.fr.IV,158;IX,92,197.

ALLOPHYLUS, Linné, amoen. acad. 124 (1747). (Schmidelia.)
A. ternatus, Loureiro, Fl. Cochinch. ed. Willd. 286 (1790) ... — — — — Q. N.A. B.fl.I,455 M.fr.IX,90.

CUPANIA, Plumier, Gen. 45, t. 19 (1703). (Cuioa, Ratonia, Arytera, Elattostachys, Mischocarpus, Euphoria partly.)
C. Wadsworthii, Harpullia, F. v. M., fragm. IV, 1, t. XXVI (1863) — — — — — Q. — M.fr.IV,1;IX,89,197.
C. anacardioides, A. Richard, sert. Astrol. 33, t. 13 (1833) ... — — — — N.S.W. Q. N.A. B.fl.I,458 M.fr.IX,91.
C. Robertsonii, F. v. M., fragm. V, 146 (1866) ... ... — — — — — Q. — M.fr.V,146;IX,94.
C. serrata, F. v. M., fragm. III, 43 (1862) ... ... — — — — N.S.W. Q. — B fl.I,458 M.fr.III,43;IX,94.
C. foveolata, F. v. M., fragm. IX, 95 (1875) ... ... — — — — — Q. — M.fr.IX,95.
C. tomentella, F. v. M. in Benth. Fl. Austr. I, 458 (1863) ... — — — — — Q. — B.fl I,458
C. pseudorhus, A. Richard, sert. Astrol. 34, t. 14 (1833) ... — — — — N.S.W. Q. — B.fl.I,459 M.fr.IX,92.
C. xylocarpa, Cunningham in Benth. Fl. Austr. I, 459 (1863) ... — — — — N.S.W. Q. — B.fl.I,459 M.fr.IX,91.
C. erythrocarpa, F. v. M., fragm. V, 7 (1865) ... ... — — — — — Q. — M.fr.V,7;IX,91.
C. Mortoniana, F. v. M., fragm. V, 177 (1866) ... ... — — — — — Q. — M.fr.V,177;IX,94.
C. lachnocarpa, F. v. M., fragm. IV, 157 (1864) ... ... — — — — — Q. — M.fr.IV,156;IX,91.
C. grandisima, F. v. M., fragm. IV, 156 (1864) ... ... — — — — — Q. — M.fr.IV,156;IX,91.
C. pyriformis, F. v. M., App. Rep. Intercol. Exhib. 25 (1867)... — — — — N.S.W. — B.fl.I,461 M.fr.I,2;II,76;IX,90.
C. anodonta, F. v. M., fragm. II, 76 (1860) ... ... — — — — — Q. — B.fl.I,461 M.fr.I,2;II,76;IX,90.
C. stipitata, F. v. M., fragm. II, 75 (1860) ... ... — — — — N.S.W. — B.fl.I,461 M.fr.II,75;IX,91.
C. punctulata, F. v. M., fragm. III, 12 (1862) ... ... — — — — — Q. — B.fl.I,458 M.fr.III,12;IX,91.
C. exangulata, F. v. M., fragm. IV, 156 (1864) ... ... — — — — — Q. — M.fr.IV,156;IX,91.
C. O'Shanesiana, F. v. M., fragm. IX, 96 (1875) ... ... — — — — — Q. — M.fr.IX,96.
C. Daemeliana, F. v. M., fragm. IX, 96 (1875) ... ... — — — — — Q. — M.fr.IX,96.
C. Martyana, F. v. M., fragm. V, 6 (1865) ... ... — — — — — Q. — M.fr.V,6;IX,04.
C. Cordieri, F. v. M., fragm. IX, 93 (1875) ... ... — — — — — Q. — M.fr.IX,93.
C. tenax, Cunningham in Benth. Fl. Austr. I, 461 (1863) ... — — — — N.S.W. Q. — B.fl.I,461 M.fr.IX,94.
C. semiglauca, F. v. M. in Benth. Fl. Austr. I, 457 (1863) ... — — — — — Q. — B-fl.I,457 M.fr.IX,97.

NEPHELIUM, Linné, mantiss. 18 (1767). (Euphoria partly, Alectryon, Spanoghea.)
N. connatum, F. v. M. in Benth. Fl. Austr. I, 465 (1863) ... — — — — N.S.W. Q. — B.fl.I,465 M.fr.IX,158.
N. semicinereum, F. v. M., fragm. IV, 158 (1864) ... ... — — — — — Q. — M.fr.IX,158.
N. subdentatum, F. v. M. in Benth. Fl. Austr. I, 465 (1863) ... — — — — N.S.W. — B.fl.I,465 M.fr.IX,99.
N. tomentosum, F. v. M. in Transact. Vict. Inst. II, 64 (1858) ... — — — — N.S.W. Q. — B.fl.I,466 M.fr.IX,90.
N. coriaceum, Bentham, Fl. Austr. I, 466 (1863)... ... ... — — — — — Q. — B.fl.I,466
N. foveolatum, F. v. M. in Benth. Fl. Austr. I, 466 (1863) ... — — — — — Q. — B.fl I,466 M.fr.IX,90.
N. leiocarpum, F. v. M. in Transact. phil. Inst. Vict. III, 23 (1858) — — — V. N.S.W. Q. — B.fl.I,467 M.fr.IX,98.
N. Beckleri, Bentham, Fl. Austr. I, 467 (1863) ... ... — — — — N.S.W. — B.fl.I,467
N. divaricatum, F. v. M. in Benth. Fl. Austr. I, 467 (1863) ... — — — — — Q. — B.fl.I,467 M.fr.IX,98.
N. microphyllum, Bentham, Fl. Austr. I, 468 (1863) ... ... — — — — — Q. — B.fl I,468
N. Leichhardtii, F. v. M., App. Rep. Intercol. Exhib. 25 (1867) ... — — — — — Q. — B.fl.I,468 M.fr.IX,00.
N. distyle, F. v. M., fragm. IX, 99 (1875)... ... ... — — — — — Q. — B.fl.I,402 M.fr.IX,99.

HETERODENDRON, Desfontaines in Mém. du Mus. IV, 8, t. 3 (1818).
H. oleaefolium, Desfontaines in Mém. du Mus. IV, 8, t. 3 (1818) W.A. S.A. — V. N.S.W. Q. N.A. B.fl.I,469 M.fr.IX,90;X 82.
H. diversifolium, F. v. M., fragm. I, 46 (1858) ... ... N.S.W. Q. — B.fl I,469 M.fr.I,46;IX,90;X,82.

HARPULLIA, Roxburgh, Fl. Ind. ed. Carey II, 44 (1826).
H. alata, F. v. M., fragm. II, 103 (1861) ... ... — N.S.W. Q. — B.fl.I,470 M.fr.II,103;IX,89.
H. Hillii, F. v. M. in Transact. Vict. Inst. III, 26 (1859) ... ... — N.S.W. Q. — B.fl.I,470
H. Leichhardtii, F. v. M. in Benth. Fl. Austr. I, 470 (1863) ... — — — N.A. B.fl.I,470
H. pendula, Planchon in Transact. Vict. Inst. III, 26 (1859) ... — N.S.W. Q. — B.fl.I,471 M.fr.IX,89&187.

AKANIA, J. Hooker in B. & H. Gen. pl. 409 (1862).
A. Hillii, J. Hooker in B. & H. Gen. pl. 409 (1862) ... ... — N.S.W. Q. — B.fl.I,471 M.fr.III,44;IX,89&107.

DIPLOPELTIS, Endlicher in Huegel, enum. 13 (1837).
D. Huegelii, Endlicher in Hueg. enum. 13 (1837)... ... ... W.A. — — — — — N.A. fl.fl.I,456 M.fr.III,12;IX,89.
D. Stuartii, F. v. M., fragm. III, 12 (1862) ... ... — S.A. — — — — N.A. B.fl.I,456 M.fr.III,12&167;IX,80.
DODONAEA, Linné, gen. pl. 341 (1737). (Empleurosma).
D. triquetra, Wendland, Bot. Beobacht. p. 44 (1798) ... ... — — V. N.S.W. Q. B.fl.I,474 M.fr.I,73;IX,88.
D. lanceolata, F. v. M., fragm. I, 73 (1858) ... ... — — N.S.W. Q. N.A. B.fl.I,475 M.fr.I,73.
D. petiolaris, F. v. M., fragm. III, 13 (1862) ... W.A. S.A. — N.S.W. — B.fl.I,475 M.fr.IX,89.
D. viscosa, Linné, mantissa X, altera 149 & 228 (1771) ... W.A. S.A. T. V. N.S.W. Q. N.A. B.fl.I,475 M.fr.IX,80.
D. peduncularis, Lindley in Mitch. Trop. Austr. 361 (1848) ... — — N.S.W. Q. — B.fl.I,478 M.fr.IX,87.
D. procumbens, F. v. M. in Transact. Vict. Inst. I, 3 (1854) ... — S.A. — V. N.S.W. — B.fl.I,478
D. cricifolia, G. Don, gen. syst. I, 674 (1831) ... ... — T. — — — B.fl.I,478 M.fr.IX,87.
D. filifolia, Hooker in Mitch. Trop. Austr. 241 (1848) ... — — — Q. — B.fl.I,478 M.fr.I,71.
D. lobulata, F. v. M. in Linnaea XXV, 372 (1852) ... W.A. S.A. — N.S.W. — — B.fl.I,479 M.fr.IX,87.
D. ptarmicifolia, Turcz. in Bull. Soc. Mosc. XXV, part III, 165(1852) W.A. — — — — B.fl.I,479 M.fr.I,07;IX,87.
D. truncatiales, F. v. M., fragm. II, 143 (1861) ... — — V. N.S.W. Q. — B.fl.I,479 M.fr.IX,90.
D. platyptera, F. v. M., fragm. I, 73 (1858) ... ... — — — — N.A. B.fl.I,480 M.fr.IX,87.
D. stenophylla, F. v. M , fragm. I, 72 (1858) ... ... — — — Q. — B.fl.I,480 M.fr.IX,87.
D. pinifolia, Miquel in Lehm. pl. Preiss. I, 227 (1844) ... ... W.A. — — — — B.fl.I,480 M.fr.IX,88.
D. ceratocarpa, Endlicher in Hueg. enum. 13 (1837) ... ... W.A. — — — — B.fl.I,481 M.fr.X,89.
D. divaricata, Bentham, Fl. Austr. I, 481 (1863)... ... ... W.A. — — — — B.fl.I,481 M.fr.IX,87.
D. triangularis, Lindley in Mitch. Trop. Austr. 219 (1848) — — — N.S.W. Q. — B.fl.I,481 M.fr.IX,87.
D. aptera, Miquel in Lehm. pl. Preiss. I, 225 (1844) ... ... W.A. — — — — D.fl.I,481 M.fr.IX,88.
D. bursarifolia, Behr & F. v. M. in Transact. Vict. Inst. I, 8 (1855) W.A. S.A. — V. N.S.W. — B.fl.I,482 M.fr.IX,87.
D. Baueri, Endlicher in Hueg. enum. 13 (1837) ... ... — S.A. — V. N.S.W. — — B.fl.I,482 M.fr.IX,88.
D. trifida, F. v. M., fragm. IX, 58 (1875) ... ... W.A. — — — — — M.fr.IX,88.
D. humifusa, Miquel in Lehm. pl. Preiss. I, 208 (1844) ... W.A. — — — — — B.fl.I,492 M f IX,88.
D. hexandra, F. v. M. in Transact. Vict. Inst. I, 117 (1855) ... — S.A. — — — — B.fl.I,483
D. ericoides, Miquel in Lehm. pl. Preiss. I, 227 (1844) ... W.A. — — — — B.fl.I,483 M.fr.IX,88.
D. polyzyga, F. v. M. in fragm. I, 74 (1858) ... ... — — — — N.A. B.fl.I,483 M.fr.I,74.
D. megazyga, F. v. M. in Benth. Fl. Austr. I, 483 (1863) ... — — — N.S.W. — B.fl.I,483 M.fr.IX,86.
D. macrozyga, F. v. M., fragm. IV, 135 (1864) ... ... — — — — Q. — M.fr.IX,86.
D. physocarpa, F. v. M., fragm. I, 74 (1858) ... ... — — — — N.A. B.fl.I,484 M.fr.IX,80.
D. vestita, Hooker in Mitch. Trop. Austr. 265 (1848) ... — — — — N.S.W. — B.fl.I,484 M.fr.IX,86.
D. pinnata, Smith in Rees' Cycl. XII (1909) ... ... — — — — — B.fl.I,484
D. oxyptera, F. v. M., fragm. I, 74 (1858)... ... ... — — — — N.A. B.fl.I,484 M.fr.I,74.
D. humilis, Endlicher, nov. stirp. dec. 26 (1839) ... ... — S.A. — — — N.A. B.fl.I,485 M.fr.IX,86.
D. Macrossani, F. v. M. & Scortechini in Melb. Chemist Jan. 1882 — — — — Q. — M.fr.XII.
D. microzyga, F. v. M. annual Rep. bot. Gard. 12 (1863) — — — — — — M.fr.IX,86.
D. boronifolia, G. Don, gen. syst. I, 674 (1831) ... ... — S.A. — V. N.S.W. Q. — B.fl.I,485 M.fr.IX,86.
D. multijuga, G. Don, gen. syst. I, 674 (1831) ... ... — — — N.S.W. — — B.fl.I,486 M.fr.IX,86.
D. larraeoides, Turcz. in Bull. Soc. Mosc. XXXI, 408 (1858) W.A. — — — — B.fl.I,486
D. inaequifolia, Turcz. in Bull. Soc. Mosc. XXXI, 408 (1858) ... W.A. — — — — B.fl.I,486 M.fr.I,210;IX,80.
D. adenophora, Miquel in Linnaea XVIII, 95 (1845) ... ... W.A. — — N.S.W. Q. — B.fl.I,486 M.fr.IX,85.
D. tenuifolia, Lindley in Mitch. Trop. Austr. 248 (1848) — — — — — — M.fr.IX,85.
D. stenozyga, F. v. M., fragm. I, 96 (1858) ... ... W.A. S.A. — V. N.S.W. — B.fl.I,486 M.fr.IX,87.
D. concinna, Bentham, Fl. Austr. I, 487 (1863) ... ... W.A. — — — — — B.fl.I,487 M.fr.IX,87.

DISTICHOSTEMON, F. v. Mueller in Hook. Kew misc. IX, 306 (1857).
D. phyllopterus, F. v. M. in Hook. Kew misc. IX, 306 (1857) ... — — — — — Q. N.A. B.fl.I,487 M.fr.IX,89.

BLEPHAROCARYA, F. v. Mueller, fragm. XI, 15 (1878).
B. involucrigera, F. v. M., fragm. XI, 16 (1878) ... ... — — — — — Q. — — M.fr.XI,16&137

MALPIGHIACEAE.

Ventenat, Tabl. III, 131 (1799).

RYSSOPTERYS, Blume in Delessert ic. sel. III, 21, t. 35 (1837).
R. Timorensis, Blume in Deless. ic. sel. III, 21, t. 35 (1837) ... — — — — — Q. — B.fl.I,285

TRISTELLATEIA, Petit-Thouars in Roemer. Collect. bot. 206 (1809).
T. Australasica, A. Richard, Sert. Astrol. 38, t. 15 (1833) ... ... — — — — — Q. — B.fl.I,286

BUESERACEAE.

Kunth in Ann. des sc. nat. II, 346 (1824).

GARUGA, Roxburgh, Pl. of Coromandel III, 5, t. 208 (1810).
G. floribunda, Descalsne, in nouv. annal. du Mus. III, 477 (1834) — — — — — N.A. B.fl.I,377
CANARIUM, Rumphius, herb. Amboin. II, 145, t. 47, 49, 54 (1741). (Sonzaya.)
C. Australianum, F. v. M., fragm. III, 15 (1800)... ... — — — — Q. N.A. B.fl.I,379 M.fr.VII,25.

ANACARDIACEAE.

R. Brown in Tuckey's Expedit. Congo 431 (1818).

RHUS, Tournefort, Inst. 611, t. 381 (1700), from Hippocr. Theophr. Diosc. & Plin. (Rhodosphaera.)
R. rhodanthema, F. v. M. in Journ. pharm. soc. Vict. pl. I, 43 (1858) — — — N.S.W. Q. — B.fl.I,460 M.fr.VII,23.
R. panaciformis, F. v. M., fragm. VII, 22 (1800)... ... — — — — Q. — — M.fr.VII,22.

EUROSCHINUS, J. Hooker in B. & H. Gen. pl. I, 422 (1862).
E. falcatus, J. Hooker in B. & H. Gen. pl. I, 422 (1862) ... — — N.S.W. Q. — B.fl.I,400 M.fr.VII,24.

BUCHANANIA, Sprengel in Schrader's Journ. II, 234 (1800).
B. angustifolia, Roxburgh, Pl. Corom. III, t. 262 (1819) ... — — — — Q. N.A. B.fl.I,400 M.fr.VII,23.
B. mangoides, F. v. M., fragm. VII, 23 (1869) ... ... — — — — — Q. — M.fr.VII,23.
B. arborescens, Blume, Mus. Bot. Lugd. Batav. I, 183 (1850) ... — — — — — Q. — M.fr.VII,23.

SEMECARPUS, Linné fil., suppl. 25 (1781).
S, Anacardium, Linné fil., suppl. 182 (1781) ...   — — — — —   Q. N.A. B.fl.I,401 M.fr.VII,23.

SPONDIAS, Linné, gen. pl. 365 (1737).
S. Solandri, Bentham, Fl. Austr. I, 492 (1863) ... ... — — — — —   Q. — B.fl.I,402
S. pleiogyna, F. v. M., fragm. IV, 78 (1864) ... ...   — — — — —   Q. —   M.fr.III,15.

CELASTRINEAE.
R. Brown in App. Flind. Voy. 554 (1814).

EUONYMUS, Tournefort, inst. 617, t. 388 (1700), from Theophr. & Plinius. (Evonymus.)
E. Australianus, F. v. M., fragm. IV, 118 (1864)... ... ... — — — — —   Q. —   — M.fr.IV 118;VI,202.

CELASTRUS, Linné, gen. pl. 59 (1737).
C. Australis, Harvey&F. v. M. in Trans. phil. Soc. Vict. I, 41 (1854) — — V. N.S.W. Q. — B.2.I,398 M.fr.III,93.
C. Muelleri, Bentham, Fl. Austr. I, 399 (1863) ... ... — — — — — N.A. B.fl.I,399
C. dispermus, F. v. M. in Transact. phil. Inst. Vict. III, 31 (1850) — — — — — Q. B.fl.I,399 M.fr.VI,203.
C. bilocularis, F. v. M. in Transact. phil. Inst. Vict. III, 31 (1859) — — — — Q. — B.fl.I,399 M.fr.VI,204.
C. Cunninghamii, F. v. M. in Trans. phil. Inst. Vict. III, 30 (1859) — — — .— — N.S.W. Q. N.A. B.fl.I,399 M.fr.VI,203.

GYMNOSPORIA, Wight & Arnott, prodr. fl. poenius. Ind. I, 159 (1834.)
G. montana, Wight & Arnott, prodr. I, 159 (1834) ... ... — — — — — Q. — B.fl.I,400

LEUCOCARPUM, A. Richard, sert. Astrolabe 46 (1834). (Denhamia, Leucocarpon.)
L. Oleaster, F. v. M. in Transact. Roy. Soc. of N.S.W. 15 (1881) — — — — — Q. — B.fl.I,401 M.fr.VI,203.
L. obscurum, A. Richard, sert. Astrol. 46, t. 18 (1834) ... — — — — Q. N.A. B.fl.I,401 M.fr.VI,203.
L. pittosporoïdes, F. v. M. in Trans. Roy. Soc. of N.S.W. 15 (1881) — — — N.S.W. Q. — B.fl.I,402 M.fr.VI,203.

ELAEODENDRON, J. F. Jacquin in Nov. Act. Helv. I, 36 (1787).
E. Australe, Ventenat, Jard. Malm. t. 117 (1804) ... ... — — N.S.W. Q. N.A. B.fl.I,402 M.fr.III,61;IV,175;VI,204.

CARYOSPERMUM, Blume, Mus. bot. Lugd. I, 175 (1850).
C. arborescens, F. v. M., fragm. VI, 202 (1868) ... ... — — — — — Q. —   M.fr.VI,202.

SIPHONODON, Griffith in Calc. journ. of Nat. Hist. IV, 150 (1843).
S. Australe, Bentham, Fl. Austr. I, 403 (1863) ... ... — — — N.S.W. Q. — B.fl.I,403 M.fr.VI,204.

HIPPOCRATEA, Linné, gen. pl. 363 (1737).
H. obtusifolia, Roxburgh, flor. Ind. edit. Carey, I, 170 (1820).. — — — N.S.W. Q. — B.fl.I,404

SALACIA, Linné, mantiss. II, 159 (1771).
S. prinoides, De Candolle, prodr. I, 571 (1824) ... — — — — — Q. — —   M.fr.VI,202.

STACKHOUSIEAE.
R. Brown in App. to Flind. Voy. 555 (1814).

STACKHOUSIA, Smith in Transact. Linn. Soc. IV, 218 (1798). (Tripterococcus, Plokiostigma.)
S. pulvinaris, F. v. M. in Transact. phil. Soc. IV, in (1855) — — — — — B.fl.I,405 M.fr.III,88.
S. linarifolia, Cunningham in Field's N.S.Wales 365 (1825) — — S.A. T. V. N.S.W. Q. — B.fl.I,406 M.fr.III,87;XI.27.
S. pubescens, A. Richard, sert. Astrol. 89, t. 33 (1834) ... W.A. — — — — — B.fl.I,407
S. Huegelii, Endlicher in Hueg. enum. 17 (1837) ... ... W.A. — — — — — B.fl.I,407
S. flava, Hooker, icon. pl. t. 209 (1840) ... ...   W.A. S.A. T. V. — — — B.fl.I,407
S. muricata, Lindley, Bot. Reg. XXII, 1917 (1837) ...   — — N S W. Q. N.A. B.fl.I,408 M.fr.VIII,36.
S. viminea, Smith in Rees' Cycl. XXXII (1816)... ... ...   — — V. N.S.W. Q. N.A. B.fl.I,408 M.fr.VIII,36.
S. elata, F. v. M., fragm. III, 86 (1862) ... ... W.A. — — — — — B.fl.I,408 M.fr.III,86.
S. scoparia, Bentham, Fl. Austr. I, 409 (1863) ... ... W.A. — — — — — B.fl.I,409
S. spathulata, Sieber in Spreng. syst. cur. post. 124 (1827) W.A. — S.A. T. V. N.S.W. Q. — B.fl.I,406 M.fr.VIII,30.
S. megaloptera, F. v. M., fragm. VIII, 35 (1873)... W.A. — — — — — —   M.fr.X,119.
S. Brunonis, Bentham, Fl. Austr. I, 409 (1863) ... W.A. — — — N.A. B.fl.I,409 M.fr.VIII,36.

MACCREGORIA, F. v. Mueller in Caruel, Giorn. Bot. Italian. 128 (1873).
M. racemigera, F. v. M. in Caruel, Giorn. Bot. Ital. 128 (1873) — S.A. — — —   M.fr.VIII,161;XI,134.

FRANKENIACEAE.
A. de St. Hilaire in Bull. de la Soc. Philom. 22 (1815).

FRANKENIA, Linné, gen. pl. 92 (1737).
F. laevis, Linné, spec. plant. 331 (1753) ... ... W.A. S.A. T. V. N.S.W. Q. N.A. B.fl.I,151
F. bracteata, Turczan. in Bull. Soc. Mosc. XXVII, 367 (1854)... W.A. — — — — — B.fl.I,150
F. glomerata, Turczan. in Bull. Soc. Mosc. XXVII, 368 (1854) W.A. — — — — — B.fl.I,151
F. parvula, Turczan. in Bull. Soc. Mosc. XXVII, 368 (1854) ... W.A. — — — .— — B.fl.I,152
F. Drummondii, Bentham, Fl. Austr. I, 152 (1863)   W.A. — — — — — B.fl.I,151
F. tetrapetala, Labillard., nov. Holl. pl. spec. I, 88, t. 114 (1804) W.A. — — — — ... — B.fl.I,152
F. punctata, Turczan. in Bull. Soc. Mosc. XXVII, 367 (1854) ... W.A. — — — — — B.fl.I,153

PLUMBAGINEAE.
A. L. de Jussieu, gen. plant. 92 (1789).

PLUMBAGO, Tournefort, inst. 140, t. 58 (1700).
P. Zeilanica, Linné, sp. pl. 151 (1753) ... ... — — — — N.S.W. Q. N.A. B.fl.IV,267 M.fr.XI,7.

STATICE, Linné, gen. pl. 88 (1737).
S. Taxanthema, Roemer & Schultes, syst. veg. VI, 798 (1820) ... — T. V. N.S.W. Q. — B.fl.IV,267 M.fr.XI,7.
S. salicorniacea, F. v. M., fragm. XI, 7 (1878) ... ... W.A. — — — — N.A. —   M.fr.XI,7.

AEGIALITIS, R. Brown, prodr. 426 (1810). (Aegialinites).
A. annulata, R. Brown, prodr. 426 (1810) ... ... — — Q. — B.fl.IV,266

PORTULACEAE.

A. L. de Jussieu, Gen. 312 (1789).

PORTULACA, Tournefort, inst. 236, t. 118 (1700), from 1 Obel (1581).
P. oleracea, Linné, spec. plant. 445 (1753)... ... ... W.A. S.A. — V. N.S.W. Q. N.A. B.fl.I,160 M.fr.X,71.
P. napiformis, F. v. M. in Benth. Fl. Austr. I, 169 (1863) ... — — — — — Q. N.A. B.fl.I,169 M.fr.X.72.
P. australis, Endlicher, Atakta 7, t. 6 (1833) ... ... — — — — V. Q. N.A. B.fl.I,169 M.fr.X,72.
P. filifolia, F. v. M., fragm. I, 169 (1859) ... ... — S.A. — — N.S.W. Q. N.A. B.fl.I,169 M.fr.III,104;X,72.
P. digyna, F. v. M., fragm. I, 170 (1859) ... ... — — — — — N.A. B.fl.I,170 M.fr.I,170.
P. oligosperma, F. v. M., fragm. I, 170 (1859) ... ... — — — — Q. N.A. B.fl.I,170 M.fr.III,164;X,72.
P. bicolor, F. v. M., fragm. I, 171 (1859) ... ... — — — — Q. N.A. B.fl.I,170 M.fr.III,164;X,72.
P. Armitti, F. v. M., fragm. X, 97 (1877) ... ... — — — — — Q. N.A. — M.fr.X,97.

CLAYTONIA, Gronovius in Linné, gen. pl. 339 (1737). (Calandriula, Talinum.)
C. spergularina; Calandrinia, F. v. M., fragm. I, 175 (1859) ... — — — — Q. N.A. B.fl.I,176 M.fr.I,175;X,71.
C. ptychosperma; Calandrinia, F. v. M., fragm. IV, 137 (1864) — S.A. — — — —•Q. — — M.fr.IV,137;X,70
C. strophiolata, F. v. M., fragm. XI, 82 (1880) ... ... W.A. — — — — N.A. — M.fr.XI,82.
C. Lehmanni, F. v. M.; Calandr., Endl. in Lehm. pl. Preiss. II,
     235 (1847) ... ... W.A. — — — — — B.fl.I,172 M.fr.X,71.
C. uniflora, F. v. M.; Calandr. in Trans. Inst. Vict. III, 41 (1859) — — — — Q. N.A. B.fl.I,172 M.fr.X,71.
C. Balonnensis, F. v. M.; Calandrinia, Lindley in Mitch. Trop.
     Austr. 148 (1848) ... ... — S.A. — — N.S.W. Q. — B.fl.I,172 M.fr.X,71.
C. polyandra, F. v. M.; Calandr., Benth. Fl. Austr. I, 172 (1863) W.A. S.A. — — N.S.W. — — B.fl.I,172 M.fr.X,71.
C. pleiopetala; Calandrinia, F. v. M., fragm. X, 70 (1877) ... — S.A. — — — Q. — — M.fr.X,70.
C. quadrivalvis; Calandrinia, F. v. M., fragm. I, 176 (1859) ... — — — — — N.A. B.fl.I,173 M.fr.I,176.
C. uniflora, F. v. M.; Calandr., Fenzl. in Hueg. enum. 52 (1837) W.A. — — — — N.A. B.fl.I,173 M.fr.X,71.
C. gracilis, F. v. M.; Calandrinia, Benth. Fl. Austr. I, 173 (1863) — — — — — N.A. B.fl.I,173
C. polypetala, F. v. M.; Calandr., Fenzl. in Hueg. enum. 51 (1837) W.A. — — — — — B.fl.I,174
C. Pickeringi, F. v. M.; Calandr., A. Gray, Bot. Wilkes' Exped.
     144 (1854) ... ... — — — N.S.W. — — — M.fr.X,69;XI,130.
C. volubilis, F. v. M.; Calandr., Benth. Fl. Austr. I, 174 (1863) W.A. S.A. — — N.S.W. Q. — B.fl.I,174 M.fr.X,68;XI,27.
C. calyptrata, F. v. M., fragm. III, 80 (1862) ... ... W.A. S.A. T. V. N.S.W. Q. — B.fl.I,174 M.fr.IV,175;X,68.
C. pumila; Calandrinia, F. v. M., fragm. X, 68 (1877) ... ... W.A. S.A. — — — Q. — — M.fr.X,68.
C. composita, F. v. M.; Calandrinia, Nees in Lehm. pl. Preiss.
     I, 247 (1844) ... ... W.A. — — — — — B.fl.I,175 M.fr.X,69.
C. corrigioloides; Calandr., F. v. M., native pl. of Vict. I, 135 (1879) W.A. S.A. — V. N.S.W. — — B.fl.I,175 M.fr.X,69.
C. brevipedata; Calandr., F. v. M., fragm. X, 69 (1877) ... W.A. S.A. — — — — — M.fr.X,69;XI,130.
C. granulifera, F. v. M.; Calandr., Benth. Fl. Austr. I, 176 (1863) W.A. — — — — — B.fl.I,176
C. pogonophora; Calandrinia, F. v. M., fragm. X, 69 (1877) ... — — — — — Q. — M.fr.X,69.
C. pygmaea, F. v. M., fragm. III, 80 (1862) ... ... W.A. S.A. — V. N.S.W. — — B.fl.I,176 M.fr.I,175;X,71.
C. Australasica, J. Hooker, icon. plant. t. 293 (1840) ... W.A. S.A. T. V. N.S.W. — B.fl.I,177 M.fr.III,80;X,71.
MONTIA, Micheli, nov. pl. gen. 17, t. 13 (1729).
M. fontana, Linné, spec. plant. 87 (1753) ... ... — T. — — — B.fl.I,177

CARYOPHYLLEAE.

Linné, philosoph. bot. 31 (1751).

STELLARIA, Linné, spec. plant. 421 (1753).
S. pungens, Brongniart, Voy. sur la Coquille t. 78 (1826) ... — S.A. T. V. N.S.W. — B.fl.I,157
S. glauca, Withering, Arr. veg. in Great Brit. I, 420 (1776) ... — S.A. T. V. N.S.W. Q. — B.fl.I,158
S. flaccida, Hooker, Comp. Bot. Mag. I, 275 (1835) ... ... — T. V. N.S.W. — B.fl.I,158
S. multiflora, Hooker, Comp. Bot. Mag. I, 275 (1835) ... ... W.A. S.A. T. V. N.S.W. — B.fl.I,159 M.fr.XI,27.
SAGINA, Linné, syst. nat. 8 (1735); Linné, gen. pl. 118 (1737).
S. procumbens, Linné, spec. plant. 128 (1753) ... ... — — V. N.S.W. — B.fl.I,100
S. apetala, Linné, mantiss. 559 (1771) ... ... — S.A. T. V. — — B.fl.I,100
COLOBANTHUS, Bartling, Reliq. Haenk. II, 13, t. 49 (1830).
C. Benthamianus, Fenzl in Ann. Wien. Mus. I, 49 (1838) ... — — V. N.S.W. — B.fl.I,160
C. Billardieri, Fenzl in Ann. des Wien. Mus. I, 48 (1838) ... — S.A. T. V. — — B.fl.I,161
SCLERANTHUS, Linné, gen. plant. 130 (1737). (Mniarum.)
S. pungens, R. Brown, prodr. 412 (1810) ... ... — S.A. — V. N.S.W. — B.fl.V,260
S. diander, R. Brown, prodr. 412 (1810) ... ... — T. V. N.S.W. — B.fl.V,260
S. biflorus, J. Hooker, Fl. N. Zeal. I, 74 (1853) ... ... — T. V. N.S.W. Q. — B.fl.V,259
S. mniaroides, F. v. M., pl. of Vict. I, 215 (1862) ... ... — — V. N.S.W. — B.fl.V,259
SAPONARIA, Linné, gen. plant. 130 (1737), from Camerarius (1586). (Gypsophila.)
S. tubulosa, F. v. M., native pl. of Vict. I, 136 (1870) ... — S.A. — V. N.S.W. — — B.fl.I,155
SPERGULARIA, Persoon, synops. II, 1 (1806). (Lepigonum, Arenaria partly.)
S. rubra, Cambessèdes in St. Hilaire, Fl. Brazil. II, 179 (1829)... W.A. S.A. T. V. N.S.W. — — B.fl.I,162 M.fr.XI,27
DRYMARIA, Willdenow in Roemer & Schultes, syst. V, 406 (1819).
D. filiformis, Bentham, Fl. Austr. I, 162 (1863) ... ... W.A. S.A. — — — — B.fl.I,162
POLYCARPON, Loefling, Iter Hispanic. 7 (1758).
P. tetraphyllum, Loefling in Linné, syst. nat. ed. X, 881 (1759) W.A. S.A. T. V. N.S.W. Q. — B.fl.I,163
POLYCARPAEA, Lamarck in Journ. d'hist. nat. II, 8, t. 25 (1792). (Aylmeyera.)
P. longiflora, F. v. M., Rep. of Babb. Exped. 8 (1858) ... — — — — — N.A. B.fl.I,164
P. spirostylis, F. v. M., Rep. of Babb. Exped. 8 (1858) ... — — — — Q. N.A. B.fl.I,165
P. synandra, F. v. M., Rep. of Babb. Exped. 8 (1858) ... — S.A. — — — N.A. B.fl.I,165
P. violacea, Bentham, Fl. Austr. I, 165 (1863) ... ... — — — — — N.A. B.fl.I,165
P. Indica, Lamarck in journ. d'hist. nat. II, 8 (1792) ... — S.A. — — Q. N.A. B.fl.I,166

P. breviflora, F. v. M., Rep. of Babb. Exped. 9 (1858) ... ... — — — — — Q. N.A. B.fl.I,166
P. spicata, Arnott in Ann. of nat. hist. III, 91 (1830) ... ... — — — — — — N.A. B.fl.I,167
P. involucrata, F. v. M., Rep. of Babb. Exped. 9 (1858) ... ... — — — — — — N.A. B.fl.I,167

AMARANTACEAE.

A. L. de Jussieu, Gen. 87 (1789).

GOMPHRENA, Linné, gen. plant. 69 (1737). (Philoxerus.)
G. canescens, R. Brown, prodr. 416 (1810) ... ... — — — — — N.A. B.fl.V,233
G. flaccida, R. Brown, prodr. 416 (1810) ... ... — — — — — Q. N.A. B.fl.V,234 M.fr.III,123.
G. affinis, F. v. M. in Benth. Fl. Austr. V, 254 (1870) ... ... — — — — — N.A. B.fl.V,254
G. humilis, R. Brown, prodr. 416 (1810) ... ... — — — — — Q. N.A. B.fl.V,234
G. Browni, Moquin in De Cand. prodr. XIII, 2, 397 (1849) ... — — — — — Q. N.A. B.fl.V,235
G. brachystylis, F. v. M., fragm. III, 124 (1862) ... ... — — — — — N.A. B fl.V,235 M.fr.III,124.
G. leptoclada, Bentham, Fl. Austr. V, 235 (1870) ... ... — — — — — N.A. B.fl.V,235
G. Maitlandi, F. v. M., fragm. III, 124, t. 23 (1862) ... ... — — — — — N.A. B.fl.V,236 M.fr.VI,234.
G. pusilla, Bentham, Fl. Austr. V, 256 (1870) ... ... — — — — — N.A. B.fl.V,236
G. tenella, Bentham, Fl. Austr. V, 256 (1870) ... ... — — — — — N.A. B.fl.V,236
G. conica, Sprengel, syst. I, 824 (1825) ... ... — — — — — N.A. B.fl.V,236 M.fr.III,125.
G. conferta, Bentham, Fl. Austr. V, 257 (1870) ... ... — — — — — Q. N.A. B.fl.V,257
G. diffusa, Sprengel, syst. I, 824 (1825) ... ... — — — — — N.A. B.fl.V,257
G. parviflora, Bentham, Fl. Austr. V, 257 (1870) ... ... — — — — — N.A. B.fl.V,257

ALTERNANTHERA, Forskael, flor. Aegypt. Arab. 28 (1775). (Telanthera.)
A. triandra, Lamarck, encycl. meth. I, 95 (1783) ... ... W.A. S.A. T. V. N.S.W. Q. N.A. B.fl.V,249
A. nana, R. Brown, prodr. 417 (1810) ... ... — S.A. — — N.S.W. Q. N.A. B.fl.V,250
A. angustifolia, R. Brown, prodr. 417 (1810) ... ... — — — — — N.A. B.fl.V,250
A. decipiens, Bentham, Fl. Austr. V, 251 (1870) ... ... — — — — — Q. B.fl.V,251
A. polycephala, Bentham, Fl. Austr. V, 251 (1870) ... ... — — — — — N.A. B.fl.V,251
A. leptophylla, Bentham, Fl. Austr. V, 251 (1870) ... ... — — — — — N.A. B.fl.V,251
A. longipes, Bentham, Fl. Austr. V, 252 (1870) ... ... — — — — — N.A. B.fl.V,252

ACHYRANTHES, Linné, gen. plant. 34 (1737).
A. aspera, Linné, spec. plant. 204 (1753) ... ... — S.A. — — N.S.W. Q. N.A. B.fl.V,246
A. arborescens, R. Brown, prodr. 417 (1810) ... ... — — — — N.S.W. — — — M.fr.IX,169.

NYSSANTHES, R. Brown, prodr. 418 (1810).
N. erecta, R. Brown, prodr. 418 (1810) ... ... — — — — N.S.W. Q. B.fl.V,247
N. diffusa, R. Brown, prodr. 418 (1810) ... ... — — — — N.S.W. Q. B.fl.V,247

PTILOTUS, R. Brown, prodr. 415 (1810). (Trichinium, Psilotrichum, Goniotriche, Hemisteirus, Arthrotrichum.)
P. conicus, R. Brown, prodr. 415 (1810) ... ... — — — — — N.A. B.fl.V,242
P. corymbosus, R. Brown, prodr. 415 (1810) ... ... — — — — — N.A. B.fl.V,243 M.fr.III,125.
P. grandiflorus, F. v. M., fragm. I, 237 (1859) ... ... W.A. — — — — B.fl.V,243 M.fr.I,237.
P. spicatus, F. v. M. in Benth. Fl. Austr. V, 243 (1870) ... ... — — — — — Q. N.A. B.fl.V,243
P. Murrayi, F. v. M., fragm. III, 145 (1863) ... ... — S.A. — — — N.A. B.fl.V,243 M.fr.III,145.
P. gomphrenoides, F. v. M., fragm. VI, 233 (1868) ... ... — — — — — N.A. B.fl.V,244
P. latifolius, R. Brown in App. Sturt's Exped. 25 (1849) ... ... — S.A. — — — N.A. B.fl.V,244 M.fr.VI,232.
P. macrotrichus, F. v. M., fragm. IV, 90 (1864) ... ... W.A. — — — — N.A. B.fl.V,244 M.fr.IV,232.
P. villosiflorus, F. v. M., fragm. III, 123 (1863) ... ... — — — — — N.A. B.fl.V,245 M.fr.III,123.
P. humilis, F. v. M., fragm. VI, 229 (1868) ... ... W.A. — — — — B.fl.V,245 M.fr.III,161;VI,229.
P. obovatus, F. v. M., fragm. VI, 229 (1868) ... ... W.A. S.A. — V. N.S.W. Q. N.A. B.fl.V,220 M.fr.XI,27.
P. incanus, Poiret, Encycl. méth. suppl. IV, 620 (1817) ... ... — — — — — N.A. B.fl.V,221 M.fr.VI,228.
P. parviflorus, Trich., Lindley in Mitch. three Exp.II,13(1838) ... ... — — — N.S.W Q. B.fl.V,222
P. astrolasius, F. v. M., fragm. VI, 233 (1869) ... ... — — — — — N.A. B.fl.V,222 M.fr.VI,233.
P. rotundifolius, F. v. M., fragm. VI, 230 (1868) ... ... W.A. — — — — — N.A. B.fl.V,223 M.fr.III,162.
P. distitiflorus, F. v. M., fragm. IV, 89 (1864) ... ... — — — — — N.A. B.fl.V,223 M.fr.IV,89.
P. distans, Poiret, Encycl. suppl. IV, 620 (1817) ... ... — — — — — Q. N.A. B.fl.V,223 M.fr.VI,228.
P. alopecuroideus, F. v. M., fragm. VI, 227 (1868) ... ... W.A. S.A. — — N.S.W. Q. N.A. B.fl.V,224 M.fr.VI,227.
P. nobilis, F. v. M., fragm. VI, 227 (1868) ... ... — S.A. — V. N.S.W — — Q. B.fl.V,224
P. macrocephalus, Poiret, Encycl. suppl. IV, 620 (1817) ... ... — S.A. — V. N.S.W. Q. — B.fl.V,225 M.fr.VI,229; XI,27.
P. hemisteirus, F. v. M., fragm. IV, 90 (1864) ... ... W.A. S.A. — — — D.B.V,241 M.fr.VI,231.
P. parvifolius, F. v. M., fragm. VI, 229 (1868) ... ... — — S.A. — — N.A. B.fl.V,226 M.fr.VI,229.
P. exaltatus, Nees in Lehm. pl. Preiss. I, 630 (1845) ... ... W.A. S.A. — V. N.S.W. Q. N.A. B.fl.V,227
P. Manglesii, F. v. M., fragm. VI, 230 (1869) ... ... W.A. — — — — N.A. B.fl.V,228 M.fr.VI,230.
P. Beckeri, F. v. M., fragm. VI, 233 (1869) ... ... — — — — — N.A. B.fl.V,228
P. gomphrenoides, Moquin in De Cand., prodr. XIII, 2, 297 (1849) — S.A. — — — N.A. B.fl.V,229
P. cæquamatus; Trichinium, Bentham, Fl. Austr. V, 229 (1870) W.A. — — — — B.fl.V,229
P. declinatus, Nees in Lehm. pl. Preiss. I, 631 (1845) ... ... W.A — — — — B.fl.V,229
P. erubescens, Schlechtendal in Linnaea XX, 575 (1847) ... ... S.A. — V. N.S.W. — — N.A. B.fl.V,230 M.fr.VI,229.
P. divaricatus, F. v. M., fragm. VI, 229 (1868) ... ... — S.A. — — — N.A. B.fl.V,230 M.fr.VI,229.
P. helipteroides, F. v. M., fragm. VI, 231 (1868) ... ... W.A. — — — — N.A. B.fl.V,231 M.fr.III,122&161).
P. Stirlingi; Trichinium, Lindley in Bot. Reg. under n. 28 (1839) W.A. — — — — N.A. B.fl.V,231
P. laxus; Trichinium, Bentham, Fl. Austr. V, 232 (1870) ... W.A. — — — — B.fl.V,232
P. axillaris, Trichinium, F. v. M. in Benth. Fl. Austr. V, 232 (1870) — — — — — N.A. B.fl.V,232
P. striatus; Trichinium, Moquin in Benth. Fl. Austr. V, 233 (1870) — — — — — N.A. B.fl.V,233
P. auriculifolius; Trichinium, Cunningham in De Cand., prodr.
XIII, part II, 287 (1849) ... ... — — — — — N.A. B.fl.V,233

P. sericostachyus, F. v. M., fragm. VI, 230 (1868) ... ... — — — — — N.A. B.fl.V,233
P. roseus; Trichin., Moq. in De Cand., prodr. XIII, p. II, 284 (1849) W.A. — — — — — B.fl.V,234 M.fr.VI,230
P. fusiformis, Poiret, Encycl. suppl. IV, 619 (1817) ... ... W.A. — — — — — N.A. B.fl.V,234
P. gracilis, Poiret, Encycl. suppl. IV, 620 (1817)... ... ... — — — — — N.A. B.fl.V,235 M.fr.VI,228.
P. Drummondii, F. v. M., fragm. VI, 229 (1868) ... ... W.A. — — — — — B.fl.V,235 M.fr.VI,229.

P. calostachyus, F. v. M., fragm. VI, 231 (1868)... ... ... — — — — — — N.A. B.fl.V.236 M.fr.VI,231.
P. Fraseri ; Trichinium, Cunningham in De Cand., prodr. XIII,
    part II, 295 (1849) ... ... — — — — N.S.W. — — B.fl.V,236
P. spathulatus, Poiret, Encycl. suppl. IV, 620 (1817) ... ... W.A. S.A. T. V. N.S.W. — — B.fl.V,236 M.fr.VI,228.
P. pyramidatus, F. v. M., fragm. VI, 230 (1868) ... .. W.A. — — — — — — B.fl.V,237 M.fr.VI,230.
P. holosericeus, F. v. M., fragm. VI, 229 (1868) ... ... W.A. — — — — — — B.fl.V,237 M.fr.VI,229.
P. lanatus, Cunningham in De Cand., prodr. XIII, p. II, 281 (1849) — — — — — N.A. B.fl.V,238
P. leucocoma ; Trichinium, Moquin in De Cand., prodr. XIII,
    part II, 292 (1849) ... ... — S.A. — — — — — B.fl.V,238
P. brachyanthus;Trichin., F. v. M. in Benth. Fl. Austr. V,639(1870) — — — — — — N.A. B.fl.V,239
P. arthrolasius, F. v. M., fragm. VI, 232 (1868) ... ... — — — — — — N.A. B.fl.V,239 M.fr.VI,232.
P. Forrestii, F. v. M., pl. of N. W. Austr. 7 (1881) ... ... — — — — — — N.A. —
P. aervoides, F. v. M., fragm. VI, 231 (1868) ... ... — — — — — Q. N.A. B.fl.V,240 M.fr.III,123;VI,231.
P. Roei; Trichinium, F. v. M. in Benth. Fl. Austr. V, 240(1870) W.A. — — — — — — B.fl.V,240
P. Hoodii, F. v. M., fragm. VIII, 232 (1874) ... ... — S.A. — — — — — M.fr.VIII,232.
P. caespitulosus, F. v. M., fragm. VI, 232 (1868)... ... W.A. — — — — — — B.fl.V,240 M.fr.VI,232.
P. helichrysoides, F. v. M., fragm. VI, 231 (1868) ... ... W.A. — — — — — — B.fl.V,241 M.fr.VI,231.
P. psilotrichoides, F. v. M., fragm. VI, 231 (1869) ... .. W.A. — — — — — — B.fl.V,209 M.fr.XI,97.

EUXOLUS, Rafinesque, Fl. Tell. 42 (1836).
E. Mitchellii, F. v. M. in Giles, geogr. trav. Centr. Austr. 214 (1875) — S.A. — — N.S.W. Q. — B.fl.V,214
E. interruptus, Moquin in De Cand., prodr. XIII, part II, 275(1849) — — — — — Q. N.A. B.fl.V,215
E. viridis, Moquin in De Cand., prodr. XIII, part II, 273 (1849) W.A. — — — — N.S.W. Q. — B.fl.V,215
E. macrocarpus; Amarantus, Benth. Fl. Austr. V, 216(1870) ... — — — — — V. N.S.W. Q. — B.fl.V,216

AMARANTUS, Dodoens, stirp. hist. pompt. 185 (1583).
A. leptostachyus, Bentham, Fl. Austr. V, 214 (1870) ... ... — — — — — Q. N.A. B.fl.V,213
A. pallidiflorus, F. v. M., fragm. I, 140 (1859) ... ... W.A. — — — — — Q. N.A. B.fl.V,214 M.fr.I,140.

POLYCNEMON, Linné, gen. pl. ed. sec. 21(1742). (Hemichroa.)
P. pentandrum, F. v. M., native pl. of Vict. I, 162 (1879) ... W.A. S.A. T. V. N.S.W. — — B.fl.V,211 M.fr.VI,234.
P. diandrum, F. v. M., native pl. of Vict. I, 163 (1879)... ... W.A. S.A. — V. N.S.W. — N.A. B.fl.V,211 M.fr.VI,234.
P. mesembrianthemum, F. v. M., fragm. VIII, 38 (1873) ... — S.A. — — — — — M.fr.VIII,38.

DEERINGIA, R. Brown, prodr. 413 (1810). (Lestibudesia partly, Celosia partly, Lagrezia.)
D. celosioides, R. Brown, prodr. 413 (1810) ... ... — — — — — N.S.W. Q. — B fl.V,210 M.fr.II,92.
D. altissima, F. v. M., fragm. II, 92 (1860) ... ... — — — — N.S.W. Q. — B.fl.V,210 M.fr.VI,251.

SALSOLACEAE.
Linné, class. pl. 507 (1738).

RHAGODIA, R. Brown, prodr. 408 (1810).
R. Billardieri, R. Brown, prodr. 408 (1810) ... ... W.A. S.A. T. V. N.S.W. Q. — B.fl.V,152 M.fr.VII,10.
R. parabolica, R. Brown, prodr. 408 (1810) ... ... — S.A. — — N.S.W. Q. — B.fl.V,153 M.fr.VII,10.
R. dioica, Nees in Lehm. pl. Preiss. I, 636 (1845)... ... W.A. — — — — — — B.fl.V,154
R. Gaudichaudiana, Moquin, Chenop. monogr. enum. 11 (1840) — S.A. — V. N.S.W. — N.A. B.fl.V,154 M.fr.VII,10.
R. crassifolia, R. Brown, prodr. 408 (1810) ... ... W.A. S.A. — V. N.S.W. — — B.fl.V,154
R. Preissü, Moquin in De Cand. prodr. XIII, part II, 40 (1849) W.A. — — — — — — B.fl.V,155 M.fr.VII,10.
R. obovata, Moquin, Chenop. monogr. enum. 10 (1840)... ... W.A. — — — — — — B.fl.V,155
R. spinescens, R. Brown, prodr. 408 (1810) ... ... — — — — V. N.S.W. Q. N.A. B.fl.V,155 M.fr.VII,10;XI,20.
R. hastata, R. Brown, prodr. 408 (1810) ... ... — — — V. N.S.W. Q. — B.fl.V,156 M.fr.VII,10.
R. nutans, R. Brown, prodr. 408 (1810) ... ... — S.A. T. V. N.S.W. Q. — B.fl.V,156 M.fr.VII,10.
R. linifolia, R. Brown, prodr. 408 (1810) ... ... — — — — N.S.W. Q. — B.fl.V,157 M.fr.VII,10.

CHENOPODIUM, Tournefort, inst. 506, t. 288 (1700). (Blitum, Ambrina.)
C. nitrariaceum, F. v. M. in Benth. Fl. Austr. V, 158 (1870) ... W.A. S.A. — V. N.S.W. — — B.fl.V,158 M.fr.XI,20&27.
C. auricomum, Lindley in Mitch. Trop. Austr. 94 (1848) ... — S.A. — V. N.S.W. Q. N.A. B.fl.V,159 M.fr.VII,11.
C. triangulare, R. Brown, prodr. 407 (1810) ... ... — — V. N.S.W. Q. — B.fl.V,160 M.fr.VII,11.
C. microphyllum, F. v. M. in Trans. phil. Inst. Vict. II, 74 (1857) — S.A. — — N.S.W. Q. — B.fl.V,161 M.fr.VII,11.
C. rhadinostachyum, F. v. M. in Wing's S. Science Rec. May (1880) — S.A. — — — — — M.fr.XII.
C. carinatum, R. Brown, prodr. 407 (1810) ... ... W.A. S.A. — V. N.S.W. Q. — B.fl.V,162 M.fr.VII,11.
C. cristatum, F. v. M., fragm. VII, 11 (1869) ... ... W.A. S.A. — V. N.S.W. Q. — B.fl.V,163 M.fr.XI,27.
C. atriplicinum, F. v. M., fragm. VII, 11 (1860) ... ... W.A. S.A. — V. N.S.W. — — B.fl.V,163 M.fr.I,140;XI,27.

DYSPHANIA, R. Brown, prodr. 411 (1810).
D. plantaginella, F. v. M., fragm. I, 61 (1858) ... ... — S.A. — — — — N.A. B.fl.V,164 M.fr.I,61.
D. litoralis, R. Brown, prodr. 411 (1810) ... ... — S.A. — — — Q. N.A. B.fl.V,164 M.fr.VII,11.
D. myriocephala, Bentham, Fl. Austr. V, 165 (1870) ... W.A. S.A. — V. N.S.W. Q. — B.fl.V,165

ATRIPLEX, Tournefort, inst. 506, t. 288 (1700), from l'Obel (1581). (Obione, Theloophyton.)
A. Moquinianum, Webb in Cand. prodr. XIII, part. II, 97 (1849) W.A. — — — — — N.A. B.fl.V,169
A. stipitatum, Bentham, Fl. Austr. V, 168 (1870) ... — S.A. — — V. N.S.W. — — B.fl.V,168 M.fr.VII,9.
A. paludosum, R. Brown, prodr. 406 (1810) ... ... — S.A. — V. N.S.W. — — B.fl.V,169 M.fr.VII,9.
A. Drummondii, Moquin in Cand. prodr. XIII, part. II, 102 (1849) W.A. — — — — — — B.fl.V,170
A. iaxidenm, Moquin, Chenop. monogr. enum. 63 (1840) ... W.A. — — — — — — B.fl.V,170
A. nummularium, Lindley in Mitch. Trop. Austr. 64 (1848) ... — S.A. — V. N.S.W. Q. — B.fl.V,170 M.fr.VII,10;XI,27.
A. cinereum, Poiret, Encycl. suppl. I, 471 (1810) ... ... — S.A. — V. N.S.W. — N.A. B.fl.V,170 M.fr.VII,0.
A. rhagodioides, F. v. M. in Trans. phil. Inst. Vict. II, 74 (1857) W.A. S.A. — V. N.S.W. — — B.fl.V,172
A. incrassatum, F. v. M., Rep. Babb. Exped. 20(1858)... ... — S.A. — — — — — B.fl.V,172
A. vesicarium, Howard in Benth. Fl. Austr. V, 172 (1870) ... — S.A. — V. N.S.W. Q. — B.fl.V,172
A. hymenothecum, Moquin in Cand. prodr. XIII, part. II, 101 (1849)W.A. — — — — — — B.fl.V,173 M.fr.VII,0.
A. humile, F. v. M., fragm. IV, 48 (1866)... ... ... — — — — — — N.A. B.fl.V,174 M.fr.IV,48.
A. fissivalve, F. v. M., fragm. IX, 123 (1875) ... ... — S.A. — — — — — M.fr.IX,123.
A. volutinellum, F. v. M., Rep. Babb. Exped. 20 (1858) ... — S.A. — — N.S.W. — — B.fl.V,174 M.fr.VII,0.
A. angulatum, Bentham, Fl. Austr. V, 174 (1870) ... ... — S.A. — — N.S.W. — — B.fl.V,174

A. acmibaccatum, R. Brown, prodr. 406 (1810) ... .. W.A. S.A. — V. N.S.W. Q. — B.fl.V,175 M.fr.VII,9.
A. exilifolium, F. v. M., fragm. VII, 9 (1870) ... . W.A. — — — — — B.fl.V,175 M.fr.VII,9.
A. Muelleri, Bentham, Fl. Austr. V, 175 (1870) ... .. — S.A. — V. N.S.W. Q. N.A. B.fl.V,175 M.fr.VII,9.
A. elachophyllum, F. v. M., fragm. VII, 8 (1870) .. — — — — — N.A. B.fl.V,176 M.fr.VII,8.
A. microcarpum, Bentham, Fl. Austr. V, 176 (1870) — S.A. — — N.S.W. — — B.fl.V,176
A. prostratum, R. Brown, prodr. 406 (1810) ... . — S.A. — — — — — B.fl.V,176
A. pumilio, R. Brown, prodr. 406 (1810) ... . — S.A. — — — — — B.fl.V,176
A. campanulatum, Bentham, Fl. Austr. V, 177 (1870) ... — S.A. — — — — — B.fl.V,177
A. leptocarpum, F. v. M. in Trans. phil. Inst. Vict. II, 74 (1857) — S.A. — V. N.S.W. Q. — B.fl.V,178 M.fr.VII,9;XI 27
A. limbatum, Bentham, Fl. Aust. V, 178 (1870) ... ... — S.A. — V. N.S.W. — — B.fl.V,178
A. halimoides, Lindley in Mitch. Three Exped. I, 282 (1838) ... — S.A. — V. N.S.W. Q. — B.fl.V,179 M.fr.VII,10;XI,2
A. holocarpum, F. v. M., Rep. Babb. Exped. 19 (1858) ... — S.A. — V. N.S.W. Q. — B.fl.V,179
A. spongiosum, F. v. M. in Trans. Vict. phil. Inst. II, 74 (1857) W.A. S.A. — V. N.S.W. Q. N.A B.fl.V,179
A. crystallinum, J. Hooker in Lond. journ. of Bot. VI, 279 (1847) — S.A. T. V. N.S.W. — — B.fl.V,180 M.fr.VII.9.

KOCHIA, Roth in Schrader's Journ. Bot. I, 303 (1799). (Maireana, Sclerochlamys.)
K. fimbriolata, F. v. M., fragm. IX, 75 (1875) ... — S.A. — — — — — M.fr.IX,75.
K. lobiflora, F. v. M. in Benth. Fl. Austr. V, 184 (1870) — S.A. — — N.S.W. — — B.fl.V,184
K. lanosa, Lindley in Mitch. Trop. Austr. 88 (1848) ... W.A. S.A. — V. N.S.W. Q. — B.fl.V,184 M.fr.VII,13.
K. triptera, Bentham, Fl. Austr. V, 185 (1870) ... W.A. S.A. — V. N.S.W. — — B.fl.V,185
K. oppositifolia, F. v. M. in Transact. Vict. Inst. 134 (1855) W.A. S.A. — V. N.S.W. Q. — B.fl.V,185 M.fr.VII,12.
K. brevifolia, R. Brown, prodr. 409 (1810) ... ... W.A. S.A. — V. N.S.W. Q. — B.fl.V,185 M.fr.VII,13.
K. pyramidata, Bentham, Fl. Austr. V, 186 (1870) — S.A. — — N.S.W. — — B.fl.V,186
K. eriantha, F. v. M., Rep. Babb. Exped. 20 (1858) — S.A. — — N.S.W. — — B.fl.V,186 M.fr.VII,13.
K. villosa, Lindley in Mitch. Trop. Austr. 91 (1848) W.A. S.A. — V. N.S.W. Q. N.A. B.fl.V,186 M.fr.VII,12.
K. planifolia, F. v. M., fragm. I, 213 (1859) ... W.A. — — — — — — B.fl.V,187 M.fr.I,213.
K. sedifolia, F. v. M. in Transact. Vict. Inst. I, 134 (1855) ... — S.A. — V. N.S.W. — — B.fl.V,187 M.fr.VII,12;VIII,37.
K. humillima, F. v. M., fragm. IX, 168 (1875) ... — S.A. — V. N.S.W. — — — M.fr.IX,168.
K. microphylla, F. v. M., fragm. VIII, 148 (1874) — — — — N.S.W. Q. — B.fl.V,181 M.fr.VIII,37.
K. dichoptera, F. v. M., fragm. VIII, 37 (1873) ... ... — — — — — Q. — M.fr.VIII,37,163.
K. decaptera, F. v. M., fragm. IX, 75 (1875) ... ... — S.A. — — — — — M.fr.IX,75.
K. ciliata, F. v. M., Rep. of Babb. Exped. 20 (1858) ... — S.A. — V. N.S.W. — — B.fl.V,189 M.fr.VII,12.
K. brachyptera, F. v. M., second gen. Rep. 15 (1854) ... — S.A. — V. N.S.W. — — B.fl.V,189 M.fr.VII,13.
K. stelligera, F. v. M., fragm. VII, 13 (1869) ... ... — S.A. — V. N.S.W. — — B.fl.V,189 M.fr.VII,13.

DIDYMANTHUS, Endlicher, nov. stirp. decad. 7 (1839).
D. Roei, Endlicher, nov. stirp. decad. 8 (1839) ... W.A. — — — — — — B.fl.V,193 M.fr.VII,11.

BABBAGIA, F. v. M., Rep. ou pl. of Babb. Exped. 21 (1858).
B. dipterocarpa, F. v. M., Rep. of Babb. Exped. 21 (1858) ... — S.A. — — N.S.W. Q. — B.fl.V,192 M.fr.VII,11.

BASSIA, Allioni in Misc. Taurin. III, 177, t. 4 (1766). (Chenolea, Sclerolaena, Anisacantha, Echinopsilon, Kentropsis, Dissocarpus, Eriochiton, Osteocarpum, Collocarpus.)
B. carnosa; Chenolea, Benth. Fl. Austr. V, 190 (1870) ... W.A. — — — — — — B.fl.V,190 M.fr.VII,13.
B. Dallachyana; Chenolea, Benth. Fl. Austr. V, 191 (1870) — S.A. — V. N.S.W. — — B.fl.V,190
B. tricornis; Chenolea, Benth. Fl. Austr. V, 191 (1870)... ... — S.A. — V. N.S.W. — — B.fl.V,191
B. eurotioides; Chenolea, F. v. M. in Benth. Fl. Austr. V, 191 (1870)W.A. — — — — — N.A. B.fl.V,191 M.fr.VII,13.
B. Muelleri; Chenolea, Benth. Fl. Austr. V, 191 (1870) ... — — — — — — N.A. B.fl.V,191
B. sclerolaenoides; Echinopsilon, F. v. M. in Tr. phil. Inst. Vict. II, 75 (1857) ... ... W.A. S.A. — V. N.S.W. — — B.fl.V,192 M.fr.VII,13.
B. uniflora; Sclerolaena, R. Brown, prodr. 410 (1810) ... W.A. S.A. — V. N.S.W. — — B.fl.V,194
B. diacantha; Anisacantha, Nees in Lehm. pl. Preiss. I, 635 (1844) W.A. S.A. — V. N.S.W. Q. N.A. B.fl.V,194 M.fr.VII,14;X,91.
B. lanicuspis; Anisacantha, F. v. M., fragm. II, 170 (1861) — S.A. — — N.S.W. Q. — B.fl.V,195 M.fr.II,140,170;VII,14.
B. bicornis; Sclerolaena, Lindl. in Mitch. Three Exp. II. 47 (1838) — S.A. — V. N.S.W. Q. N.A. B.fl.V,195 M.fr.VII,14.
B. biflora; Sclerolaena, R. Brown, prodr. 410 (1810) ... — S.A. — V. N.S.W. — — B.fl.V,196 M.fr.VII,11;X,91.
B. paradoxa; Sclerolaena, R. Brown, prodr. 410 (1810) ... W.A. S.A. — V. N.S.W. — — B.fl.V,196 M.fr.VII,11;X,91.
B. quinquecuspis; Anisac., F. v. M. in Tr. Vict. Inst. I, 134 (1855) — S.A. — V. N.S.W. Q. — B.fl.V,199 M.fr.VII,14;X,91.
B. Drummondii; Anisacantha, Benth. Fl. Austr. V, 199 (1870) W.A. — — — — — — B.fl.V,199
B. Birchii; Anisacantha, F. v. M., fragm. VIII, 163 (1874) .. — — — — — Q. — M.fr.VIII,163.
B. divaricata; Anisacantha, R. Brown, prodr. 410 (1810) ... — S.A. — V. N.S.W. — — B.fl.V,200 M.fr.VII,14;X,92.
B. bicuspis; Anisac., F. v. M. in Trans. Vict. Inst. I, 133 (1855) — S.A. — — N.S.W. Q. — B.fl.V,200 M.fr.VII,13.
B. glabra; Kentropsis, F. v. M., fragm. I, 139 (1859) ... — — — — — — N.A. B.fl.V,200
B. echinopsila; Anisacantha, F. v. M., fragm. VII, 14 (1870) ... — S.A. — V. N.S.W. — — B.fl.V,201 M.fr.I,139;X,92.
B. enchylaenoides; Chenolea, F. v. M., fragm. X, 92 (1876) ... — S.A. — V. N.S.W. — — B.fl.V,182 M.fr.X,92.
B. micrantha; Enchylaena, Benth. Fl. Austr. V, 181 (1870) W.A. — — — — — — B.fl.V,181
B. brevicuspis; Anisacantha, F. v. M., fragm. IV, 150 (1864) ... — — — — N.S.W. — — B.fl.V,198 M.fr.VII,14;VIII,36.
B. salsuginosa; Osteocarpum, F. v. M., second gen. Rep. 15 (1854) — S.A. — V. N.S.W. — — B.fl.V,197 M.fr.VII,12;X,92.

ENCHYLAENA, R. Brown, prodr. 408 (1810).
E. tomentosa, R. Brown, prodr. 408 (1810) ... ... W.A. S.A. — V. N.S.W. Q. N.A. B.fl.V,181 M.fr.VII,12;XI,27.

THRELKELDIA, R. Brown, prodr. 409 (1810).
T. diffusa, R. Brown, prodr. 410 (1810) ... ... W.A. S.A. T. V. N.S.W. — — B.fl.V,197 M.fr.VII,12.
T. proceriflora, F. v. M., fragm. VIII, 38 (1873)... ... — S.A. — V. N.S.W. — — — M.fr.VII,38.

SALICORNIA, Tournefort, coroll. 51 (1703). (Halocnemum, Arthrocnemum, Tecticornia, Pachycornia.)
S. robusta, F. v. M., fragm. VI, 251 (1868) ... — S.A. — V. N.S.W. — — B.fl.V,202 M.fr.I,139;VII,14.
S. pruinosa, R. Brown, prodr. 411 (1810)... ... — S.A. — V. N.S.W. Q. — B.fl.V,203 M.fr.VII,15.
S. cinerea, F. v. M., fragm. VI, 251 (1868) ... — — — V. N.S.W. Q. N.A. B.fl.V,203 M.fr.VII,15.
S. leiostachya, Bentham, Fl. Austr. V, 203 (1870) ... — S.A. — V. N.S.W. — — B.fl.V,203
S. bidens, Bentham, Fl. Austr. V, 204 (1870) ... W.A. — — — — — — B.fl.V,204
S. tenuis, Bentham, Fl. Austr. V, 204 (1870) ... ... — S.A. — — N.S.W. — — B.fl.V,204
S. australis, Solander in G. Forster, prodr. 88 (1786) ... — S.A. — — N.S.W. — — B.fl.V,205 M.fr.VII,15;XI,20,27.

SUAEDA, Forskael, Fl. Aegypt. Arab. 69 (1770). (Chenopodium partly, Schoberia, Chenopodina.)
S. maritima, Dumortier, Fl. Belg. 22 (1827) ... ... W.A. S.A. T. V. N.S.W. Q. N.A. B.fl.V,206 M.fr.XI,20.

SALSOLA, Linné, gen. pl. 67 (1737), from Cisalpini.
S. Kali, Linné, spec. plant. 222 (1753) ... . W.A, S.A. — V. N.S.W. Q. N.A. B.fl.V,207 M.fr.XI,20,27.

FICOIDEAE.
A. L. de Jussieu, Gen. 315 (1789).

MESEMBRIANTHEMUM, Breyne, prodr. fasc. rar. pl. II, 67 (1680).
M. aequilaterale, Haworth, misc. nat. 77 (1803) ... ... W.A. S.A. T. V. N.S.W. Q. N.A. B.fl.III,324 M.fr.XI,20,27.
M. australe, Solander in G. Forster, prodr. 90 (1786) ... W.A. S.A. T. V. N.S.W. Q. N.A. B.fl.III,324 M.fr.V,158;X,83;XI,20,27

TETRAGONIA, Linné, syst. nat. 9 (1735); Linné, gen. pl. 144 (1737). (Tetragonella.)
T. expansa, Murray in Comm. Goetting. VI, 13, t. 5 (1783) ... — S.A. T. V. N.S.W. Q. — B.fl.III,325 M.fr.VII,129;X,84.
T. diptera, F. v. M., fragm. XI, 8 (1878) ... ... W.A. — — — — — N.A. — M.fr.XI,8.
T. implexicoma, J. Hooker, Fl. Tasman. I, 148 (1859) ... ... W.A. S.A. T. V. N.S.W. — N.A. B fl.III,326 M.fr.X,84;XI,8.

AIZOON, Linné, gen. pl. 161 (1737). (Aizonm.)
A. quadrifidum, F. v. M., fragm. II, 148 (1861) ... ... W.A. S.A. — — N.S.W. — — B.fl.III,327 M.fr.VII,129;X,83.
A. zygophylloides, F. v. M., fragm. VII, 129 (1871) ... W.A. S.A. — — — — — — M.fr.V,83.

GUNNIA, F. v. M., Rep. on pl. of Babb. Exped. 9 (1858).
G. septifraga, F. v. M., Rep. of Babb. Exped. 9 (1858) ... ... — S.A. — — — — — B.fl.III,327
G. Drummondii, Bentham, Fl. Austr. III, 327 (1866) ... W.A. — — — — — — B.fl.III,327

SESUVIUM, Linné, syst. ed. decim. 1058 (1759).
S. portulacastrum, Linné, syst. ed. dec. 1058 (1759) ... ... — — — N.S.W. Q. N.A. B.fl.III,328 M.fr.VII,120;X,83.

ZALEYA, N. L. Burmann, Fl. Indic. 110, t. 31 (1768).
Z. decandra, N. L. Burmann, Fl. Ind. 110, t. 31 (1768) ... — S.A. — — N.S.W. Q. N.A. B.fl.III,329 M.fr.I,172;X,83.

TRIANTHEMA, Sauvage, Méth. fol. 127 (1751). (Anciatrostigma.)
T. turgidifolia, F. v. M., fragm. X, 83 (1876) ... ... — — — — — N.A. — M.fr.X,83.
T. crystallina, Vahl, symbol. I, 32 (1790) ... ... — S.A. — — — Q. N.A. B.fl.III,330 M.fr.I,171-2.
T. galericulata, F. v. M., fragm. I, 172 (1859) ... ... — — — — — Q. — — M.fr.I,172.
T. pilosa, F. v. M., fragm. I, 174 (1859) ... ... — S.A. — — — — N.A. B.fl.III,330 M.fr.X,83.
T. oxycalyptra, F. v. M., fragm. I, 173 (1859) ... ... — — — — — — N.A. B.fl.III,330 M.fr.I,173.
T. rhynchocalyptra, F. v. M., fragm. I, 174 (1859) ... ... — — — — — — N.A. B.fl.III,330 M.fr.I,174.
T. cypseloides, Bentham, Fl. Austr. III, 331 (1866) ... ... — — — N.S.W. Q. N.A. B.fl.III,331 M.fr.X,83.

POMATOTHECA, F. v. M., fragm. X, 72 (1876).
P. humillima, F. v. M., fragm. X, 72 (1876) ... ... — S.A. — — N.S.W. — — — M.fr.X,72.

MACARTHURIA, Huegel, enum. pl. Nov. Holl. Austr. occ. 11 (1837).
M. apetala, Harvey in Hook. Kew. misc. VII, 55 (1855) ... W.A. — — — — — — B.fl.III,332 M.fr.XI,20.
M. australis, Huegel, enum. pl. 11 (1837) ... ... W.A. — — — — — — B.fl.III,332
M. Neo-Cambrica, F. v. M., fragm. V, 28 (1866)... ... — — — N.S.W. Q. — B.fl.III,332 M.fr.V,28.

MOLLUGO, Linné, gen. plant. 336 (1737). (Glinus, Trigastrotheca.)
M. Glinus, A. Richard, tentam. fl. Abyss. I, 48 (1847) ... W.A. S.A. T. V. N.S.W. Q. N.A. B.fl.III,333
M. oryzloides, F. v. M. in Benth. Fl. Austr. III, 333 (1866) ... — S.A. — N.S.W. — Q. N.A. B.fl.III,333
M. Spergula, Linné, spec. pl. ed. alt. 131 (1762) ... ... W.A. S.A. — T. V. N.S.W. Q. N.A. B.fl.III,333
M. trignatrotheca, F. v. M., pl. of Vict. I, 201 (1860) ... ... — — — — — — N.A. B.fl.III,334
M. Cervana, Scringe in De Cand., prodr. I, 392 (1824) ... ... — — — — — — — B.fl.III,334 M.fr.II,148.

POLYGONACEAE.
A. L. de Jussieu, gen. pl. 82 (1789), from B. de Jussieu (1759).

EMEX, Neeker, elem. bot. II, 214 (1790). (Perhaps immigrated.)
E. australis, Steinheil in Ann. des sc. nat. ser. 2, IX, 195, t. 7 (1839) W.A. S.A. — — — — — B.fl.V,262

RUMEX, Linné, gen. pl. 105 (1737), from Plinius.
R. Brownii, Campdera, Monogr. Rum. 81 (1819)... ... ... — S.A. T. V. N.S.W. Q. — B.fl.V,263
R. flexuosus, Solander in G. Forst., fl. ins. Austr. prodr. 90 (1786) — S.A. — N.S.W. — — B.fl.V,264
R. crystallinus, Lange, ind. sem. hort. Hafn. 23 (1861) ... ... — S.A. T. V. N.S.W. Q. — B.fl.V,265 M.fr.IV,48.
R. bidous, R. Brown, prodr. 421 (1810) ... ... — S.A. T. V. N.S.W. Q. — B.fl.V,265

POLYGONUM, l'Obel, stirp. hist. 228 (1576), from Dioscorides and Plinius.
P. plebejum, R. Brown, prodr. 420 (1810) ... ... — — — N.S.W. Q. N.A. B.fl.V,267
P. strigosum, R. Brown, prodr. 420 (1810)... ... . — — T. V. N.S.W. Q. — B.fl.V,268
P. prostratum, R. Brown, prodr. 419 (1810) ... ... — S.A. — N.S.W. — — B.fl.V,268
P. hydropiper, Linné, spec. plant. 361 (1753) ... ... — — T. V. N.S.W. Q. — B.fl.V,269
P. minus, Hudson, Fl. Anglica I, 148 (1762) ... ... W.A. S.A. — N.S.W. — — B.fl.V,269
P. subsessile, R. Brown, prodr. 419 (1810) ... ... — — T. V. N.S.W. Q. — B.fl.V,269
P. barbatum, Linné, spec. plant. 362 (1753) ... ... — — — N.S.W. — — B.fl.V,270
P. articulatum, R. Brown, prodr. 420 (1810) ... ... — — — — — — B.fl.V,270
P. lapathifolium, Linné, spec. plant. 360 (1753) ... ... — S.A. T. V. N.S.W. Q. — B.fl.V,270
P. orientale, Linné, spec. plant. 362 (1753)... ... ... — — — N.S.W. — — B.fl.V,271
P. attenuatum, R. Brown, prodr. 420 (1810) ... ... — — — N.S.W. Q. N.A. B.fl.V,272

MUEHLENBECKIA, Meissner, gen. pl. vasc. 316 (1840).
M. adpressa, Meissner, gen. pl. comm. 227 (1843) ... W.A. S.A. T. V. N.S.W. — — B.fl.V,273 M.fr.XI,20.
M. australis, Meissner, gen. pl. comm. 227 (1843) ... ... — — — N.S.W. Q. — B.fl.V,274 M.fr.IX,169.
M. gracillima, Meissner in De Cand., prodr. XIV, 145 (1856) ... — — — N.S.W. Q. — B.fl.V,274
M. rhyticarya, F. v. M., fragm. V, 62 (1866) ... ... — — — — — Q. — B.fl.V,274
M. axillaris, J. Hooker in Lond. Journ. VI, 278 (1847) ... ... — T. V. N.S.W. — — B.fl.V,275
M. polybotrya, Meissner in Lehm. pl. Preiss. I, 623 (1845) W.A. — — — — — B.fl.V,475 M.fr.IV,130.
M. polygonoides, F. v. M., fragm. V, 73 (1865) ... ... — S.A. — T. V. N.S.W. — B.fl.V,275 M.fr.V,73.
M. stenophylla, F. v. M., fragm. I, 138 (1859) ... ... — — V. — — — B.fl.V,275 M.fr.I,138.
M. Cunninghamii, F. v. M., fragm. V, 91 (1865) ... ... W.A. S.A. — V. N.S.W. Q. N.A. B.fl.V,275 M.fr.V,95.

PHYTOLACCEAE.
R. Brown in Tuck. Exped. Cong. App. V, 454 (1818).

MONOCOCCUS, F. v. M., fragm. I, 47 (1858).
M. echinophorus, F. v. M., fragm. I, 47 (1858) ... ... — — — — N.S.W. Q. — B.fl.V,144 M.fr.I,47.

DIDYMOTHECA, J. Hooker in Lond. Journ. Bot. VI, 278 (1847). (Cyclotheca.)
D. thesioides, J. Hooker in Lond. Journ. V, 279 (1847)... ... W.A. S.A. T. — — — — B.fl.V,145
D. pleiococca, F. v. M., fragm. I, 202 (1859) ... ... W.A. S.A. — V. N.S.W. — — B.fl.V,146 M.fr.I,202.

GYROSTEMON, Desfontaines in Mém. du Mus. VI, 16, t. 6 (1820).
G. brachystigma, F. v. M. in Benth. Fl. Austr. V, 146 (1870) ... W.A. — — — — — — B.fl.V,146
G. ramulosus, Desfontaines in Mém. du Mus. V, 17, t. 6 (1820) W.A. S.A. — — — — N.A. B.fl.V,147

CODONOCARPUS, Cunningham in Hook. Bot. Misc. I, 244 (1830). (Hymenotheca.)
C. pyramidalis, F. v. M., plants of Vict. I, 201 (1862) ... ... — S.A. — — — — B.fl.V,148
C. australis, Cunningham in Hook. bot. misc. I, 244 (1830) ... — — — N.S.W. Q. — B.fl.V,148
C. cotinifolius, F. v. M., pl. of Vict. I, 200 (1862) ... ... W.A. S.A. — V. N.S.W. — N.A. B.fl.V,149

TERSONIA, Moquin in De Cand., prodr. XIII, part sec. 40 (1849).
T. brevipes, Moquin in De Cand., prod. XIII, part II, 40 (1849) W.A. — — — — — — B.fl.V,149
T. subvolubilis, Bentham, Fl. Austr. V, 150 (1870) ... ... W.A. — — — — — — B.fl.V,150

CYPSELOCARPUS, F. v. M., fragm. VIII, 36 (1873).
C. haloragoides, F. v. M., fragm. VIII, 36 (1873) ... W.A. — — — — — — B.d.V,108 M.fr.VIII,36.

NYCTAGINEAE.
A. L. de Jussieu, Gen. 90 (1789).

BOERHAAVIA, Vaillant, serm. de struct. fl. 50 (1718).
B. diffusa, Linné, spec. fl. Zeylan. 4 (1747) ... ... W.A. S.A. — V. N.S.W. Q. N.A. B.fl.V,277
B. repanda, Willdenow, sp. pl. I, 22 (1797) ... ... W.A. S.A. — — — Q. N.A. B.fl.V,278

PISONIA, Plumier, nov. pl. Amer. gen. 7, t. 11 (1703).
P. aculeata, Linné, sp. pl. 1026 (1753) ... ... — — — — N.S.W. Q. — B.fl.V,279 M.fr.VI,197.
P. inermis, G. Forster, prodr. 75 (1786) ... ... — — — — — Q. N.A. B.fl.V,280 M.fr.VI,197.
P. Brunoniana, Endlicher, prodr. fl. Norf. 43 (1833) ... ... — — — N.S.W. Q. — B.fl.V,280 M.fr.I,20;IV,169.

CHORIPETALEAE PERIGYNAE.
F. v. M. in Woolls pl. of the neighb. of Sydney 18 (1880).

CONNARACEAE.
R. Brown, Narr. exped. Cong. App. V, 431 (1878).

ROUREA, Aublet, Hist. des pl. de la Guian. I, 467, t. 187 (1775).
R. brachyandra, F. v. M., fragm. VIII, 6 (1872) ... ... — — — — — Q. — — M.fr.VIII,6.

TRICHOLOBUS, Blume, Mus. Bot. Lugd. I, 237 (1850).
T. connaroides, F. v. M., fragm. VIII, 224 (1874) ... ... — — — — — — M.fr.V,224;X,120.

LEGUMINOSAE.
Haller, enum. stirp. Helv. II, 565 (1742).

JANSONIA, Kippist in Gard. Chron. 19 (1847). (Cryptosema.)
J. formosa, Kippist in Transact. Linn. Soc. XX, 334, t. 16 (1851) W.A. — — — — — — B.fl.II,8

BRACHYSEMA, R. Brown in Ait. hort. Kew. sec. edit. III, 10 (1811). (Leptosema, Kaleniczenkia, Burgesia.)
B. praemorsum, Meissner in Lehm. pl. Preiss. I, 23 (1844) ... W.A. — — — — — — B.fl.II,10
B. lanceolatum, Meissner in Lehm. pl. Preiss. I, 24 (1844) ... W.A. — — — — — — B.fl.II,10
B. latifolium, R. Brown in Ait. hort. Kew. sec. ed. III, 10 (1811) W.A. — — — — — — B.fl.II,10
B. undulatum, Ker in Bot. Regist. t. 642 (1922) ... ... W.A. — — — — — — B.fl.II,11 M.fr.IV,11.
B. subcordatum, Bentham, Fl. Austr. II, 11 (1864) ... ... W.A. — — — — — — B.fl.II,11
B. bracteolosum, F. v. M., fragm. IV, 10 (1864) ... ... W.A. — — — — — — B.fl.II,11 M.fr.IV,10.
B. bossiaeoides, Bentham, Fl. Austr. II, 12 (1864) ... ... — — — — — N.A. B.d.II.12
B. oxylobioides, Bentham, Fl. Austr. II, 12 (1864) ... ... — — — — — Q. N.A. B.fl.II,12
B. uniflorum, R. Brown in Benth., Fl. Austr. II, 12 (1864) ... — — — — — N.A. B.fl.II,12
B. aphyllum, Hooker, Bot. Mag. t. 4481 (1849) ... ... W.A. — — — — — — B.fl.II,12 M.fr.I,222.
B. macrocarpum, Bentham, Fl. Austr. II, 13 (1864) ... ... W.A. — — — — — — B.fl.II,13
B. tomentosum, Beutham, Fl. Austr. II, 13 (1864) ... ... W.A. — — — — — — B.fl.II,13
B. Chambersii, F. v. M. in Benth., Fl. Austr. II, 13 (1864) ... — S.A. — — — N.A. B.fl.II,13
B. daviesioides, Bentham, Fl. Austr. II, 13 (1864) ... ... W.A. — — — — — — B.fl.II,13

OXYLOBIUM, Andrews, Bot. Reposit. t. 492 (1909). (Chorizematis subgenus, Callistachys, Podolobium.)
O. Callistachys, Bentham, Fl. Austr. II, 16 (1864) ... ... W.A. — — — — — — B.fl.II,16 M.fr.IV,18.
O. ellipticum, R. Brown in Ait hort. Kew. sec. ed. III, 10 (1811) — — T. V. N.S.W. Q. — B.fl.II,16
O. alpestre, F. v. M. in Transact. Bot. Soc. Vict. I, 38 (1854) ... — — — V. N.S.W. — — B.fl.II,17
O. lineare, Bentham, Fl. Austr. II, 17 (1864) ... ... W.A. — — — — — — B.fl.II,17 M.fr.IV,17.
O. carinatum, Bentham, Fl. Austr. II, 18 (1864) ... ... W.A. — — — — — — B.fl.II,18
O. spathulatum, Bentham, Fl. Austr. II, 18 (1864) ... ... W.A. — — — — — — B.fl.II,18
O. obtusifolium, Sweet, Fl. Austr. t. 5 (1827) ... ... W.A. — — — — — — B.fl.II,18
O. cordifolium, Andrews, Bot. Regist. t. 492 (1809) ... ... — — — — N.S.W. — — B.fl.II,19 M.fr.IV,17.
O. microphyllum, Bentham, Fl. Austr. II, 19 (1864) ... ... W.A. — — — — — — B.fl.II,19
O. Pultenaeae, De Candolle, prodr. II, 104 (1825) ... ... — — — — N.S.W. — — B.fl.II,19 M.fr.IV,19.
O. hamulosum, Bentham in A. Gray, Bot. Wilk. Exped. t. 379 — — — — N.S.W. — — B.d.II,20
O. scandens, Bentham in Ann. Wien. Mus. II, 70 (1839)... ... — — — — N.S.W. Q. — B.d.II,20
O. procumbens, F. v. M. in Transact. phil. Soc. Vict. I, 37 (1854) — — — V. N.S.W. — — B.fl.II,20
O. tricuspidatum, Meissner in Lehm. pl. Preiss. I, 30 (1844) ... W.A. — — — — — — B.fl.II,21

O. spectabile, Endlicher, nov. stirp. dec. 2 (1839)	W.A.	— — — — — —	B.fl.II,21	
O. atropurpureum, Turczanin. in Bull. Soc. Mosc. XXVI, 250 (1853)	W.A.	— — — — — —	B.fl.II,22	
O. retusum, R. Brown in Bot. Regist. t. 913 (1823)	V.A.	— — — — — —	B.fl.II,22	
O. virgatum, Bentham, Fl. Austr. II, 22 (1864)	W.A.	— — — — — —	B.fl.II,22	
O. reticulatum, Meissner in Lehm. pl. Preiss. I, 29 (1844)	W.A.	— — — — — —	B.fl.II,23	
O. capitatum, Bentham in Hueg. enum. 28 (1837)	W.A.	— — — — — —	B.fl.II,23	
O. cuneatum, Bentham in Bot. Regist. XXV, App. XII (1839)	W.A.	— — — — — —	B.fl.II,23	
O. acutum, Bentham, Fl. Austr. II, 24 (1864)	W.A.	— — — — — —	B.fl.II,24	
O. parviflorum, Bentham in Bot. Regist. XXV, App. XII (1839)	W.A.	— — — — — —	B.fl.II,24	
O. heterophyllum, Bentham, Fl. Austr. II, 25 (1864)	W.A.	— — — — — —	B.fl.II,25	
O. aciculiferum, F. v. M., fragm. I, 75, implied (1859)		— — — — — Q.	B.fl.II,25	M.fr.I,75.
O. trilobatum, F. v. M., annual report 23 (1853)		— — — — N.S.W. Q.	B.fl.II,25	M.fr.IV,19.
CHORIZEMA, Labillardière, Voy. I, 405 (1799). (Orthotropis.)				
C. Dicksonii, Graham in Maund's Botanist, t. 100 (1841)	W.A.	— — — — — —	B.fl.II,27	
C. nervosum, Th. Moore in Gard. Companion (1852)	W.A.	— — — — — —	B.fl.II,27	
C. varium, Bentham in Bot. Regist. XXV, t. 40 (1839)	W.A.	— — — — — —	B.fl.II,28	
C. cordatum, Lindley, Bot. Regist. t. 10 (1838)	W.A.	— — — — — —	B.fl.II,28	
C. ilicifolium, Labillardière, Voy. I, 405, t. 21 (1799)	W.A.	— — — — — —	B.fl.II,28	
C. rhombeum, R. Brown in Ait. hort. Kew. sec. ed. III, 9 (1811)	W.A.	— — — — — —	B.fl.II,28	
C. diversifolium, A. de Candolle, Pl. rar. jard. Genèv. 44, t. 8 (1836)	W.A.	— — — — — —	B.fl.II,29	
C. angustifolium, Bentham in Hueg. enum. 28 (1837)	W.A.	— — — — — —	B.fl.II,29	
C. reticulatum, Meissner in Lehm. pl. Preiss. I, 34 (1844)	W.A.	— — — — — —	B.fl.II,30	
C. trigonum, Turcz. in Bull. de la Soc. Mosc. XXVI, 254 (1853)	W.A.	— — — — — —	B.fl.II,30	
C. humile, Turcz. in Bull. de la Soc. Mosc. XXVI, 254 (1853)	W.A.	— — — — — —	B.fl.II,30	
C. parviflorum, Bentham in Ann. des Wien. Mus. II, 71 (1838)		— — — — N.S.W. Q.	B.fl.II,30	
C. cytisoides, Turczaninow in Bull. Soc. Mosc. XXVI, 256 (1853)	W.A.	— — — — — —	B.fl.II,31	
C. Henchmanni, R. Brown in Bot. Regist. t. 986 (1826)	W.A.	— — — — — —	B.fl.II,31	
C. ericifolium, Meissner in Lehm. pl. Preiss. II, 209 (1847)	W.A.	— — — — — —	B.fl.II,32	
GASTROLOBIUM, R. Brown in Ait. hort. Kew. sec. edit. III, 16 (1811).				
G. pyramidale, Th. Moore in Gard. Comp. I, 81 (1852)	W.A.	— — — — — —	B.fl.II,98	
G. Lehmanni, Meissner in Lehm. pl. Preiss. I, 70 (1844)	W.A.	— — — — — —	B.fl.II,98	M.fr.X,35.
G. pulchellum, Turcz. in Bull. de la Soc. Mosc. XXVI, 274 (1853)	W.A.	— — — — — —	B.fl.II,98	
G. stipulare, Meissner in Lehm. pl. Preiss. II, 218 (1847)	W.A.	— — — — — —	B.fl.II,99	
G. Brownii, Meissner in Lehm. pl. Preiss. I, 71 (1844)	W.A.	— — — — — —	B.fl.II,99	M.fr.X,35.
G. reticulatum, Bentham, Fl. Austr. II, 99 (1864)	S.A.	— — — — — —		M.fr.IX,67.
G. clachistum, F. v. M., fragm. IX, 67 (1875)	S.A.	— — — — — —		
G. truncatum, Bentham, Fl. Austr. II, 99 (1864)	W.A.	— — — — — —	B.fl.II,99	
G. spathulatum, Bentham in Bot. Regist. XX, App. XIV (1839)	W.A.	— — — — — —	B.fl.II,100	
G. plicatum, Turczanin. in Bull. Soc. Mosc. XXVI, 274 (1853)	W.A.	— — — — — —	B.fl.II,100	
G. tricuspidatum, Meissner in Lehm. pl. Preiss. I, 66 (1844)	W.A.	— — — — — —	B.fl.II,100	
G. obovatum, Bentham in Bot. Regist. XXV, App. XIV (1839)	W.A.	— — — — — —	B.fl.II,101	M.fr.X,35.
G. epacridioides, Meissner in Lehm. pl. Preiss. I, 72 (1844)	W.A.	— — — — — —	B.fl.II,101	
G. trilobum, Bentham in Bot. Regist. XXV, App. XIII (1839)	W.A.	— — — — — —	B.fl.II,101	
G. ilicifolium, Meissner in Lehm. pl. Preiss. I, 67 (1844)	W.A.	— — — — — —	B.fl.II,102	
G. villosum, Bentham in Bot. Regist. XXV, App. XIII (1839)	W.A.	— — — — — —	B.fl.II,102	
G. polystachyum, Meissner in Lehm. pl. Preiss. II, 217 (1847)	W.A.	— — — — — —	B.fl.II,102	
G. ovalifolium, Henfrey in Gard. Comp. I, 41 (1852)	W.A.	— — — — — —	B.fl.II,102	
G. grandiflorum, F. v. M., fragm. III, 17 (1862)		— — — — Q. N.A.	B.fl.II,103	M.fr.III,17;IV,174;X,35.
G. pycnostachyum, Bentham, Fl. Austr. II, 103 (1864)	W.A.	— — — — — —	B.fl.II,103	M.fr.X,35.
G. spinosum, Bentham in Bot. Regist. XXV, App. XIII (1839)	W.A.	— — — — — —	B.fl.II,103	
G. rotundifolium, Meissner in Lehm. pl. Preiss. II, 216 (1847)	W.A.	— — — — — —	B.fl.II,104	
G. microcarpum, Meissner in Lehm. pl. Preiss. I, 70 (1844)	W.A.	— — — — — —	B.fl.II,104	
G. oxylobioides, Bentham in Bot. Regist. XXV, App. XIII (1839)	W.A.	— — — — — —	B.fl.II,104	M.fr.X,35.
G. calycinum, Bentham in Bot. Regist. XXV, App. XIII (1839)	W.A.	— — — — — —	B.fl.II,104	
G. Callistachys, Meissner in Lehm. pl. Preiss. II, 216 (1846)	W.A.	— — — — — —	B.fl.II,105	
G. stenophyllum, Turczanin. in Bull. Soc. Mosc. XXVI, 275 (1853)	W.A.	— — — — — —	B.fl.II,105	
G. crassifolium, Bentham, Fl. Austr. II, 105 (1864)	W.A.	— — — — — —	B.fl.II,105	
G. parvifolium, Bentham, Fl. Austr. II, 106 (1864)	W.A.	— — — — — —	B.fl.II,106	
G. hamulosum, Meissner in Lehm. pl. Preiss. II, 218 (1847)	W.A.	— — — — — —	B.fl.II,106	M.fr.X,35.
G. velutinum, Lindley & Paxton, Flower Gard. III, 76 (1853)	W.A.	— — — — — —	B.fl.II,106	
G. bidens, Meissner in Regensb. Bot. Zeit. 29 (1855)	W.A.	— — — — — —	B.fl.II,106	M.fr.X,35.
G. bilobum, R. Brown in Ait. hort. Kew. sec. edit. III, 16 (1811)	W.A.	— — — — — —	B.fl.II,107	M.fr.X,35.
G. scorsifolium, F. v. M., fragm. X, 35 (1876)	W.A.	— — — — — —		M.fr.X,35.
ISOTROPIS, Bentham in Hueg. enum. 28 (1837).				
I. striata, Bentham in Hueg. enum. 28 (1837)	W.A.	— — — — — —	B.fl.II,39	M.fr.III,16.
I. Drummondii, Meissner in Lehm. pl. Preiss. I, 31 (1844)	W.A.	— — — — — —	B.fl.II,39	
I. juncea, Turczanin. in Bull. Soc. Mosc. XXVI, 251 (1853)	W.A.	— — — — — —	B.fl.II,39	M.fr.X,51.
I. canescens, F. v. M., fragm. X, 51 (1876)	W.A.	— — — — — —		M.fr.X,51.
I. atropurpurea, F. v. M., fragm. III, 16 (1862)		— — — — — N.A.	B.fl.II,40	M.fr.III,16;X,51.
I. filicaulis, Bentham in Ann. des Wien. Mus. II, 71 (1838)		— — — — — Q.	B.fl.II,40	M.fr.IV,20.
I. parviflora, Bentham in Ann. des Wien. Mus. II, 71 (1838)		— — — — — N.A.	B.fl.II,40	
I. Wheeleri, F. v. M. in Benth. Fl. Austr. II, 40 (1864)		— S.A. — — N.S.W. —	B.fl.II,40	
MIRBELIA, Smith in Koenig & Sims, Ann. of Bot. I, 511 (1805). (Dichosema, Oxycladium.)				
M. dilatata, R. Brown in Ait. hort. Kew. sec. ed. III, 16 (1811)		— — — — — —	B.fl.II,33	M.fr.IV,18.
M. racemosa, Turczanin. in Bull. Soc. Mosc. XXVI, 282 (1853)	W.A.	— — — — — —	B.fl.II,33	
M. grandiflora, Alton in Bot. Mag. t. 2771 (1827)		— — — — N.S.W. —	B.fl.II,34	
M. subcordata, Turczanin. in Bull. Soc. Mosc. XXVI, 282 (1853)	W.A.	— — — — — —	B.fl.II,34	
M. ovata, Meissner in Lehm. pl. Preiss. I, 77 (1844)	W.A.	— — — — — —	B.fl.II,34	

M. oxylobioides, F. v. M., fragm. II, 154 (1801) ... ... — — — V. N.S.W. — — B.fl.II,34 M.fr.IV 12.
M. reticulata, Smith in Koen. & Sims. Ann. of Bot. I, 511 (1805) — — — — N.S.W. Q. — B.fl.II,35
M. aotoides, F. v. M. in Transact. phil. Inst. Vict. III, 53 (1850) — — — — N.S.W. Q. — B.fl.II,35
M. pungens, Cunningham in G. Don, gen. syst. II, 126 (1832) ... — — — — N.S.W. Q. — B.fl.II,35
M. speciosa, Sieber in De Cand. prodr. II, 115 (1824) ... ... — — — — N.S.W. — — B.fl.II,36
M. floribunda, Bentham in Bot. Regist. XXV, App. XII (1839) W.A. — — — — — — B.fl.II,36
M. spinosa, Bentham, Fl. Austr. II, 36 (1864) ... ... W.A. — — — — — — B.fl.II,36
M. multicaulis, Bentham, Fl. Austr. II, 37 (1864) ... ... W.A. — — — — — — B.fl.II,37
M. microphylla, Bentham, Fl. Austr. II, 37 (1864) ... ... W.A. — — — — — — B.fl.II,37
M. daviesioides, Bentham, Fl. Austr. II, 37 (1864) ... ... W.A. — — — — — — B.fl.II,37 M.fr.IV,11.
M. oxyclada, F. v. M., fragm. IV, 12 (1864) ... ... — S.A. — — — — N.A. B.fl.II,38 M.fr.IV,12.

GOMPHOLOBIUM, Smith in Transact. Linn. Soc. IV, 220 (1798).
G. ovatum, Meissner in Lehm. pl. Preiss. I, 35 (1844) ... W.A. — — — — — — B fl.II,42
G. amplexicaule, Meissner in Lehm. pl. Preiss. I, 36 (1844) ... W.A. — — — — — — B.fl.II,42
G. latifolium, Smith in Koenig and Sims, Ann. of Bot. I, 505 (1805) — — — — N.S.W. Q. — B.fl.II,42
G. Huegelii, Bentham in Hueg. enum. 29 (1837) ... ... — T. V. N.S.W. — — B.fl.II,43
G. polymorphum, R. Brown in Ait. hort. Kew, sec. ed. III, 11 (1811)W.A. — — — — — — B.fl.II,43
G. obcordatum, Turczanin. in Bull. Soc. Mosc. XXVI, 258 (1853) W.A. — — — — — — B.fl.II,44
G. marginatum, R. Brown in Ait. hort. Kew. sec. ed. III, 11 (1811)W.A. — — — — — — B.fl.II,44
G. grandiflorum, Smith, Exot. Bot. t. 5 (1804) ... ... — — — — N.S.W. — — B.fl.II,44
G. virgatum, Sieber in De Cand. prodr. II, 105 (1824) ... — — — — N.S.W. Q. — B.fl.II,45
G. minus, Smith in Transact. Linn. Soc. IX, 251 (1808) ... S.A. — V. N.S.W. — — B.fl.II,45
G. uncinatum, Cunningham in Annal. des Wien. Mus. II, 72 (1838) — — — — N.S.W. — — B fl.II,46
G. Baxteri, Bentham, Fl. Austr. II, 46 (1864) ... ... W.A. — — — — — — B.fl.II,46
G. aristatum, Bentham in Ann. des Wien. Mus. II, 72 (1839) ... W.A. — — — — — — B.fl.II,46
G. burtonioides, Meissner in Lehm. pl. Preiss. I, 37 (1844) ... W.A. — — — — — — B.fl.II,46
G. capitatum, Cunningham in Bot. Regist. t. 1563 (1830) ... W.A. — — — — — — B.fl.II,47
G. tomentosum, Labillard., Nov. Holl. pl. sp. I. 106, t. 134 (1804) W.A. — — — — — — B.fl.II,47
G. Preissii, Meissner in Lehm. pl. Preiss. I, 40 (1844) ... W.A. — — — — — — B.fl.II,17
G. viscidulum, Meissner in Lehm. pl. Preiss. I, 39 (1844) ... W.A. — — — — — — B.fl.II,48
G. glabratum, De Candolle, prodr. II, 106 (1824) ... ... — — — — N.S.W. — — B.fl.II,48
G. nitidum, Solander in Benth. Fl. Austr. II, 48 (1864) ... — — — — — Q. — B.fl.II,48
G. pinnatum, Smith in Transact. Linn. Soc. IX, 251 (1808) ... — — — — N.S.W. Q. — B.fl.II,48
G. Shuttleworthii, Meissner in Lehm. pl. Preiss. I, 39 (1844) ... W.A. — — — — — — B.fl.II,49
G. venustum, R. Brown in Ait. hort. Kew. sec. ed. III, 12 (1811) W.A. — — — — — — B.fl.II,49
G. Knightianum, Lindley, Bot. Regist. t. 1468 (1837) ... ... W.A. — — — — — — B.fl.II,49

BURTONIA, R. Brown in Ait. hort. Kew, ed. sec. III, 13 (1811).
B. subulata, Bentham, Fl. Austr. II, 50 (1864) ... ... — — — — — N.A. B.fl.II,50 M.fr.III,30;X,35.
B. foliolosa, Bentham, Fl. Austr. II, 50 (1864) ... ... — — — — — Q. — B.fl.II,50
B. polyzyga, Bentham, Fl. Austr. II, 51 (1864) ... ... S.A. — — — — N.A. B.fl.II,51 M.fr.III,29;X,34.
B. villosa, Meissner in Lehm. pl. Preiss. I, 41 (1844) ... ... W.A. — — — — — — B.fl.II,51 M.fr.X,35.
B. Hendersonii, Bentham, Fl. Austr. II, 51 (1864) ... ... W.A. — — — — — — B.fl.II,51
B. gompholobioides, F. v. M., fragm. X, 34 (1876) ... ... W.A. — — — — — — — M.fr.X,34.
B. scabra, R. Brown in Ait. hort. Kew. sec. ed. III, 12 (1811) ... W.A. — — — — — — B.fl.II,51 M.fr.X,35.
B. conferta, De Candolle, prodr. II, 106 (1825) ... ... W.A. — — — — — — B.fl.II,52 M.fr.X,35.

JACKSONIA, R. Brown in Ait. hort. Kew. sec. ed. III, 12 (1811). (Piptomeris.)
J. dilatata, Bentham in Ann. des Wien. Mus. II, 74 (1838) ... — — — — — N.A. B.fl.II,54 M.fr.X,37.
J. densiflora, Bentham in Bot. Regist. XXV, App. XIII (1839) W.A. — — — — — — B.fl.II,54 M.fr.X,38.
J. carduacea, Meissner in Bot. Zeit. 25 (1855) ... ... W.A. — — — — — — B.fl.II,55
J. floribunda, Endlicher in Ann. des Wien. Mus. II, 197 (1838) W.A. — — — — — — B.fl.II,55
J. odontoclada, F. v. M. in Benth. Fl. Austr. II, 55 (1864) ... — — — — — N.A. B.fl.II,55 M.fr.X,37.
J. ramosissima, Bentham in Mitch. Trop. Austr. 258 (1848) ... — — — — — Q. — B.fl.II,56 M.fr.X,37.
J. foliosa, Turcz. in Bull. de la Soc. Imp. de Mosc. XXVI, 260 (1853)W.A. — — — — — — B.fl.II,56
J. spinosa, R. Brown in Ait. hort. Kew. sec. ed. III, 13 (1811) W.A. — — — — — — B.fl.II,56 M.fr.X,38&39.
J. stricta, Meissner in Regensb. Bot. Zeit. 27 (1855) ... ... W.A. — — — — — — B.fl.II,57
J. hakeoides, Meissner in Lehm. pl. Preiss. I, 45 (1844) ... W.A. — — — — — — B.fl.II,57 M.fr.X,39.
J. furcellata, De Candolle, prodr. II, 107 (1824) ... ... W.A. — — — — — — B.fl.II,57 M.fr.X,39.
J. horrida, De Candolle, prodr. II, 107 (1824) ... ... W.A. — — — — — — B.fl.II,57
J. sericea, Bentham in Hueg. enum. 31 (1837) ... ... W.A. — — — — — — B.fl.II,58
J. Sternbergiana, Huegel, Bot. Archiv. t. 3 (1837) ... ... W.A. — — — — — — B.fl.II,58 M.fr.X,39.
J. vernicosa, F. v. M. in Benth. Fl. Austr. II, 58 (1864) ... — — — — — N.A. B.fl.II,58 M.fr.X,37.
J. rhadinoclona, F. v. M., fragm. X, 37 (1876) ... ... — — — — N.S.W. Q. — M fr.X,37.
J. thesioides, Meissner in Lehm. pl. Preiss. I, 44 (1844) ... — — — — Q. N.A. B.fl.II,59 M.fr.IV,161;X,37.
J. compressa, Turcz. in Bull. de la Soc. de Mosc. XXVI, 260 (1853)W.A. — — — — — — B.fl.II,59 M.fr.X,38.
J. scoparia, R. Brown in Ait. hort. Kew. sec. ed. III, 13 (1811) W.A. — N.S.W. Q. — B.fl.II,59 M.fr.X,37.
J. cupulifera, Meissner in der Bot. Zeitung 27 (1855) ... W.A. — — — — — — — M.fr.X,36.
J. restioides, Meissner in Lehm. pl. Preiss. I, 46 (1844) ... W.A. — — — — — — B.fl.II,60
J. velutina, Bentham, Fl. Austr. II, 60 (1864) ... ... W.A. — — — — — — B.fl.II,60
J. Lehmanni, Meissner in Lehm. pl. Preiss. I, 46 (1844) ... W.A. — — — — — — B.fl.II,60
J. racemosa, Meissner in Lehm. pl. Preiss. II, 212 (1847) ... W.A. — — — — — — B.fl.II,60
J. umbellata, Turczaninow in Bull. Soc. Mosc. XXVI, 261 (1853) W.A. — — — — — — B.fl.II,61
J. nematoclada, F. v. M., fragm. X, 50 (1876) ... ... — — — — — — — M.fr.X,50.
J. rhadinoclada, F. v. M., fragm. X, 38 (1876) ... ... — — — — — — — M.fr.X,38.
J. capitata, Meissner in Lehm. pl. Preiss. I, 45 (1844) ... W.A. — — — — — — B.fl.II,61 M.fr.X,38.
J. alata, Bentham in Hueg. enum. 30 (1837) ... ... W.A. — — — — — — B.fl.II,61
J. angulata, Bentham, Fl. Austr. II, 62 (1864) ... ... W.A. — — — — — — B.fl.II,62
J. Stackhousii, F. v. M. in proc. Linn. Soc. N.S.W., VI, 791 (1881) — — — — N.S.W. Q. — — M.fr.XI,138
J. pteroclada, F. v. M., fragm. X, 37 (1876) ... ... W.A. — — — — — — — M.fr.X,37.
J. macrocalyx, Meissner in der Bot. Zeit. 26 (1855) ... ... W.A. — — — — — — B.fl.II,62

J. purpurascens, F. v. M., fragm. IV, 161 (1864)... ... ... — — — — — Q. — B.fl.II,62 M.fr.IV,161.
J. piptomeris, Bentham, Fl. Austr. II, 62 (1864)... ... ... W.A. — — — — — — —

SPHAEROLOBIUM, Smith in Koenig & Sims, Ann. of Bot. I, 509 (1806). (Roca.)
S. linophyllum, F. v. M., fragm. I, 167 (1859) ... ... W.A. — — — — — B.fl.II,64 M.fr.I,167.
S. foliosum, F. v. M., fragm. I, 166 (1830)... ... ... W.A. — — — — — B.fl.II,64 M.fr.I,166.
S. gracile, Bentham, Fl. Austr. II, 64 (1864) ... ... W.A. — — — — — B.fl.II,64
S. racemulosum, Bentham, Fl. Austr. II, 64 (1864) ... ... W.A. — — — — — B.fl.II,65
S. alatum, Bentham in Hueg. enum. 32 (1837) ... ... W.A. — — — — — B.fl.II,65
S. vimineum, Smith in Koen. & Sims, Ann. of Bot. I, 509 (1805)... — S.A. T. V. N.S.W. Q. — B.fl.II,65
S. grandiflorum, R. Brown in Benth. Fl. Austr. II, 66 (1864) ... W.A. — — — — — B.fl.II,66
S. fornicatum, Bentham in Hueg. enum. 32 (1837) ... ... W.A. — — — — — B.fl.II,66
S. medium, R. Brown in Ait. hort. Kew. sec. ed. III, 14 (1811) W.A. — — — — — B.fl.II,66
S. scabriusculum, Meisner in Lehm. pl. Preiss. II, 214 (1849)... W.A. — — — — — M.fl.II,66
S. macranthum, Meisner in Lehm. pl. Preiss. II, 213 (1847) ... W.A. — — — — — B.fl.II,67
S. daviesioides, Turczaninow in Bull. Soc. Mosc. XXVI,266 (1853) W.A. — — — — — B.fl.II,67

VIMINARIA, Smith, Exotic Botany 51, t. 27 (1804).
V. denudata, Smith, Exot. Bot. 51, t. 27 (1804) ... ... W.A. S.A. T. V. N.S.W. Q. — B.fl.II,68

DAVIESIA, Smith in Transact. Linn. Soc. IV, 222 (1798).
D. cordata, Smith in Transact. Linn. Soc. IX, 259 (1808) ... W.A. — — — — — B.fl.II,72
D. ovata, Bentham, Fl. Austr. II, 72 (1864) ... ... W.A. — — — — — B.fl.II,72
D. crenulata, Turczanin. in Bull. Soc. de Mosc. XXVI, 265 (1853) W.A. — — — — — B.fl.II,72 M.fr.IV,10.
D. oppositifolia, Endl. in den. Ann. des Wien. Mus. II, 199 (1838) W.A. — — — — — B.fl.II,73
D. alternifolia, Endlicher in Ann. Wien. Mus. II, 199 (1838) ... W.A. — — — — — B.fl.II,73
D. elongata, Bentham, Fl. Austr. II, 74 (1864) ... ... W.A. — — — — — B.fl.II,74
D. pedunculata, Beutham in Bot. Regist. XXV, App. XIV (1839) W.A. — — — — — B.fl.II,74
D. arthropoda, F. v. M., fragm. VIII, 223 (1874)... ... ... W.A. S.A. — — — — — M.fr.VIII,223.
D. mollis, Turczaninow in Bull. Soc. Mosc. XXVI, 263 (1853)... — — — — — — B.fl.II,74
D. concinna, R. Brown in Benth. Fl. Austr. II, 75 (1864) ... — — — — N.S.W. Q. — B.fl.II,75
D. umbellulata, Smith in Koen. & Sims, Ann. of Bot. I, 507 (1805) — — — — N.S.W. Q. — B.fl.II,75
D. Wyattii, Bailey in pap. Gardn. Assoc. of S. Austr. (1880) ... — — — V. N.S.W. — — M.fr.X,138.
D. latifolia, R. Brown in Ait. hort. Kew. sec. ed. III, 507 (1805) — T. V. N.S.W. — B.fl.II,76
D. corymbosa, Smith in Koen. & Sims, Ann. of Bot. I, 507 (1805) — S.A. — V. N.S.W. Q. — B.fl.II,76
D. horrida, Meisner in Lehm. pl. Preiss. I, 54 (1844) ... W.A. — — — — — B.fl.II,77
D. reclinata, Cunningham in Benth. Fl. Austr. II, 77 (1864) ... — — — — — N.A. B.fl.II,77
D. obtusifolia, F. v. M., fragm. II, 104 (1861) ... ... W.A. — — — — — B.fl.II,77 M.fr.II,104.
D. obovata, Turczaninow in Bull. Soc. Mosc. XXVI, 261 (1853) W.A. — — — — — B.fl.II,78
D. longifolia, Bentham in Bot. Regist. XXV, App. XIV (1839) W.A. — — — — — B.fl.II,78
D. chordophylla, Meisner in Lehm. plant. Preiss. I, 48 (1844)... W.A. — — — — — B.fl.II,78
D. nematophylla, F. v. M. in Benth. Fl. Austr. II, 78 (1864) ... W.A. — — — — — B.fl.II,78
D. daphnoides, Meisner in Lehm. plant. Preiss. I, 54 (1844) ... W.A. — — — — — B.fl.II,79
D. nudiflora, Meisner in Lehm. plant. Preiss. I, 53 (1844) ... W.A. — — — — — B.fl.II,79
D. rhombifolia, Meisner in Lehm. plant. Preiss. I, 56 (1844) ... W.A. — — — — — B.fl.II,79
D. Drummondii, Meisner in Lehm. plant. Preiss. I, 53 (1844)... W.A. — — — — — B.fl.II,80
D. filipes, Bentham in Mitch. Trop. Austr. 363 (1848) ... ... — — — — N.S.W. Q. — B.fl.II,80
D. cardiophylla, F. v. M., fragm. II, 103 (1861) ... ... W.A. — — — — — B.fl.II,80 M.fr.II,103.
D. squarrosa, Smith in Koen. & Sims, Ann. of Bot. I, 507 (1805) — — — N.S.W. Q. — B.fl.II,80
D. ulicina, Smith in Koen. & Sims, Ann. of Bot. I, 506 (1805)... — S.A. T. V. N.S.W. Q. — B.fl.II,81
D. acicularis, Smith in Koen. & Sims, Ann. of Bot. I, 506 (1805) — — — N.S.W. — B.fl.II,81
D. pachyphylla, F. v. M., fragm. IV, 15 (1864) ... ... W.A. — — — — — B.fl.II,82 M.fr.IV,15.
D. teretifolia, R. Brown in Benth. Fl. Austr. II, 82 (1864) ... W.A. — — — — — B.fl.II,82
D. genistifolia, Cunningham in Annal. Wien. Mus. II, 75 (1838) — S.A. — V. N.S.W. Q. — B.fl.II,82
D. hakeoides, Meisner in Lehm. plant. Preiss. I, 47 (1844) ... W.A. — — — — — B.fl.II,83
D. colletioides, Meisner in Lehm. plant. Preiss. I, 48 (1844) ... W.A. — — — — — B.fl.II,83
D. reversifolia, F. v. M., fragm. I, 145 (1850) ... ... W.A. — — — — — B.fl.II,83 M.fr.I,145.
D. incrassata, Smith in Transact. Linn. Soc. IX, 255 (1808) ... W.A. S.A. — — — — B.fl.II,84
D. brevifolia, Lindley in Mitch. Three Exped. II, 201 (1838) ... W.A. S.A. — V. N.S.W. — B.fl.II,84
D. acanthoclona, F. v. M., fragm. X, 32 (1876) ... ... W.A. — — — — — B.fl.II,84 M.fr.X,32.
D. Preissii, Meisner in Lehm. plant. Preiss. I, 50 (1844) ... W.A. — — — — — B.fl.II,84
D. spinosissima, Meisner in Lehm. plant. Preiss. I, 51 (1844) ... W.A. — — — — — B.fl.II,85
D. pachylina, Turczaninow in Bull. Soc. Mosc. XXVI, 263 (1853) W.A. — — — — — B.fl.II,85
D. quadrilatera, Bentham in Bot. Regist. XXV, App. XIV (1839) W.A. — — — — — B.fl.II,85
D. striata, Turczaninow in Bull. Soc. Mosc. XXVI, 264 (1853) W.A. — — — — — B.fl.II,85 M.fr.II,105.
D. polyphylla, Bentham in Hueg. enum. 32 (1837) ... ... W.A. — — — — — B.fl.II,86
D. microphylla, Bentham, Fl. Austr. II, 86 (1864) ... ... W.A. — — — — — B.fl.II,86
D. flexuosa, Bentham in Hueg. enum. 32 (1837) ... ... W.A. — — — — — B.fl.II,86
D. pectinata, Lindley in Mitch. Three Exped. II, 151 (1838) ... W.A. S.A. — V. N.S.W. — B.fl.II,87
D. trigonophylla, Meisner in Lehm. plant. Preiss. II, 213 (1847) W.A. — — — — — B.fl.II,87
D. epiphylla, Meisner in Regensb. Bot. Zeit. 27 (1855)... ... W.A. — — — — — B.fl.II,87
D. euphorbioides, Bentham, Fl. Austr. II, 88 (1864) ... ... W.A. — — — — — B.fl.II,88
D. divaricata, Bentham in Hueg. enum. 31 (1837) ... ... W.A. — — — — — B.fl.II,88
D. paniculata, Bentham in Hueg. enum. 31 (1837) ... ... W.A. — — — — — B.fl.II,88
D. aphylla, F. v. M. in Benth. Fl. Austr. 88 (1864) ... ... W.A. — — — — — B.fl.II,88
D. juncea, Smith in Transact. Linn. Soc. IX, 260 (1808) ... — — — — N.S.W. — B.fl.II,89
D. alata, Smith in Transact. Linn. Soc. IX, 259 (1808) ... ... — — — — N.S.W. — B.fl.II,89
D. anceps, Turczaninow in Bull. Soc. Mosc. XXVI, 266 (1853)... W.A. — — — — — B.fl.II,89

AOTUS, Smith in Koenig & Sims, Ann. of Bot. I, 504 (1805).
A. gracillima, Meisner in Lehm. pl. Preiss. I, 50 (1844) ... W.A. — — — — — B.fl.II,90
A. villosa, Smith in Koen. & Sims, Ann. of Bot. I, 504 (1805) ... — — T. V. N.S.W. Q. — B.fl.II,90
A. mollis, Bentham in Mitch. Trop. Austr. 236 (1848) ... ... — — — N.S.W. Q. — B.fl.II,91

A. Proissii, Meissner in Lehm. plant. Preiss. II, 214 (1848) ... W.A. — — — — — — B.fl.II,91
A. phylicoides, F. v. M. in Benth. Fl. Austr. II, 72 (1864) ... W.A. — — — — — — B.fl.II,92
A. lanigera, Cunningham in Annal. des Wien. Mus. II, 78 (1838) — — — — N.S.W. Q. — B.fl.II,92
A. genistoides, Turczaninow in Bull. Soc. Mosc. XXVI, 268 (1853) W.A. — — — — — — B.fl.II,92
A. carinata, Meissner in Lehm. plant. Preiss. II, 215 (1849) ... W.A. — — — — — — B.fl.II,93
A. passerinoides, Meissner in Lehm. plant. Preiss. I, 61 (1844)... W.A. — — — — — — B.fl.II,93
A. Tietkensii, F. v. M., fragm. X, 33 (1876) ... ... W.A. — — — — — — — M.fr.X,33.
A. cordifolia, Bentham in Hueg. enum. 33 (1837)... ... ... W.A. — — — — — — B.fl.II,93

PHYLLOTA, De Candolle, prodr. II, 113 (1825). (Urodon.)
P. barbata, Bentham in Hueg. enum. 33 (1837) ... ... W.A. — — — — — — B.fl.II,94
P. gracilis, Turczaninow in Bull. Soc. Mosc. XXVI, 267 (1853) W.A. — — — — — — B.fl.II,94
P. Sturtii, Bentham, Fl. Austr. II, 95 (1864) ... ... — S.A. — — — — — B.fl.II,95
P. phylicoides, Bentham in Ann. des Wien. Mus. II, 77 (1838)... — — — — N.S.W. Q. — B.fl.II,95 M.fr.X,33.
P. humifusa, Bentham, Fl. Austr. II, 95 (1864) ... ... — — — — N.S.W. — — B.fl.II,95
P. diffusa, F. v. M., fragm. I, 8 (1858) ... ... — T. — — — — B.fl.II,119 M.fr.I,8.
P. pleurandroides, F. v. M. in Transact. phil. Inst. Vict. I, 38 (1854) — S.A. — V. — — — B.fl.II,96 M.fr.X,33.
P. Lachmanni, F. v. M., fragm. X, 33 (1876) ... ... W.A. — — — -- — — — M.fr.X,33.
P. Urodon, F. v. M. in papers of the Roy. Soc. of Tasm. 120 (1877) W.A. — — — — — — B.fl.II,124

PULTENAEA, Smith, Specim. of the Bot. of New Holl. 35, t. 12 (1793). (Euchilus, Spadostyles, Bartlingia.)
P. daphnoides, Wendland, Bot. Beob. 49 (1798) ... — S.A. T. V. N.S.W. — — B.fl.II,112
P. stricta, Sims, Bot. Mag. 1588 (1813) ... — S.A. T. V. N.S.W. — — B.fl.II,112
P. retusa, Smith in Koen. & Sims, Ann. of Bot. I, 502 (1805) ... — — — V. N.S.W. Q. — B.fl.II,113
P. Benthamii, F. v. M. in Transact. phil. Inst. Vict. I, 38 (1854) — — — V. — — — B.fl.II,113
P. pycnocephala, F. v. M. in Benth. Fl. Austr. II, 114 (1864) ... — — — — N.S.W. — — B.fl.II,114
P. myrtoides, Cunningham in Annal. des Wien. Mus. II, 81 (1838) — — — — — Q. — B.fl.II,114
P. mucronata, F. v. M., fragm. I, 8 (1858)... ... ... — S.A. — V. — — — B.fl.II,114 M.fr.I,8.
P. polifolia, Cunningham in Field's N. S. Wales 346 (1825) ... — — — — N.S.W. — — B.fl.II,115
P. petiolaris, Cunningham in Ann. des Wien. Mus. II, 82 (1838) — — — — — Q. — B.fl.II,115
P. paleacea, Willdenow, spec. plant. II, 1340 (1799) ... — — V. N.S.W. — — B.fl.II,115
P. Gunnii, Bentham in Ann. des Wien. Mus. II, 82 (1838) ... — — T. V. N.S.W. — — B.fl.II,116
P. scabra, R. Brown in Ait. hort. Kew. sec. ed. III, 18 (1811)... — — — V. N.S.W. — — B.fl.II,116
P. Hartmanni, F. v. M., fragm. VIII, 166 (1874)... ... — — — — N.S.W. Q. — — M.fr.VIII,166.
P. microphylla, Sieber in De Cand. prodr. II, 11 (1825) ... — — — — N.S.W. Q. — B.fl.II,117
P. Skinneri, F. v. M., fragm. VIII, 166 (1874) ... W.A. — — — — — — — M.fr.VIII,166.
P. Drummondii, Meissner in Lehm. plant. Preiss. II, 219 (1847) W.A. — — — — — — B.fl.II,117
P. pinifolia, Meissner in Lehm. plant. Preiss. II, 220 (1847) ... W.A. — — — — — — B.fl.II,118
P. pedunculata, Hooker, Bot. Magaz. t. 2859 (1828) ... — S.A. T. V. N.S.W. — — B.fl.II,118
P. conferta, Bentham, Fl. Austr. II, 118 (1864) ... W.A. — — — — — — B.fl.II,118
P. aciphylla, Bentham in Hueg. enum. 35 (1837) ... W.A. — — — — — — B.fl.II,119
P. ochreata, Meissner in Lehm. plant. Preiss. I, 73 (1844) W.A. — — — — — — B.fl.II,120
P. aspalathoides, Meissner in Lehm. plant. Preiss. I, 73 (1844) W.A. — — — — — — B.fl.II,120
P. obcordata, Bentham, Fl. Austr. II, 120 (1864)... ... W.A. — — — — — — B.fl.II,121 M.fr.I,145.
P. rotundifolia, Bentham, Fl. Austr. II, 121 (1864) ... W.A. — — — — — — B.fl.II,121
P. calycina, Bentham, Fl. Austr. II, 121 (1864) ... W.A. — — — — — — B.fl.II,121
P. spinulosa, Bentham, Fl. Austr. II, 121 (1864) ... W.A. — — — — — — B.fl.II,122
P. tenella, Bentham, Fl. Austr. II, 122 (1864) ... — — — V. N.S.W. Q. — B.fl.II,122 M.fr.I,8;IV,10.
P. ternata, F. v. M., fragm. I, 6 (1858) ... — — — V. N.S.W. Q. — B.fl.II,122
P. styphelioides, Cunningham in G. Don, gen. syst. II, 124 (1832) — — — V. N.S.W. — — B.fl.II,123
P. altissima, F. v. M. in Bentham, Fl. Austr. II, 123 (1864) ... — — — V. N.S.W. — — B.fl.II,123
P. obovata, Bentham, Fl. Austr. II, 123 (1864) ... — — — — N.S.W. — — B.fl.II,123
P. incurvata, Cunningham in Field's N.S. Wales 346 (1825) ... — — — — N.S.W. — — B.fl.II,124
P. subumbellata, Hooker, Bot. Mag. t. 3254 (1833) ... — — T. V. N.S.W. — — B.fl.II,125
P. stipularis, Smith, Specim. of the Bot. of New Holl. 35, t. 12 (1793) — — — — N.S.W. — — B.fl.II,125
P. glabra, Bentham, Fl. Austr. II, 124 (1864) ... — — — — N.S.W. — — B.fl.II,125
P. dentata, Labillardière, Nov. Holl. pl. spec. I, 103, t. 131 (1804) — — T. V. N.S.W. — — B.fl.II,125
P. aristata, Sieber in De Cand. prodr. II, 112 (1825) ... — — — — N.S.W. — — B.fl.II,126
P. plumosa, Sieber in De Cand. prodr. II, 111 (1825) ... — — — — N.S.W. — — B.fl.II,126
P. viscosa, R. Brown in Benth. Fl. Austr. II, 127 (1864) ... — — — — N.S.W. — — B.fl.II,127
P. echinula, Sieber in De Cand. prodr. II, 112 (1825) ... — — — — N.S.W. Q. — B.fl.II,127
P. hibbertioides, J. Hooker, Fl. Tasm. I, 89 (1860) ... — — T. V. N.S.W. — — B.fl.II,127
P. rosea, F. v. M., fragm. II, 15 (1860) ... — — — V. — — — B.fl.II,128 M.fr.II,15.
P. mollis, Lindley in Mitch. Three Exped. II, 260 (1838) ... — S.A. — V. — — — B.fl.II,128
P. strobilifera, Meissner in Lehm. pl. Preiss. I, 75 (1844) W.A. — — — — — — B.fl.II,128
P. ericifolia, Bentham in Bot. Regist. XXV, App. XIII (1839) W.A. — — — — — — B.fl.II,129
P. verruculosa, Turcz. in Bull. Soc. Mosc. XXVI, 278 (1853) W.A. — — — — — -- B.fl.II,129
P. empetrifolia, Meissner in Lehm. plant. Preiss. I, 76 (1844) W.A. — — — — — — B.fl.II,129
P. adunca, Turczaninow in Bull. Soc. Mosc. XXVI, 279 (1853) ... W.A. — — — — — — B.fl.II,130
P. neurocalyx, Turcz. in Bull. Soc. Mosc. XXVI, 281 (1853) ... W.A. — — — — — — B.fl.II,130
P. rigida, R. Brown in Benth. Fl. Austr. II, 130 (1864) ... — S.A. — V. — — — B.fl.II,130
P. juniperina, Labillardière, Nov. Holl. pl. spec. I, 102, t. 130 (1804) — — T. V. — — — B.fl.II,131
P. acerosa, R. Brown in Benth. Fl. Austr. II, 131 (1864) ... — S.A. — V. — — — B.fl.II,131
P. humilis, Bentham in J. Hook. Fl. Tasm. I, 91 (1860)... ... — S.A. T. V. N.S.W. — — B.fl.II,131
P. parvifora, Sieber in De Caud. prodr. II, 111 (1825) ... — — — V. N.S.W. — — B.fl.II,132
P. setulosa, Bentham, Fl. Austr. II, 132 (1864) ... — — — — — Q. — B.fl.II,132
P. vestita, R. Brown in Ait. hort. Kew. sec. ed. III, 19 (1811)... W.A. S.A. — — — — — B.fl.II,132
P. procumbens, Cunningham in Field's N. S. Wales 347 (1838)... — — — — N.S.W. — — B.fl.II,133
P. hispidula, R. Brown in Bentham, Fl. Austr. II, 133 (1864) ... — S.A. — V. — — — B.fl.II,133
P. laxiflora, Bentham, Fl. Austr. I, 133 (1864) ... — S.A. — V. — — — B.fl.II,133
P. largiflorens, F. v. M. in Benth. Fl. Austr. II, 134 (1864) ... — S.A. — V. — — — B.fl.II,134

P. villosa, Willdenow, spec. plant. II, 507 (1799)... — — — V. N.S.W. Q. — B.fl.II,134
P. foliolosa, Cunningham in Ann. des Wien. Mus. II, 83 (1838) — — — V. N.S.W. — B.fl.II,135
P. flexilis, Smith in Koen. & Sims, Ann. of Bot. I, 502 (1805) ... — — — N.S.W. — B.fl.II,135
P. euchila, De Candolle, prodr. II, 112 (1825) ... — — — N.S.W. Q. — B.fl.II,135
P. selaginoides, J. Hooker, Fl. Tasm. I, 87 (1860) — — T. — — — B.fl.II,136
P. densifolia, F. v. M. in Transact. Vict. Inst. 119 (1854) ... S.A. — — — — B.fl.II,136
P. elliptica, Smith in Transact. Linn. Soc. IX, 246 (1808) — — — N.S.W. — B.fl.II,136
P. subspicata, Bentham, Fl. Austr. II, 137 (1864) ... — — — N.S.W. — B.fl.II,137
P. villifera, Sieber in De Cand. prodr. II, 111 (1825) ... S.A. — V. N.S.W. — B.fl.II,137
P. involucrata, Bentham, Fl. Austr. II, 138 (1864) ... S.A. — — — — B.fl.II,138
P. Muelleri, Bentham, Fl. Austr. II, 138 (1864) ... — — V. — — B.fl.II,138
P. prostrata, Bentham in J. Hook. Fl. Tasm. I, 89 (1860) — S.A. T. V. N.S.W. — B.fl.II,138
P. canaliculata, F. v. M. in Transact. Vict. Inst. 119 (1854) — S.A. — V. — — B.fl.II,139
P. fasciculata, Bentham in Ann. des Wien. Mus. II, 82 (1838) ... — T. V. N.S.W. — B.fl.II,139
P. tenuifolia, R. Brown in Bot. Mag. t. 2086 (1819) ... W.A. S.A. T. V. — — B.fl.II,130 M.fr.I,9.

LATROBEA, Meissner in Lehm. plant. Preiss. II, 219 (1847). (Leptocytisus.)
L. pungens, Bentham, Fl. Austr. II, 140 (1864) ... W.A. — — — — — B.fl.II,140 M.fr.II,106.
L. genistoides, Meissner in Lehm. plant. Preiss. II, 219 (1847)... W.A. — — — — — B.fl.II,141
L. Brunonis, Meissner in Lehm. plant. Preiss. II, 219 (1847) ... W.A. — — — — — B.fl.II,141
L. tenella, Bentham, Fl. Austr. II, 141 (1864) ... ... W.A. — — — — — B.fl.II,141
L. hirtella, Bentham, Fl. Austr. II, 142 (1864) ... ... W.A. — — — — — B.fl.II,142
L. diosmifolia, Bentham, Fl. Austr. II, 142 (1864) ... W.A. — — — — — B.fl.II,142

EUTAXIA, R. Brown in Ait. hort. Kew. sec. ed. III, 16 (1811). (Sclerothamnus.)
E. cuneata, Meissner in Lehm. plant. Preiss. I, 65 (1844) ... W.A. — — — — — B.fl.II,143
E. myrtifolia, R. Brown in Ait. hort. Kew. sec. ed. III, 16 (1811) W.A. — — — — — B.fl.II,144
E. epacridoides, Meissner in Lehm. plant. Preiss. I, 64 (1844) ... W.A. — — — — — B.fl.II,144
E. virgata, Bentham in Hueg. enum. 34 (1837) ...... W.A. — — — — — B.fl.II,144
E. densifolia, Turczaninow in Bull. Soc. Mosc. XXVI, 271 (1853) W.A. — — — — — B.fl.II,145
E. dillwynioides, Meissner in Lehm. plant. Preiss. I, 63 (1844) W.A. — — — — — B.fl.II,145
E. parvifolia, Bentham in Hueg. enum. 34 (1837)... ... W.A. — — — — — B.fl.II,145
E. empetrifolia, Schlechtendal, Linnaea XX, 607 (1847) ... W.A. S.A. — V. N.S.W. — B.fl.II,145 M.fr.I,7.

DILLWYNIA, Smith in Koenig & Sims, Ann. of Bot. I, 510 (1805).
D. hispida, Lindley in Mitch. Three Exped. II, 251 (1838) ... — S.A. — V. — — B.fl.II,147
D. ericifolia, Smith in Koen. & Sims, Ann. of Bot. I, 510 (1805) — S.A. T. V. N.S.W. Q. — B.fl.II,147
D. floribunda, Smith in Koen. & Sims, Ann. of Bot. I, 510 (1805) — S.A. T. V. N.S.W. Q. — B.fl.II,149
D. Preissii, Bentham, Fl. Austr. II, 149 (1864) ... ... W.A. — — — — — B.fl.II,150
D. brunioides, Meissner in Lehm. pl. Preiss. I, 62 (1844) ... — — — N.S.W. — B.fl.II,150
D. juniperina, Sieber in Hueg. enum. pl. N. Holl. occ. 33 (1837) — — V. N.S.W. Q. — B.fl.II,150
D. pungens, Mackay in den Annal. des Wien. Mus. II, 79 (1838) W.A. — — — — — B.fl.II,150
D. cinerascens, R. Brown in Bot. Mag. t. 2247 (1821) ... — S.A. T. V. N.S.W. — B.fl.II,151
D. divaricata, Bentham, Fl. Austr. II, 150 (1864) ... W.A. — — — — — B.fl.II,151
D. patula, F. v. M., fragm. IV, 10 (1864) ... ... W.A. S.A. — V. N.S.W. — B.fl.II,151 M.fr.IV,10.

EUCHILOPSIS, F. v. M. in Melbourne Chemist, June (1882).
E. linearis, F. v. M. in Melb. Chem., June (1882) ... W.A. — — — — — B.fl.II,67

PLATYLOBIUM, Smith in Transact. Linn. Soc. II, 350 (1794). (Cheilococca.)
P. formosum, Smith in Transact. Linn. Soc. II, 350 (1794) ... — T. V. N.S.W. Q. — B.fl.II,153 M.fr.IX,137.
P. obtusangulum, Hooker, Bot. Mag. t. 3258 (1633) ... — S.A. T. V. — — B.fl.II,153
P. triangulare, R. Brown in Ait. hort. Kew. sec. ed. III, 266 (1811) — S.A. T. V. — — B.fl.II,152 M.fr.IX,158.

BOSSIAEA, Ventenat, Jardin de Cels I, 7, t. 7 (1800). (Lalage, Scottia.)
B. dentata, Bentham, Fl. Austr. II, 156 (1864) ... W.A. — — — — — B.fl.II,156
B. Aquifolium, Bentham, Fl. Austr. II, 157 (1864) ... W.A. — — — — — B.fl.II,157
B. strigillosa, Bentham, Fl. Austr. II, 157 (1864)... ... W.A. — — — — — B.fl.II,157
B. cordigera, Bentham in J. Hook. Fl. Tasm. I, 95, t. 16 (1860)... — T. V. — — — B.fl.II,157 M.fr.III,100;IX,45.
B. lenticularis, Sieber in De Cand. prodr. II, 117 (1825) — — — N.S.W. — B.fl.II,157
B. Kiamensis, Bentham, Fl. Austr. II, 158 (1864) — — — N.S.W. — B.fl.II,158
B. ornata, F. v. M., fragm. IV, 13 (1863) ... ... W.A. — — — — — B.fl.II,158 M.fr.IX,45.
B. eriocarpa, Bentham in Hueg. enum. 36 (1837) ... ... W.A. — — — — — B.fl.II,159 M.fr.IX,45.
B. biloba, Bentham in Hueg. enum. 36 (1837) ... ... W.A. — — — — — B.fl.II,160 M.fr.IX,45.
B. Preissii, Meissner in Lehm. pl. Preiss. I, 82 (1844) ... W.A. — — — — — B.fl.II,161 M.fr.IX,45.
B. concinna, Bentham, Fl. Austr. II, 161 (1864) ... ... W.A. — — — — — B.fl.II,161
B. carinalis, Bentham in Mitch. Trop. Austr. 260 (1848) ... — — — — Q. — B.fl.II,162 M.fr.IX,45.
B. rupicola, Cunningham in Benth. Fl. Austr. II, 162 (1864) ... — — — — Q. — B.fl.II,162
B. linophylla, R. Brown in Ait. hort. Kew. sec. ed. IV, 268 (1812) W.A. — — — — — B.fl.II,162 M.fr.IX,45.
B. disticha, Lindley, Bot. Regist. t. 55 (1841) ... W.A. — — — — — B.fl.II,162
B. prostrata, R. Brown in Ait. hort. Kew. sec. ed. IV, 268 (1812) — S.A. T. V. N.S.W. Q. — B.fl.II,162 M.fr.III,97.
B. Neo-Anglica, F. v. M., fragm. V, 106 (1866) ... ... — — — N.S.W. — — M.fr.V,106.
B. buxifolia, Cunningham in Field's N. S. Wales, 348 (1825) ... — — — N.S.W. — B.fl.II,163 M.fr.I,9;IX,45.
B. Brownii, Bentham, Fl. Austr. II, 163 (1864) ... ... — — — — Q. — B.fl.II,163
B. rhombifolia, Sieber in De Cand., prodr. II, 117 (1825) ... — — — N.S.W. — B.fl.II,164 M.fr.IX,45.
B. pulchella, Meissner in Lehm. pl. Preiss. I, 84 (1844) W.A. — — — — — B.fl.II,164
B. microphylla, Smith in Transact. Linn. Soc. IX, 303 (1808) ... — — — N.S.W. — B.fl.II,164 M.fr.III,99.
B. pedunularis, Turczaninow in Bull. Soc. Mosc. XXVI, 287 (1853) W.A. — — — — — B.fl.II,165
B. heterophylla, Ventenat, Jardin de Cels t. 7 (1800) ... — — V. N.S.W. Q. — B.fl.II,165 M.fr.III,97.
B. rufa, R. Brown in Ait. hort. Kew. sec. ed. III, 267 (1812) ... W.A. — — — — — B.fl.II,165 M.fr.IX,45.

B. bracteosa, F. v. M. in Benth. Fl. Austr. II, 166 (1864) ... — — — V. — — — B.fl.II,166
B. riparia, Cunningham in Benth. Fl. Austr. II, 166 (1864) ... — S.A. T. V. N.S.W. — — B.fl.II,166
B. ensata, Sieber in De Cand. prodr. II, 117 (1825) ... ... — — — V. N.S.W. Q. — B.fl.II,167   M.fr.XI,27&123.
B. scolopendria, Smith in Transact. Linn. Soc. IX, 307 (1808) ... — — — — N.S.W. — — B.fl.II,167   M.fr.III,95.
B. Walkeri, F. v. M., fragm. II, 120 (1861) ... ... — S.A. — — N.S.W. — — B.fl.II,167   M.fr.II,120;IX,45;XI,27.
B. Armitii, F. v. M., fragm. IX, 44 (1875)... ... ... — — — — — — Q. — M.fr.IX,44.
B. phylloclada, F. v. M. in Transact. phil. Inst. Vict. III, 52 (1859) — — — — — — — N.A. B.fl.II,168

TEMPLETONIA, R. Brown in Ait. hort. Kew. sec. ed. 260 (1812).
T. retusa, R. Brown in Ait. hort. Kew. sec. ed. IV, 269 (1872)... W.A. S.A. — — — — — B.fl.II,169   M.fr.IX,158.
T. Drummondii, Bentham, Fl. Austr. II, 169 (1864) ... W.A. — — — — — — B.fl.II,169
T. Muelleri, Bentham, Fl. Austr. II, 169 (1864) ... ... — — — V. N.S.W. Q. — B.fl.II,169   M.fr.I,0;II,178;IX,158.
T. aculeata, Bentham. Fl. Austr. II, 170 (1864) ... ... W.A. — — — — — — B.fl.II,170   M.fr.II,120.
T. egena, Bentham, Fl. Austr. II, 170 (1864) ... ... W.A. S.A. — V. N.S.W. — N.A. B.fl.II,170   M.fr.III,04;IX,158.
T. sulcata, Bentham, Fl. Austr. II, 171 (1864) ... W.A. S.A. — V. N.S.W. — — B.fl.II,171   M.fr.III,04.

HOVEA, R. Brown in Ait. hort. Kew. sec. ed. IV, 275 (1812).   (Poiretia, Plagiolobium, Platychilum.)
H. linearis, R. Brown in Ait. hort. Kew. sec. ed. IV, 275 (1812) — — — N.S.W. — — B.fl.II,172
H. heterophylla, Cunningh. in J. Hook. Fl. Tasm. I, 93, t. 15 (1860) — S.A. T. V. N.S.W. Q. — B.fl.II,172   M.fr.IX,158.
H. longifolia, R. Brown in Ait. hort. Kew. sec. ed. IV, 275 (1812) — S.A. T. V. N.S.W. Q. N.A. B.fl.II,172   M.fr.IX,158.
H. acutifolia, Cunningham in G. Don, gen. syst. II, 126 (1832)... — — — — N.S.W. Q. — B.fl.II,174   M.fr.IX,158.
H. longipes, Bentham in Hueg. enum. 37 (1837) ... ... — — — — N.S.W. Q. — B.fl.II,174   M.fr.IX,158.
H. acanthoclada, F. v. M., fragm. IV, 15 (1864) ... W.A. — — — — — — B.fl.II,174   M.fr.IV,15.
H. chorizemifolia, De Candolle, prodr. II, 116 (1825) ... W.A. — — — — — — B.fl.II,174   M.fr.IX,158.
H. Celsi, Boupland, plant. cult. à Malmais. t. 51 (1813) ... W.A. — — — — — — B.fl.II,175   M.fr.IX,158.
H. trisperma, Bentham in Hueg. enum. 37 (1837) ... ... W.A. — — — — — — B.fl.II,176   M.fr.IX,158.
H. stricta, Meissner in Lehm. pl. Preiss. I, 79 (1844) ... W.A. — — — — — — B.fl.II,176   M.fr.IX,158.
H. pungens, Bentham in Hueg. enum. 37 (1837) ... ... W.A. — — — — — — B.fl.II,176

NEMATOPHYLLUM, F. v. M. in Hook. Kew miscell. IX, 20 (1857).
N. Hookeri, F. v. M. in Hook. Kew misc. IX, 20 (1857) ... ... — — — — — N.A. B.fl.II,170   M.fr.I,169;IV,170;IX,157

GOODIA, Salisbury, paradis. Londin. t. 41 (1806).
G. lotifolia, Salisbury, parad. Lond. t. 41 (1806) ... W.A. S.A. T. V. N.S.W. Q. — B.fl.II,177   M.fr.I,10;IX,157.
G. medicaginea, F. v. M., Second Gen. Rep. 11 (1854) ... W.A. S.A. — V. — — — M.fr.I.10;IX,157.

CROTALARIA, Hermann, hort. acad. Lugd. Bat. 196—202, t. 197—203 (1687).   (Pentadynamis.)
C. verrucosa, Linné, spec. plant. 715 (1753) ... ... — — — — — Q. N.A. B.fl.II,179   M.fr.III,54;IX,156.
C. crispata, F. v. M. in Benth. Fl. Austr. II, 179 (1864) ... — — — — — — N.A. B.fl.II,179   M.fr.III,55.
C. juncea, Linné, spec. plant. 714 (1753) ... ... — — — — — Q. N.A. B.fl.II,179   M.fr.III,51;IX,156.
C. linifolia, Linné, fil. suppl. 322 (1781) ... ... — — — — N.S.W. Q. N.A. B.fl.II,180   M.fr.III,55;IX,156.
C. humifusa, Graham in Wall. numeric. list 5421 (1828) ... — — — — — — Q. — M.fr.VI,225;IX,156.
C. calycina, Schranck, pl. rar. hort. Monac. t. 12 (1819) — — — — — Q. N.A. B.fl.II,180   M.fr.IX,156.
C. retusa, Linné, spec. plant. 715 (1753) ... ... — — — — N.S.W. Q. N.A. B.fl.II,181   M.fr.III,51;IX,156.
C. Mitchelli, Bentham in Mitch. Trop. Austr. 120 (1848) — — — — N.S.W. Q. — B.fl.II,181   M.fr.IX,156.
C. Novae-Hollandiae, De Candolle, prodr. II, 127 (1825) ... — — — — — Q. N.A. B.fl.II,181   M.fr.IX,156.
C. crassipes, Hooker, icon. plant. t. 830 (1852) ... ... — — — — — N.A. B.fl.II,182
C. Cunninghamii, R. Brown, App. to Sturt's Centr. Austr. 8 (1849) W.A. S.A. — — — — Q. N.A. B.fl.II,182   M.fr.III,52;IX,156.
C. medicaginea, Lamarck, encycl. meth. II, 201 (1786) ... ... — — — — — Q. N.A. B.fl.II,183   M.fr.III,56;IX,157.
C. incana, Linné, spec. plant. 716 (1753) ... ... — — — — N.S.W. — — B.fl.II,183   M.fr.IX,157.
C. dissitiflora, Bentham in Mitch. Trop. Austr. 386 (1848) — S.A. — — N.S.W. Q. N.A. B.fl.II,184   M.fr.IX,156.
C. laburnifolia, Linné, spec. plant. 715 (1753) ... ... — — — — — Q. — B.fl.II,184   M.fr.III,52;IX,157.
C. quinquefolia, Linné, spec. plant. 716 (1753) ... ... — — — — — Q. N.A. B.fl.II,184   M.fr.IX,157.
C. alata, Hamilton in D. Don, prodr. Fl. Nepal. 241 (1825) ... — — — — — — N.A. — M.fr.XI,100

WESTONIA, Sprengel, syst. veg. III, 152 (1826).
W. trifoliata, Sprengel, syst. veg. III, 230 (1826) ... ... — — — — — — N.A. B.fl.II,185

TRIGONELLA, Linné, gen. pl. 351 (1737).
T. suavissima, Lindley in Mitch. Three Exped. I, 255 (1839) ... W.A. S.A. — V. N.S.W. — — B.fl.II,187

LOTUS, Tournefort, inst. 402, t. 227 (1700), from C. Bauhin (1623).
L. corniculatus, Linné, spec. plant. 775 (1753) ... — — S.A. T. V. N.S.W. — — B.fl.II,188
L. australis, Andrews, Botanist's Reposit. t. 624 (1810) ... W.A. S.A. T. V. N.S.W. Q. N.A. B.fl.II,188   M.fr.XI,20&27.

PSORALEA, Linné, gen. ed. sec. 358 (1742).
P. badocana, Bentham, Fl. Austr. II, 190 (1864) ... ... — — — — — Q. N.A. B.fl.II,190   M.fr.IX,156.
P. cephalantha, F. v. M., fragm. IV, 35 (1863) ... ... — — — — — Q. — M.fr.IV,35.
P. Martini, F. v. M., fragm. V, 11 (1865) ... ... — — — — — — N.A. — M.fr.V,11.
P. Archeri, F. v. M., fragm. IV, 21 (1865) ... ... — — — — — — N.A. B.fl.II,190   M.fr.IV,21.
P. balsamica, F. v. M., fragm. III, 55 (1858) ... In Transact. Vict. Inst. III, 55 (1858) ... — — — — — N.A. B.fl.II,191
P. plumosa, F. v. M., fragm. IV, 22 (1864) ... ... — — — — — — N.A. B.fl.II,191   M.fr.IV,22.
P. pustulata, F. v. M. in Transact. Vict. Inst. III, 54 (1858) ... — — — — — — N.A. B.fl.II,191
P. leuchnotachys, F. v. M., fragm. III, 105 (1863) ... ... — — — — — — N.A. B.fl.II,191   M.fr.III,105.
P. eriantha, Bentham in Transact. Vict. Inst. I, 40 (1854)... W.A. S.A. — V. N.S.W. Q. — B.fl.II,192
P. patens, Lindley in Mitch. Three Exped. II, 9 (1838)... W.A. — — V. N.S.W. Q. N.A. B.fl.II,192   M.fr.IX,155.
P. cinerea, Lindley in Mitch. Three Exped. II, 85 (1839) ... W.A. — — — — Q. N.A. B.fl.II,192
P. leucantha, F. v. M. in Transact. Vict. Inst. I, 40 (1854) ... S.A. — — V. N.S.W. Q. N.A. B.fl.II,193   M.fr.IX,155.
P. parva, F. v. M. in Transact. Vict. Inst. I, 40 (1854) ... — — — — — Q. N.A. B.fl.II,193   M.fr.IX,156.
P. tenax, Lindley in Mitch. Three Exped. II, 10 (1838)... — S.A. — — N.S.W. Q. N.A. B.fl.II,194   M.fr.IX,155.
P. adscendens, F. v. M. in Transact. Vict. Inst. I, 40 (1854) ... — — — — N.S.W. — N.A. B.fl.II,193   M.fr.IX,155.
P. Schultzii, F. v. M., fragm. IX, 155 (1875) ... ... — S.A. T. V. N.S.W. — — B.fl.II,193   M.fr.IX,155.
P. Testariae, F. v. M., fragm. V, 45 (1865) ... ... — — — — — Q. — M.fr.V,45,189

INDIGOFERA, Royen, pl. hort. Lugd. 372 (1740), from Linné (1737).
I. llulfolia, Retzius, observ. bot. IV, 29 (1786) ... ... — S.A. — — N.S.W. Q. N.A. B.fl.II,195 M.fr.III,101;IX,44.
I. cordifolia, Heyne in Roth, nov. pl. spec. 357 (1821) ... ... — — — — — N.A. B.fl.II,196
I. cuneaphylla, Linné, mantissa altera 272 (1771) .. ... — S.A. — — N.S.W. Q. N.A. B.fl.II,196 M.fr.III,102;IX,44. [44.
I. glandulosa, Willdenow, spec. plant. III, 1227 (1800) ... — — — — — Q. — B.fl.II,196 M.fr.IV,22;VII,106;IX,
I. haplophylla, F. v. M., fragm. III, 102 (1863) ... ... — — — — — — N.A. B.fl.II,196 M.fr.IX,44.
I. trifoliata, Linné, amoen. acad. IV, 327 (1750) ... ... — — — — — Q. — B.fl.II,197 M.fr.III,104.
I. trita, Linné fil., suppl. plant. 335 (1781) ... ... — — — — N.S.W. Q. N.A. B.fl.II,197 M.fr.III,103.
I. parviflora, Heyne in Wall. numeric. list 5457 (1828) .. ... — — — — — Q. N.A. B.fl.II,197 M.fr.III,103;IX,44.
I. viscosa, Lamarck, Encycl. meth. III, 247 (1789) ... — S.A. — — — Q. N.A. B.fl.II,198 M.fr.III,104;IX,44.
I. hirsuta, Linné, spec. plant. 751 (1753) ... ... — S.A. — — N.S.W. Q. N.A. B.fl.II,198 M.fr.III,105;IX,44.
I. pratensis, F. v. M., essay on plants of Burdek. Exped. 10 (1800) — — — — — Q. — B.fl.II,198
I. Schultziana, F. v. M., fragm. VII, 105 (1870) ... ... — — — — — — N.A. M.fr.VII,105. [44.
I. monophylla, De Candolle, prodr. II, 222 (1825) ... ... — S.A. — — — — N.A. B.fl.II,199 M.fr.III,44;VII,106;IX,
I. rugosa, Bentham, Fl. Austr. II, 199 (1864) ... ... — — — — — — N.A. B.fl.II,199
I. saxicola, F. v. M. in Benth. Fl. Austr. II, 199 (1864)... ... — — — — — — N.A. B.fl.II,199
I. australis, Willdenow, spec. plant. III, 1235 (1800) ... W.A. S.A. T. V. N.S.W. Q. — B.fl.II,199
I. brevidens, Bentham in Mitch. Trop. Austr. 385 (1848) ... W.A. S.A. — — N.S.W. Q. N.A. B.fl.II,200
I. Baileyi, F. v. M., fragm. IX, 43 (1875)... ... ... — — — — — Q. — — M.fr.IX,43.
I. coronillifolia, Cunningham in Benth. Fl. Austr. II, 201 (1864) — S.A. — — N.S.W. — — B.fl.II,201

PTYCHOSEMA, Bentham in Lindl. Bot. Regist. XXV, App. XVI (1839).
P. pusillum, Bentham in Lindl. Bot. Regist. XXV, App. XVI (1839) W.A. — — — — — — B.fl.II,201
P. anomalum, F. v. M., fragm. IX, 62 (1875) ... ... — S.A. — — — — — M.fr.IX,62.
P. trifoliolatum, F. v. M. in Wing's S. Science Record II, 72 (1892) W.A. — — — — — ... — ..

LAMPROLOBIUM, Bentham, Fl. Austr. II, 202 (1864).
L. fruticosum, Bentham, Fl. Austr. II, 202 (1864) ... ... — — — — — Q. — B.fl.II,202 M.fr.IX,68.
L. megalophyllum, F. v. M., fragm. IX, 67 (1875) ... ... — — — — — — N.A. — M.fr.IX,67.

TEPHROSIA, Persoon, synops. pl. II, 328 (1807).
T. coriacea, Bentham, Fl. Austr. II, 204 (1864) ... ... — — — — — — N.A. B.fl.II,204 M.fr.IX,65&66.
T. lamprolobioides, F. v. M., fragm. IX, 64 (1875) ... ... — — — — — — N.A. — M.fr.IX,64.
T. Forrestiana, F. v. M., fragm. XI, 98 (1880) ... ... — — — — — — N.A. — M.fr.XI,98.
T. flammea, F. v. M. in Benth. Fl. Austr. II, 204 (1864) ... — — — — — Q. — B.fl.II,204
T. reticulata, Bentham, Fl. Austr. II, 205 (1864) ... ... — — — — — Q. N.A. B.fl.II,205 M.fr.IX,64.
T. crocea, Bentham, Fl. Austr. II, 205 (1864) ... ... — — — — — — N.A. B.fl.II,205
T. oblongata, Bentham, Fl. Austr. II, 205 (1864) ... ... — — — — — — N.A. B.fl.II,205
T. porrecta, Bentham, Fl. Austr. II, 206 (1864) ... ... — — — — — — N.A. B.fl.II,206
T. polyzyga, F. v. M. in Benth. Fl. Austr. II, 206 (1864) ... — — — — — — N.A. B.fl.II,206 M.fr.IX,66.
T. grammifolia, F. v. M. in Benth. Fl. Austr. II, 206 (1864) ... — — — — — — N.A. B.fl.II,206
T. simplicifolia, F. v. M. in Benth. Fl. Austr. II, 206 (1864) ... — — — — — — N.A. B.fl.II,206
T. uematophylla, F. v. M., fragm. IX, 63 (1879) ... ... — — — — — — N.A. — M.fr.IX,63.
T. leptoclada, Bentham, Fl. Austr. II, 207 (1864)... ... ... — — — — — Q. N.A. B.fl.II,207
T. brachycarpa, F. v. M. in Benth. Fl. Austr. II, 207 (1864) ... — — — — — — N.A. B.fl.II,207
T. uniovulata, F. v. M., fragm. XI, 70 (1879) ... ... — — — — — — N.A. — M.fr.XI,70.
T. Stuartii, Bentham, Fl. Austr. II, 207 (1864) ... ... — — — — — — N.A. B.fl.II,207
T. conferta, F. v. M., fragm. IX, 65 (1879) ... ... — — — — — — N.A. — M.fr.IX,75.
T. eriocarpa, Bentham, Fl. Austr. II, 207 (1864) ... ... — — — — — — N.A. B.fl.II,207
T. phaeosperma, F. v. M. in Benth. Fl. Austr. II, 208 (1864) ... — — — — — — N.A. B.fl.II,208
T. astragaloides, Bentham, Fl. Austr. II, 208 (1864) ... ... — — — — — Q. — B.fl.II,208
T. juncea, Bentham, Fl. Austr. II, 208 (1864) ... ... — — — — — — N.A. B.fl.II,208 M.fr.IX,64.
T. filipes, Bentham, Fl. Austr. II, 208 (1864) ... ... — — — — — Q. N.A. B.fl.II,208
T. remotiflora, F. v. M. in Benth. Fl. Austr. II, 209 (1864) ... — — — — — — N.A. B.fl.II,209
T. oligophyllum, Bentham, Fl. Austr. II, 209 (1864) ... ... — — — — — — N.A. B.fl.II,209
T. macrocarpa, Bentham, Fl. Austr. II, 209 (1864) ... ... — — — — — — N.A. B.fl.II,209
T. purpurea, Persoon, synops. plant. II, 329 (1807) ... — S.A. — — N.S.W. Q. N.A. B.fl.II,209
T. Bidwilli, Bentham, Fl. Austr. II, 210 (1864) ... ... — — — — N.S.W. Q. — B.fl.II,210
T. rosea, F. v. M. in Benth. Fl. Austr. II, 211 (1864) ... — — — — — — N.A. B.fl.II,211

WISTARIA, Nuttall, Gen. of N. Amer. pl. II, 115 (1818). (Milletia.)
W. megasperma, F. v. M., fragm. I, ad hist. natur. 10 (1859) ... — — — — N.S.W. Q. — B.fl.II,211 M.fr.I,10.

SESBANIA, Scopoli, Introd. 308 (1777). (Agati.)
S. grandiflora, Persoon, synops. pl. II, 310 (1807) ... ... — — — — — N.A. B.fl.II,212 M.fr.II,88,XI,33.
S. Aegyptiaca, Persoon, synops. pl. II, 316 (1807) ... ... — — — — — — N.A. B.fl.II,212 M.fr.XI,33.
S. aculeata, Persoon, synops. pl. II, 310 (1807) ... ... — S.A. — — N.S.W. Q. N.A. B.fl.II,213 M.fr.XI,33.
S. simpliciuscula, F. v. M. in Benth. Fl. Austr. II, 213 (1864)... — — — — — — N.A. B.fl.II,213
S. brachycarpa, F. v. M., fragm. XI, 32 (1878) ... ... — — — — — — N.A. M.fr.IX,32.

CARMICHAELLA, R. Brown in Bot. Regist. XI, 912 (1825).
C. exul, F. v. M., fragm. VII, 126 (1871)... ... ... — — N.S.W. — — — M.fr.VII,126,193.

CLIANTHUS, Banks & Solander in G. Don, dichlam. pl. II, 468; indicative (1832). (Donia.)
C. Dampieri, Cunning. in Trans. hort. soc. Lond. sec. ser. I, 522 (1835) W.A. S.A. — — N.S.W. — N.A. B.fl.II,214

STREBLORRHIZA, Endlicher, prodr. fl. Norfolk. 92 (1833).
S. speciosa, Endlicher, prodr. fl. Norf. 93 (1833) ... ... — — N.S.W. — — — M.fr.IX,109.

SWAINSONIA, Salisbury, parad. Londin. t. 28 (1806). (Cyclogyne, Diplolobium.)
S. Greyana, Lindley, Bot. Regist. t. 66 (1846) ... ... — S.A. — — N.S.W. Q. — B.fl.II,216 M.fr.IX,154;X,7.
S. galegifolia, R. Brown in Ait. hort. Kew. sec. ed. III, 327 (1811) — — — N.S.W. Q. N.A. B.fl.II,216
S. coronillifolia, Salisbury, paradis. Lond. t. 28 (1806) ... — — N.S.W. — — B.fl.II,217 M.fr.XI,70.
S. brachycarpa, Bentham, Fl. Austr. II, 217 (1864) ... — — — N.S.W. Q. — B.fl.II,217 M.fr.IX,155.
S. phacoides, Bentham in Mitch. Trop. Austr. 363 (1848) ... W.A. S.A. — V. N.S.W. Q. N.A. B.fl.II,217
S. Burkittii, F. v. M. in Benth. Fl. Austr. II, 218 (1864) ... — S.A. — — N.S.W. — — B.fl.II,218

S. Burkei, F. v. M. in Benth. Fl. Austr. II, 218 (1864) ..	— S.A. — — —	— N.A.	B.fl.II,218	M.fr.IX,155.
S. oligophylla, F. v. M. in Benth. Fl. Austr. II, 219 (1864)	— S.A. — — N.S.W.	— N.A.	B.fl.II,219	
S. campylantha, F. v. M. in Gregory's pap. rel. to Leichh. Exp.6(1858)	— S.A. — — N.S.W. Q.	—	R.fl.II,219	M.fr.IX,154.
S. occidentalis, F. v. M., fragm. III, 46 (1862)	W.A. — — — —	— N.A.	B.fl.II,219	M.fr.III,46.
S. stenodonta, F. v. M., fragm. XI, 70 (1879)	— — — — —	— N.A.		M.fr.XI,70.
S. gracilis, Bentham, Fl. Austr. II, 220 (1864)	W.A. — — — —	— —	B.fl.II,220	
S. plagiotropis, F. v. M., fragm. IX, 153 (1875)	— — — V. N.S.W.	— —		M.fr.IX,143.
S. procumbens, F. v. M., fragm. III, 46 (1863)	— S.A. — V. N.S.W. Q.	—	R.fl.II,220	M.fr.IX,154;XI,27.
S. Drummondii, Bentham, Fl. Austr. II, 220 (1864)	W.A. — — — —	—	B.fl.II,220	
S. canescens, F. v. M., fragm. III, 46 (1863)	W.A. — — — —	—	B.fl.II,221	
S. phacifolia, F. v. M. in South Austr. Register (1850)	— S.A. — V. N.S.W. —	—	D.fl.II,221	M.fr.IX,154.
S. oroboides, F. v. M. in Benth. Fl. Austr. II, 222 (1864)	— S.A. — — N.S.W. Q.	—	B.fl.II,222	
S. unifoliolata, F. v. M., fragm. VIII, 226 (1874)	W.A. S.A. — — —	—	—	M.fr.VIII,226.
S. lessertiifolia, De Candolle, prodr. II, 271 (1825)	— S.A. T. V. N.S.W. —	—	B.fl.II,222	M.fr.IX,154.
S. luteola, F. v. M., fragm. I, 75 (1858)	— — — — — Q.	—	B.fl.II,223	M.fr.I,75;XI,70.
S. parviflora, Bentham, Fl. Austr. II, 223 (1864) ..	— — — — — Q.	—	B.fl.II,223	
S. monticola, Cunningham in A. Gray's Bot. Wilk. Exp. I, 411(1854)	— — — — N.S.W. Q.	—	B.fl.II,223	M.fr.IX,155.
S. microphylla, A. Gray, Bot. Wilk. Exped. I, 410 (1854)	— S.A. — V. N.S.W. Q.	—	B.fl.II,223	M.fr.IX,155.
S. laxa, R. Brown in Sturt's Centr. Austr. App. 18 (1849)	— S.A. — V. N.S.W. —	—	B.fl.II,224	
S. colutoides, F. v. M., fragm. X, 6 (1876)...	— S.A. — — — —	—	—	M.fr.X,6.
S. Fraseri, Bentham, Fl. Austr. II, 224 (1864)	— — — — N.S.W. Q.	—	B.fl.II,224	M.fr.IX,154.
S. Macullochiana, F. v. M., fragm. VII, 25 (1869)	W.A. — — — —	— N.A.		M.fr.IX,154.

GLYCYRRHIZA, Tournefort, inst. 389, t. 210 (1700), from Dioscorides and Plinius. (Clidanthera.)
G. psoraleoides, Bentham, Fl. Austr. II, 225 (1864)	— S.A. — V. N.S.W. —	—	B.fl.II,225	M.fr.III,45.

ORMOCARPUM, Palisot, Fl. d'Ovar. I, 95, t. 53 (1805).
O. sennoides, De Candolle, prodr. II, 315 (1825)	— — — — — — Q.	—	B.fl.II,226	

AESCHYMOMENE, Linné, gen. pl. 350 (1737).
A. Indica, Linné, spec. plant. 713 (1753)	W.A. — — — — Q. N.A.	B.fl.II,227	
A. falcata, De Candolle, prodr. II, 322 (1825)	— — — — — — Q.	—	B.fl.II,227
A. Americana, Linné, spec. plant. 713 (1753)	— — — — — — Q.	—	

SMITHIA, Aiton, hort. Kew. III, 496 (1789).
S. conferta, Smith in Rees, Cyclop. XXXIII (1816)	— — — — — — Q.	—	B.fl.II,228	
S. sensitiva, Aiton, hort. Kew. III, 496, t. 13 (1769)	— — — — — — Q.	—		M.fr.VII,27;IX,193.

ZORNIA, Gmelin, Syst. Nat. 1076 (1791).
Z. diphylla, Persoon, synops. II, 318 (1807)	— — — — N.S.W. Q. N.A.	B.fl.II,228	
Z. chaetophora, F. v. M. in Transact. phil. Inst. Vict. III, 56 (1857)	— — — — — — N.A.	B.fl.II,229	

DESMODIUM, Desvaux, Journ. de Bot. III, 122 (1813). (Dicerma, Phyllodium, Pteroloma.)
D. umbellatum, De Candolle, prodr. II, 325 (1825)	— — — — — — Q.	B.fl.II,230	M.fr.IX,66.
D. pulchellum, Bentham, Fl. Hongk. 83 (1861)	— — — — — — N.A.	B.fl.II,231	
D. biarticulatum, F. v. M., fragm. II, 121 (1861)	— — — — — — Q. N.A.	B.fl.II,231	M.fr.IX,66.
D. acanthocladum, F. v. M., fragm. II, 122 (1861)	— — — — V. N.S.W. Q.	—	B.fl.II,231	M.fr.IX,66.
D. triquetrum, De Candolle, prodr. II, 326 (1825)	— — — — — — Q.	—	
D. Gangeticum, De Candolle, prodr. II, 327 (1825)	— — — — — — Q. N.A.	B.fl.II,232	
D. brachypodum, A. Gray, Bot. Wilk. Exped. 434 (1854)	— — — — — N.S.W. Q.	—	B.fl.II,232	
D. varians, Endlicher in Ann. des Wien. Mus. II, 185 (1839)	— — — — T. V. N.S.W. Q.	—	B.fl.II,232	
D. flagellare, Bentham, Fl. Austr. II, 233 (1864)	— — — — — — N.A.	B.fl.II,233	
D. rhytidophyllum, F. v. M. in Benth. Fl. Austr. II, 233 (1864)	— — — — N.S.W. Q.	—	B.fl.II,233	
D. campylocaulon, F. v. M. in Benth. Fl. Austr. II, 233 (1864)	— — — — — — N.A.	B.fl.II,233	
D. nemorosum, F. v. M. in Benth. Fl. Austr. II, 234 (1864)	— — — — N.S.W. Q.	—	B.fl.II,234	
D. neurocarpum, Bentham, Fl. Austr. II, 234 (1864)	— — — — — — N.A.	B.fl.II,234	
D. trichostachyum, Bentham, Fl. Austr. II, 234 (1864)	— — — — — — N.A.	B.fl.II,234	
D. polycarpum, De Candolle, prodr. II, 334 (1825)	— — — — N.S.W. Q. N.A.	B.fl.II,235	M.fr.IX,66.
D. trichocaulon, De Candolle, prodr. II, 335 (1825)	— — — — N.S.W. Q.	—	B.fl.II,235	
D. Muelleri, Bentham, Fl. Austr. II, 335 (1864) ...	— — — — — — Q. N.A.	B.fl.II,235	
D. reniforme, De Candolle, prodr. II, 327 (1825) ...	— — — — — — N.A.	—	M.fr.XII.
D. parvifolium, De Candolle, prodr. II, 334 (1825)	— — — — N.S.W. Q. N.A.	B.fl.II,235	M.fr.IX,66. [X,120.
D. dependens, Blume in Miq. Fl. Ind. Batav. I, 249 (1855)	— — — — — — N.A.	—	M.fr.VIII,225;IX,66&195

PYCNOSPORA, R. Brown in Wight & Arnott, prodr. I, 197 (1834).
P. hedysaroides, R. Brown in Wight & Arn., prodr. 197 (1834)	— — — — — — Q. N.A.	B.fl.II,236	

URARIA, Desvaux, Journ. de bot. III, 122 (1813).
U. picta, Desvaux, Journ. de Bot. III, 122 (1813)	— — — — N.S.W. Q.	—	B fl.II,237
U. cylindracea, Bentham, Fl. Austr. II, 237 (1864)	— — — — — — Q. N.A.	B.fl.II,237	
U. lagopoides, De Candolle, prodr. II, 324 (1325)	— — — — — — Q.	—	B.fl.II,237

LOUREA, Necker, elem. bot. III, 17 (1790).
L. obcordata, Desvaux, Journ. de Bot. III, 122 (1813) ...	— — — — — — N.A.	B.fl.II,238	

ALYSICARPUS, Necker, elem. bot. III, 15 (1790).
A. vaginalis, De Candolle, prodr. II, 353 (1825)	— — — — — — Q.	—	B.fl.II,239
A. longifolius, Wight & Arnott, prodr. 223 (1834)	— — — — — — N.A.	B.fl.II,239	
A. cylindricus, Desvaux in Annal. Soc. Linn. Par. 301 (1825)	— — — — — — Q. N.A.	B.fl.II,239	

LESPEDEZA, Cl. Richard in Michaux, fl. bor. am. II, 70 (1803).
L. cuneata, G. Don, gen. syst. II, 307 (1832)	— — — V. N.S.W. Q.	—	B.fl.II,240
L. lanata, Bentham, Fl. Austr. II, 241 (1864)	— S.A. — — — —	—	B.fl.II,241

CLITORIA, Linné, gen. plant. 216 (1737), from Petiver.
C. australis, Bentham, Fl. Austr. II, 242 (1864) ...	— — — — — — N.A.	B.fl.II,242	

GLYCINE, Linné, gen. pl. 349 (1737). (Leptolobium, Leptocyamus.)
G. falcata, Bentham, Fl. Austr. II, 243 (1864) ... ... — S.A. — — N.S.W. Q. N.A. 11.fl.II,243
G. clandestina, Wendland, Bot. Reob. 54 (1798) ... . W.A. S.A. T. V. N.S.W. Q. — B.fl.II,243
G. Latrobeana, Bentham, Fl. Austr. II, 244 (1864) ... ... — S.A. T. V. — — B.fl.II,244
G. tabacina, Bentham, Fl. Austr. II, 244 (1864) ... ... W.A. S.A. — V. N.S.W. Q. — B.fl.II,244
G. sericea, Bentham, Fl. Austr. II, 245 (1864) ... ... — S.A. — V. N.S.W. — B.fl.II,245
G. tomentosa, Bentham, Fl. Austr. II, 245 (1864) ... ... — S.A. — N.S.W. Q. N.A. B.fl.II,245

KENNEDYA, Ventenat, Jardin de la Malmaison II, t. 104, 106 (1804). (Caulinia, Moonch 1802, Hardenbergia, Physolobium, Zichya, Amphodus.)
K. nigricans, Lindley, Bot. Regist. t. 1715 (1835) ... ... W.A. — — — — — B.fl.II,249 M.fr.VII,127.
K. Beckxiana, F. v. M., fragm. XI, 98 (1880) ... . W.A. — — — — — — M.fr.XI,98.
K. rubicunda, Ventenat, Jard. de la Malm. t. 104 (1804) — — — V. N.S.W. Q. — B.fl.II,249 M.fr.VII,127.
K. procurrens, Bentham in Mitch. trop. Austr. 305 (1848) —. — — N.S.W. Q. — B.fl.II,249
K. prorepens, F. v. M., fragm. VIII, 225 (1874) ... ... — S.A. — — — — — M.fr.VIII,225.
K. prostrata, R. Brown in Ait. hort. Kew. sec. ed. IV, 299 (1812) W.A. S.A. T. V. N.S.W. Q. — B.fl.II,250 M.fr.VII,128.
K. bracteata, Gaudichaud in Freyc. voy. 286, t. 113 (1826) .. W.A. — — — — — B.fl.II,230
K. eximia, Lindley in Paxt. Mag. XVI, 35 ... . W.A. — — — — — B.fl.II,230
K. microphylla, Meissner in Lehm. pl. Preiss. I, 91 (1844) .. W.A. — — — — — B.fl.II,231
K. parviflora, Meissner in Lehm. pl. Preiss. I, 91 (1844) .. W.A. — — — — — B.fl.II,231
K. Stirlingii, Lindley, Bot. Regist. t. 1843 (1837) ... ... W.A. — — — — — B.fl.II,252
K. glabrata, Lindley, Bot. Regist. t. 1838 (1837)... ... W.A. — — — — — B.fl.II,232 M.fr.VII,128.
K. macrophylla, F. v. M., fragm. IV, 79 (1864) ... ... W.A. — — — — — B.fl.II,252 M.fr.IV,79.
K. coccinea, Ventenat, Jard. de la Malm. t. 105 (1804) ., .. W.A. — — — — — B.fl.II,230 M.fr.VII,128.
K. monophylla, Ventenat, Jard. de la Malm. t. 106 (1804) — S.A. T. V. N.S.W. Q. — B.fl.II,246 M.fr.VII,128.
K. Comptoniana, Link, enum. pl. hort. bot. Berol. II, 235 (1822) W.A. — — — — — B.fl.II,247 M.fr.VII,128.
K. retusa, F. v. M., fragm. V, 106 (1866) ... ... — — — — Q. — B.fl.II,247 M.fr.VII,128.

ERYTHRINA, Linné, gen. pl. 216 (1737).
E. vespertilio, Bentham in Mitch. Trop. Austr. 218 (1848) ... — S.A. — — — Q. N.A. B.fl.II,253
E. Indica, Lamarck, Encyclop. méthod. II, 391 (1786) ... — — — — — Q. N.A. B.fl.II,253

STRONGYLODON, T. Vogel in Linnaea X, 585 (1836).
S. ruber, T. Vogel in Linnaea X, 585 (1836) ... ... — — — — — Q. —

MUCUNA, Marcgraf, hist. nat. Brazil. 18 (1648).
M. gigantea, De Candolle, prodr. II, 405 (1825) ... ... — — — N.S.W. Q. N.A. B.fl.II,254

GALACTIA, P. Browne, Civil & Nat. Hist. of Jamaica 298 (1756).
G. tenuiflora, Wight & Arnott, prodr. fl. penins. Ind. 206 (1834) — S.A. — — N.S.W. Q. N.A. B.fl.II,255
G. Muelleri, Bentham, Fl. Austr. II, 255 (1864) ... ... — — — — — — N.A. B.fl.II,255

CANAVALIA, De Candolle, prodr. II, 403 (1825). (Canavali.)
C. obtusifolia, De Candolle, prodr. II, 404 (1825)... ... — — — — N.S.W. Q. N.A. B.fl.II,256

PHASEOLUS, Tournefort, inst. 412, t. 232 (1700), from Dioscorides & Columella.
P. vulgaris, de l'Obel, stirp. icon. 59 (1581) ... ... — — — — — N.A. B.fl.II,257
P. Truxillensis, Humb. Bonpl. & Kunth, nov.gen.Amer.VI,451 (1823) — — — — — Q. N.A. B.fl.II,257
P. Mungo, Linné, mantissa plantar. 101 (1767) ... . — — — — — Q. N.A. B.fl.II,257

VIGNA, Savi, Dissert. 6 (1824). (Callicystus.)
V. vexillata, Bentham in Mart. Fl. Bras. Papil. III, t. 50 (1862) — — — — N.S.W. Q. N.A. B.fl.II,258
V. lutea, A. Gray, Bot. Wilk. Exped. I, 454 (1854) ... ... — — — — N.S.W. Q. N.A. B.fl.II,259
V. luteola, Bentham in Mart. Fl. Bras. Papil. 194, t. 50 (1862) — — — — N.S.W. Q. — B.fl.II,260
V. lanceolata, Bentham in Mitch. trop. Austr. 350 (1848) ... — S.A. — — N.S.W. Q. N.A. B.fl.II,260

DOLICHOS, Linné, gen. plant. 222 (1737).
D. biflorus, Linné, spec. plant. 727 (1753) ... ... — — — — — Q. N.A. B.fl.II,261

DUNBARIA, Wight & Arnott, prodr. I, 258 (1834).
D. conspersa, Bentham in plant. Junghuhn. (1851) ... ... — — — — — Q. — B.fl.II,262

CAJANUS, De Candolle, Catal. hort. Monsp. 85 (1813). (Cajan, Atylosia.)
C. marmoratus, F. v. M.; Atylosia, Benth., Fl. Austr. II, 263 (1864) — — — — — Q. N.A. B.fl.II,263
C. scarabeoides, F. v. M.; Atyl., Benth. in pl. Jungh. I, 242 (1851) — — — — — Q. — B.fl.II,263
C. reticulatus, F. v. M.; Atylosia, Benth., Fl. Austr. II, 263 (1864) — — — — — Q. N.A. B.fl.II,263
C. grandifolius, F. v. M., Essay on pl. Burdekin Exped. 9 (1860) — — — — — Q. — B.fl.II,264
C. confertiflorus, F. v. M., Essay on pl. Burdekin Exped. 9 (1860) — — — — — Q. — B.fl.II,264
C. cinereus; Atylosia, F. v. M. in Benth. Fl. Austr. II, 264 (1864) — — — — — — N.A. B.fl.II,264

RHYNCHOSIA, Loureiro, Fl. Cochinch. II, 460 (1790).
R. rhomboidea, F. v. M. in Benth. Fl. Austr. II, 265 (1864) ... — — — — — N.A. B.fl.II,265
R. acutifolia, Bentham, Fl. Austr. II, 266 (1864)... ... — — — — — N.A. B.fl.II,266
R. rostrata, Bentham, Fl. Austr. II, 266 (1864) ... ... — — — — — — N.A. B.fl.II,266
R. Cunninghamii, Bentham, Fl. Austr. II, 266 (1864) ... — — — — — Q. — B.fl.II,266
R. minima, De Candolle, prodr. II, 385 (1825) ... ... — — — N.S.W. Q. N.A. B.fl.II,267

ERIOSEMA, De Candolle, prodr. II, 388 (1825).
E. Chinense, T. Vogel in Meyen's Beitr. zur Bot. 31 (1843) ... — — — — — Q. N.A. B.fl.II,268

FLEMINGIA, Roxburgh in Ait. hort. Kew. sec. ed. IV, 349 (1812).
F. lineata, Roxburgh in Ait. hort. Kew. sec. ed. IV, 350 (1812) — — — — — Q. N.A. B.fl.II,268
F. pauciflora, Bentham, Fl. Austr. II, 269 (1864)... ... ... — — — — — — N.A. B.fl.II,269
F. parviflora, Bentham, Fl. Austr. II, 269 (1864) ... ... — — — — — Q. — B.fl.II,269
F. involucrata, Bentham in pl. Jungh. I, 246 (1851) ... ... — — — — — Q. — B.fl.II,269

ABRUS, Rauwolf in Dalechamps, hist. general. plant. append. 193 (1587).
A. precatorius, Linné, syst. natur. ed. XII, 533 (1767) ... — — — — — Q. N.A. B.fl.II,270

G

DALBERGIA, Linné, fil. suppl. 52 (1781).
D. densa, Bentham in Hook.'Lond. Journ. II, 217 (1843) ... — — — — — Q. — B.fl.II,271

LONCHOCARPUS, Humboldt, Bonpland & Kunth, nov. gen. VI, 383 (1823).
L. Blackii, Bentham, Fl. Austr. II, 271 (1864) ... ... — — — N.S.W. Q. — B.fl.II,272

DERRIS, Loureiro, Fl. Cochinch. II, 432 (1790). (Brachypterum.)
D. scandens, Bentham in Journ. Linn. soc. IV, suppl. 103 (1860) — — — N.S.W. Q. — B.fl.II,272
D. uliginosa, Bentham in pl. Jungh. I, 252 (1851) ... ... — — — — Q. N.A. B.fl.II,272

PONGAMIA, Lamarck in Ventenat. Jardin de la Malmaison, t. 28 (1803).
P. glabra, Ventenat, Jard. Malm. t. 28 (1803) ... ... — — — — Q. N.A. B.fl.II,273

SOPHORA, Linné, gen. plant. 125 (1737).
S. tomentosa, Linné, spec. plant. 373 (1753) ... ... — — — N.S.W. Q. N.A. B.fl.II,274 M.fr.VII,27.
S. Fraseri, Bentham, Fl. Austr. II, 274 (1864) ... ... — — — N.S.W. Q. — B.fl.II,274 M.fr.V,31;VII,27.
S. tetraptera, J. Miller, icon. plant. t. 1 (1780) ... ... — — — N.S.W. — — M.fr.VII,26.

PODOPETALUM, F. v. M. in Melbourne Chemist, June (1882).
P. Ormondi, F. v. M., fragm. XII (inedit.) ... ... — — — — — Q. — —

CASTANOSPERMUM, Cunningham & Fraser in Hook. bot. misc. I, 241, t. 51 (1830).
C. Australe, Cunningh. & Fraser in Hook. bot. misc. I, 241, t. 51 (1830) — — — N.S.W. Q. — B.fl.II,275 M.fr.VII,27.

BARKLYA, F. v. M., fragm. I, 109 (1859).
B. syringifolia, F. v. M. in Journ. Linn. Soc. III, 159 (1859) ... — — — — N.S.W. Q. — B.fl.II,275 M.fr.I,109;III,163;VIII,27.

CAESALPINA, Plumier, nov. pl. Amer. gen. 28, t. 0 (1703), from Ammand. (Guilandina, Caesalpinia.)
C. Bonducella, Fleming in Asiat. Research XI, 159 (1801) — — — N.S.W. N.A. B.fl.II,276
C. Nuga, W. T. Aiton, hort. Kew., sec. ed. III, 32 (1811) ... — — — — Q. — B.fl.II,277 M.fr.X,7.
C. sepiaria, Roxburgh, Fl. Indic. II, 360 (1832) ... ... — — — — Q. — B.fl.II,277 M.fr.X,7.

MEZONEURON, Desfontaines in Mém. du Mus. IV, 245, t. 10 et 11 (1818).
M. brachycarpum, Bentham, Fl. Austr. II, 278 (1864) ... — — — N.S.W. Q. — B.fl.II,278 M.fr.X,7.
M. Scortechinii, F. v. M. in Wing's South. Sc. Record'II, 73 (1882) — — — N.S.W. Q. — —

PTEROLOBIUM, R. Brown in Salt. Abyssin. 65 (1814).
P. nitens, F. v. M. in Bentham, Fl. Austr. II, 279 (1864) ... — — — Q. — B.fl.II,279 M.fr.X,7.

PELTOPHORUM, Vogel in Linnaea XI, 406 (1837).
P. ferrugineum, Bentham, Fl. Austr. II, 279 (1864) ... ... — — — — — N.A. B.fl.II,279 M.fr.X,7.

CASSIA, Tournefort, inst. 619, t. 392 (1700), from Plumier (1693). (Cathartocarpus.)
C. Brewsteri, F. v. M., fourth ann. Rep. 17 (1858) ... ... — N.S.W. Q. — B.fl.II,282 M.fr.X,8.
C. laevigata, Willdenow, enum. pl. hort. bot. Berol. I, 441 (1809) — — N.S.W. Q. — B.fl.II,282 M.fr.IV,14;X,8.
C. Sophera, Linné, spec. plant. 379 (1853) ... ... — S.A. — N.S.W. Q. — B.fl.II,283 M.fr.IV,14;X,8.
C. laxiflora, Bentham, Fl. Austr. II, 283 (1864) ... ... — — — — N.A. B.fl.II,283
C. magnifolia, F. v. M., fragm. I, 166 (1859) ... ... — — — — Q. N.A. B.fl.II,283 M.fr.I,166;X,0.
C. venusta, F. v. M., fragm. I, 165 (1859)... ... ... — — — — Q. N.A. B.fl.II,284 M.fr.I,165;X,9.
C. notabilis, F. v. M., fragm. III, 28 (1862) ... ... W.A. S.A. — — — B.fl.II,284 M.fr.III,28;X,9.
C. pleurocarpa, F. v. M., fragm. I, 223 (1859) ... W.A. S.A. — N.S.W. Q. — B.fl.II,284 M.fr.I,223;X,9.
C. glauca, Lamarck, Encycl. meth. I, 647 (1783) ... — — — N.S.W. Q. N.A. B.fl.II,285 M.fr.IV,13;X,9.
C. retusa, Solander in Linnaea XI, 72 (1842) ... ... — — — — Q. — B.fl.II,285
C. australis, Sims in Bot. Mag. t. 2676 (1826) ... ... — — — V. N.S.W. Q. N.A. B.fl.II,285 M.fr.X,9.
C. Chatelainiana, Gaudichaud in Freyc. voy. 485, t. 111 (1826) W.A. — — — N.A. B.fl.II,286
C. glutinosa, De Candolle, prodr. II, 495 (1825) ... ... — S.A. — — Q. N.A. B.fl.II,286
C. pruinosa, F. v. M., fragm. III, 48 (1862) ... ... — S.A. — N.S.W. Q. N.A. B.fl.II,286 M.fr.III,48.
C. circinata, Bentham in Mitch. trop. Austr. 384 (1848)... ... — S.A. — N.S.W. Q. — B.fl.II,287 M.fr.I,165.
C. phyllodinea, R. Brown, App. to Sturt's Centr. Austr. 15 (1849) — S.A. — V. N.S.W. Q. — B.fl.II,287
C. eremophila, Cunningh. in T. Vogel, gen. Cass. synops. 47 (1837) W.A. S.A. — V. N.S.W. Q. — B.fl.II,287
C. artemisioides, Gaudichaud in De Cand. prodr. II, 495 (1825) W.A. S.A. — V. N.S.W. Q. — B.fl.II,288
C. Sturtii, R. Brown, App. to Sturt's Centr. Austr. 14 (1849) ... W.A. S.A. — V. N.S.W. Q. — B.fl.II,288
C. desolata, F. v. M. in Linnaea XXV, 387 (1852) ... ... W.A. S.A. — V. N.S.W. Q. — B.fl.II,289 M.fr.III,49.
C. oligophylla, F. v. M., fragm. III, 49 (1862) ... ... — — — — N.A. B.fl.II,289 M.fr.III,49.
C. cardiosperma, F. v. M., fragm. X, 50 (1877) ... ... W.A. — — — — M.fr.X,50.
C. oligoclada, F. v. M., fragm. III, 49 (1862) ... ... — — — — N.A. B.fl.II,289 M.fr.III,40;X,9.
C. heptanthera, F. v. M., fragm. X, 8 (1876) ... ... — — — — N.A. — M.fr.X,8.
C. leptoclada, Bentham, Fl. Austr. II, 290 (1864)... ... — — — — N.A. B.fl.II,290
C. Absus, Linné, spec. plant. 376 (1753) ... ... — — — — Q. N.A. B.fl.II,290 M.fr.III,50;X,9.
C. pumila, Lamarck, Encycl. meth. I, 620 (1783) ... — — — Q. N.A. B.fl.II,290 M.fr.III,47;X,10.
C. concinna, Bentham, Fl. Austr. II, 291 (1864) ... ... — — N.S.W. Q. N.A. B.fl.II,291 M.fr.X,10.
C. mimosoides, Linné, spec. plant. 379 (1753) ... ... — — N.S.W. Q. N.A. B.fl.II,291 M.fr.III,48;X,10.

PETALOSTYLIS, R. Brown, App. to Sturt's Centr. Austr. 79 (1849).
P. labichoides, R. Brown, App. to Sturt's Centr. Austr. 17 (1849) W.A. S.A. — N.S.W. Q. N.A. B.fl.II,292 M.fr.X,7.

LABICHEA, Gaudichaud in De Cand. prodr. II, 507 (1825).
L. cassioides, Gaudichaud in De Cand. prodr. II, 507 (1825) ... W.A. — — — — B.fl.II,292
L. nitida, Bentham, Fl. Austr. II, 293 (1864) ... ... W.A. — — — Q. N.A. B.fl.II,293 M.fr.X,7.
L. lanceolata, Bentham in Hueg. enum. 41 (1837)... ... W.A. — — — — B.fl.II,293 M.fr.X,7.
L. rupestris, Bentham in Mitch. Trop. Austr. 342 (1848) ... — — — — Q. B.fl.II,293 M.fr.X,7.
L. punctata, Bentham in Bot. Regist. XXV, App. XV (1839) ... W.A. — — — — B.fl.II,294 M.fr.X,8.
L. Buettneriana, F. v. M. in Melb. Chemist, June (1882) ... — — — — Q. — —

TAMARINDUS, Tournefort, inst. 660, t. 445. App. (1700) from C. Bauhin (1623).
T. Indicus, Linné, spec. plant. 34 (1853) ... ... — — — N.A. B.fl.II,294

BAUHINIA, Plumier, nov. gen. pl. Amer. 22, t. 13 (1703).
B. Leichhardtii, F. v. M. in Transact. Vict. Inst. III, 50 (1858) — — — N.S.W. Q. — B.fl.II,295 M.fr.X,8.
B. Carronii, F. v. M. in Transact. Vict. Inst. III, 49 (1855) ... S.A. — — N.A. B.fl.II,295 M.fr.X,8.
B. Hookeri, F. v. M. in Transact. Vict. Inst. III, 51 (1858) ... — — — N.S.W. Q. — B.fl.II,295 M.fr.X,8.

B. Gilesii, F. v. M. and Bailey in Wing's S. Sc. Record, Jul. (1882) — — — — — — N.A. —

    AFZELIA, Smith in Transact. Linn. Soc. IV, 221 (1798).
A. bijuga, A. Gray, Bot. Wilk. Expl. Exped. 467, t. 51 (1854)... — — — — — Q. — —

    CYNOMETRA, Linné in Act. Soc. Ups. 78 (1741).
C. ramiflora, Linné, spec. plant. 382 (1753)   ...   . — — — — — Q. — B.fl.II,296

    ERYTHROPHLAEUM, Afzelius in Tuckey's Congo 430 (1818). (Laboucheria.)
E. Laboucherii, F. v. M. in Benth. Fl. Austr. II, 297 (1864) ... — — — — — Q. N.A. B.fl.II,297

    ENTADA, Adanson, Familles des plantes II, 318 (1763).
E. Pursaetha, De Cand., mémoir. sur la fam. des Légum. 12 (1825) — — — — — Q. — B.fl.II,298

    ADENANTHERA, Royen in Linné, coroll. 7 (1737).
A. pavonina, Linné, spec. plant. 384 (1753) ... ... — — — — — Q. — — M.fr.V,30.
A. abrosperma, F. v. M., fragm. V, 30 (1865) ... ... — — — — — Q. — B.fl.II,298 M.fr.V,31.

    NEPTUNIA, Loureiro, Fl. Cochinch. II, 653 (1790). (Dichrostachys.)
N. cinerea,F.v.M.; Dichrostachys,Wight&Arnott,prodr.271(1834) — — — — — — N.A. B.fl.II,299
N. spicata, F. v. M., fragm. III, 151 (1862) ... — — — — — — N.A. B.fl.II,299 M.fr.III,151.
N. gracilis; Bentham in Hook. journ. of Bot. IV, 355 (1842) ... — — — — N.S.W. Q. N.A. B.fl.II,300
N. monosperma, F. v. M. in Benth. Fl. Austr. II, 299 (1864) ... — — — — — Q. N.A. B.fl.II,300

    ACACIA, Tournefort, inst. 605, t. 375 (1700) from Dioscorides & Plinius. (Vachellia, Chithonanthus, Tetracheilus, Mimosa partly.)

    I. ALATAE.
A. glaucoptera, Bentham in Schlecht. Linnaea XXVI, 604 (1853) W.A. — — — — — B.fl.II,320
A. alata, R. Brown in Ait. hort. Kew. sec. ed. V, 464 (1813) ... W.A. — — — — — B.fl.II,320
A. diptera, Lindley, Bot. Regist. XXV. App. XV (1830) ... W.A. — — — — — B.fl.II,321
A. stenoptera, Bentham in Hook. Lond. Journ. I, 325 (1842) ... W.A. — — — — — B.fl.II,321

    II. CONTINUAE.
A. incurva, Bentham in Hook. Lond. Journ. I, 325 (1842) ... W.A. — — — — — B.fl.II,322
A. trigonophylla, Meissner in Lehm. pl. Preiss. II, 199 (1846) ... W.A. — — — — — B.fl.II,322 M.fr.IV,3.
A. continua, Bentham, Fl. Austr. II, 322 (1864) ... ... S.A. — V. N.S.W. — B.fl.II,322
A. Peuce, F. v. M., fragm. III, 151 (1863)... ... ... S.A. — — — Q. — B.fl.II,323 M.fr.III,151.

    III. PUNGENTEAE.
A. spinescens, Bentham in Hook. Lond. Journ. I, 323 (1842) ... — S.A. — V. N.S.W. — B.fl.II,323
A. cochlearis, H. L. Wendland, comm. de Acac. aphyll. 15 (1820) W.A. S.A. — — — — B.fl.II,324
A. lanigera, Cunningham in Field's N.S. Wales, 345 (1825) ... — — — V. N.S.W. — B.fl.II,324
A. phlebocarpa, F. v. M. in Journ. Linn. Soc. III, 119 (1858) ... — — — — — N.A. B.fl.II,325
A. trinervata, Sieber in De Cand. prodr. II, 451 (1825) ... — — — — N.S.W. — B.fl.II,325
A. colletioides, Cunningham in Hook. Lond. Journ. I, 336 (1842) — S.A. — V. N.S.W. — B.fl.II,325 M.fr.IV,4.
A. striatula, Bentham in Hook. Lond. Journ. I, 336 (1842) W.A. — — — — — B.fl.II,326
A. campylophylla, Bentham in Schlecht. Linnaea XXVI, 605 (1853) W.A. — — — — — B.fl.II,326
A. teretifolia, Bentham in Hook. Lond. Journ. I, 326 (1842) ... W.A. — — — — — B.fl.II,326
A. sulcata, R. Brown in Ait. hort. Kew. sec. ed. V, 460 (1813) W.A. — — — — — B.fl.II,327
A. costata, Bentham in Hook. Lond. Journ. I, 339 (1842) ... W.A. — — — — — B.fl.II,327
A. barbinervis, Bentham in Hook. Lond. Journ. I, 326 (1842) ... W.A. — — — — — B.fl.II,328
A. ataxiphylla, Bentham in Schlecht. Linnaea XXVI, 605 (1853) W.A. — — — — — B.fl.II,328
A. Baxteri, Bentham in Hook. Lond. Journ. I, 327 (1842) ... W.A. — — — — — B.fl.II,328
A. aureonitens, Lindley, Bot. Regist. XXV, App. XV (1839) ... W.A. — — — — — B.fl.II,328 M.f.III,127.
A. quadrisulcata, F. v. M., fragm. III, 127 (1863) ... W.A. — — — — — B.fl.II,329
A. orioclada, Bentham in Schlecht. Linnaea XXVI, 606 (1853) W.A. — — — — — B.fl.II,329
A. siculiformis, Cunningham in Hook. Lond. Journ. I, 337 (1842) — — T. V. N.S.W. — B.fl.II,329
A. patens, F. v. M. in Journ. Linn. Soc. III, 120 (1858)... — — — — — N.A. B.fl.II,330 M.fr.III,46.
A. laricina, Meissner in Lehm. pl. Preiss. II, 198 (1846) ... W.A. — — — — — B.fl.II,330
A. tetragonophylla, F. v. M. in Journ. Linn. Soc. III, 121 (1858) S.A. — — N.S.W. — B.fl.II,330 M.fr.IV,3.
A. genistoides, Cunningham in Benth. Fl. Austr. II, 330 (1864) W.A. — — — — — B.fl.II,331
A. spaclata, Bentham in Hook. Lond. Journ. I, 338 (1842) ... W.A. — — — — — B.fl.II,331
A. ingrata, Bentham, Fl. Austr. II, 331 (1864) ... ... W.A. — — — — — B.fl.II,331
A. juniperina, Willdenow, spec. plant. IV, 1049 (1805) ... — — T. V. N.S.W. Q. — B.fl.II,331
A. asparagoides, Cunningham in Field's N.S. Wales 343 (1825)... — — — N.S.W. — B.fl.II,332
A. tenuifolia, F. v. M. in Transact. phil. Soc. Vict. I, 37 (1854) — — — V. — — B.fl.II,332
A. diffusa, Edwards, Bot. Regist. t. 634 (1822) ... — — T. V. N.S.W. — B.fl.II,332
A. rupicola, F. v. M. in Schlecht. Linnaea XXVI, 610 (1853) ... — S.A. — V. — — B.fl.II,333

    IV. CALAMIFORMES.
A. volubilis, F. v. M., fragm. X, 98 (1877)... ... W.A. — — — — — M.fr.X,98.
A. tetragonocarpa, Meissner in Lehm. pl. Preiss. I, 4 (1844) ... W.A. — — — — — B.fl.II,336
A. restiacea, Bentham in Hook. Lond. Journ. I, 3 (1842) ... W.A. — — — — — B.fl.II,336
A. squamata, Lindley, Bot. Regist. XXV, App. XV (1839) ... W.A. — — — — — B.fl.II,336
A. brachyphylla, Bentham in Schlecht. Linnaea XXVI, 615 (1853) W.A. — — — — — B.fl.II,337
A. Bynoeana, Bentham in Schlecht. Linnaea XXVI, 614 (1853) — — — — — N.A. B.fl.II,337 M.fr.IV,9.
A. triptycha, F. v. M. in Benth. Fl. Austr. II, 337 (1864) ... W.A. — — — — — B.fl.II,337
A. leptoneura, Bentham in Hook. Lond. Journ. I, 341 (1842) ... W.A. — — — — — B.fl.II,337
A. rigens, Cunningham in Loudon, hort. 406 (1830) ... — S.A. — V. N.S.W. — B.fl.II,338
A. papyrocarpa, Bentham, Fl. Austr. II, 338 (1864) ... W.A. S.A. — — — — B.fl.II,338
A. Gilesiana, F. v. M. in Melb. Chemist, July (1882) ... — S.A. — — — — B.fl.II,338
A. sessiliceps, F. v. M. in Melb. Chemist, July (1882) ... — S.A. — — — —
A. pugioniformis, Wendl., comm. de Acac. aphyll. 38, t. 9 (1820) — — — N.S.W. Q. — B.fl.II,339
A. juncifolia, Bentham in Hook. Lond. Journ. I, 341 (1842) ... — — — N.S.W. Q. N.A. B.fl.II,339
A. calamifolia, Sweet in Bot. Regist. t. 839 (1824) ... S.A. — V. N.S.W. — B.fl.II,339
A. sclrpifolia, Meissner in Bot. Zeitung 10 (1855)... ... W.A. — — — — — B.fl.II,340
A. extensa, Lindley, Bot. Regist. XXV, App. XV (1839) ... W.A. — — — — — B.fl.II,340

44

A. gonophylla, Bentham in Schlecht. Linnaea XXVI, 613 (1853) W.A. — — — — — — — B.fl.II,340
A. ericifolia, Bentham in Hook. Lond. Journ. I, 345 (1842) ... W.A. — — — — — — — B.fl.II,340
A. uncinella, Bentham in Schlecht Linnaea XXVI, 613 (1853)... W.A. — — — — — — — B.fl.II,341
A. oxyclada, F. v. M. in Benth. Fl. Austr. II, 341 (1864) ... W.A. — — — — — — — B.fl.II,341

V. BRUNIOIDAE.
A. cedroides, Bentham in Schlecht. Linnaea XXVI, 615 (1853) W.A. — — — — — — B.fl.II,341
A. lycopodifolia, Cunningham in Hook. icon. plant. t. 172 (1837) — S.A. — — Q. N.A. B.fl.II,342
A. spondylophylla, F. v. M., fragm. VIII, 243 (1874) ... ... — S.A. — — — N.A. — M.fr.243.
A. galioides, Bentham in Hook. Lond. Journ. I, 344 (1842) ... — — — — — Q. N.A. B.fl.II,342
A. Baueri, Bentham in Hook. Lond. Journ. I, 344 (1842) ... — — — — N.S.W. Q. — B.fl.II,342 M.fr.XI,33.
A. subternata, F. v. M. in Journ. Linn. Soc. III, 124 (1858) ... — — — — — — N.A. B.fl.II,343
A. minutifolia, F. v. M., fragm. VIII, 243 (1874) ... ... — S.A. — — — — — B.fl.II,343 M.fr.X,243.
A. brunioides, Cunningham in G. Don, gen. syst. II, 404 (1832) — — — — — Q. B.fl.II,343
A. conferta, Cunningham in Hook. Lond. Journ. I, 345 (1842)... — — — N.S.W. Q. — B.fl.II,343

VI. UNINERVES.
A. scabra, Bentham in Schlecht. Linnaea XXVI, 605 (1853) ... W.A. — — — — — — B.fl.II,344
A. nodiflora, Bentham in Schlecht. Linnaea XXVI, 621 (1853) ... W.A. — — — — — — B.fl.II,344
A. spinosissima, Bentham in Schlecht. Linnaea XXVI, 621 (1853) W.A. — — — — — — B.fl.II,344
A. ulicina, Meissner in Lehm. pl. Preiss. II, 202 (1846) ... W.A. — — — — — — B.fl.II,345
A. erinacea, Bentham in Hook. Lond. Journ. I, 360 (1842) ... W.A. — — — — — — B.fl.II,345
A. Huegelii, Bentham in Hueg. enum. 42 (1837) ... ... W.A. — — — — — — B.fl.II,345
A. nervosa, De Candolle, Mém. sur la fam. des Légum. 444 (1825) W.A. — — — — — — B.fl.II,346
A. obovata, Bentham in Hook. Lond. Journ. I, 329 (1842) ... W.A. — — — — — — B.fl.II,346
A. congesta, Bentham in Hook. Lond. Journ. I, 327 (1842) ... W.A. — — — — — — B.fl.II,346
A. dermatophylla, Bentham, Fl. Austr. II, 346 (1864) ... ... W.A. — — — — — — B.fl.II,346
A. aspera, Lindley in Mitch. Three Exped. II, 139 (1838) ... — S.A. — V. N.S.W. — B.fl.II,347
A. armata, R. Brown in Ait. hort. Kew. sec. ed. V, 463 (1813) W.A. S.A. — V. N.S.W. Q. — B.fl.II,347
A. idiomorpha, Cunningham in Hook. Lond. Journ. I, 329 (1842) W.A. — — — — — — B.fl.II,348
A. Shuttleworthii, Meissner in Lehm. pl. Preiss. I, 7 (1844) ... W.A. — — — — — — B.fl.II,348
A. Gregorii, F. v. M., fragm. III, 47 (1862) ... ... W.A. — — — — — N.A B.fl.II,348 M.fr.III,47.
A. pilosa, Bentham in Schlecht. Linnaea XXVI, 607 (1853) ... W.A. — — — — — — B.fl.II,348
A. crispula, Bentham in Schlecht. Linnaea XXVI, 607 (1853) .. W.A. — — — — — — B.fl.II,349
A. crassistipulea, Bentham in Hook. Lond. Journ. I, 326 (1842) W.A. — — — — — — B.fl.II,349
A. hastulata, Smith in Rees' Cyclopaed. XXXIX, Suppl. (1819) W.A. — — — — — — B.fl.II,350
A. horridula, Meissner in Lehm. pl. Preiss. I, 9 (1844) ... W.A. — — — — — — B.fl.II,350
A. divergens, Bentham in Hook. Lond. Journ. I, 331 (1842) ... W.A. — — — — — — B.fl.II,350
A. vomeriformis, Cunningham in Hook. Lond. Journ. I, 332 (1842) — S.A. T. V. N.S.W. — B.fl.II,350
A. plagiophylla, F. v. M. in journ. Linn. Soc. III, 131 (1858) ... — — — N.S.W. Q. — B.fl.II,351
A. biflora, R. Brown in Ait. hort. Kew. sec. ed. V, 463 (1813) W.A. — — — — — — B.fl.II,351
A. decipiens, R. Brown in Ait. hort. Kew. sec. ed. V, 463 (1813) W.A. — — — — — — B.fl.II,351
A. cuneata, Bentham in Hueg. enumer. 42 (1837)... ... ... W.A. — — — — — — B.fl.II,352
A. dilatata, Bentham in Schlecht. Linnaea XXVI, 608 (1853) ... W.A. — — — — — — B.fl.II,352
A. bidentata, Bentham in Hook. Lond. Journ. I, 333 (1842) .. W.A. — — — — — — B.fl.II,352
A. acanthoclada, F. v. M., fragm. III, 127 (1863) ... ... W.A. S.A. — V. N.S.W. — B.fl.II,352 M.fr.III,127.
A. obliqua, Cunningham in Hook. Lond. Journ. I, 334 (1842) ... — S.A. — V. N.S.W. — B.fl.II,353
A. acinacea, Lindley in Mitch. Three Exped. II, 267 (1838) ... — S.A. — V. N.S.W. — B.fl.II,353
A. lineata, Cunningham in G. Don, gen. syst. II, 403 (1832) ... — S.A. — V. N.S.W. — B.fl.II,353 M.fr.I,5;II,177.
A. lachnophylla, F. v. M. in Wing's S. Sc. Record, July (1882) W.A. — — — — — — B.fl.II,354
A. triquetra, Bentham in Hook. Lond. Journ. I, 358 (1842) ... W.A. — — — — — — B.fl.II,354
A. ligustrina, Meissner in Lehm. pl. Preiss. II, 203 (1846) ... W.A. — — — — — — B.fl.II,354
A. Meissneri, Lehmann, delect. sem. hort. bot. Hamb. (1842) ... W.A. — — — — — — B.fl.II,355
A. anceps, De Candolle, Mém. sur la fam. des Légum. 446 (1825) W.A. S.A. — — — — B.fl.II,355
A. bispidula, Willdenow, spec. plant. IV, 1054 (1805) ... — — — N.S.W. Q. — B.fl.II,355
A. undulifolia, Fraser in Loddiges, Bot. Cabinet XVI, 1544 (1829) — — — N.S.W. Q. — B.fl.II,355
A. flexifolia, Cunningham in Hook. Lond. Journ. I, 359 (1842)... — — — N.S.W. — B.fl.II,356
A. dura, Bentham in Schlecht. Linnaea XXVI, 622 (1853) ... W.A. — — — — N.A. B.fl.II,356
A. spathulata, F. v. M. in Benth. Fl. Austr. II, 356 (1864) ... W.A. — — — — — — B.fl.II,357 M.fr.I,6.
A. microcarpa, F. v. M., second gen. Report 7 (1854) ... ... — S.A. — V. N.S.W. — B.fl.II,357
A. montana, Bentham in Hook. Lond. Journ. I, 360 (1842) ... — S.A. — V. N.S.W. — B.fl.II,358
A. verniciflua, Cunningham in Field's N.S.Wales 344 (1825) ... — S.A. T. V. N.S.W. — B.fl.II,358
A. leprosa, Sieber in De Cand. prodr. II, 450 (1825) ... — — — V. N.S.W. — B.fl.II,358
A. stricta, Willdenow, spec. plant. IV, 1052 (1805) ... — — T. V. N.S.W. — B.fl.II,358
A. dodonaeifolia, Willdenow, enum. pl. hort. Berol. suppl. 68 (1813) — S.A. — — — — B.fl.II,359
A. Gnidium, Bentham, Fl. Austr. II, 359 (1864) ... ... — S.A. — — — — B.fl.II,359
A. ramosissima, Bentham in Hook. Lond. Journ. I, 356 (1842)... W.A. — — — — — — B.fl.II,360
A. Dempsteri, F. v. M., fragm. XI, 65 (1879) ... ... W.A. — — — — — — B.fl.II,360 M.fr.XI,65.
A. Sentis, F. v. M., second gen. Report 11 (1854) ... ... W.A. S.A. — V. N.S.W. Q. N.A. B.fl.II,360 M.fr.XI,65.
A. dentifera, Bentham in Maund, Botanist IV, t. 179 (1839) ... W.A. — — — — — N.A. B.fl.II,361
A. sclerosperma, F. v. M. in Wing's South. Sc. Rec., July (1882) W.A. — — — — — N.A. B.fl.II,361
A. fasciculifera, F. v. M. in Hook. Lond. Journ. II, 361 (1864) ... — — — N.S.W. Q. — B.fl.II,361
A. falcata, Willdenow, spec. plant. IV, 1053 (1805) ... — — — N.S.W. Q. — B.fl.II,361
A. macradenia, Bentham in Mitch. Trop. Austr. 300 (1848) ... — — — — Q. — B.fl.II,362
A. penninervis, Sieber in De Cand. prodr. II, 452 (1825) ... — — T. V. N.S.W. — B.fl.II,362
A. amblyphylla, F. v. M. in Wing's Sc. Record, July (1882) W.A. — — — — — — B.fl.II,362
A. retinodes, Schlechtendal, Linnaea XX, 664 (1847) ... — S.A. — V. — — B.fl.II,362
A. neriifolia, Cunningham in Hook. Lond. Journ. I, 353 (1842)... — — — N.S.W. Q. — B.fl.II,363
A. microbotrya, Bentham in Hook. Lond. Journ. I, 353 (1842)... W.A. — — — — — — B.fl.II,363
A. leiophylla, Bentham in Hook. Lond. Journ. I, 353 (1842) ... — S.A. — — — — B.fl.II,363
A. cyanophylla, Lindley, Bot. Regist. Misc. 49 (1835) ... W.A. — — — — — — B.fl.II,364
A. pycnantha, Bentham in Hook. Lond. Journ. I, 351 (1842) ... — S.A. — V. N.S.W. — B.fl.II,365

45

A. notabilis, F. v. M., fragm. I, 6 (1858) ... ... — S.A. — — N.S.W. — — B.fl.II,365 M.fr.I,6.
A. gladiiformis, Cunningham in Hook. Lond. Journ. I, 354 (1842) — — — — N.S.W. — — B.fl.II,365
A. obtusata, Sieber in De Cand. prodr. II, 453 (1825) ... ... — — — — N.S.W. — — B.fl.II,366
A. rubida, Cunningham in Field's N.S.Wales, 344 (1825) ... — — — — N.S.W. — — B.fl.II,366
A. amoena, II. L. Wendland, comm. de Acac. aphyll. 10, t. 4 (1820) — — — V. N.S.W. — B.fl.II,366
A. bakeoides, Cunningham in Hook. Lond. Journ. I, 354 (1842) — S.A. — V. N.S.W. — — B.fl.II,367
A. salicina, Lindley in Mitch. Three Exped. II, 20 (1838) ... W.A. S.A. — V. N.S.W. Q. N.A. B.fl.II,367
A. rostellifera, Bentham in Hook. Lond. Journ. I, 356 (1842) ... W.A. — — — — — — B.fl.II,368
A. pycnophylla, Bentham, Fl. Austr. II, 368 (1864) ... ... W.A. — — — — — — B.fl.II,368
A. Harveyi, Bentham, Fl. Austr. II, 368 (1864) ... ... W.A. — — — — — — B.fl.II,368
A. iteaphylla, F. v. M. in Schlecht. Linnaea XXVI, 617 (1853) — — — — — — —
A. suaveolens, Willdenow, spec. plant. IV, 1050 (1805)... — S.A. T. V. N.S.W. Q — B.fl.II,369
A. subcoerulea, Lindley, Bot. Regist. t. 1075 (1827) ... W.A. — — — — — — B.fl.II,369
A. Lindleyi, Meissner in Lehm. pl. Preiss. I, 14 (1844) ... W.A. — — — — — — B.fl.II,370
A. leptopetala, Bentham in Schlecht. Linnaea XXVI, 619 (1853) W.A. — — — — — — B.fl.II,370
A. Dietrichiana, F. v. M. in Melb. Chemist, July (1882) ... — — — — — — —
A. Murrayana, F. v. M. in Benth. Fl. Austr. II, 370 (1864) ... — S.A. — — — Q. — B.fl.II,370
A. subulata, Bonpland, pl. rar. cult. à Malm. 110 t. 43 (1813)... — — — — N.S.W. Q. — B.fl.II,370
A. linifolia, Willdenow, spec. plant. IV, 1051 (1805) ... — — — — N.S.W. Q. — B.fl.II,371
A. Leichhardtii, Bentham, Fl. Austr. II, 372 (1864) ... — — — — — Q. — B.fl.II,372
A. crassiuscula, H. L. Wendland, comm. de Acac. 31, t. 8 (1820) — — T. — N.S.W. Q. — B.fl.II,372
A. buxifolia, Cunningham in Field's N.S.Wales, 344 (1825) ... — — — V. N.S.W. Q. — B.fl.II,372
A. lunata, Sieber in De Cand. prodr. II, 452 (1825) ... — — — V. N S W. Q. — B.fl.II,373
A. brachybotrya, Bentham in Hook. Lond. Journ. I, 347 (1842) — S.A. — V. N.S.W. — — B.fl.II,373
A. Wattsiana, F. v. M. in Benth. Fl. Austr. II, 374 (1864) ... — S.A. — — — — — B.fl.II,374
A. podalyriifolia, Cunningham in Loudon, hort. Brit. 707 (1830) — — — — N.S.W. Q. — B.fl.II,374
A. uncifera, Bentham in Mitch. Trop. Austr. 341 (1848) ... — — — — — Q. — B.fl.II,374
A. vestita, Ker in Bot. Regist. t. 698 (1823) ... ... — — — — N.S.W. — — B.fl.II,375
A. cultriformis, Cunningham in G. Don. gen. syst. II, 406 (1832) — — — — N.S.W. — — B.fl.II,375 M.fr.XI,123.
A. pravissima, F. v. M., first gen. report 12 (1853) ... ... — — — — N.S.W. — — B.fl.II,375 M.fr.I,5.
A. strongylophylla, F. v. M., fragm. VIII, 220 (1874) ... ... — S.A. — — — — — M.fr.VIII,226
A. pyrifolia, De Candolle, Mém. sur la fam. des Légum. 447 (1825) W.A. — — — — — N.A. B.fl.II,376 M.fr.III,17.
A. myrtifolia, Willdenow, spec. plant. IV, 1054 (1805) ... ... W.A. S.A. T. V. N.S.W. Q. — B.fl.II,376

VII. PLURINERVES.

A. scalpelliformis, Meissner in Lehm. pl. Preiss. II, 200 (1847)... W.A. — — — — — — B.fl.II,377
A. urophylla, Bentham in Bot. Reg. Misc. 24 (1841) ... W.A. — — — — — — B.fl.II,377
A. Luehmanni, F. v. M., fragm. XI, 116 (1881) ... ... — S.A. — — — — N.A. — M.fr.XI,116.
A. sublanata, Bentham in Hucg. enum. 42 (1837)... ... ... — S.A. — — — — — B.fl.II,378 M.fr.XI,117.
A. pravifolia, F. v. M., fragm. I, 4 (1858) ... ... — S.A. — — N.S.W. Q. — B.fl.II,378 M.fr.I,4.
A. amblygona, Cunningham in Hook. Lond. Journ. I, 332 (1842) — — — — N.S.W. Q. — B.fl.II,378 M.fr.IV,3.
A. adnata, F. v. M. in Melb. Chemist, July (1882) ... W.A. — — — — — — B.fl.II,378
A. deltoidea, Cunningham in G. Don. gen. syst. II, 405 (1832)... — — — — — N.A. B.fl.II,378
A. stipulosa, F. v. M. in Journ. Linn. Soc. III, 119 (1858)... W.A. — — — — — N.A. B.fl.II,379 M.fr.XI,117.
A. loxophylla, Bentham in Schlecht. Linnaea XXVI, 625 (1853) W.A. — — — — — N.A. B.fl.II,379
A. setulifera, Bentham in Schlecht. Linnaea XXVI, 625 (1853) — — — — — N.A. B.fl.II,379
A. translucens, Cunningham in Hook. icon. plant. t. 100 (1837) — — — — — N.A. B.fl.II,380
A. improcera, F. v. M. in Journ. Linn. Soc. III, 133 (1858) — — — — — N.A. B.fl.II,380
A. bivenosa, De Candolle, prodr. II, 452 (1825) ... ... W.A. — — — — — N.A. B.fl.II,380
A. trineura, F. v. M., pl. of Vict. II, 25 (1860) ... ... — S.A. — V. — — N.A. B.fl.II,381 M.fr.IV,5.
A. nitidula, Bentham, Fl. Austr. II, 381 (1864) ... W.A. — — — — — — B.fl.II,381
A. ostrophiolata, F. v. M. in Wing's South. Sc. Rec. July (1882) — S.A. — — — — — B.fl.II,381
A. heteroclita, Meissner in Lehm. pl. Preiss. I, 318 (1845) ... W.A. — — — — — — B.fl.II,381 M.fr.IV,6.
A. elongata, Sieber in De Cand. prodr. II, 451 (1825) ... — — — V. N.S.W. Q. — B.fl.II,381
A. subporosa, F. v. M., pl. of Vict. II, 24 (1860)... ... — — T. V. N.S.W. — — B.fl.II,382 M.fr.IV,5.
A. Simsii, Cunningham in Hook. Journ. I, 368 (1842) ... — — — — — Q. N.A. B.fl.II,382
A. dissoneura, F. v. M. in Wing's S. Sc. Record, July (1882) ... — — — — — — —
A. leptospermoides, Bentham in Schlecht. Linnaea XXVI (1853) W.A. — — — — — — B.fl.II,383
A. homalophylla, Cunningham in Hook. Lond. Journ. I, 365 (1842) — S.A. — V. N.S.W. — — B.fl.II,383
A. pendula, Cunningham in Loudon, hort. Brit. 490 (1830) — — — — N.S.W. Q. — B.fl.II,384
A. Oswaldi, F. v. M., pl. of Vict. II, 27 (1860) ... ... W.A. S.A. — V. N.S.W. Q. — B.fl.II,384 M.fr.IV,5.
A. lineolata, Bentham in Schlecht. Linnaea XXVI, 626 (1853) ... W.A. — — — — — — B.fl.II,384
A. quadrimarginea, F. v. M., fragm. X, 31 (1876) ... W.A. — — — — — — B.fl.II,385 M.fr.X,31.
A. coriacea, De Candolle, Mém. sur la fam. des Légum. 446 (1825) W.A. — — — — — N.A. B.fl.II,385
A. stenophylla, Cunningham in Hook. Lond. Journ. I, 366 (1842) — S.A. — V. N.S.W. Q. N.A. B.fl.II,385
A. hemignosta, F. v. M. in Journ. Linn. Soc. III, 134 (1858) ... — S.A. — V. N.S.W. — — B.fl.II,385 M.fr.XI,67
A. sclerophylla, Lindley in Mitch. Three Exped. II, 138 (1838) — S.A. — V. N.S.W. — — B.fl.II,386
A. farinosa, Lindley in Mitch. Three Exped. II, 145 (1838) ... — S.A. — V. N.S.W. — — B.fl.II,386
A. Whanii, F. v. M. in Benth. Fl. Austr. II, 386 (1864) ... — — — — — — — B.fl.II,386
A. heteroneura, Bentham in Schlecht. Linnaea XXVI, 624 (1853) W.A. — — — — — — B.fl.II,387
A. viscidula, Cunningham in Hook. Lond. Journ. I, 366 (1842) — — — V. N.S.W. Q. — B.fl.II,387
A. ixiophylla, Bentham in Hook. Lond. Journ. I, 364 (1842) ... W.A. — — — — — — B.fl.II,387 M.fr.IV,6.
A. dictyophleba, F. v. M., fragm. III, 128 (1863)... ... — S.A. — — — Q. N.A. B.fl.II,388 M.fr.III,128.
A. venulosa, Bentham in Hook. Lond. Journ. I, 366 (1842) — — — — N.S.W. Q. — B.fl.II,388
A. cyclops, Cunningham in Loudon, hort. Brit. 407 (1830) W.A. S.A. — — — — — B.fl.II,388
A. melanoxylon, R. Brown in Ait. hort. Kew. sec. ed. V, 462 (1813) — S.A. T. V. N.S.W. — — B.fl.II,388
A. oraria, F. v. M., fragm. XI, 66 (1879) ... ... — — — — — Q. — M.fr.XI,66.
A. implexa, Bentham in Hook. Lond. Journ. I, 368 (1842) ... — — — V. N.S.W. Q — B.fl.II,389
A. harpophylla, F. v. M. in Benth. Fl. Austr. II, 389 (1864) ... — — — — — Q. — B.fl.II,389
A. excelsa, Bentham in Mitch. Trop. Austr. 225 (1848) ... ... — — — — — Q. — B.fl.II,390 M.fr.IV,6.
A. homaloclada, F. v. M., fragm. XI, 34 (1878) ... ... — — — — — — — M.fr.XI,34.

A. complanata, Cunningham in Hook. Lond. Journ. I, 380 (1842) — — — — N.S.W. Q. — B.fl.II,390 M.fr.XI,34.
A. dineura, F. v. M. in Journ. Linn. Soc. III, 130 (1858) ... — — — — — — N.A. B.fl.II,391 M.fr.XI,65.
A. binervata, De Candolle, prodr. II, 452 (1825) ... ... — — — — N.S.W. Q. — B.fl.II,390 M.fr.XI,66.
A. latescens, Bentham in Hook. Lond. Journ. I, 380 (1842) ... — — — — — — N.A. B.fl.II,391
A. platycarpa, F. v. M. in Journ. Linn. Soc. III, 145 (1858) ... — — — — — Q. N.A. B.fl.II,391 M.fr.XI,67.
A. flavescens, Cunningham in Hook. Lond. Journ. I, 381 (1842) — — — — — Q. — B.fl.II,391 M.fr.XI,68.
A. retivonea, F. v. M., fragm. III, 128 (1863) ... ... — S.A. — — — N.A. B.fl.II,392 M.fr.III,128.
A. triptera, Bentham in Hook. Lond. Journ. I, 325 (1842) — — — — N.S.W. Q. — B.fl.II,323
A. rhigiophylla, F. v. M. in Schlecht. Linnaea XXVI, 611 (1853) — S.A. — — — — — B.fl.II,333
A. Oxycedrus, Sieber in De Cand. prodr. II, 453 (1825) ... ... — S.A. T. V. N.S.W. — — B.fl.II,334
A. verticillata, Willdenow, spec. plant. IV, 1049 (1805) ... — S.A. T. V. N.S.W. — — B.fl.II,334
A. Riceana, Henslow in Maund, Botanist III, t. 135 (1839) ... — — T. — — — — B.fl.II,335

VIII. JULIFERAE.
A. amentifera, F. v. M. in Journ. Linn. Soc. III, 141 (1858) ... — — — — — — N.A. B.fl.II,392
A. Wickhami, Bentham in Hook. Lond. Journ. I, 379 (1842) ... — — — — — — N.A. B.fl.II,392
A. lysiphloia, F. v. M. in Journ. Linn. Soc. III, 137 (1858) ... — S.A. — — — — N.A. B.fl.II,393
A. linarioides, Bentham in Hook. Lond. Journ. I, 371 (1842) ... — — — — — — N.A. B.fl.II,393
A. stipuligera, F. v. M. in Journ. Linn. Soc. III, 144 (1858) ... — — — — — — N.A. B.fl.II,393
A. ptychophylla, F. v. M. in Journ. Linn. Soc. III, 142 (1858)... — — — — — — N.A. B.fl.II,394
A. stigmatophylla, Cunningham in Hook. Lond. Journ. I, 377 (1842) — — — — — — N.A. B.fl.II,394
A. umbellata, Cunningham in Hook. Lond. Journ. I, 378 (1842) — — — — — — N.A. B.fl.II,394
A. leptophleba, F. v. M. in Journ. Linn. Soc. III, 143 (1858) ... — — — — — — N.A. B.fl.II,395
A. limbata, F. v. M. in Journ. Linn. Soc. III, 145 (1858) ... — — — — — — N.A. B.fl.II,395
A. brevifolia, Bentham, Fl. Austr. II, 395 (1864) ... ... — — — — — Q. — B.fl.II,395
A. megalantha, F. v. M. in Journ. Linn. Soc. III, 143 (1858) .. — — — — — — N.A. B.fl.II,395
A. gonoclada, F. v. M. in Journ. Linn. Soc. III, 140 (1858) ... — — — — — — N.A. B.fl.II,396
A. pycnostachya, F. v. M., pl. of Vict. II, 33 (1860) ... ... — — — — N.S.W. — B.fl.II,396
A. subtilinervis, F. v. M., pl. of Vict. II, 32 (1860) ... ... — — — V. N.S.W. — B.fl.II,396
A. cochllocarpa, Meissner in Bot. Zeitung 10 (1855) ... ... W.A. — — — — — B.fl.II,397 M.fr.IV,8.
A. Dallachiana, F. v. M., fragm. I, 7 (1858) ... ... — — — V. — — B.fl.II,397 M.fr.I,7.
A. alpina, F. v. M., fragm. III, 129 (1863)... ... ... — — — V. — — B.fl.II,397 M.fr.III,129.
A. longifolia, Willdenow, spec. pl. IV, 1052 (1805) ... ... — S.A. T. V. N.S.W. Q. — B.fl.II,390
A. linearis, Sims, Bot. Mag. t. 2156 (1820)... ... ... — — T. V. N.S.W. — — B.fl.II,390
A. aciphylla, Bentham in Schlecht. Linnaea, XXVI, 627 (1853) W.A. — — — — — — B.fl.II,399
A. ephedroides, Bentham in Hook. Lond. Journ. I, 370 (1842)... W.A. — — — — — — B.fl.II,400
A. Burkittii, F. v. M. in Benth. Fl. Austr. II, 400 (1864) ... — S.A. — — — — — B.fl.II,400
A. microneura, Meissner in Lehm. pl. Preiss. I, 19 (1844) ... W.A. — — — — — — B.fl.II,400
A. cyperophylla, F. v. M. in Benth. Fl. Austr. II, 400 (1864) ... — S.A. — — N.S.W. Q. — B.fl.II,400
A. multispicata, Bentham, Fl. Austr. II, 400 (1864) ... W.A. — — — — — — B.fl.II,400
A. pityoides, F. v. M. in Journ. Linn. Soc. III, 135 (1858) — — — — — Q. N.A. B.fl.II,400
A. xylocarpa, Cunningham in Hook. Lond. Journ. I, 370 (1842) — — — — — — N.A. B.fl.II,401
A. gonocarpa, F. v. M. in Journ. Linn. Soc. III, 136 (1858) ... — — — — — — N.A. B.fl.II,401
A. oncinophylla, Lindley, Bot. Regist. XXV, App. XV (1839)... W.A. — — — — — — B.fl.II,401
A. conjunctifolia, F. v. M., fragm. XI, 68 (1879) ... ... — — — — — N.A. — M.fr.XI,68.
A. drepanocarpa, F. v. M. in Journ. Linn. Soc. III, 137 (1858)... — — — — — N.A. B.fl.II,402
A. arida, Bentham in Hook. Lond. Journ. I, 370 (1842) ... — — — — — — N.A. B.fl.II,402 M.fr.IV,8.
A. aneura, F. v. M. in Schlecht. Linnaea, XXVI, 627 (1853) ... W.A. S.A. — V. N.S.W. Q. — B.fl.II,402
A. cibaria, F. v. M. in Melb. Chemist, July (1882) ... W.A. — — — — — — —
A. Kempeana, F. v. M. in Melb. Chemist, July (1882) ... — S.A. — — N.S.W. Q. — —
A. conspersa, F. v. M. in Journ. Linn. Soc. III, 140 (1858) ... — — — — — — N.A. B.fl.II,403
A. Doratoxylon, Cunningham in Field's N.S. Wales, 345 (1825) — S.A. — V. N.S.W. Q. — B.fl.II,403
A. acuminata, Bentham in Hook. Lond. Journ. I, 373 (1862) ... W.A. — — — — — — B.fl.II,404 M.fr.IV,7.
A. stereophylla, Meissner in Lehm. pl. Preiss. II, 203 (1847) ... W.A. — — — — — — B.fl.II,404
A. signata, F. v. M., fragm. IV, 7 (1864) ... ... W.A. — — — — — — B.fl.II,404 M.fr.IV,7.
A. delibrata, Cunningham in Hook. Lond. Journ. I, 374 (1842) — — — — — — N.A. B.fl.II,405
A. oligoneura, F. v. M. in Journ. Linn. Soc. III, 130 (1858) ... — — — — — — N.A. B.fl.II,405
A. julifera, Bentham in Hook. Lond. Journ. I, 374 (1842) ... — — — — — — Q. — B.fl.II,405
A. Solandri, Bentham, Fl. Austr. II, 406 (1864) ... ... — — — — — Q. N.A. B.fl.II,406
A. leptostachya, Bentham, Fl. Austr. II, 406 (1864) ... ... — — — — — Q. N.A. B.fl.II,406
A. glaucescens, Willdenow, spec. plant. IV, 1052 (1805)... — — — — N.S.W. Q. — B.fl.II,406
A. Cunninghamii, Hooker, icon. plant. t. 165 (1837) ... — — — — — Q. — B.fl.II,407
A. leptocarpa, Cunningham in Hook. Lond. Journ. I, 376 (1842) — — — — — Q. N.A. B.fl.II,407
A. polystachya, Cunningham in Hook. Lond. Journ. I, 376 (1842) — — — — — Q. N.A. B.fl.II,408
A. holocarpa, Bentham, Fl. Austr. II, 408 (1864) ... ... — — — — — — N.A. B.fl.II,408 M.fr.XI,69.
A. plectocarpa, Cunningham in Hook. Lond. Journ. I, 375 (1842) — — — — — — N.A. B.fl.II,408
A. pachycarpa, F. v. M. in Journ. Linn. Soc. III, 139 (1858) ... — — — — — — N.A. B.fl.II,408
A. tumida, F. v. M. in Journ. Linn. Soc. III, 144 (1858) ... — — — — — — N.A. B.fl.II,409
A. loxocarpa, Bentham in Hook. Lond. Journ. I, 377 (1842)... — — — — — — N.A. B.fl.II,409
A. oncinocarpa, Bentham in Hook. Lond. Journ. I, 378 (1842) ... — — — — — — N.A. B.fl.II,409
A. retinervis, Bentham in Hook. Lond. Journ. I, 379 (1842) ... — — — — — — N.A. B.fl.II,410
A. aulacocarpa, Cunningham in Hook. Lond. Journ. I, 379 (1842) — — — — — Q. — B.fl.II,410
A. calyculata, Cunningham in Hook. Lond. Journ. I, 379 (1842) — — — — — — N.A. B.fl.II,410
A. crassicarpa, Cunningham in Hook. Lond. Journ. I, 379 (1842) — — — — — Q. N.A. B.fl.II,411 M.fr.XI,69.
A. auriculiformis, Cunningham in Hook. Lond. Journ. I, 377 (1842) — — — — — Q. N.A. B.fl.II,411
A. latifolia, Bentham in Hook. Lond. Journ. I, 382 (1805) ... — — — — — Q. N.A. B.fl.II,411
A. Mangium, Willdenow, spec. plant. IV, 1053 (1805) ... — — — — — Q. — —
A. holosericea, Cunningham in G. Don, gen. syst. II, 407 (1832) — — — — — Q. N.A. B.fl.II,411 M.fr.XI,36.
A. cincinnata, F. v. M., fragm. XI, 35 (1878) ... ... — — — — — Q. — M.fr.XI,35.
A. dimidiata, Bentham in Hook. Lond. Journ. I, 381 (1842) ... — — — — — — N.A. B.fl.II,412

A. humifusa, A. Cunningham in Lond. Journ. I, 382 (1842) ... — — — — — Q. N.A. B.fl.II,412
A. denticulosa, F. v. M., fragm. X, 32 (1876) ... ... W.A. — — — — — — — M.fr.X,32

IX. Pipinnatae.
A. Drummondii, Lindley, Bot. Regist. XXV, App. XV (1839)... W.A. — — — — — — B.fl.II,410
A. pulchella, R. Brown in Ait. hort. Kew. sec. ed. V, 464 (1813) W.A. — — — — — — B.fl.II,416
A. Mitchelli, Bentham in Hook. Lond. Journ. I, 387 (1842) ... — S.A. — V. — — — B.fl.II,417
A. pentadenia, Lindley, Bot. Regist. t. 1521 (1832) ... ... W.A. — — — — — — B.fl.II,417
A. Gilberti, Meissner in Lehm. pl. Preiss. II, 204 (1847) ... W.A. — — — — — — B.fl.II,417
A. nigricans, R. Brown in Ait. hort. Kew. sec. ed. V, 465 (1813) W.A. — — — — — — B.fl.II,418
A. Tayloriana, F. v. M. in Wing's S. Sc. Record, July (1882) ... W.A. — — — — — — —
A. strigosa, Link, enum. pl. hort. Berol. 444 (1822) ... ... W.A. — — — — — — B.fl.II,418
A. elata, Cunningham in Hook. Lond. Journ. I, 383 (1842) ... — — — — N.S.W. — B.fl.II,413
A. pruinosa, Cunningham in Hook. Lond. Journ. I, 383 (1842)... — — — — N.S.W. — B.fl.II,413
A. spectabilis, Cunningham in Hook. Lond. Journ. I, 383 (1842) — — — — N.S.W. Q. — B.fl.II,413
A. polybotrya, Bentham in Hook. Lond. Journ. I, 384 (1842) ... — — — — N.S.W. — B.fl.II,414
A. discolor, Willdenow, spec. plant. IV, 1068 (1805) ... ... — — — T. V. N.S.W. — B.fl.II,414
A. decurrens, Willdenow, spec. plant. IV, 1072 (1805) ... — S.A. T. V. N.S.W. Q. — B.fl.II,414
A. dealbata, Link, cnum. pl. hort. Berol. 445 (1822) ... — S.A. T. V. N.S.W. — B.fl.II,415
A. cardiophylla, Cunningham in Hook. Lond. Journ. I, 385 (1842) — — — — N.S.W. — B.fl.II,415
A. leptoclada, Cunningham in Hook. Lond. Journ. I, 385 (1842) — — — — N.S.W. Q. — B.fl.II,416
A. pubescens, R. Brown in Ait. hort. Kew. sec. ed. V, 464 (1813) — — — — N.S.W. — B.fl.II,416
A. Farnesiana, Willdenow, spec. plant. IV, 1083 (1805) ... — S.A. — — N.S.W. Q. N.A. B.fl.II,419
A. suberosa, Cunningham in Hook. Lond. Journ. I, 490 (1842)... — — — — — — N.A. B.fl.II,420 M.fr.XI,36.
A. Bidwilli, Bentham in Schlecht. Linnaea XXVI, 620 (1853)... — — — — — Q. N.A. B.fl.II,420
A. pallida, F. v. M. in Journ. Linn. Soc. III, 14 (1859) ... — — — — — — N.A. B.fl.II,421

ALBIZZIA, Durazzini in Magazzino Toscano III, part IV, 13 (1772). (Zygia 1756, Pithccolobium, Cathormion, Calliandra, Enterolobium, Serianthes.)
A. lophantha, Bentham in Hook. Lond. Journ. III, 86 (1844) ... W.A. — — — — — — B.fl.II,421
A. Sutherlandi, F. v. M., fragm. VI, 22 (1867) ... ... — — — — — — N.A. M.fr.IX,179.
A. basaltica, F. v. M., fragm. Fl. Austr. II, 422 (1864) ... — — — — — Q. — B.fl.II,422 M.fr.IX,179.
A. Thozetiana, F. v. M. in Benth. Fl. Austr. II, 422 (1864) ... — — — — — Q. — B.fl.II,422 M.fr.IV,0&176
A. amoenissima, F. v. M., fragm. VIII, 165 (1874) ... ... — — — V. N.S.W. Q. — — M.fr.IX,179.
A. procera, Bentham in Hook. Lond. Journ. III, 86 (1844) ... — — — — — — N.A. B.fl.II,422
A. pruinosa, F. v. M. in Trimen's Journ. of Bot. 9 (1872) ... — — — — N.S.W. Q. — B.fl.II,423 M.fr.IX,179.
A. canescens, Bentham, Fl. Austr. II, 423 (1864)... ... — — — — — Q. — B.fl.II,423 M.fr.IX,179.
A. Tozeri, F. v. M. in Trimen's Journ. of Bot. 10 (1872) ... — — — — — Q. — B.fl.II,424 M.fr.X,178.
A. Hendersoni, F. v. M. in Trimen's Journ. of Bot. 10 (1872)... — — — — N.S.W. Q. — — M.fr.X,178.
A. monilifera, F. v. M. in Trimen's Journ. of Bot. 10 (1872) ... — — — — — — N.A. B.fl.II,424
A. ramiflora, F. v. M., fragm. IX, 178 (1875) ... ... — — — — — — — M.fr.V,10;IX,178.

ARCHIDENDRON, F. v. M., fragm. V, 50 (1865).
A. Vaillantii, F. v. M., fragm. V, 60 (1865) ... ... — — — — — — — M.fr.V,0;IX,178.
A. Lucyi, F. v. M., fragm. VI, 201 (1868)... ... ... — — — — — — — M.fr.IX,178.

ROSACEAE.
A. L. de Jussieu, gen. pl. 334 (1789), from B. de Jussieu (1730).
PARINARIUM, A. L. de Jussieu, gen. pl. 342 (1789).
P. Nonda, F. v. M. in Benth. Fl. Austr. II, 428 (1864) ... — — — — — Q. N.A. B.fl.II,426
P. Griffithianum, Bentham in Hook. Niger-Flora 334 (1849) ... — — — — — Q. N.A. B.fl.II,426 M.fr.VIII,34
STYLOBASIUM, Desfontaines in Mém. du Mus. V, 37, t. 2 (1819). (Macrostigma.)
S. spathulatum, Desfontaines in Mém. du Mus. V, 37, t. 2 (1819) W.A. S.A. — — — — N.A. B.fl.II,427
S. lineare, Nees in Lehm. pl. Preiss. I, 95 (1844) ... ... W.A. — — — — — — B.fl.II,427
GEUM, Linné, gen. plant. 148 (1737) from Plinius. (Sieversia.)
G. urbanum, Linné, spec. plant. 501 (1753) ... ... — S.A. T. V. N.S.W. — — B.fl.II,428
G. renifolium, F. v. M. in Transact. phil. Inst. Vict. II, 66 (1857) — — T. — — — — B.fl.II,428
POTENTILLA, Linné, gen. plant. 147 (1737) from Camerarius (1586).
P. anserina, Linné, spec. plant. 495 (1753) ... ... — S.A. T. V. — — — B.fl.II,429
RUBUS, Tournefort, Inst. 614, t. 385 (1700) from Plinius.
R. Gunnianus, Hooker, Icon. plant. t. 291 (1840) ... ... — — T. — — — — B.fl.II,430
R. parvifolius, Linné, spec. plant. 1197 (1753) ... ... — S.A. T. V. N.S.W. Q. — B.fl.II,430 M.fr.IV,30.
R. rosifolius, Smith, plantar. icon. ex herb. Linn. t. 60 (1790) ... — — — V. N.S.W. Q. N.A. B.fl.II,430 M.fr.IV,31;VIII,34.
R. Moluccanus, Linné, spec. plant. 1197 (1753) ... ... — — — V. N.S.W. Q. N.A. B.fl.II,430 M.fr.IV,31;VIII,34.
R. Moorei, F. v. M. in Transact. phil. Inst. Vict. II, 67 (1857)... — — — V. N.S.W. Q. — B.fl.II,431 M.fr.IV,29&176;VIII,34.
AGRIMONIA, Tournefort, inst. 301, t. 155 (1700) from C. Bauhin (1623).
A. Eupatoria, Linné, spec. plant. 448 (1753) ... ... — — — — N.S.W. — — M.fr.VIII,34.
ALCHEMILLA, Linné, gen. pl. 30 (1737) from C. Bauhin (1623). (Alchimilla.)
A. vulgaris, C. Bauhin, pinax 319 (1623) ... ... — — — V. N.S.W. — — B.fl.II,432
ACAENA, Mutis in Linné, mantiss. II, 200 (1771). (Ancistrum.)
A. ovina, Cunningham in Field's N.S. Wales, 358 (1825) ... W.A. S.A. T. V. N.S.W. Q. — B.fl.II,433
A. sanguisorbae, Vahl, enum. plant. I, 294 (1804) ... ... — S.A. T. V. N.S.W. Q. — B.fl.II,434
A. montana, J. Hooker in Hook. Lond. Journ. VI, 276 (1847) ... — — — T. — — — B.fl.II,434

SAXIFRAGEAE.
Ventenat, Tabl. III, 277 (1799).
ARGOPHYLLUM, R. et G. Forster, charact. gen. 29, t. 15 (1776).
A. Lejourdanii, F. v. M., fragm. IV, 33 (1864) ... ... — — — — N.S.W. Q. — B.fl.II,437 M.fr.VI,168.

ABROPHYLLUM, J. Hooker in Benth. fl. Austr. II, 437 (1864).   (Brachynema.)
A. ornans, J. Hooker in Benth. pl. Austr. II, 437 (1864)   ...   — — — — N.S.W. Q.   —   B.fl.II,437   M.fr.VI,189;VII,150

CUTTSIA, F. v. M., fragm. V, 47 (1865).
C. viburnea, F. v. M., fragm. V, 47 (1865)... ...   ...   — — — — N.S.W. Q.   —   —   M.fr.V,47&189;VI,189

COLMEIROA, F. v. M., fragm. VII, 149 (1871).
C. carpodetoides, F. v. M., fragm. VII, 149 (1871)   ...   ...   — — — — N.S.W. — — —   M.fr.VII,149.

QUINTINIA, A. de Candolle, Mou. Campan. 92 (1830).
Q. Sieberi, A. De Candolle, Monogr. Camp. 92 (1830) ...   ...   — — — — N.S.W. Q.   —   B.fl.II,438   M.fr.III,126;VI,189.
Q. Verdonii, F. v. M., fragm. II, 225 (1861)   ...   ...   — — — — N.S.W. Q.   —   B.fl.II,438   M.fr.VI,180.
Q. Fawkneri, F. v. M., fragm. VI, 92 (1867)   ...   .   — — — — — Q.   —   —   M.fr.VII,150.

POLYOSMA, Blume, Bijdr. 658 (1825).
P. Cunninghamii, J. J. Bennett, pl. Jav. rar. 196 (1840)   ...   — — — — N.S.W. —   B.fl.II,438   M.fr.VI.189.
P. alangiacea, F. v. M., fragm. VIII, 8 (1872)   ...   .   — — — — — Q.   —   —   M.fr.VIII,8.

ANOPTERUS, Labillardière, Nov. Holl. pl. spec. I, 86, t. 112 (1804.)
A. glandulosus, Labillardière, Nov. Holl. pl. spec. I, 86, t. 112 (1804)   — — T. — — — —   B.fl.II,439
A. Macleayanus, F. v. M. in Journ. Pharm. Soc. Vict. 43 (1858)   — — — — N.S.W. Q.   —   B.fl.II,439   M.fr.VI,188;VII,150.

CALLICOMA, Andrews, Bot. Reposit. t. 566 (1809).   (Calycomis partly.)
C. serratifolia, Andrews, Bot. Reposit. t. 566 (1809)   ...   — — — — N.S.W. Q.   —   B.fl.II,440   M.fr.V,32.
C. Stutzeri, F. v. M., fragm. V, 34 (1865) ...   —   —   —   —   —   —   Q.   —   —   M.fr.VI,188&232;VII,150.

ANODOPETALUM, Cunningham in Endl. gen. 818 (1838).
A. biglandulosum, Cunningham in J. Hook. Fl. Tasm. I, 148 (1860) —   — T. — — — —   B.fl.II,440   M.fr.VI,189.

APHANOPETALUM, Endlicher, nov. stirp. decad. I, 34 (1839).   (Platyptelea.)
A. resinosum, Endlicher, nov. stirp. dec. I, 35 (1839)   ...   ... — V. N.S.W. Q.   —   B.fl.II,441   M.fr.I,228.
A. occidentale, F. v. M., fragm. I, 228 (1859)   ...   ...   W.A. — — — — —   —   B.fl.II,441   M.fr.XI,20.

CERATOPETALUM, Smith, specim. of the Bot. of New Holland, 9 (1793).
C. gummiferum, Smith, specim. of the Bot. of New Holl. t. 3 (1793)   — — — — N.S.W. — — —   B.fl.II,442   M.fr.VI,189;VII,150.
C. apetalum, D. Don in Edinb. new phil. journ. (1830) ..   — — — — N.S.W. — — —   B.fl.II,442   M.fr.VI,189;VII,150.

SCHIZOMERIA, D. Don in Edinb. new phil. journ. IX, 94 (1830).
S. ovata, D. Don in Edinb. new phil. journ. IX, 94 (1830)   — — — — N.S.W. Q.   —   B.fl.II,443   M.fr.VI,189.

ACROPHYLLUM, Bentham in Maund & Henslow, Botanist II, 95 (1840).   (Calycomis partly.)
A. venosum, Bentham in Maund & Henslow, Bot. II, 95 (1840)   — — — — N.S.W. —   —   B.fl.II,443   M.fr.VI,189;VII,150.

WEINMANNIA, Linné, syst. ed. X, 1005 (1759).   (Ackama.)
W. paniculosa. F. v. M., fragm. II, 175 (1861)   ...   ...   — — — — N.S.W. Q.   —   B.fl.II,444   M.fr.II,83; VI,188; VII,
W. Biagiana, F. v. M., fragm. V, 16 (1865)   ...   .   — — — — — Q.   —   —   M.fr.VII,150.   [150.
W. lachnocarpa, F. v. M., fragm. VIII, 7 (1872)...   ...   ...   — — — — N.S.W. Q.   — —   M.fr.VIII,7.
W. rubifolia, Bentham, Fl. Aust. II, 445 (1864) ...   .   — — — — N.S.W. Q.   —   B.fl.II,445   M.fr.II,82&175;VI,181.
W. Benthami, F. v. M., Docum. intercol. Exhibit 31 (1866) ...   — — — — — —   —   B.fl.II,446   M.fr.V,180;VI,188.

GILLBEEA, F. v. M., fragm. V, 17 (1865).
G. adenopetala, F. v. M., fragm. V, 17 (1865)   ...   ...   — — — — — — Q.   —   —   M.fr.V,17;VI,188.

DAVIDSONIA, F. v. M., fragm. VI, 4 (1867).
D. pruriens, F. v. M., fragm. VI, 4 (1867)...   ...   ...   — — — — — — Q.   —   —   M.fr.VI,249.

TETRACARPAEA, J. Hooker, icon. plant. t. 264 (1840).
T. Tasmanica, J. Hooker, icon. plant. t. 264 (1840)   ...   ... — — T. — — — —   B.fl.II,445

EUCRYPHIA, Cavanilles, icon. IV, 48, t. 372 (1797).   (Carpodontos.)
E. Billardieri, Spach, veg. phan. V, 344 (1836)   ...   ...   — — T. — — — —   B.fl.II,446   M.fr.VI,189;VII,150.
E. Moorei, F. v. M., fragm. IV, 2 (1864)   ...   .   — V. N.S.W. — —   B.fl.II,447   M.fr.IV,176;VI,189;VII   [150

BAUERA, Banks & Kennedy in Andrews, Bot. Rep. t. 198 (1793).
B. rubioides, Andrews, Bot. Reposit. t. 198 (1793)   ...   — S.A. T. V. N.S.W. —   —   B.fl.II,447   M.fr.IV,23.
B. capitata, Seringe in De Cand. prodr. IV, 13 (1830)   ...   — — V. N.S.W. Q.   —   B.fl.II,448   M.fr.IV,24;VII,150
B. sessiliflora, F. v. M. in Transact. phil. Soc. Vict. I, 41 (1854)   — — — V. — — —   B.fl.II,448   M.fr.IV,24;VII,150

EREMOSYNE, Endlicher in Hueg. enum. pl. Nov. Holl. austr. occ. 53 (1837).
E. pectinata, Endlicher in Hueg. enum. pl.53 (1837)   ...   W.A. — — — — —   B.fl.II,449   M.fr.VI,189.

CEPHALOTUS, Labillardière, Nov. Holl. pl. spec. II, 7, t. 145 (1806).
C. follicularis, Labillardière, Nov. Holl. pl. spec. II, 7, t. 145 (1806) W.A. — — — — —   B.fl.II,449   M.fr.VI,189.

CRASSULACEAE.
De Candolle, Bull. de la soc. philom. n. 49 (1801).

TILLAEA, Micheli, nov. gen. 22, t. 20 (1729).   (Bulliarda, Crassula subgentis).
T. verticillaris, De Candolle, prodr. III, 382 (1828)   ...   W.A. S.A. T. V. N.S.W. Q.   —   B.fl.II,451   M.fr.XI,118.
T. purpurata, J. Hooker in Lond. Journ. of Bot. VI, 472 (1847)   W.A. S.A. T. V. N.S.W. —   B.fl.II,451   M.fr.XI,118
T. macrantha, J. Hooker in Hook. icon. plant. t. 310 (1841)   ...   — S.A. T. V. N.S.W. —   B.fl.II,452   M.fr.XI,117.
T. pedicellosa, F. v. M., fragm. XI, 118 (1881)   ...   W.A. — — — — — —   —   —   M.fr.XI,118.
T. intricata, Nees in Lehm. pl. Preiss. I, 278 (1844)   ...   W.A. — — — — — —   —   —   M.fr.XI,117.
T. recurva, J. Hooker, Fl. Tasm. I, 146 (1860)   ...   ...   — — T. — — — —   B.fl.II,452   M.fr.XI,118.

HAMAMELIDEAE.
R. Brown in Abel, Narr. journ. Chin. 374 (1818).

Genus and species undetermined   ...   ...   — — — — — — Q.   —   —

ONAGREAE.

Adanson, Familles des plantes II, 81 (1763), from B. de Jussieu (1759).

OENOTHERA, Linné, syst. nat. 8 (1735); Linné, gen. plant. 112 (1737).
O. Tasmanica, J. Hooker, Fl. Tasm. I, 119 (1860) ... ... — T. — — — B.fl.III,303

EPILOBIUM, Dillenius in Linné, syst. nat. 8 (1735), from Goener (1542).
E. tetragonum, Linné, spec. plant. 348 (1753) ... ... W.A. S.A. T. V. N.S.W. Q. — B.fl.III,305

JUSSIEUA, Linné, gen. plant. 126 (1737). (Jussiaea.)
J. repens, Linné, spec. plant. 388 (1753) ... ... — S.A. — V. N.S.W. Q. — B.fl.III,300
J. suffruticosa, Linné, spec. plant. 388 (1753) ... ... — — N.S.W. Q. N.A. B.fl.III,307 M.fr.III,13.

LUDWIGIA, Linné, coroll. 3 (1737).
L. parviflora, Roxburgh, Fl. Ind. edit. Carey. I, 440 (1820) — — — — — Q. N.A. B.fl.III,307 M.fr.III,120;IV,175.

SALICARIEAE.

Adanson, Familles des plantes II, 232 (1763), from B. de Jussieu (1759).

ROTALA, Linné, mantissa altera 143 (1771).
R. verticillaris, Linné, mantissa altera 143 (1771) ... ... — — — — — N.A. B.fl.III,295 M.fr.III,108.
R. Roxburghiana, Wight, Icon. t. 260 (1840) ... ... — — — — — Q. N.A. B.fl.III,296 M.fr.XI,118.
R. diandra, F. v. M. in Transact. Roy. Soc. of N.S.Wales 89 (1880) — — — — — — N.A. B.fl.III,296

AMMANNIA, Houston in Linné, gen. plant. 337 (1737).
A. crinipes, F. v. M. in Transact. phil. Soc. Vict. III, 49 (1858) — — — — — — N.A. B.fl.III,296
A. triflora, R. Brown in Benth. Fl. Austr. III, 297 (1866) — — — — — — N.A. B.fl.III,297
A. baccifera, Linné, spec. plant. ed. sec. 175 (1762) ... ... — S.A. — — — Q. N.A. B.fl.III,297
A. auriculata, Willdenow, hort. Berol. t. 7 (1806) ... ... — — — — — Q. — B.fl.III,297
A. multiflora, Roxburgh, Fl. Ind. edit. Carey. I, 447 (1820) ... — S.A. — V. N.S.W. Q. N.A. B.fl.III,298

LAGERSTROEMIA, Linné, syst. ed. X, 1076 (1759).
L. Indica, Linné, spec. plant. 734 (1753) ... ... — — — — Q. — M.fr.VIII,35.
L. Flos Reginae, Retzius, observ. botan. V, 25 (1789) ... ... — — — — Q. — M.fr.XII.

LYTHRUM, Linné, syst. nat. 8 (1737).
L. Salicaria, Linné, spec. plant. 446 (1753) ... ... — S.A. T. V. N.S.W. Q. — B.fl.III,298 M.fr.VII,147.
L. Hyssopifolia, Linné, spec. plant. 447 (1753) ... ... — S.A. T. V. N.S.W. Q. — B.fl.III,299 M.fr.VII,147.

NESÆA, Commerçon in A. L. de Jussieu, gen. plant. 332 (1789).
N. Robertsii, F. v. M., fragm. VII, 145 (1871) ... ... — — — — — Q. — M.fr.VII,145.
N. Aruhemica, F. v. M., fragm. VII, 146 (1871) ... ... — — — — — N.A. B.fl.III,299 M.fr.III,109;VII,146.

PEMPHIS, R. et G. Forster, charact. gen. 67, t. 34 (1776).
P. acidula, R. & G. Forster, charact. gen. 67, t. 34 (1776) ... — — — — — Q. N.A. B.fl.III,300 M.fr.VII,147.

LAWSONIA, Linné, gen. plant. 111 (1737).
L. alba, Lamarck, Encycl. méth. III, 106 (1789) ... ... — — — — — N.A. B.fl.III,301

HALORAGEAE.

R. Brown in Flind. voy. II, 549 (1814).

LOUDONIA, Lindley, Bot. Regist. XXV, App. XLII (1839). (Glischrocaryon.)
L. aurea, Lindley, Bot. Regist. XXV, App. XLII (1839) ... W.A. — — — — B.fl.II,472 M.fr.VIII,162;XI,20.
L. Behrii, Schlechtendal, Linnaea XX, 648 (1847) ... ... — S.A. — V. N.S.W. — B.fl.II,472 M.fr.VIII,162.
L. Roei, Schlechtendal, Linnaea XX, 648 (1847) ... ... W.A. — — — — B.fl.II,472 M.fr.VIII,162.

HALORAGIS, R. et G. Forster, charact. gen. 61, t. 31 (1776). (Cercodia, Goniocarpus.)
H. Gossei, F. v. M., fragm. VIII, 161 (1874) ... ... — S.A. — — — N.A. — M.fr.XI,134.
H. trigonocarpa, F. v. M., fragm. X, 64 (1876) ... ... W.A. S.A. — — — — M.fr.X,84.
H. digyna, Labillardière, Nov. Holl. pl. spec. I, 101, t. 129 (1804) W.A. — — — — B.fl.II,475
H. mucronata, Bentham, Fl. Austr. II, 475 (1864) ... W.A. S.A. — V. N.S.W. — B.fl.II,475
H. pithyoides, Bentham, Fl. Austr. II, 476 (1864) ... W.A. — — — — B.fl.II,476
H. digyna, Fenzl in Hueg. enum. 45 (1837) ... ... W.A. — — — — B.fl.II,476
H. elata, Cunningham in Hueg. enum. 45 (1837) ... ... — S.A. — V. N.S.W. Q. — B.fl.II,476 M.fr.VIII,162.
H. tenuifolia, Bentham, Fl. Austr. II, 477 (1864) ... ... W.A. — — — — B.fl.II,477
H. scoparia, Fenzl in Hueg. enum. 45 (1837) ... ... W.A. — — — — B.fl.II,477 M.fr.X,54.
H. aculeata, Bentham, Fl. Austr. II, 477 (1864) ... ... W.A. — — — — B.fl.II,477 M.fr.X,54.
H. foliosa, Bentham, Fl. Austr. II, 477 (1864) ... ... W.A. — — — — B.fl.II,477
H. confertifolia, F. v. M., fragm. X, 53 (1876) ... ... W.A. — — — — B.fl.II,478 M.fr.X,54.
H. polygarpa, Bentham, Fl. Austr. II, 478 (1864) ... ... W.A. — — — — B.fl.II,478
H. ceratophylla, Zahlbruckner in Endl. Atakta, 16, t. 15 (1833) — S.A. T. V. N.S.W. Q. N.A. B.fl.II,478
H. acutangula, F. v. M. in Transact. Vict. Inst. 125 (1854) ... — S.A. — — — — B.fl.II,478
H. hexandra, F. v. M., fragm. III, 31 (1862) ... ... W.A. — — — — B.fl.II,478 M.fr.III,31.
H. odontocarpa, F. v. M., fragm. I, 108 (1859) ... ... — S.A. — V. N.S.W. — B.fl.II,479 M.fr.I,108;X,54.
H. serra, Brongniart in Duperr. Voy. sur la Coquill. t. 69 (1829) ... — — — N.S.W. — B.fl.II,479
H. glauca, Lindley in Mitch. Trop. Austr. 91 (1848) ... — — — N.S.W. — B.fl.II,479
H. alata, N. J. Jacquin, Icon. plant. rar. I, 7, t. 69 (1786) ... — — — N.S.W. — B.fl.II,479
B. ramosa, Labillardière, Nov. Holl. pl. spec. I, 100, t. 128 (1804) W.A. — — — — B.fl.II,480 M.fr.VIII,162.
H. rotundifolia, Bentham, Fl. Austr. II, 480 (1864) ... W.A. — — — — B.fl.II,480 M.fr.VIII,162.
H. rudis, Bentham, Fl. Austr. II, 480 (1864) ... ... W.A. — — — — B.fl.II,480
H. paniculata, R. Brown in Benth. Fl. Austr. II, 481 (1864) ... W.A. — — — — B.fl.II,481
H. pusilla, R. Brown in Benth. Fl. Austr. II, 481 (1864) ... W.A. — — — — B.fl.II,481
H. intricata, Bentham, Fl. Austr. II, 481 (1864) ... ... W.A. — — — — B.fl.II,481
H. nodulosa, Walpers, repert. V, 672 (1846) ... ... W.A. — — — — B.fl.II,481
H. trichostachya, Bentham, Fl. Austr. II, 481 (1864) ... W.A. — — — — B.fl.II,482
H. lanceolata, R. Brown in Benth. Fl. Austr. II, 481 (1864) ... W.A. — — — — B.fl.II,482
H. micrantha, R. Brown in Flind. Voy. II, 550 (1814) ... — S.A. T. V. N.S.W. Q. — B.fl.II,482 M.fr.VIII,162;X,54.

H. stricta, R. Brown in Benth. Fl. Austr. II, 482 (1864) ... — — — — — Q. — B.fl.II,482
H. heterophylla, Brongniart in Duperr. Voy. t. 68 (1829) ... — S.A. — V. N.S.W. Q. — B.fl.II,483
H. pinnatifida, A. Gray, Bot. Wilk. Expl. Exped. I, 627 (1854) — — — — N.S.W. — — B.fl.II,483
H. leptothoca, F. v. M., fragm. III, 32 (1862) ... ... — — — — — — N.A. B.fl.II,483 M.fr.VIII,162.
H. tetragyna, R. Brown in Flind. Voy. II, 550 (1814) ... — S.A. T. V. N.S.W. Q. — B.fl.II,484 M.fr.IV,26.
H. touerioides, A. Gray, Bot. Wilk. Expl. Exped. I, 625 (1854) W.A. S.A. T. V. N.S.W. — — B.fl.II,484
H. scordioides, Bentham, Fl. Austr. II, 485 (1864) ... W.A. — — — — — — B.fl.II,485
H. depressa, Walpers, repertor. II, 99 (1843) ... ... — T. V. — — B.fl.II,485
H. salsoloides, Bentham, Fl. Austr. II, 485 (1864) ... — — — — N.S.W. — — B.fl.II,485

MEIONECTES, R. Brown in Flind. Voy. II, 550 (1814).
M. Brownii, J. Hooker in Hook. icon. plant. t. 306 (1841) ... W.A. S.A. T. V. N.S.W. — — B.fl.II,486 M.fr.VIII,163.

MYRIOPHYLLUM, l'Ecluse, rar. stirp. hist. II, 252 (1583) from Dioscorides. (Pelonastes).
M. variifolium, J. Hooker in Hook. icon. plant. t. 289 (1840) ... W.A. S.A. T. V. N.S.W. Q. — B.fl.II,487
M. elatinoides, Gaudichaud in Ann. des sc. nat. V, 105 (1825) ... — S.A. T. V. N.S.W. — — B.fl.II,487
M. verrucosum, Lindley in Mitch. Trop. Austr. 384 (1848) ... W.A. S.A. — V. N.S.W. Q. N.A. B.fl.II,488 M.fr.VIII,163.
M. latifolium, F. v. M., fragm. II, 87, (1860) ... ... — — — — N.S.W. Q. — B.fl.II,488 M.fr.II,87;IV,172;X,54.
M. Muelleri, Sonder in Linnaea, XXVIII, 233 (1855) ... W.A. S.A. — V. — — — B.fl.II,488 M.fr.VIII,162.
M. amphibium, Labillardière, Nov. Holl. pl. spec. II,70, t. 220(1806) — S.A. T. V. — — — B.fl.II,489 M.fr.VIII,163.
M. pedunculatum, J. Hooker in Lond. Journ. VI, 474 (1847) ... W.A. S.A. T. V. — — — B.fl.II,489 M.fr.VIII,162.
M. dicoccum, F. v. M. in Transact. phil. Inst. Vict. III, 41 (1858) — — — — — N.A. B.fl.II,489 M.fr.VIII,162.
M. trachycarpum, F. v. M., fragm. II, 87 (1860) ... — — — — — N.A. B.fl.II,488 M.fr.II,87.
M. gracile, Bentham, Fl. Austr. II, 489 (1864) ... ... — — — — N.S.W. Q. — B.fl.II,489
M. filiforme, Bentham, Fl. Austr. II, 489 (1864) ... ... — — — — — N.A. B.fl.II,489
M. integrifolium, J. Hooker, Fl. Tasm. I, 123, t. 23 (1860) ... W.A. S.A. T. V. N.S.W. — — B.fl.II,490 M.fr.VIII,162.
M. Drummondii, J. Hooker, Fl. Tasm. I, 123, implied (1860) ... W.A. — — — — — — B.fl.II,490

GUNNERA, Linné, mantiss. plant. 16 (1767). (Milligania.)
G. cordifolia, J. Hooker, Fl. Tasm. I, 125 (1860) ... ... — — T. — — — — B.fl.II,491

CERATOPHYLLUM, Linné, syst. nat. 9 (1735); Linné, gen. plant. 290 (1737).
C. demersum, Linné, spec. plant. 992 (1753) ... ... — S.A. — V. N.S.W. Q. — B.fl.II,491 N.fr.VIII,163.

CALLITRICHE, Linné, syst. nat. ed. sext. 82 (1748).
C. verna, Linné, flor. Suec. edit. sec. (1755) ... ... W.A. S.A. T. V. N.S.W. Q. — B.fl.II,492
C. Muelleri, Sonder in Linnaea XXVIII, 229 (1855) ... ... — — V. N.S.W. — — — M.fr.VIII,163.

RHIZOPHOREAE.
R. Brown in Flind. Voy. II, 549 (1814).

RHIZOPHORA, Linné, gen. plant. 137 (1737).
R. mucronata, Lamarck, Encycl. méth. VI, 189 (1804) ... ... — — — N.S.W. Q. N.A. B.fl.II,493 M.fr.IX,159.

CERIOPS, Arnott in Jardine, Ann. of nat. hist. I, 363 (1838).
C. Candolleana, Arnott in Ann. of nat. hist. I, 364 (1838) ... — — — — — Q. N.A. B.fl.II,494 M.fr.IX,159.

BRUGUIERA, Lamarck, Encycl. méth. IV, 606, t. 397 (1796).
B. Rheedii, Blume, enum. pl. Jav. 92 (1827) ... ... — — — — — Q. N.A. B.fl.II,494 M.fr.IX,159.
B. caryophylloides, Blume, enum. pl. Jav. 93 (1827) ... ... — — — — — — M.fr.IX,159.
B. gymnorrhiza, Lamarck, Encycl. méth. IV, 696, t. 397 (1796) — — — — — Q. N.A. B.fl.II,495 M.fr.IX,159.

CARALLIA, Roxburgh, in Flind. Voy. II, 549 (1814).
C. integerrima, De Candolle, prodr. III, 33 (1828) ... ... — — — — — Q. N.A. B.fl.II,495 M.fr.IX,159.

COMBRETACEAE.
R. Brown, prodr. 351 (1810).

TERMINALIA, Linné, mantiss. plant. 21 (1767). (Chuncoa.)
T. platyptera, F. v. M., fragm. II, 151 (1861) ... ... — — — — — N.A. B.fl.II,498 M.fr.II,151;IX,160.
T. volucris, R. Brown in Benth. Fl. Austr. II, 498 (1864) ... — — — — — N.A. B.fl.II,498
T. oblongata, F. v. M., fragm. II, 152 (1861) ... ... — — — — — Q. — B.fl.II,499 M.fr.III,167;IX,160.
T. bursarina, F. v. M., fragm. II, 149 (1861) ... ... — — — — — N.A. B.fl.II,499 M.fr.II,149.
T. circumalata, F. v. M., fragm. III, 91 (1862) ... ... — — — — — N.A. B.fl.II,499 M.fr.III,91.
T. pterocarya, F. v. M., fragm. II, 152 (1861) ... ... — — — — — N.A. B.fl.II,499 M.fr.II,152.
T. Thozetii, Bentham, Fl. Austr. II, 500 (1864) ... ... — — — — — Q. — B.fl.II,500 M.fr.IX,160.
T. Catappa, Linné, mantiss. pl. 128 (1767)... ... ... — — — — — Q. — B.fl.II,500 M.fr.II,100.
T. melanocarpa, F. v. M., fragm. II, 02 (1802) ... ... — — — — — Q. N.A. B.fl.II,500 M.fr.II,02;IX,160.
T. Muelleri, Bentham, Fl. Austr. II, 500 (1864) ... ... — — — — — N.A. B.fl.II,500 M.fr.IX,160.
T. latipes, Bentham, Fl. Austr. II, 501 (1864) ... ... — — — — — N.A. B.fl.II,501
T. edulis, F. v. M., fragm. II, 151 (1861) ... ... — — — — — N.A. B.fl.II,501 M.fr.II,151;IX,160.
T. discolor, F. v. M., fragm. III, 92 (1862) ... ... — — — — — N.A. B.fl.II,501 M.fr.III,92.
T. porphyrocarpa, F. v. M. in Benth. Fl. Austr. II, 501 (1864) — — — — — N.A. B.fl.II,501
T. platyphylla, F. v. M., fragm. II, 150 (1861) ... ... — — — — — N.A. B.fl.II,502 M.fr.II,150.
T. sericocarpa, F. v. M., fragm. IX, 159 (1875) ... ... — — — — — Q. — M.fr.IX,150.
T. microcarpa, Decaisne in Ann. du Mus. II, 150 (1866) ... — — — — — N.A. B.fl.II,502
T. petiolaris, Cunningham in Benth. Fl. Aust. II, 502 (1864) ... — — — — — N.A. B.fl.II,502 M.fr.II,150.
T. erythrocarpa, F. v. M., fragm. II, 150 (1861) ... ... — — — — — N.A. B.fl.II,503 M.fr.IX,100.
T. grandiflora, Bentham, Fl. Austr. II, 503 (1864) ... ... — — — — — N.A. B.fl.II,503 M.fr.IX,100.

LUMNITZERA, Willdenow, in den Verh. der Ges. Nat. Freunde zu Berl. IV, 186 (1803).
L. coccinea, Wight & Arnott, prodr. 316 (1834) ... ... — — — — — Q. — B.fl II,503 M.fr.IX,160.
L. racemosa, Willd. in Verh. der Nat. Freunde zu Berl. IV,186(1803) — — — — — Q. N.A. B.fl.II,304 M.fr.IX,160.

MACROPTERANTHES, F. v. M., fragm. III, 91 (1862).
M. montana, F. v. M., fragm. III, 91 (1862) ... ... — — — — — Q. — B.fl.II,504 M.fr.II,149;III,91&166.
M. Kekwickii, F. v. M., fragm. III, 151 (1863) ... ... — — — — — N.A. B.fl.II,504 M.fr.III,151.
M. Leichhardtii, F. v. M., fragm. III, 91 (1862) ... ... — — — — — Q. — B.fl.II,505 M.fr.III,91.
M. Fitzalani, F. v. M., fragm. VIII, 160 (1874) ... ... — — — — — Q. — — M.fr.VIII,160.
GYROCARPUS, N. Jacquin, select. stirp. Amer. hist. 282 (1763).
G. Americanus, N. Jacquin, select. stirp.Amer. hist. 282, t.178(1763) — — — — — Q. N.A. B.fl.II,505 M.fr.IX,160.

MYRTACEAE.

Adanson, Familles des plantes II, 86 (1763), from B. de Jussieu (1759).

ACTINODIUM, Schauer in Schlecht. Linnaea X, 311 (1835). (Triphelia.)
A. Cunninghamii, Schauer in Lindl. introd. to the nat. syst., scc.
ed. 440 (1835) ... ... W.A. — — — — — — B.fl.III,5
DARWINIA, Rudge in Transact. Linn. Soc. XI, 299 (1813). (Genetyllis, Chamaelaucium, Homoranthus, Hedaroma, Polyzone, Francisia, Decalophium, Schuermannia, Cryptostemon.)
D. macrostegia, Bentham in Journ. Linn. Soc. IX, 170 (1865) ... W.A. — — — — — — B.fl.III,8
D. Hookeriana, Bentham in Journ. Linn. Soc. IX, 179 (1865) ... W.A. — — — — — — B.fl.III,0 M.fr.VIII,182.
D. fimbriata, Bentham in Journ. Linn. Soc. IX, 179 (1865) ... W.A. — — — — — — B.fl.III,0 M.fr.VIII,182;IX,177.
D. speciosa, Bentham in Journ. Linn. Soc. IX, 179 (1865) ... W.A. — — — — — — B.fl.III,9 M.fr.XI,10.
D. Meisneri, Bentham in Journ. Linn. Soc. IX, 179 (1865) ... W.A. — — — — — — B.fl.III,9 M.fr.VIII,182;IX,177.
D. helichrysoides, Bentham in Journ. Linn. Soc. IX, 170 (1865) W.A. — — — — — — B.fl.III,10
D. Neildiana, F. v. M., fragm. IX, 177 (1875) ... ... W.A. — — — — — — M.fr.IX,177.
D. oederoides, Bentham in Journ. Linn. Soc. IX, 170 (1865) ... W.A. — — — — — — B.fl.III,10 M.fr.IX,177;XI,9.
D. virescens, Bentham in Journ. Linn. Soc. IX, 179 (1865) ... W.A. — — — — — — B.fl.III,10 M.fr.XI,10.
D. Oldfieldii, Bentham in Journ. Linn. Soc. IX, 180 (1865) ... W.A. — — — — — — B.fl.III,10
D. purpurea, Bentham in Journ. Linn. Soc. IX, 180 (1865) ... W.A. — — — — — — B.fl.III,11 M.fr.XI,9. {177;XI,9.
D. citriodora, Bentham in Journ. Linn. Soc. IX, 180 (1865) ... W.A. — — — — — — B.fl.III,11 M.fr.I,169;IV,174;IX,
D. thymoides, Bentham in Journ. Linn. Soc. IX, 180 (1865) ... W.A. — — — — — — B.fl.III,11 M.fr.VIII,182;IX,176.
D. taxifolia, Cunningham in Field's N.S.Wales 352 (1825) ... — — — V. N.S.W. — B.fl.III,12 M.fr.IX,176.
D. vestita, Bentham in Journ. Linn. Soc. IX, 180 (1865) ... W.A. — — — — — — B.fl.III,12 M.fr.VIII,182;IX,176.
D. pauciflora, Bentham in Journ. Linn. Soc. IX, 180 (1865) ... W.A. — — — — — — B.fl.III,12 M.fr.XI,9.
D. diosmoides, Bentham in Journ. Linn. Soc. IX, 180 (1865) ... W.A. — — — — — — B.fl.III,13 M.fr.IX,176.
D. fascicularis, Rudge in Transact. Linn. Soc. XI, 290, t. 22 (1813) — — — N.S.W. — B.fl.III,13 M.fr.IX,176.
D. pinifolia, Bentham in Journ. Linn. Soc. IX, 181 (1865) ... W.A. — — — — — — B.fl.III,14
D. rhadinophylla, F. v. M., fragm. IX, 175 (1875) ... W.A. — — — — — — M.fr.IX,175.
D. sanguinea, Bentham in Journ. Linn. Soc. IX, 181 (1865) ... W.A. — — — — — — B.fl.III,14 M.fr.IX,176.
D. micropetala, Bentham in Journ. Linn. Soc. IX, 181 (1865) ... — S.A. — — — — B.fl.III,14 M.fr.I,12;IX,176.
D. Schuermanni, Bentham in Journ. Linn. Soc. IX, 181 (1865)... — S.A. — — — — B.fl.III,14 M.fr.I,12.
D. verticordina, Bentham in Journ. Linn. Soc. IX, 181 (1865) ... W.A. — — — — — — B.fl III,15 M.fr.IV,57.
D. Thomasii, Bentham in Journ. Linn. Soc. IX, 181 (1865) ... — — — — — Q. — B.fl.III,15 M.fr.V,137.
D. virgata, F. v. M., fragm. IX, 176 (1875) ... ... — — — N.S.W. Q. — B.fl.III,15 M.fr.IX,176.
D. ciliata, F. v. M.; Chamaelaucium, Desfontaines in Mém. du
Mus. Par. V, 40, t. 3 (1821) ... ... W.A. — — — — — — B.fl.III,36 M.fr.VIII,182.
D. gracilis; Chamaelaucium, F. v. M., fragm. IV, 62 (1864) ... W.A. — — — — — — B.fl.III,36 M.fr.IV,62;VIII,182.
D. Forrestii, F. v. M., fragm. XI, 9 (1878)... ... ... W.A. — — — — — — M.fr.XI,9.
D. heterandra, F. v. M.; Cham., Benth., Fl. Austr. III, 36 (1866)... W.A. — — — — — — B.fl.III,36
D. Drummondii, F. v. M.; Chamaelaucium, Meisener in Journ.
Linn. Soc. I, 44 (1857) ... ... W.A. — — — — — — B.fl.III,37
D. Endlicheri, F. v. M.; Chamaelaucium virgatum, Endlicher in
Ann. Wien. Mus. II, 193 (1838) ... ... W.A. — — — — — — B.fl.III,37
D. brevifolia, F. v. M.; Cham., Benth., Fl. Austr. III, 37 (1866)... W.A. — — — — — — B.fl.III,37
D. uncinata, F. v. M.; Cham., Schauer in Pl. Preiss. I, 97 (1844) W.A. — — — — — — B.fl.III,37
D. megaloptala; Cham., F. v. M. in Benth. Fl. Austr. III,38 (1866) W.A. — — — — — — B.fl.III,38 M.fr.VIII,182.
D. Turczaninowii, F. v. M.; Chamaelaucium pauciflorum, Benth.,
Fl. Austr. III, 38 (1866) ... ... W.A. — — — — — — B.fl.III,38 M.fr.IX,176;XI,9.
D. axillaris, Cham., F. v. M. in Benth. Fl. Austr. III, 38 (1866) W.A. — — — — — — B.fl.III,38
VERTICORDIA, De Candolle in Dict. class. XI, 400 (1826). (Chrysorrhoea.)
V. Wilhelmii, F. v. M. in Transact. Vict. Inst. 122 (1855) ... — S.A. — — — — — B.fl.III,19 M.fr.X,28.
V. densiflora, Lindley in Bot. Regist. XXV, App. VI (1839) ... W.A. — — — — — — B.fl.III,20
V. stelluligera, Meisner in Journ. Linn. Soc. I, 36 (1857) ... W.A. — — — — — — B.fl.III,20
V. minutiflora, F. v. M., fragm. IV, 58 (1864) ... ... W.A. — — — — — — B.fl.III,20 M.fr.IV,38.
V. Fontanesii, De Candolle, prodr. III, 209 (1828) ... ... W.A. — — — — — — B.fl.III,21 M.fr.VIII,182;X,28.
V. helichrysantha, F. v. M. in Benth. Fl. Austr. III, 21 (1866) W.A. — — — — — — B.fl.III,21
V. Brownii, De Candolle, prodr. III, 209 (1828) ... ... W.A. — — — — — — B.fl.III,21
V. fastigiata, Turcz. in Bullet. de l'Ac. de St. Petersb. 402 (1852) W.A. — — — — — — B.fl.III,22 M.fr.VIII,122;X,28.
V. Harveyi, Bentham, Fl. Austr. III, 22 (1864) ... ... W.A. — — — — — — B.fl.III,22
V. fimbrilepis, Turczaninow in Bull. Mosc. XX, 158 (1847) ... W.A. — — — — — — B.fl.III,22 M.fr.X,28.
V. serrata, Schauer in nov. act. acad. Caes. XIX, suppl. II,70 (1841) W.A. — — — — — — B.fl.III,23
V. nitens, Schauer in nov. act. acad.Caes.XIX,suppl.II,71,t.4(1841) W.A. — — — — — — B.fl.III,23 M.fr.X,28.
V. grandiflora, Endlicher in Ann. Wien. Mus. II, 195 (1838) ... W.A. — — — — — — B.fl.III,24
V. chrysantha, Endlicher in Ann. Wien. Mus. II, 195 (1838) ... W.A. — — — — — — B.fl.III,24 M.fr.X,28.
V. Preissii, Schauer in Lehm. pl. Preiss. I, 101 (1844) ... W.A. — — — — — — B.fl.III,24 M.fr.VIII,182.
V. acerosa, Lindley in Bot. Regist. XXV, App. VI (1839) ... W.A. — — — — — — B.fl.III,25
V. polytricha, Bentham, Fl. Austr. III, 25 (1866) ... ... W.A. — — — — — — B.fl.III,25
V. oxylepis, Turcz. in Bullet. de l'Ac. de St. Petersb. 402 (1852) W.A. — — — — — — B.fl.III,25 M.fr.VIII,182.
V. humilis, Bentham, Fl. Austr. III, 26 (1866) ... ... W.A. — — — — — — B.fl.III,26
V. penicillaris, F. v. M., fragm. I, 226 (1859) ... ... W.A. — — — — — — B.fl.III,26 M.fr.I,226.
V. multiflora, Turczaninow in Bull. Mosc. XX, 159 (1847) ... W.A. — — — — — — B.fl.III,26 M.fr.X,28.
V. Huegelii, Endlicher in Hueg. enum. 16 (1837)... ... ... W.A. — — — — — — B.fl.III,27

V. insignis, Endlicher in Hueg. enum. 47 (1837) ... ... W.A. — — — — — — — B.fl.III,27 M.fr.X,28.
V. habrantha, Schauer in Lehm. pl. Preiss. I, 100 (1844) ... W.A. — — — — — — — B.fl.III,28 M.fr.I,104.
V. monadelpha, Turczaninow in Bull. Mosc. XX, 158 (1847) ... W.A. — — — — — — — B.fl.III,28 M.fr.X,23.
V. Lehmanni, Schauer in Lehm. pl. Preiss. I, 99 (1844) .. .. W.A. — — — — — — — B.fl.III,29
V. Cunninghamii, Schauer in nov. act. acad. Caes. XIX, suppl. II,
    55 (1841) ... ... — — — — — — N.A. D.fl.III,29 M.fr.X,28.
V. picta, Endlicher in Ann. des Wien. Mus. II, 194 (1838) ... W.A. — — — — — — — B.fl.III,30 M.fr.X,28.
V. pennigera, Endlicher in Hueg. enum. 46 (1837) ... ... W.A. — — — — — — — B.fl.III,30 M.fr.X,28.
V. Drummondii, Schauer in nov. act. acad. Caes. XIX, suppl.
    II, 56 (1841) ... ... W.A. — — — — — — — B.fl.III,31
V. Hughani, F. v. M., fragm. XI, 10 (1878) ... ... W.A. — — — — — — — — M.fr.XI,10.
V. pholidophylla, F. v. M., fragm. I, 227 (1859) ... ... W.A. — — — — — — — B.fl.III,31 M.fr.I,227.
V. apicata, F. v. M., fragm. I, 226 (1859) ... ... W.A. — — — — — — — B.fl III,32 M.fr.I,226.
V. lepidophylla, F. v. M., fragm. I, 228 (1859) ... ... W.A. — — — — — — — B.fl.III,32 M.fr.I,228.
V. ovalifolia, Meissner in Journ. Linn. Soc. I, 49 (1857)... ... W.A. — — — — — — — B.fl.III,32 M.fr.X,28.
V. chrysostachya, Meissner in Journ. Linn. Soc. I, 41 (1857) .. W.A. — — — — — — — R.fl.III,33 M.fr.X,28.
V. oculata, Meissner in Journ. Linn. Soc. I, 41 (1857) ... W.A. — — — — — — — B.fl.III,33
V. grandis, Drummond in Hook. Kew Misc. V, 119 (1853) ... W.A. — — — — — — — B.fl.III,33 M.fr.X,28;XI,10.

PILEANTHUS, Labillardière, Nov. Holl. pl. specim. II, 11 (1806).
P. peduncularis, Endl. in Ann. des Wien. Mus. II, 196 (1838)... W.A. — — — — — — — B.fl.III,34 M.fr.I,225.
P. Limacis, Labillardière, Nov. Holl. pl. specim. II, 11, t. 149(1806) W.A. — — — — — — — B.fl.III,34
P. filifolius, Meissner in Journ. Linn. Soc. I, 45 (1857) ... W.A. — — — — — — — B.fl.III,35

CALYCOTHRIX, Labillardière, Nov. Holl. pl. specim. II, 8, t. 146 (1806). (Calytrix, Calythrix.)
C. aurea, Lindley, Bot. Regist. XXV, App. V, t. 3 (1839) ... W.A. — — — — — — — B.fl.III,41
C. puberula, Meissner in Journ. Linn. Soc. I, 48 (1857) ... ... W.A. — — — — — — — B.fl.III,42
C. flavescens, Cunningham in Bot. Mag. 3323 (1834) ... ... W.A. — — — — — — — D.fl.III,42
C. asperula, Schauer in Lehm. pl. Preiss. I, 106 (1844) ... ... W.A. — — — — — — — B.fl.III,43
C. sapphirina, Lindley, Bot. Regist. XXV, App. V (1839) ... W.A. — — — — — — — B.fl.III,43 M.fr.I,224.
C. breviseta, Lindley, Bot. Regist. XXV, App. V (1839) ... W.A. — — — — — — — B.fl.III,43 M.fr.VIII,182.
C. simplex, Lindley, Bot. Regist. XXV, App. V (1839)... ... W.A. — — — — — — — B.fl.III,43
C. empetroidea, Schauer in nov. act. acad. Caes. XIX, suppl. II,
    102 (1841) ... ... W.A. — — — — — — — B.fl.III,44
C. variabilis, Lindley, Bot. Regist. XXV, App. V (1839) ... W.A. — — — — — — — B.fl.III,44
C. muricata, F. v. M., fragm. I, 224 (1859) ... ... W.A. — — — — — — — B.fl.III,44 M.fr.I,224.
C. gracilis, Bentham, Fl. Austr. III, 45 (1866) ... ... W.A. — — — — — — — B.fl.III,45
C. brevifolia, Meissner in Journ. Linn. Soc. I, 46 (1857) ... W.A. — — — — — — — B.fl.III,45
C. brachyphylla, Turczaninow in Bull. Mosc. XX, 161 (1847) ... W.A. — — — — — — — B.fl.III,45
C. Leschenaultii, Schauer in Lehm. pl. Preiss. I, 104 (1844) ... W.A. — — — — — — — B.fl.III,46
C. Oldfieldii, Bentham, Fl. Austr. III, 46 (1866)... ... ... W.A. — — — — — — — B.fl.III,46
C. glutinosa, Lindley, Bot. Regist. XXV, App. V (1839) ... W.A. — — — — — — — B.fl.III,47
C. angulata, Lindley, Bot. Regist. XXV, App. VI (1839) ... W.A. — — — — — — — B.fl.III,47
C. depressa, Turczaninow in Bull. Soc. Mosc. XX, 162 (1847) ... W.A. — — — — — — — B.fl.III,47
C. tenuifolia, Meissner in Journ. Linn. Soc. I, 46 (1857)... ... W.A. — — — — — — — B.fl.III,47
C. sirigosa, Cunningham in Bot. Mag. 3323 (1834) ... ... W.A. — — — — — — — — M.fr.X,27.
C. plumulosa, F. v. M., fragm. X, 27 (1876) ... ... W.A. — — — — — — — B.fl.III,48 M.fr.I,146.
C. decandra, R. Brown in Can. Prodr. III, 208 (1828) ... ... W.A. — — — — — — — — M.fr.X,26.
C. Birdii, F. v. M., fragm. X, 26 (1876) ... ... W.A. — — — — — — — B.fl.III,48
C. tenuiramea, Turczaninow in Bull. Mosc. XXII, 20 (1849) ... W.A. — — — — — — — B.fl.III,48 **
C. Fraseri, Cunningham in Bot. Mag. 3323 (1834) ... ... W.A. — — — — — — — —
C. granulosa, Bentham, Fl. Austr. III, 49 (1866)... ... ... — — — — — — — — M.fr.X,27.
C. Creswelli, F. v. M., fragm. X, 27 (1876) ... ... W.A. — — — — — — — —
C. microphylla, Cunningham in Bot. Mag. 3323 (1834) ... ... — — — — — — N.A. B.fl.III,49
C. longiflora, F. v. M., fragm. I, 12 (1858) ... ... — S.A. — — — — — Q. B.fl.III,49 M.fr.I,12.
C. leptophylla, Bentham, Fl. Austr. III, 50 (1866) ... ... — — — — — — Q. B.fl.III,50
C. megaphylla, F. v. M., fragm. I, 146 (1858) ... ... — — — — — — N.A. D.fl.III,50 M.fr.I,146.
C. tetragona, Labillardière, Nov. Holl. pl.specim. II,8, t. 146(1806) W.A. S.A. T. V. N.S.W. Q. — B.fl.III,50 M.fr.IV,36;VIII,183.
C. Sullivani, F. v. M., fragm. IX, 1 (1875)... ... ... — — — — V. — — — M.fr.IX,1.
C. conferta, Cunningham in Bot. Mag. 3323 (1834) ... ... — — — — — — N.A. B.fl.III,51
C. arborescens, F. v. M. in Transact. phil. Inst. Vict. III, 42 (1858) — — — — — — N.A. B.fl.III,52
C. brachychaeta, F. v. M. in Transact. phil. Inst. Vict. III, 43(1858) — — — — — — N.A. B.fl.III,52
C. achacta, F. v. M. in Transact. phil. Inst. Vict. III, 43 (1858) — — — — — — N.A. B.fl.III,52
C. laricina, R. Brown in Benth. Fl. Austr. III, 52 (1866) ... — — — — — — N.A. B.fl.III,52

LHOTZKYA, Schauer in Linnaea X, 309 (1835).
L. glaberrima, F. v. M., fragm. I, 13 (1858) ... ... — S.A. — — — — — B.fl.III,53 M.fr.I,13.
L. genetylloides, F. v. M. in Transact. phil. Soc. Vict. I, 10 (1854) — S.A. — V. — — — B.fl.III,54
L. violacea, Lindley, Bot. Regist. XXV, App. VII (1839) ... W.A. — — — — — — B.fl.III,54 M.fr.XI,8.
L. ciliata, F. v. M. in Benth. Fl. Austr. III, 54 (1866) ... ... W.A. — — — — — — B.fl.III,54
L. brevifolia, Schauer in Lehm. pl. Preiss. I, 103 (1844) ... W.A. — — — — — — B.fl.III,54
L. purpurea, F. v. M., fragm. I, 224 (1859) ... ... W.A. — — — — — — B.fl.III,55 M.fr.I,224.
L. ericoides, Schauer in Lindl. Introd. acc. ed. 439 (1835) ... W.A. — — — — — — B.fl.III,55 M.fr.XI,9.
L. acutifolia, Lindley, Bot. Regist. XXV, App. VII (1839) ... W.A. — — — — — — B.fl.III,55 M.fr.XI,8.
L. Harvestiana, F. v. M., fragm. XI, 8 (1878) ... ... W.A. — — — — — — — M.fr.XI,8.

THRYPTOMENE, Endlicher in den Ann. des Wien. Mus. II, 192 (1838). (Homalocalyx, Paryphantha, Astraea, Eremopyxis, Micromyrtus.)
T. mucronulata, Turczaninow in Bull. Mosc. XX, 156 (1847) ... W.A. — — — — — — B.fl.III,58
T. australis, Endlicher in den Ann. des Wien. Mus. II, 192 (1838) W.A. — — — — — — B.fl.III,58
T. tenella, Bentham, Fl. Austr. III, 59 (1866) ... ... W.A. — — — — — — B.fl.III,59 M.fr.VIII,182;X,24.
T. prolifera, Turczaninow in Bull. Mosc. XXXV, 324 (1862) ... W.A. — — — — — — B.fl.III,59

T. saxicola, Schauer in Lehm. pl. Preiss. I, 102 (1844) ... ... W.A. — — — — — — B.fl.III,59 M.fr.IV,75.
T. Johnsonii, F. v. M., fragm. IV, 77 (1864) ... ... W.A. — — — — — — B.fl.III,00 M.fr.IV,77.
T. racemulosa, Turczaninow in Bull. Mosc. XX, 156 (1847) ... W.A. — — — — — — B.fl.III,00
T. denticulata, Bentham in Fl. Austr. III, 60 (1866) ... ... W.A. — — — — — — B.fl.III,00 M.fr.IV,75.
T. baeckeacea, F. v. M., fragm. IV, 65 (1864) ... ... W.A. — — — — — — B.fl.III,60 M.fr.IV,65.
T. strongylophylla, F. v. M. in Benth. Fl. Austr. III, 61 (1866) W.A. — — — — — — B.fl.III,61
T. hyporhytis, Turczaninow in Bull. Mosc. XXXV, 324 (1862) W.A. — — — — — — B.fl.III,61
T. Maisonneuvii, F. v. M., fragm. IV, 64 (1864) ... ... — S.A. — — — — N.A. B.fl.III,61 M.fr.IV,64.
T. urceolaris, F. v. M., fragm. X, 25 (1876) ... ... W.A. — — — — — — — M.fr.X,25.
T. Mitchelliana, F. v. M., fragm. I, 11 (1858) ... ... — S.A. — V. — — — B.fl.III,61 M.fr.I,11;IX,62.
T. Miqueliana, F. v. M., fragm. I, 11 (1858) ... ... — S.A. — — N.S.W. — — B.fl.III,02 M.fr.I,11.
T. micrantha, J. Hooker in Hook. Kew Misc. V, 299, t. 8 (1853) — — T. — — — — B.fl.III,02
T. stenocalyx, F. v. M., fragm. X, 23 (1876) ... ... W.A. — — — — — — — M.fr.X,23.
T. ericoea, F. v. M., fragm. I, 12 (1858) ... ... — S.A. — — — — — B.fl.III,02 M.fr.I,12.
T. oligandra, F. v. M., fragm. I, 11 (1858)... ... ... W.A. — — — — Q. N.A. B.fl.III,63 M.fr.I,11;IV,169.
T. elobata, F. v. M., fragm. IV, 63 (1864)... ... ... W.A. — — — — — — B.fl.III,64 M.fr.IV,63.
T. racemosa, F. v. M. in Benth. Fl. Austr. III, 64 (1866) ... W.A. — — — — — — B.fl.III,64
T. trachycalyx, F. v. M., fragm. X, 25 (1876) ... ... W.A. — — — — — — — M.fr.X,25.
T. imbricata, F.v.M.; Micromyrtus, Benth., Fl. Aus. III, 64 (1866) W.A. — — — — — — B.fl.III,64
T. flaviflora, F. v. M., fragm. VIII, 13 (1873) ... ... W.A. S.A. — — — — — M.fr.X,25.
T. Drummondii, F. v. M.; Microm., Benth., Fl. Aus. III, 64 (1866) W.A. — — — — — — B.fl.III,64 M.fr.VIII,183.
T. Elliotii, F. v. M., fragm. IX, 62 (1875)... ... ... — S.A. — — — — — — M.fr.IX,62.
T. auriculata, F. v. M., fragm. X, 24 (1876) ... ... — S.A. — — — — — — M.fr.X,24.
T. hymenonema, F. v. M., fragm. X, 26 (1876) ... ... W.A. — — — — — — — M.fr.X,26.
T. ciliata, F. v. M. in Woolls, pl. of the neighb. of Sydney 23 (1880) — S.A. — V. N.S.W. Q. — B.fl.III,65 M.fr.I,30&242;IV,63.
T. minutiflora, F.v.M. in Woolls, pl. of the neighb. of Sydn. 23 (1880) — — — — N.S.W. — — B.fl.III,65
T. leptocalyx, F. v. M.; Microm., Benth., Fl. Austr. III, 65 (1866) — — — — — Q. — B.fl.III,65 M.fr.I,30.
T. bomalocalyx, F. v. M., fragm. IV, 63 (1864) ... ... — — — — — — N.A. B.fl.III,56 M.fr.IV,63.
T. polyandra, F. v. M., fragm. IV, 77 (1864) ... ... — — — — — — N.A. B.fl.III,56 M.fr.IV,77.

WEHLIA, F. v. M., fragm. X, 22 (1876).
W. thryptomenoides, F. v. M., fragm. X, 22 (1876) ... ... W.A. — — — — — — — M.fr.X,22.
W. coarctata, F. v. M., fragm. X, 23 (1876) ... ... W.A. — — — — — — — M.fr.X,23.

BAECKEA, Linné, spec. plant. 358 (1753). (Jungia, Imbricaria, Scholtzia, Schidiomyrtus, Rinzia, Euryomyrtus, Camphoromyrtus,
Tetrapora, Harmogia, Oxymyrrhine, Babingtonia, Ericomyrtus, Piptandra, Anticoryne.)
B. platystemonea, Bentham, Fl. Austr. III, 74 (1866) ... ... W.A. — — — — — — B.fl.III,74 M.fr.VIII,183.
B. Fumana, F. v. M., fragm. IV, 68 (1864) ... ... W.A. — — — — — — B.fl.III,74
B. dimorphandra, F. v. M. in Benth. Fl. Austr. III, 74 (1866)... W.A. — — — — — — B.fl.III,74
B. scholtzifolia, Lehmann in Lehm. pl. Preiss. II, 360 (1847) ... W.A. — — — — — — B.fl.III,75 M.fr.VIII,183.
B. oxycoccoides, Bentham, Fl. Austr. III, 75 (1866) ... ... W.A. — — — — — — B.fl.III,75 M.fr.VIII,183.
B. Drummondii, Bentham, Fl. Austr. III, 75 (1866) ... ... W.A. — — — — — — B.fl.III,75 M.fr.VIII,183.
B. diffusa, Sieber in De Cand. prodr. III, 230 (1828) ... ... — S.A. T. V. N.S.W. — — B.fl.III,76 M.fr.IV,67.
B. crassifolia, Lindley in Mitch. Three Exped. II, 115 (1838) ... W.A. S.A. — V. N.S.W. — — B.fl.III,76 M.fr.IV,66.
B. cryptandroides, F. v. M., fragm. X, 29 (1876) ... ... W.A. — — — — — — — M.fr.X,29.
B. tetragona, F. v. M. in Benth. Fl. Austr. III, 77 (1866) ... W.A. — — — — — — B.fl.III,77
B. ericoea, F. v. M., fragm. I, 31 (1858) ... ... — S.A. — V. N.S.W. — — B.fl.III,77
B. polystemonea, F. v. M., fragm. II, 124 (1861)... ... ... — S.A. — — — — — B.fl.III,77 M.fr.II,124.
B. orculata, R. Brown in Flind. voy. 548 (1814)... ... ... — — — N.S.W. Q. — B.fl.III,78 M.fr.IV,65.
B. brevifolia, De Candolle, prodr. III, 230 (1828)... ... ... — — — — — — — B.fl.III,78
B. Gunniana, Schauer in Walp. Rep. II, 920 (1843) ... ... — — T. V. N.S.W. — — B.fl.III,78 M.fr.IV,66.
B. diosmifolia, Rudge in Transact. Linn. Soc. VIII, 299, t. 13 (1807) — — — N.S.W. — — B.fl.III,79 M.fr.I,29.
B. leptocaulis, J. Hooker, icon. pl. t. 298 (1840) ... ... — — T. — — — — B.fl.III,79
B. arbuscula, R. Brown in Benth. Fl. Austr. III, 79 (1866) ... W.A. — — — — — — B.fl.III,79
B. astartcoides, Bentham, Fl. Austr. III, 80 (1866) ... ... W.A. — — — — — — B.fl.III,80
B. linifolia, Rudge in Transact. Linn. Soc. VIII, 297, t. 12 (1807) — — — V. N.S.W. — — B.fl.III,80 M.fr.IV,71.
B. stenophylla, F. v. M., fragm. I, 13 (1858) ... ... — — — N.S.W. Q. — B.fl.III,80 M.fr.I,13.
B. camphorata, R. Brown in Bot. Mag. t. 2094 (1826) ... ... — — — N.S.W. — — B.fl.III,81 M.fr.IV,70;IV,177.
B. virgata, Andrews, Bot. Reposit. t. 598 (1810) ... ... — — — V. N.S.W. Q. N.A. B.fl.III,81 M.fr.I,60.
B. crenatifolia, F. v. M., fragm. IV, 70 (1864) ... ... — — — V. — — — B.fl.III,82 M.fr.IV,70.
B. Cunninghamii, Bentham, Fl. Austr. III, 82 (1866) ... ... — — — N.S.W. — — B.fl.III,82
B. densifolia, Smith in Transact. Linn. Soc. III, 260 (1797) ... — — — N.S.W. — — B.fl.III,82 M.fr.IV,71.
B. Behrii, F. v. M., fragm. IV, 68 (1864) ... ... W.A. S.A. — V. N.S.W. — — B.fl.III,83 M.fr.IV,68.
B. uncinella, Bentham, Fl. Austr. III, 84 (1866) ... ... W.A. — — — — — — B.fl.III,84
B. polyandra, F. v. M., fragm. IV, 72 (1864) ... ... W.A. — — — — — — B.fl.III,84 M.fr.II,30;IV,72.
B. corynophylla, Bentham, Fl. Austr. III, 85 (1866) ... ... W.A. — — — — — — B.fl.III,84
B. pachyphylla, Bentham, Fl. Austr. III, 85 (1866) ... ... W.A. — — — — — — H.fl.III,85
B. crispiflora, F. v. M., fragm. IV, 72 (1864) ... ... W.A. — — — — — — B.fl.III,85 M.fr.II,31;IV,72.
B. Blackettii, F. v. M., fragm. VIII, 181 (1874) ... ... W.A. — — — — — — — M.fr.VIII,181.
B. camphoromae, Endlicher in Hueg. enum. 51 (1837) ... ... W.A. — — — — — — B.fl.III,86
B. pulchella, De Candolle, prodr. III, 230 (1828) ... ... W.A. — — — — — — B.fl.III,86
B. ochropetala, F. v. M., fragm. X, 29 (1876) ... ... W.A. — — — — — — — M.fr.X,29.
B. pygmaea, R. Brown in Benth. Fl. Austr. III, 86 (1866) ... W.A. — — — — — — B.fl.III,86 M.fr.VIII,183.
B. corymbulosa, Bentham, Fl. Austr. III, 87 (1866) ... ... W.A. — — — — — — B.fl.III,87 M.fr.VIII,183.
B. floribunda, Bentham, Fl. Austr. III, 87 (1866) ... ... W.A. — — — — — — B.fl.III,87 [183.
B. pentandra, F. v. M., fragm. II, 31; IV, 72 (1864) ... ... W.A. — — — — — — B.fl.III,87 M.fr.II,31;IV,72;VIII,
B. pentagonantha, F. v. M., fragm. IV, 73 (1864) ... ... W.A. — — — — — — B.fl.III,87 M.fr.IV,73.
B. robusta, F. v. M., fragm. IV, 72 (1864) ... ... W.A. — — — — — — B.fl.III,88 M.fr.IV,72. [183.
B. ovalifolia, F. v. M., fragm. IV, 72 (1864) ... ... W.A. — — — — — — B.fl.III,88 M.fr.II,32;IV,72;VIII,
B. subcuneata, F. v. M., fragm. IV, 73 (1864) ... ... — — — — — — — B.fl.III,88 M.fr.IV,73.
B. grandiflora, Bentham, Fl. Austr. III, 89 (1866) ... ... W.A. — — — — — — B.fl.III,89
B. uberiflora; Scholtzia, F. v. M., fragm. IV, 74 (1864) ... ... W.A. — — — — — — B.fl.III,07 M.fr.IV,74.

64

B. involucrata, Endlicher in Hueg. enum. 51 (1837) ... ... W.A. — — — — — — B.fl.III,68 M.fr.IV,74.
B. apathulata, F. v. M.; Scholtzia, Benth., Fl. Austr. III, 68 (1860) W.A. — — — — .. — B.fl.III,68
B. ciliata; Scholtzia, F. v. M., fragm. IV, 76 (1864) ... ... W.A. — — — — — — B.fl.III,68 M.fr.IV,76.
B. capitata; Scholtzia, F. v. M. in Benth. Fl. Austr. III, 69 (1866) W.A. — — — — — — B.fl.III,69
B. umbellifera; Scholtzia, F. v. M., fragm. IV, 75 (1864) ... W.A. — — — — — — B.fl.III,69 M.fr.IV,75.
B. laxiflora, F. v. M.; Scholtzia, Benth., Fl. Austr. III, 69 (1866) W.A. — .. — — — — B.fl.III,69
B. leptantha, F. v. M.; Scholtzia, Benth., Fl. Austr. III, 69 (1866) W.A. — .— — — — — B.fl.III,69
B. parviflora; Scholtzia, F. v. M., fragm. IV, 76 (1864) ... ... W.A. — .— —. — — — B.fl.III,70 M.fr.IV,76.
B. oligandra; Scholtzia, F. v. M. in Benth. Fl. Austr. III, 70 (1866) W.A. — — — — — .. — B.fl.III,70
B. scrpillifolia, F. v. M., fragm. X, 30 (1876) ... ... W.A. — — — .. — — B.fl.III,70 M.fr.VIII,183.
B. teretifolia, F. v. M.; Scholtzia, Benth., Fl. Austr. III, 70 (1866) W.A. — — — .. - .. — B.fl.III,70

ASTARTEA, De Candolle in Dict. class. XI, 400 (1826).
A. ambigua, F. v. M., fragm. II, 32 (1860) ... ... W.A. — — — — — — B.fl.III,89 M.fr.II,32.
A. fascicularis, De Candolle, prodr. III, 210 (1828) ... ... W.A. — — — — —. — B.fl.III,00 M.fr.VIII,183.
A. intratropica, F. v. M., fragm. I, 83 (1858) ... ... — — — — — — N.A. B.fl.III,90 M.fr.I,83.

HYPOCALYMMA, Endlicher in Hueg. enum. pl. Nov. Holl. Austr. occ. 51 (1837).
H. xanthopetalum, F. v. M., fragm. II, 29 (1860) ... ... W.A. — — — — — B.fl.III,92 M.fr.II,29.
H. robustum, Endlicher in Hueg. enum. 50 (1837) ... ... W.A. — — — — — B.fl.III,92
H. longifolium, F. v. M., fragm. II, 28 (1860) ... ... W.A. — — — — — B.fl.III,92 M.fr.II,28.
H. strictum, Schauer in Lehm. pl Preiss. I, 111 (1844) ... ... W.A. — — — — — B.fl.III,93
H. tetrapterum, Turczaninow in Bull. Mosc. XXXV, 325 (1862) W.A. — -— — — .. B.fl.III,93
H. linifolium, Turczaninow in Bull. Mosc. XXXV, 325 (1862) ... W.A. — — — — — B.fl.III,93
H. angustifolium, Endlicher in Hueg. enum. 50 (1837) ... ... W.A. — — — — — B.fl.III,94
H. ericifolium, Bentham, Fl. Austr. III, 94 (1866) ... ... W.A. — — — — — B.fl.III,94
H. cordifolium, Lehmann, plant. Preiss. I, 112 (1844) ... ... W.A. — — — — — B.fl.III,94
H. speciosum, Turcz. in Bull. de l'Acad. de St. Pet. 409 (1852) W.A. — — — — — B.fl.III,95 M.fr.VIII,183.
H. Phillipsii, Harvey in nat. hist. Rev. V, 206, t. 22 (1865) ... W.A. — .. — — — B.fl.III,95
H. myrtifolium, Turcz. in Bullet. de St. Petersb. 410 (1852) ... W.A. — — — .— B.fl.III,95 M.fr.VIII,183.

BALAUSTION, Hooker, Icones plantarum 852 (1852). (Punicella, Cheynia.)
B. pulcherrimum, Hooker, icon. plant. 852 (1852) ... ... W.A. — — — .— — B.fl.III,96 M.fr.VIII,183.

AGONIS, Lindley, Bot. Regist. XXV, p. X (1839). (Billiotia, Billotia.)
A. spathulata, Schauer in Lehm. pl. Preiss. I, 117 (1844) ... W.A. — — — — B.fl.III,97 M.fr.XI,120.
A. floribunda, Turczaninow in Bull. Mosc. II, 20 (1849) ... W.A. — — — — B.fl.III,97
A. marginata, Schauer in Lehm. pl. Preiss. I, 117 (1844) ... W.A. — — — — B.fl.III,98 M.fr.XI,120.
A. linearifolia, Schauer in Lehm. pl. Preiss. I, 118 (1844) ... W.A. — — — — B.fl.III,98 M.fr.VIII,183;XI,120.
A. juperina, Schauer in Lehm. pl. Preiss. I, 118 (1844) ... W.A. — — — — B.fl.III,98 M.fr.XI,120.
A. parviceps, Schauer in Lehm. pl. Preiss. I, 119 (1844) ... W.A. — — — — B.fl.III,99 M.fr.XI,119.
A. obtusissima, F. v. M., fragm. XI, 119 (1881) ... ... W.A. — — — — M.fr.XI,119.
A. flexuosa, Schauer in Lehm. pl. Preiss. I, 116 (1844) ... W.A. — — — — B.fl.III,99 M.fr.XI,120.
A. undulata, Schauer in Lehm. pl. Preiss. I, 119 (1844) ... W.A. — — — — B.fl.III,100
A. hypericifolia, Schauer in Lehm. pl. Preiss. I, 117 (1844) ... W.A. — — — — B.fl.III,100 M.fr.XI,120.
A. Scortechiniana, F. v. M., fragm. XI, 118 (1881) ... — — — — Q. — M.fr.XI,118.
A. grandiflora, Bentham, Fl. Austr. III, 100 (1866) ... ... W.A. — — — — B.fl.III,100 M.fr.XI,120.

LEPTOSPERMUM, R. et G. Forster, char. gen. 71, t. 36 (1776). (Fabricia, Homalospermum, Pericalymma.)
L. Fabricia, Bentham, Fl. Austr. III, 102 (1866) ... ... — — — — Q. — B.fl.III,102
L. laevigatum, F. v. M., Annual Report 22 (1858) ... ... — S.A. T. V. N.S.W. — B.fl.III,103 M.fr.IV,60.
L. firmum, Bentham, Fl. Austr. III, 104 (1866) ... ... W.A. — — — — B.fl.III,104
L. flavescens, Smith in Transact. Linn. Soc. III, 262 (1797) ... — T. V. N.S.W. Q. — B.fl.III,104
L. scoparium, R. & G. Forster, char. gen. 48 (1776) ... — S.A. T. V. N.S.W. Q. — B.fl.III,105
L. arachnoideum, Smith in Transact. Linn. Soc. III, 263 (1797) ... — — — — N.S.W. — B.fl.III,105
L. spinescens, Endlicher in Hueg. enum. 51 (1837) ... W.A. — — — — B.fl.III,106
L. lanigerum, Smith in Transact. Linn. Soc. III, 263 (1797) ... — S.A. T. V. N.S.W. Q. — B.fl.III,106
L. parvifolium, Smith in Transact. Linn. Soc. III, 263 (1797) ... — — — — N.S.W. — B.fl.III,107
L. stellatum, Cavanilles, icon. pl. IV, 16, t. 330 ... ... — — — N.S.W. Q. — B.fl.III,107
L. attenuatum, Smith in Transact. Linn. Soc. III, 262 (1797) ... — — V. N.S.W. Q. — B.fl.III,108
L. myrtifolium, Sieber in De Cand., prodr. III, 238 (1828) ... — — V. N.S.W. Q. — B.fl.III,108
L. rupestre, J. Hooker in Hook. icon. pl. t. 308 (1841) ... — T. — — — B.fl.III,109
L. myrsinoides, Schlechtendal in Linnaea XX, 653 (1847) ... S.A. — V. N.S.W. — — B.fl.III,109
L. erubescens, Schauer in Lehm. pl. Preiss. I, 121 (1844) ... W.A. — — — — B.fl.III,109 M.fr.VIII,183.
L. abnorme, F. v. M. in Benth. Fl. Austr. III, 109 (1866) ... W.A. — — N.S.W. Q. N.A. B.fl.III,109 M.fr.II,27.
L. Roei, Bentham, Fl. Austr. III, 110 (1866) ... ... W.A. — — — — B.fl.III,110
L. floridum, Bentham, Fl. Austr. III, 110 (1866)... ... W.A. — — — — B.fl.III,110
L. ellipticum, Endlicher in Hueg. enum. 51 (1837) ... W.A. — — — — B.fl.III,110
L. crassipes, Lehmann, ind. sem. Hort. Hamb. (1842) ... W.A. — — — — B.fl.III,110

KUNZEA, Reichenbach, conspect. regn. veget. 175 (1828). (Salisia, Pentagonaster.)
K. micrantha, Schauer in Lehm. pl. Preiss. I, 125 (1844) ... W.A. — — — — B.fl.III,112 M.fr.VIII,183.
K. ericcalyx, F. v. M., fragm. II, 28 (1860) ... ... W.A. — — — — B.fl.III,112 M.fr.II,28.
K. Muelleri, Bentham, Fl. Austr. III, 113 (1866) ... ... — — V. N.S.W. — B.fl.III,113
K. ericifolia, Reichenbach, consp. 175 (1828) ... ... W.A. — — — — B.fl.III,113
K. Preissiana, Schauer in Lehm. pl. Preiss. I, 125 (1844) ... W.A. — — — — B.fl.III,114
K. recurva, Schauer in Lehm. pl. Preiss. I, 125 (1844) ... W.A. — — — — B.fl.III,114 M.fr.VIII,183.
K. micromera, Schauer in Lehm. pl. Preiss. III, 227 (1847) ... W.A. — — — — B.fl.III,114
K. pauciflora, Schauer in Lehm. pl. Preiss. I, 124 (1844) ... W.A. — — — — B.fl.III,114 M.fr.VIII,183.
K. calida, F. v. M., fragm. VI, 23 (1867) ... ... W.A. — — — Q. — M.fr.VI,23.
K. parvifolia, Schauer in Lehm. pl. Preiss. I, 124 (1844) ... — — — N.S.W. — B.fl.III,115 M.fr.VIII,13.
K. peduncularis, F. v. M. in Transact. Vict. Inst. 144 (1855) ... — — V. N.S.W. — B.fl.III,115 M.fr.X,55.
K. corifolia, Reichenbach, consp. 175 (1828) ... ... — — T. V. N.S.W. — B.fl.III,116
K. capitata, Reichenbach, consp. 175 (1828) ... ... — — — N.S.W. — B.fl.III,116
K. pomifera, F. v. M. in Transact. Vict. Inst. 124 (1855) ... S.A. — V. — — B.fl.III,116

K. sericea, Turczaninow in Bull. Mosc. XX, 162 (1847)... ... W.A. — — — — — — B.fl.III,117
K. Baxteri, Schauer in Lehm. pl. Preiss. I, 123 (1844) .. ... W.A. — — — — — — B.fl.III,117 M.fr.III,153.
K. opposita, F. v. M., fragm. VI, 24 (1867) ... ... — — — N.S.W. — — M.fr.VI,24.
. CALLISTEMON, R. Brown in Flind. voy. II, 517 (1814).
C. speciosus, De Candolle, prodr. III, 224 (1828)... ... W.A. — — — — — B.fl.III,119 M.fr.I,14.
C. phoeniceus, Lindley, Bot. Regist. XXV, App. X (1839) ... W.A. — — — — — B.fl.III,119 M.fr.IV,53.
C. lanceolatus, De Candolle, prodr. III, 224 (1828) ... ... — — V. N.S.W. Q. — B.fl.III,120 M.fr.IV,53.
C. coccineus, F. v. M., fragm. I, 13 (1858)... ... ... — S.A. — V. N.S.W. — B.fl.III,120 M.fr.I.13.
C. salignus, De Candolle, prodr. III, 223 (1828) .. .. S.A. T. V. N.S.W. Q. — B.fl.III,120 M.fr.IV,54.
C. paludosus, F. v. M., fragm. I, 14 (1858) ... ... — — — — Q. — M.fr.I,14.
C. rigidus, R. Brown in Bot. Regist. t. 393 (1819) ... ... — — — N.S.W. — — B.fl.III,121
C. linearis, De Candolle, prodr. III, 223 (1828) ... ... — — — N.S.W. — — B.fl.III,122
C. pinifolius, De Candolle, prodr. III, 223 (1828)... ... ... — — — N.S.W. — — B.fl.III,122
C. teretifolius, F. v. M. in Schlecht. Linnaea XXV, 387 (1852) — S.A. — — — — — B.fl.III,122
C. brachyandrus, Lindley in Journ. Hort. Soc. IV, 112 (1849) ... — S.A. — V. N.S.W. — — B.fl.III,122 M.fr.IV,52.
MELALEUCA, Linné, mantiss. plant. 14 (1767). (Gymnagathis, Asteromyrtus.)
M. macronychia, Turcz. in Bullet. de St. Petersb. 420 (1852) ... W.A. — — — — — B.fl.III,129 M.fr.VIII,183.
M. lateritia, Otto in allgem. Gart. Zeit. II, 257 (1834) ... ... W.A. — — — — — B.fl.III,130
M. calothamnoides, F. v. M., fragm. III, 114 (1862) ... ... W.A. — — — — — B.fl.III,130 M.fr.III,114.
M. blaeriaefolia, Turczaninow in Bull. Mosc. XX, 165 (1847) ... W.A. — — — — — B.fl.III,130
M. dicosmifolia, Andrews, Bot. Reposit. t. 476 (1808) ... ... W.A. — — — — — B.fl.III,130
M. elliptica, Labillardière, Nov. Holl. pl. spec. II, 31, t. 173 (1806) W.A. — — — — — B.fl.III,131 M.fr.VIII,183.
M. hypericifolia, Smith in Transact. Linn. Soc. III, 279 (1797)... — — — N.S.W. — — B.fl.III,131
M. fulgens, R. Brown in Ait. hort. Kew. sec. ed. IV, 415 (1812) W.A. — — — — — B.fl.III,131
M. acuminata, F. v. M., fragm. I, 15 (1858) ... ... W.A. S.A. — V. N.S.W. — B.fl.III,132 M.fr.I,15.
M. leptoclada, Bentham, Fl. Austr. III, 132 (1866) ... ... W.A. — — — — — B.fl.III,132
M. basicephala, Bentham, Fl. Austr. III, 133 (1866) ... ... W.A. — — — — — B.fl.III,133
M. gibbosa, Labillardière, Nov. Holl. pl. spec. II, 30, t. 172 (1806) — S.A. T. V. — — B.fl.III,133
M. decussata, R. Brown in Ait. hort. Kew. sec. ed. IV, 415 (1812) — S.A. — V. — — B.fl.III,133
M. Wilsonii, F. v. M., fragm. II, 124, t. 15 (1861) ... ... — S.A. — V. — — B.fl.III,134 M.fr.II,124.
M. thymifolia, Smith in Transact. Linn. Soc. III, 276 (1797) ... — — — N.S.W. Q. — B.fl.III,134 M.fr.X,55.
M. violacea, Lindley, Bot. Regist. XXV, App. VIII (1839) ... W.A. — — — — — B.fl.III,135
M. cardiophylla, F. v. M., fragm. I, 225 (1859) ... ... W.A. — — — — — B.fl.III,135 M.fr.I,225;III,164.
M. undulata, Bentham, Fl. Austr. III, 135 (1866) ... ... W.A. — — — — — B.fl.III,135
M. depauperata, Turcz. in Bullet. de St. Petersb. 424 (1852) ... W.A. — — — — — B.fl.III,136 M.fr.III,120;VIII,183.
M. lateriflora, Bentham, Fl. Austr. III, 136 (1866) ... ... W.A. — — — — — B.fl.III,136
M. exarata, F. v. M., fragm. III, 114 (1862) ... ... W.A. — — — — — B.fl.III,136 M.fr.III,114;VIII,184.
M. fasciculiflora, Bentham, Fl. Austr. III, 137 (1866) ... ... W.A. — — — — — B.fl.III,137 M.fr.VIII,184.
M. teretifolia, Endlicher in Hueg. enum. 49 (1837) ... ... W.A. — — — — — B.fl.III,137 M.fr.III,117.
M. alsophila, Cunningham in Benth. Fl. Austr. III, 137 (1866)... — — — — — N.A. B.fl.III,138 M.fr.III,116.
M. acacioides, F. v. M., fragm. III, 116 (1862) ... ... — — — — — N.A. B.fl.III,138
M. Baxteri, Bentham, Fl. Austr. III, 138 (1866) ... ... W.A. — — — — — B.fl.III,138
M. symphyocarpa, F. v. M. in Transact. phil. Inst. Vict. III, 44 (1858) — — — — — N.A. B.fl.III,138
M. angustifolia, Gaertner de fruct. I, 172, t. 35 (1788) ... ... — — — — Q. — B.fl.III,139
M. pauciflora, Turczaninow in Bull. Mosc. XX, 166 (1847) ... — — — N.S.W. — — B.fl.III,139
M. squarrosa, Donn, Hort. Cantabr. ed. sec. 101 (1800)... ... — S.A. T. V. N.S.W. — — B.fl.III,140 M.fr.X,55.
M. adnata, Turcz. in Bullet. de St. Petersb. 425 (1852) ... W.A. — — — — — B.fl.III,140 M.fr.III,117;VIII,184.
M. linariifolia, Smith in Transact. Linn. Soc. III, 278 (1797) ... — — — N.S.W. Q. N.A. B.fl.III,140 M.fr.X,55.
M. trichostachya, Lindley in Mitch. Trop. Austr. 277 (1848) ... — S.A. — — Q. N.A.
M. radula, Lindley, Bot. Regist. XXV, App. 167 (1840) ... W.A. — — — — — B.fl.III,141 M.fr.X,55.
M. pulchella, R. Brown in Ait. hort. Kew. sec. ed. IV, 414 ... W.A. — — — — — B.fl.III,141
M. conferta, Bentham, Fl. Austr. III, 142 (1866)... ... ... W.A. — — — — — B.fl.III,142
M. Leucadendra, Linné, mantiss. 105 (1767) ... ... — — — N.S.W. Q. N.A. B.fl.III,142
M. lasiandra, F. v. M., fragm. III, 115 (1862) ... ... — — — — — N.A. B.fl.III,143 M.fr.III,115.
M. genistifolia, Smith in Transact. Linn. Soc. III, 277 (1797) ... — — — N.S.W. Q. N.A. B.fl.III,143 M.fr.I,15;X,55.
M. styphelioides, Smith in Transact. Linn. Soc. III, 275 (1797) ... — — — N.S.W. Q. — B.fl.III,144
M. Huegelii, Endlicher in Hueg. enum. 48 (1837) ... ... W.A. — — — — — B.fl.III,144
M. dissitiflora, F. v. M., fragm. III, 153 (1863) ... ... — — — — — N.A. B.fl.III,144 M.fr.III,153.
M. linophylla, F. v. M., fragm. III, 115 (1863) ... ... — — — — — N.A. B.fl.III,145 M.fr.III,115.
M. parviflora, Lindley, Bot. Regist. XXV, p. VIII (1839) ... W.A. S.A. — V. N.S.W. Q. B.fl.III,145
M. laxiflora, Turczaninow in Bullet. de St. Petersb. 421 (1852) W.A. — — — — — B.fl.III,146 M.fr.VIII,184;X,55.
M. armillaris, Smith in Transact. Linn. Soc. III, 277 (1797) ... — — V. N.S.W. — — B.fl.III,146
M. bamulosa, Turczaninow in Bull. Mosc. XX, 165 (1847) ... W.A. — — — — — B.fl.III,146 M.fr.III,119;VIII,184.
M. subfalcata, Turczaninow in Bullet. de St. Petersb. 422 (1852) W.A. — — — — — B.fl.III,146 M.fr.III,119;VIII,184.
M. glaberrima, F. v. M., fragm. III, 119 (1862) ... ... W.A. — — — — — B.fl.III,147 M.fr.III,119.
M. rhaphiophylla, Schauer in Lehm. pl. Preiss. I, 143 (1844) ... W.A. — — — — — B.fl.III,147
M. cymbifolia, Bentham, Fl. Austr. III, 148 (1866) ... ... W.A. — — — — — B.fl.III,148
M. cuticularis, Labillard., Nov. Holl. pl. spec. II, 30, t. 171 (1806) W.A. — — — — — B.fl.III,148
M. sparsiflora, Turczaninow in Bull. Mosc. XX, 167 (1847) ... W.A. — — — — — B.fl.III,148
M. calycina, R. Brown in Ait. hort. Kew. sec. ed. IV, 416 (1812) W.A. — — — — — B.fl.III,149 M.fr.VIII,184.
M. cordata, Turcz. in Bullet. de St. Petersb. 418 (1852)... ... W.A. — — — — — B.fl.III,149 M.fr.III,184.
M. globifera, R. Brown in Ait. hort. Kew. sec. ed. IV, 411 (1812) W.A. — — — — — B.fl.III,149
M. megacephala, F. v. M., fragm. III, 117 (1862) ... ... W.A. — — — — — B.fl.III,149 M.fr.III,117.
M. nesophila, F. v. M., fragm. III, 113 (1862) ... ... W.A. — — — — — B.fl.III,150 M.fr.III,113.
M. Oldfieldii, F. v. M., fragm. III, 118 (1862) ... ... W.A. — — — — — B.fl.III,150
M. uncinata, R. Brown in Ait. hort. Kew. sec. ed. IV, 414 (1812) W.A. S.A. — V. N.S.W. — — B.fl.III,151 M.fr.III,118.
M. concreta, F. v. M., fragm. III, 118 (1862) ... ... W.A. — — — — — B.fl.III,151 M.fr.III,119.
M. filifolia, F. v. M., fragm. III, 119 (1862) ... ... W.A. — — — — — B.fl.III,151
M. hakeoides, F. v. M. in Benth. Fl. Austr. III, 151 (1866) ... — S.A. — — N.S.W. — — B.fl.III,151
M. glomerata, F. v. M., Rep. Babb. Exped. 8 (1858) ... ... W.A. S.A. — — — N.A. B.fl.III,151

Species	W.A.	S.A.	T.	V.	N.S.W.	Q.	N.A.	Benth.	Mueller
C. rupestris, Schauer in nov. act. acad. Cæs. XXI, 26 (1844) ...	W.A.	—	—	—	—	—	—	B.fl.III,179	
C. quadrifidus, R. Brown in Ait. hort. Kew. sec. ed. IV, 418(1812)	W.A.	—	—	—	—	—	—	M.fl.III,179	M.fr.X,31.
C. asper, Turczaninow in Bull. Mosc. XXII, 25 (1849)	W.A.	—	—	—	—	—	—	B.fl.III,180	M.fr.X,31.
C. homalophyllus, F. v. M., fragm. III, 111 (1862)	W.A.	—	—	—	—	—	—	B.fl.III,180	M.fr.X,31.

LAMARCHEA. Gaudichaud in Freyc. Voy. Bot. 483, t. 110 (1826).

Species	W.A.	S.A.	T.	V.	N.S.W.	Q.	N.A.	Benth.	Mueller
L. hakeæfolia, Gaudichaud in Freyc. Voy. 484, t. 110(1826) ...	W.A.	—	—	—	—	—	—	B.fl.III,123	

EREMAEA. Lindley, Bot. Regist. XXV, App. XI (1839).

Species	W.A.	S.A.	T.	V.	N.S.W.	Q.	N.A.	Benth.	Mueller
E. fimbriata, Lindley, Bot. Regist. XXV, App. XI (1839) ...	W.A.	—	—	—	—	—	—	B.fl.III,181	M.fr.XI,11.
E. acutifolia, F. v. M., fragm. II, 30 (1860)	W.A.	—	—	—	—	—	—	B.fl.III,181	M.fr.II,30.
E. violacea, F. v. M., fragm. XI, 10 (1878)	W.A.	—	—	—	—	—	—	—	M.fr.XI,10.
E. pilosa, Lindley, Bot. Regist. XXV, App. XI (1839)	W.A.	—	—	—	—	—	—	B.fl.III,182	M.fr.XI,11.
E. ebracteata, F. v. M., fragm. II, 29 (1860)	W.A.	—	—	—	—	—	—	B.fl.III,182	M.fr.XI,11.
E. beaufortioides, Bentham, Fl. Austr. III, 182 (1866) ...	W.A.	—	—	—	—	—	—	B.fl.III,182	M.fr.XI,11.

ANGOPHORA, Cavanilles, icon. IV, 21, t. 331 (1797).

Species	W.A.	S.A.	T.	V.	N.S.W.	Q.	N.A.	Benth.	Mueller
A. cordifolia, Cavanilles, icon. IV, 21, t. 339 (1797) ...	—	—	—	—	N.S.W.	—	—	B.fl.III,183	
A. subvelutina, F. v. M., fragm. I, 31 (1858) ...	—	—	—	—	N.S.W.	Q.	—	D.fl.III,184	M.fr.I,31.
A. intermedia, De Candolle, prodr. III, 222 (1828)	—	—	—	V.	N.S.W.	Q.	—	B.fl.III,184	
A. lanceolata, Cavanilles, icon. IV, 22, t. 339 (1797) ...	—	—	—	—	N.S.W.	Q.	—	B.fl.III,184	

EUCALYPTUS, L'Heritier, sertum Anglicum 18 (1788). (Eudesmia, Symphyomyrtus.)

I. RENANTHERAE.

Species	W.A.	S.A.	T.	V.	N.S.W.	Q.	N.A.	Benth.	Mueller
E. stellulata, Sieber in De Cand., prodr. III, 217 (1828)... ...	—	—	—	V.	N.S.W.	—	--	B.fl.III,200	M.fr.II,45;IX,172.
E. pauciflora, Sieber in Sprengel, cur. poster. 195 (1827) ...	—	S.A.	T.	V.	N.S.W.	—	--	B.fl.III,201	M.fr.III,52.
E. rognans, F. v. M. in Report Acclim. Soc. Vict. 20 (1870) ...	—	—	T.	V.	N.S.W.	—	—	B.fl.III,202	M.fr.II,53;VII,42&44.
E. amygdalina, Labill., Nov. Holl. pl. spec. II, 14, t. 154 (1806) ...	—	—	T.	V.	N.S.W.	—	—	B.fl.III,202	
E. coccifera, J. Hooker in Lond. Journ. of Bot. VI, 477 (1847)...	—	—	T.	—	—	—	—	B.fl.III,204	
E. obliqua, L'Heritier, sertum Anglicum 18, t. 20 (1788) ...	—	S.A.	T.	V.	N.S.W.	—	—	B.fl.III,204	M.fr.II,44&171;XI,33.
E. stricta, Sieber in De Candolle, prodr. III, 218 (1828)... ...	—	—	—	—	N.S.W.	—	—	B.fl.III,217	M.fr.XI,38.
E. Planchoniana, F. v. M., fragm. XI, 43 (1878)	—	—	—	—	—	Q.	—	—	M.fr.XI,43.
E. Baileyana, F. v. M., fragm. XI, 37 (1878)	—	—	—	—	—	Q.	—	—	M.fr.XI,37.
E. sepulcralis, F. v. M., Eucalyptographia eight dec. (1882)	W.A.	—	—	—	—	—	—	B.fl.III,205	M.fr.III,57. [XI,33.
E. buprestium, F. v. M., fragm. III, 57 (1862) ...	W.A.	—	—	—	—	—	—	B.fl.III,201	M.fr.II,60&175.
E. marginata, Smith in Transact. Linn. Soc. VI, 302 (1802)	W.A.	—	—	—	—	—	—	B.fl.III,209	M.fr.II,40&174;VII,43;
E. santalifolia, F. v. M. in Transact. Vict. Inst. I, 35 (1854)	W.A.	S.A.	—	—	—	—	—	B.fl.III,206	
E. macrorrhyncha, F. v. M., First general Report, 12 (1853) ...	—	—	—	V.	N.S.W.	—	—	B.fl.III,207	M.fr.XI,45.
E. capitellata, Smith in White, Journ. Voy. N.S.W. 216 (1790)	—	S.A.	—	V.	N.S.W.	—	—	B.fl.III,206	M.fr.II,173;XI,38.
E. eugenioides, Sieber in Spreng. cur. poster. 195 (1827) ...	—	—	—	V.	N.S.W.	—	—	B.fl.III,207	
E. piperita, Smith in White, Journ. Voy. N.S.Wales 220 (1790)	—	—	—	V.	N.S.W.	—	—	B.fl.III,207	M.fr.II,173;VII,44;XI,38
E. pilularis, Smith in Transact. Linn. Soc. III, 284 (1797) ...	—	—	—	—	N.S.W.	Q	—	B.fl.III,208	M.fr.II,61&172;VII,44;
E. triantha, Link, enum. pl. hort. bot. Berol. II, 20 (1822) ...	—	—	—	—	N.S.W.	Q.	—	—	[XI,38.
E. haemastoma, Smith in Transact. Linn. Soc. III, 285 (1797) ...	—	—	T.	V.	N.S.W.	Q.	—	B.fl.III,212	M.fr.II,51&175;VII,43;
E. Sieberiana, F. v. M., Eucalyptogr. sec. decade (1879) ...	—	S.A.	T.	V.	N.S.W.	Q.	—	D.fl.III,212	M.fr.XI,37. [XI,38.
E. microcorys, F. v. M., fragm. II, 50 (1860)	—	—	—	—	N.S.W.	Q.	—	B.fl.III,212	M.fr.II,50.

II. PORANTHERAE.

Species	W.A.	S.A.	T.	V.	N.S.W.	Q.	N.A.	Benth.	Mueller
E. paniculata, Smith in Transact. Linn. Soc. III, 287 (1797) ...	—	S.A.	—	V.	N.S.W.	—	—	B.fl.III,211	M.fr.II,174;VII,43.
E. Leucoxylon, F. v. M. in Transact. Vict. Inst. I, 33 (1854) ...	—	S.A.	—	V.	N.S.W.	Q.	—	B.fl.III,210	M.fr.II,60&175.
E. melliodora, Cunningham, in Walp. rep. bot. syst. II, 924 (1843)	—	—	—	V.	N.S.W.	Q.	—	B.fl.III,210	M.fr.VII,43.
E. polyanthema, Schauer in Walpers, rep. bot. syst. II, 924 (1843)	—	—	—	V.	N.S.W.	Q.	-	B.fl.III,214	M.fr.VII,44.
E. populifolia, Hooker, icon. plant. 279 (1852)	—	—	—	—	N.S.W.	Q.	-	B.fl.III,214&243	
E. ochrophloia, F. v. M., fragm. XI, 36 (1878)	—	—	—	—	N.S.W.	Q.	—	—	M.fr.XI,36. [XI,36.
E. gracilis, F. v. M. in Transact. Vict. Inst. I, 35 (1854) ...	W.A.	S.A.	—	V.	N.S.W.	—	—	B.fl.III,222	M.fr.VII,43;VIII,184;
E. uncinata, Turczaninow in Bull. de Mosc. XXII, 23 (1849) ...	W.A.	S.A.	—	V.	N.S.W.	—	—	B.fl.III,215	M.fr.II,66.
E. odorata, Behr in Schlecht. Linnaea XX, 657 (1847) ...	—	S.A.	—	V.	N.S.W.	—	—	B.fl.III,215	M.fr.VII,43.
E. largiflorens, F. v. M. in Transact. Vict. Inst. I, 34 (1854) ...	—	S.A.	—	V.	N.S.W.	Q.	—	B.fl.III,215	M.fr.VII,43.
E. Behriana, F. v. M. in Transact. Vict. Inst. I, 34 (1854) ...	—	S.A.	—	V.	N.S.W.	—	—	B.fl.III,214	
E. hemiphloia, F. v. M., fragm. II, 62 (1860)	—	S.A.	—	V.	N.S.W.	Q.	—	B.fl.III,216	M.fr.II,62.
E. pruinosa, Schauer in Walp. rep. bot. syst. II, 926 (1843) ...	—	—	—	—	—	—	N.A.	B.fl.III,213	M.fr.II,132.
E. melanophloia, F. v. M. in Transact. Linn. Soc. III, 93 (1858) ...	—	—	—	—	N.S.W.	Q.	—	B.fl.III,220	

III. PARALLELANTHERAE.

Species	W.A.	S.A.	T.	V.	N.S.W.	Q.	N.A.	Benth.	Mueller
E. Clœziana, F. v. M., fragm. XI, 44 (1878)	—	—	—	—	—	Q.	—	—	M.fr.XI,44.
E. Howittiana, F. v. M. in Wing's S. Sc. Record Aug. (1882) ...	—	—	—	—	—	Q.	—	—	
E. drepanophylla, F. v. M. in Benth. Fl. Austr. III, 221 (1866) ...	—	—	—	—	—	Q.	—	B.fl.III,221	
E. crebra, F. v. M. in Journ. Linn. Soc. III, 87 (1858)	—	—	—	—	N.S.W.	Q.	N.A.	B.fl.III,221	
E. decipiens, Endlicher in Hueg. enum. 49 (1837)	W.A.	—	—	—	—	—	—	B.fl.III,247	
E. concolor, Schauer in Lehm. pl. Preiss. I, 129 (1844)	W.A.	—	—	—	—	—	—	B.fl.III,247	
E. siderophloia, Bentham, Fl. Austr. III, 220 (1866)	—	—	—	—	N.S.W.	Q.	—	B.fl.III,222	M.fr.II,61;VII,44.
E. microtheca, F. v. M. in Journ. Linn. Soc. III, 87 (1859) ...	—	S.A.	—	—	N.S.W.	Q.	—	B.fl.III,223	
E. Raveretiana, F. v. M., fragm. X, 90 (1877)	—	—	—	—	—	Q.	—	B.fl.III,223	M.fr.X,90.
E. brachyandra, F. v. M. in Journ. Linn. Soc. III, 97 (1858) ...	—	—	—	—	—	—	N.A.	B.fl.III,237	
E. pachyphylla, F. v. M. in Journ. Linn. Soc. III, 98 (1858) ...	—	S.A.	—	—	—	—	N.A.	B.fl.III,237	
E. pyriformis, Turczaninow in Bull. Mosc. XXII, 22 (1849) ...	W.A.	S.A.	—	—	—	—	—	B.fl.III,224	M.fr.II,32;X,5&84.
E. macrocarpa, Hooker, icones plant. t. 405-407 (1842)	W.A.	—	—	—	—	—	—	B.fl.III,224	M.fr.II,41;XI,38.
E. Preissiana, Schauer in Lehm. pl. Preiss. I, 131 (1844) ...	W.A.	—	—	—	—	—	—	B.fl.III,225	M.fr.II,68;VII,42.
E. alpina, Lindley in Mitch. Three Exped. II, 175 (1838) ...	—	—	—	V.	—	—	—	B.fl.III,225	M.fr.II,68.
E. Globulus, Labillardière, Voy. I, 153, t. 13 (1799) ...	—	—	T.	V.	N.S.W.	—	—	B.fl.III,225	M.fr.II,68.
E. megacarpa, F. v. M., fragm. II, 70 (1860)	W.A.	—	—	—	—	—	—	B.fl.III,232	M.fr.VII,43.
E. cosmophylla, F. v. M. in Transact. Vict. Inst. I, 35 (1854) ...	—	S.A.	—	—	—	—	—	B.fl.III,225	M.fr.VII,43.
E. cordata, Labillardière, Nov. Holl. pl. spec. II, 13, t. 152(1806) ...	—	—	T.	—	—	—	—	B.fl.III,224	
E. urnigera, J. Hooker in Lond. Journ. of Bot. VI, 477 (1847)...	—	—	T.	—	—	—	—	D.fl.III,227	

58

E. longifolia, Link, enum. pl. hort. bot. Borol. II, 29 (1822) ... — — V. N.S.W. — — B.fl.III,226 M.fr.II,50;VII,43.
E. erythronema, Turcz. in Bull. de l'acad. de St. Petersb. 415 (1852) W.A. — — — — — B.fl.III,227 M.fr.VIII,184.
E. caesia, Bentham, Fl. Austr. III, 227 (1866) ... ... W.A. — — — — — B.fl.III,227 M.fr.XI,40. [184.
E. tetraptera, Turczaninow in Bull. de Mosc. XXII, 22 (1849)... W.A. — — — — — B.fl.III,228 M.fr.II,34;IV,172;VIII,
E. miniata, A. Cunningham in Walp. rep. bot. syst. II, 925(1843) — — — — — Q. N.A. B.fl.III,228 M.fr.XI,38&42.
E. phoenicea, F. v. M. in Journ. Linn. Soc. III, 91 (1858) ... — — — — — — N.A. B.fl.III,251
E. robusta, Smith, Specim. Bot. New Holl. 40, t. 13 (1793) ... — — — — N.S.W. Q. — B.fl.III,228 M.fr.II,43&171;VII,44.
E. botryoides, Smith in Transact. Linn. Soc. III, 286 (1797) ... — — — V. N.S.W. Q. — B.fl.III,229 M.fr.IV,52.
E. goniocalyx, F. v. M., fragm. II, 48 (1860) ... ... S.A. — V. N.S.W. — — B.fl.III,220 M.fr.IV,52.
E. pallidifolia, F. v. M., fragm. III, 131 (1863) ... ... — — — — — — N.A. B.fl.III,236 M.fr.III,131.
E. incrassata, Labillard., Nov. Holl. pl. spec. II, 12, t. 150 (1806) W.A. S.A. — V. N.S.W. — — B.fl.III,231
E. oleosa, F. v. M. in Nederl. Kruidk. Arch. IV, 127 (1850) ... W.A. S.A. — V. N.S.W. — — B.fl.III,248 M.fr.II,56;XI,14.
E. goniantha, Turczaninow in Bullet. de Mosc. XX, 163 (1847) W.A. — — — — — — B.fl.III,248
E. falcata, Turczaninow in Bullet. de Mosc. XX, 163 (1847) ... W.A. — — — — — — B.fl.III,248
E. salmonophloia, F. v. M., fragm. XI, 11 (1878)... ... ... W.A. — — — — — — M.fr.XI,11.
E. leptopoda, Bentham, Fl. Austr. III, 238 (1866) ... ... W.A. — — — — — B.fl.III,238 M.fr.XI,11&13.
E. salubris, F. v. M., fragm. X, 54 (1876) ... ... W.A. — — — — — — M.fr.XI,12.
E. angustissima, F. v. M., fragm. IV, 25 (1863) ... ... W.A. S.A. — — — — — B.fl.III,238 M.fr.IV,25.
E. Doratoxylon, F. v. M., fragm. II, 55 (1860) ... ... W.A. — — — — — — B.fl.III,249 M.fr.II,55.
E. decurva, F. v. M., fragm. III, 130 (1863) ... ... W.A. — — — — — — B.fl.III,249 M.fr.III,130.
E. Cooperiana, F. v. M., fragm. XI, 63 (1880) ... ... W.A. — — — — — — M.fr.XI,63.
E. corynocalyx, F. v. M., fragm. II, 43 (1860) ... ... — S.A. — V. — — — M.fr.VII,43.
E. gomphocephala, De Candolle, prodr. III, 220 (1828) ... ... W.A. — — — — — B.fl.III,231 M.fr.II,36.
E. Oldfieldii, F. v. M., fragm. II, 37 (1860) ... ... W.A. — — — — — B.fl.III,237 M.fr.II,37.
E. orbifolia, F. v. M., fragm. V, 50 (1865)... ... ... W.A. — — — — — B.fl.III,238 M.fr.V,50.
E. diversicolor, F. v. M., fragm. III, 131 (1863) ... ... W.A. — — — — — B.fl.III,251 M.fr.III,131.
E. patellaris, F. v. M. in Journ. Linn. Soc. III, 84 (1858) ... — — — — — — B.fl.III,251
E. tessellaris, F. v. M. in Journ. Linn. Soc. III, 88 (1858) ... — — — — — Q. N.A B.fl.III,251
E. clavigera, A. Cunningham in Walp. rep. bot. syst. II, 926 (1843) — — — — — — N.A. B.fl.III,250
E. ferruginea, Schauer in Walp. repert. II, 926 (1843) ... ... — — — — — — N.A. B.fl.III,254
E. grandifolia, R. Brown in Benth. Fl. Austr. III, 250 (1866) ... — — — — — — N.A. B.fl.III,250
E. aspera, F. v. M. in Journ. Linn. Soc. III, 95 (1858) ... ... — — — — — — N.A. B.fl.III,250
E. patens, Bentham, Fl. Austr. III, 247 (1866) ... ... W.A. — — — — — — B.fl.III,247
E. Todtiana, F. v. M., in Wing's S. Sc. Record Aug. (1882) ... W.A. — — — — — — B.fl.III,232
E. vernicosa, J. Hooker in Lond. Journ. of Bot. VI, 478 (1847)... — — — T. — — — B.fl.III,232
E. Gunnii, J. Hooker in White's Journ. Bot. III, 499 (1844) ... — — — S.A. T. V. N.S.W. — B.fl.III,240 M.fr.II,62;VII,43;XI,38.
E. resinifera, Smith in White's Journ. Voy. N.S. Wales 231 (1790) — — — — N.S.W. Q. — B.fl.III,245 M.fr.II,172;VII,43.
E. saligna, Smith in Transact. Linn. Soc. III, 285 (1797) ... — — — — N.S.W. — — B.fl.III,245 M.fr.II,173;VII,44.
E. punctata, De Candolle, prodr. III, 217 (1828) ... ... — — — — N.S.W. — — B.fl.III,242 M.fr.VII,43.
E. oligantha, Schauer in Walp. rep. II, 926 (1843) ... ... — — — — — — N.A B.fl.III,213
E. alba, Reinwardt in Blume, Bijdr. 1101 (1826) ... ... — — — — — Q. N.A B.fl.III,243
E. pulverulenta, Sims, Bot. Mag. t. 2089 (1819) ... ... — — — V. N.S.W. — — B.fl.III,224 M.fr.II,70.
E. Stuartiana, F. v. M. in Benth. Fl. Austr. III, 243 (1866) ... — S.A. T. V. N.S.W. — — B.fl.III,243 M.fr.II,62.
E. viminalis, Labillardière, Nov. Holl. pl. spec. II, 12, t. 151 (1806) — S.A. T. V. N.S.W. — — B.fl.III,239 M.fr.II,64;VII,44.
E. rostrata, Schlechtendal, Linnaea XX, 655 (1847) ... W.A. S.A. — V. N.S.W. Q. N.A. B.fl.III,240
E. tereticornis, Smith, Spec. Bot. New Holl. 41 (1793) ... ... — — — V. N.S.W. Q. — B.fl.III,241
E. rudis, Endlicher in Huog. enum. 49 (1837) ... ... W.A. — — — — — — B.fl.III,244 M.fr.XI,14.
E. foecunda, Schauer in Lehm. pl. Preiss. I, 130 (1844) ... ... W.A. — — — — — — B.fl.III,252
E. redunca, Schauer in Lehm. pl. Preiss. I, 127 (1844) ... ... W.A. — — — — — — B.fl.III,253 M.fr.XI,15.
E. grossa, F. v. M. in Benth. Fl. Austr. III, 232 (1866)... ... W.A. — — — — — — B.fl.III,232 M.fr.VII,41.
E. obcordata, Turczaninow in Bull. Acad. Petersb. 416 (1852) ... W.A. — — — — — — B.fl.III,234 M.fr.VII,44;VIII,184.
E. occidentalis, Endlicher in Huog. enum. 49 (1837) ... ... W.A. — — — — — — B.fl.III,235 M.fr.II,39;VII,43.
E. cornuta, Labillardière, Voy. I, 403, t. 20 (1790) ... ... W.A. — — — — — — B.fl.III,234 M.fr.II,39;XI,38.
E. gamophylla, F. v. M , fragm. XI, 40 (1878) ... ... — S.A. — — — — N.A. — M.fr.XI,40.
E. perfoliata, R. Brown in Benth. Fl. Austr. III, 253 (1866) ... — — — — — — N.A. B.fl.III,253 M.fr.II,123;X,106.
E. setosa, Schauer in Walpers, rep. bot. syst. II, 926 (1843) ... — — — — — Q. N.A. B.fl.III,254 M.fr.II,123;X,106.
E. Torelliana, F. v. M., fragm. X, 106 (1877) ... ... — — — — — Q. — — M.fr.X,106.
E. peltata, Bentham, Fl. Austr. III, 254 (1866) ... ... — — — — — Q. — —
E. latifolia, F. v. M. in Journ. Linn. Soc. III, 94 (1858)... ... — — — — — — N.A. B.fl.III,255
E. ptychocarpa, F. v. M. in Journ. Linn. Soc. III, 90 (1858) ... — — — — — — N.A. B.fl.III,255 M.fr.VII,44.
E. Abergiana, F. v. M., fragm. XI, 41 (1878) ... ... — — — — — Q. — — M.fr.XI,41.
E. calophylla, R. Brown in Journ. Geogr. Soc. 20 (1831) ... W.A. — — — — — — B.fl.III,255 M.fr.II,33&171. [XI,15.
E. ficifolia, F. v. M., fragm. II, 85 (1860) ... ... W.A. — — — — — — B.fl.III,256 M.fr.II,85&176;VI,23;
E. corymbosa, Smith. Specim. Bot. New Holl. 43 (1793)... — — — — N.S.W. Q. — B.fl.III,256 M.fr.II,46&174.
E. terminalis, F. v. M. in Journ. Linn. Soc. III, 89 (1858) ... — S.A. — — — Q N.A B.fl.III,257
E. dichromophloia, F. v. M. in Journ. Linn. Soc. III, 89 (1858) — — — — — — N.A. B.fl.III,257
E. trachyphloia, F. v. M. in Journ. Linn. Soc. III, 90 (1858) ... — — — — — — — B.fl.III,257
E. eximia, Schauer in Walpers, rep. bot. syst. II, 925 (1843) ... — — — — N.S.W. — — B.fl.III,258 M.fr.II,47;VII,43.
E. maculata, Hooker, icones plant. t. 619 (1844) ... ... — — — — — Q. — — M.fr.X,98.
E. Watsoniana, F. v. M., fragm. X, 98 (1877) ... ... — — — — — Q. — —
E. odontocarpa, F. v. M., fragm. II, 33 (1860) ... ... W.A. — — — — — — B.fl.III,259 M.fr.II,33.
E. tetrodonta, F. v. M. in Journ. Linn. Soc. III, 97 (1858) ... — — — — — Q. N.A. B.fl.III,260
E. erythrocorys, F. v. M., fragm. II, 33 (1860) ... ... W.A. — — — — — — B.fl.III,259 M.fr.II,33.
E. tetragona, F. v. M., fragm. VI, 51 (1804) ... ... W.A. — — — — — — B.fl.III,259 M.fr.II,37.
E. oudemsioides, F. v. M., fragm. II, 35 (1860) ... ... W.A. — — — — — — B.fl.III,260 M.fr.II,35.

TRISTANIA, R. Brown in Ait. hort. Kew. sec. ed. IV, 417 (1812). (Lophostemon, Tristaniopsis.)
T. nerifolia, R. Brown in Ait. hort. Kew. sec. ed. IV, 417 (1812) — — — — N.S.W. — — B.fl.III,261 M.fr.VII,56.
T. suaveolens, Smith in Rees, Cycl. XXXVI (1817) ... ... — — — — N.S.W. Q. N.A. B.fl.III,261 M.fr.I,81.
T. conferta, R. Brown in Ait. hort. Kew. sec. ed. IV, 417 (1812) — — — — N.S.W. Q. N.A. B.fl.III,263 M.fr.I,82.
T. lactiflua, F. v. M., fragm. I, 82 (1858)... ... ... — — — — — — N.A. B.fl.III,263 M.fr.I,82.

T. exiliflora, F. v. M., fragm. V, 11 (1865)... ... ... — — — — Q. — B.fl.III,264 M.fr.IX,100.
T. laurina, R. Brown in Ait. hort. Kew. sec. ed. IV, 417 (1812) — — — V. N.S.W. Q. — B.fl.III,264 M.fr.I,81.
T. psidioides, Cunningham in Bot. Regist. XXII, n. 1839 (1837) — — — — — N.A. B.fl.III,264
T. umbrosa, Cunningham in Bot. Regist. XXII, n. 1830 (1837)... — — — — — N.A. B.fl.III,265

METROSIDEROS, Banks in Gaertner, de fruct. I, 170, t. 34 (1788). (Syncarpia, Kamptzia, Lysicarpus, Xanthostemon, Nania, Freinya, Cloezia, Spermalepis.)
M. glomulifera, Smith in Transact. Linn. Soc. III, 269 (1797) ... — — — N.S.W. Q. — B.fl.III,265 M.fr.I,79.
M. leptopetala, F. v. M., Docum. Intercol. Exhib. 30 (1867) ... — — — N.S.W. Q. — B.fl.III,266 M.fr.I,79.
M. ternifolia, F. v. M., Docum. Intercol. Exhib. 30 (1867) ... — — — — Q. — N.A. B.fl.III,267
M. eucalyptoides, F. v. M., fragm. I, 243 (1858) ... ... — — — — — N.A. B.fl.III,267 M.fr.I,81&243.
M. polymorpha, Gaudichaud, Freyc. Voy. Bot. 99 t. 108 (1826) ... — — — N.S.W. — — M.fr.IX,14.
M. nervulosa, C. Moore & F. v. M., fragm. VIII, 15 (1873) ... — — — N.S.W. — — M.fr.I,80.
M. chrysantha, F. v. M., fragm. IV, 150 (1864) ... ... — — — — Q. — B.fl.III,268 M.fr.IV,150.
M. paradoxa, F. v. M., fragm. I, 243 (1859) ... ... — — — — — N.A. B.fl.III,269 M.fr.I,80.
M. tetrapetala, F. v. M., fragm. VII, 41 (1870) ... ... — — — — — N.A. — M.fr.VII,41.

BACKHOUSIA, Hooker and Harvey in Bot. Mag. t. 4133 (1845).
B. myrtifolia, Hooker & Harvey in Bot. Mag. t. 4133 (1845) ... — — N.S.W. Q. — B.fl.III,269 M.fr.I,78;IX,43.
B. anguetifolia, F. v. M., fragm. I, 79 (1858) ... ... — — — Q. — B.fl.III,270 M.fr.I,79;IX,43.
B. sciadophora, F. v. M., fragm. II, 26 (1860) ... ... — — — N.S.W. Q. — B.fl.III,270 M.fr.II,171.
B. citriodora, F. v. M., fragm. I, 78 (1858) ... ... — — — — Q. — B.fl.III,270 M.fr.I,78.

OSBORNIA, F. v. Mueller, fragm. III, 30 (1862).
O. octodonta, F. v. M., fragm. III, 31 (1862) ... ... — — — — Q. N.A. B.fl.III,271 M.fr.IX,143.

RHODOMYRTUS, De Candolle, prodr. III, 240 (1828).
R. psidioides, Bentham, Fl. Austr. III, 272 (1866) ... ... — — — N.S.W. Q. — B.fl.III,272 M.fr.II,86;IX,142.
R. trineura, F. v. M. in Benth. Fl. Austr. III, 272 (1866) ... — — — — Q. — B.fl.III,272 M.fr.IV,117;IX,142.
R. cymiflora, F. v. M. in Benth. Fl. Austr. III, 273 (1866) ... — — — — Q. — B.fl.III,273 M.fr.V,12;IX,140.
R. macrocarpa, Bentham, Fl. Austr. III, 273 (1866) ... ... — — — — Q. — B.fl.III,273 M.fr.IX,142.

MYRTUS, Tournefort, inst. 640, t. 409 (1700), from Plinius.
M. rhytisperma, F. v. M., fragm. I, 77 (1858) ... ... — — N.S.W. Q. — B.fl.III,274 M.fr.IX,147.
M. tenuifolia, Smith in Transact. Linn. Soc. III, 280 (1797) ... — — N.S.W. Q. — B.fl.III,274 M.fr.IX,147.
M. gonoclada, F. v. M. in Benth. Fl. Austr. III, 275 (1866) ... — — — — Q. — B.fl.III,275 M.fr.IX,149.
M. lasioclada, F. v. M., fragm. IX, 148 (1875) ... ... — — — — Q. — — M.fr.IX,148.
M. Hillii, Bentham, Fl. Austr. III, 275 (1866) ... ... — — — — Q. — B.fl.III,275 M.fr.IX,149.
M. Beckleril, F. v. M., fragm. II, 85 (1861) ... ... — — — — Q. — B.fl.III,275 M.fr.IX,147.
M. Bidwillii, Bentham, Fl. Austr. III, 275 (1866) ... ... — — N.S.W. Q. — B.fl.III,275 M.fr.IX,147.
M. racemulosa, Bentham, Fl. Austr. III, 276 (1866) ... ... — — — — Q. — B.fl.III,276 M.fr.IX,147.
M. acmenioides, F. v. M., fragm. I, 77 (1858) ... ... — — N.S.W. Q. — B.fl.III,276 M.fr.IX,147.
M. fragrantissima, F. v. M. in Benth. Fl. Austr. III, 277 (1866) — — — N.S.W. Q. — B.fl.III,277
M. Shepherdi, F. v. M., fragm. IX, 148 (1875) ... ... — — — — Q. — — M.fr.IX,148.

RHODAMNIA, Jack in Hook. Comp. to the Bot. Mag. I, 153 (1835).
R. sessiliflora, Bentham, Fl. Austr. III, 275 (1866) ... ... — — — — Q. — B.fl.III,277 M.fr.IX,142.
R. trinervia, Blume, Mus. Bot. Lugd. I, 79 (1849) ... ... — — N.S.W. Q. — B.fl.III,278 M.fr.I,76;IX,142.
R. argentea, Bentham, Fl. Austr. III, 276 (1866) .. .. — — N.S.W. Q. — B.fl.III,278 M.fr.IX,142.
R. Blairiana, F. v. M., fragm. IX, 141 (1875) ... ... — — — — Q. — — M.fr.IX,141.

FENZLIA, Endlicher, Atacta 19, t. 17 & 18 (1833).
F. obtusa, Endlicher, Atacta 19, t. 17 (1833) ... ... — — — — Q. — B.fl.III,279 M.fr.IX,142.
F. retusa, Endlicher, Atacta 20, t. 18 (1833) ... ... — — — — — N.A. B.fl.III,279

DECASPERMUM, R. & G. Forster, char. gen. 73, t. 37 (1776). (Nelitris.)
D. paniculatum, Baillon, hist. des plant. 350 (1871) ... ... — — — — Q. — B.fl.III,279 M.fr.IV,56;IX,149.

EUGENIA, Micheli, nov. pl. gen. 226, t. 108 (1729). (Plinia, Plumier 1703, Acmene, Jambosa, Syzygium.)
E. rariflora, Bentham in Hook. Lond. Journ. II, 221 (1843) ... — — — — — B.fl.III,282 M.fr.III,130;V,15;IX,143
E. Smithii, Poiret, Encycl. méthod. Suppl. III, 126 (1813) ... — — V. N.S.W. Q. N.A. B.fl.III,282 M.fr.IV,50;IX,145.
E. hemilampra, F. v. M., fragm. IX, 145 (1875) ... ... — — N.S.W. Q. — — M.fr.IX,145.
E. Ventenatii, Bentham, Fl. Austr. III, 283 (1866) ... ... — — N.S.W. Q. — B.fl.III,283 M.fr.IV,58;IX,145.
E. leptantha, Wight, Illustr. II, 15 (1850)... ... ... — — — N.S.W. Q. — B.fl.III,283 M.fr.IX,143.
E. Moorei, F. v. M., fragm. V, 33 (1865) ... ... — — N.S.W. — — B.fl.III,283 M.fr.IX,143.
E. corynantha, F. v. M., fragm. IX, 144 (1875) ... ... — — — N.S.W. Q. — — M.fr.IX,144.
E. ventriflora, F. v. M., fragm. V, 32 (1865) ... ... — — — — Q. — B.fl.III,284 M.fr.V,32.
E. Hodgkinsoniae, F. v. M., fragm. IX, 145 (1875) ... ... — — N.S.W. Q. — — M.fr.IX,145.
E. Tierneyana, F. v. M., fragm. V, 14 (1865) ... ... — — — — Q. — B.fl.III,284 M.fr.IX,143.
E. grandis, Wight, Illustr. II, 17 (1850) ... ... — — — — Q. — B.fl.III,285 M.fr.V,19;IX,143.
E. suborbicularis, Bentham, Fl. Austr. III, 285 (1866) ... ... — — — — Q. — B.fl.III,285 M.fr.IX,143.
E. Wilsonii, F. v. M., fragm. V, 12 (1865)... ... ... — — — — Q. — B.fl.III,285 M.fr.IX,143.
E. eucalyptoides, F. v. M., fragm. IV, 55 (1863)... ... ... — — — — — N.A. B.fl.III,285 M.fr.IX,143.
E. myrtifolia, Sims, Bot. Mag. t. 2230 (1821) ... ... — — N.S.W. Q. — B.fl.III,286 M.fr.I,225;IX,143.
E. angophoroides, F. v. M., fragm. V, 33 (1865) ... ... — — — — Q. — B.fl.III,286 M.fr.IX,143.
E. Armstrongii, Bentham, Fl. Austr. III, 286 (1866) ... ... — — — — — N.A. B.fl.III,286
E. oleosa, F. v. M., fragm. V, 15 (1865) ... ... — — — — Q. — B.fl.III,287 M.fr.IX,146.
E. Dallachiana, F. v. M. in Benth. Fl. Austr. III, 287 (1866) ... — — — — Q. — B.fl.III,287 M.fr.IX,143.
E. cryptophlebia, F. v. M., fragm. IX, 144 (1875) ... ... — — — — — — M.fr.IX,144.

ACICALYPTUS, A. Gray, New Gen. Un. St. Exped. 9 (1853). (Calyptranthes partly.)
A. Fullageri, F. v. M., fragm. VIII, 15 (1873) ... ... — — N.S.W. — — M.fr.VIII,15.

BARRINGTONIA, R. & G. Forster, char. gen. 75, t. 38 (1776). (Stravadium.)
B. speciosa, R. & G. Forster, char. gen. 76, t. 38 (1776)... ... — — — — Q. — B.fl.III,288 M.fr.IX,118.
B. racemosa, Gaudichaud in Freyc. Voy. Bot. 483, t. 107 (1826) — — — — Q. — — M.fr.IX,118.
B. acutangula, Gaertner, de fruct. II, 97, t. 101 (1791) .. ... — — — — — N.A. B.fl.III,288 M.fr.IX,119.

CAREYA, Roxburgh, pl. Corom. III, 13, t. 217 (1816).
C. australis, F. v. M., fragm. V, 183 (1866) ... . — — — — — Q. N.A. B.fl.III,289 M.fr.V,183.
SONNERATIA, Linné fil., suppl. plant. 38[(1781).
S. acida, Linné fil., suppl. plant. 38 (1781)... ... ... — — — — — N.A. B.fl.III,301

MELASTOMACEAE.
A. L. de Jussieu, gen. plant. 328 (1789).

MEMECYLON, Linné, Fl. Zeyl. app. 9 (1747).
M. umbellatum, N. L. Burmann, Fl. Ind. 87 (1768) ... — — — — — Q. N.A. B.fl.III,293

OSBECKIA, Linné, spec. plant. I, 345 (1753).
O. Chinensis, Linné, spec. plant. 345 (1753) ... ... — — Q. — B.fl.III,291 M.fr.IV,160.
O. Australiana, Naudin in Ann. sc. nat. trois. sér. XIV, 59 (1850) — — — — — N.A. B.fl.III,291
O. perangusta, F. v. M., fragm. V, 181 (1866) ... ... — — — — N.A. — M.fr.V,181.

OTANTHERA, Blume in Flora, Regensb. IV, 488 (1831).
O. bracteata, Korthals in Verh. Nat. Gesch. der Ned. Bes. 235,
t. 51 (1841) ... ... — — — — Q. — B.fl.III,292

MELASTOMA, J. Burmann, thesaur. Zeyl. 72 (1737).
M. Malabathricum, Linné, spec. plant. 390 (1753) ... ... — — — — N.S.W. Q. N.A. B.fl.III,292

RHAMNACEAE.
A. L. de Jussieu, gen. plant. 376 (1789), from B. de Jussieu (1759).

VENTILAGO, Gaertner, de fruct. I, 223, t. 49 (1788).
V. viminalis, Hooker in Mitch. Trop. Austr. 309 (1848) ... — S.A. — — N.S.W. Q. N.A. B.fl.I,411 M.fr.IX,140.

ZIZYPHUS, Tournefort, inst. 627, t. 403 (1700), from Plinius (Ziziphus).
. Oenoplia, Miller, Gardeners Dictionary n. 3 (1731) ... ... — — — — — N.A. B.fl.I,412
. Jujuba, Lamarck, Encycl. méth. III, 318 (1789) ... ... — — — — Q. — B.fl.I,412
. quadrilocularis, F. v. M., fragm. III, 57 (1862) ... .— — — — — N.A. B.fl.I,412 M.fr.III,57.

DALLACHYA, F. v. M., fragm. IX, 140 (1875). (Rhamnus partly.)
D. Vitiensis, F. v. M., fragm. IX, 140 (1875) ... ... — — — — Q. — M.fr.XI,134.

BERCHEMIA, Necker, elem. bot. II, 122 (1790).
B. corollata, F. v. M., fragm. IX, 141 (1875) ... ... — — — — Q. — M.fr.IX,141.

COLUBRINA, L. C. Richard in Ann. des sc. nat. X, 368 (1827).
C. Asiatica, Brogniart in Ann. des sc. nat. X, 369 (1827) ... — — — — Q. N.A. B.fl.I,413 M.fr.IX,130.

ALPHITONIA, Reisseck in Endl. gen. plant. 1098 (1840).
A. excelsa, Reisseck in Endl. gen. plant. 1098 (1840) ... — — — N.S.W. Q. N.A. B.fl.I,414 M.fr.IX,140.

EMMENOSPERMUM, F. v. M., fragm. III, 63 (1862).
E. alphitonioides, F. v. M., fragm. III, 63 (1862)... ... — — — N.S.W. Q. — B.fl.I,415 M.fr.IX,140.
E. Cunninghamii, Bentham, Fl. Austr. I, 415 (1863) ... ... — — — — — N.A. B.fl.I,415

POMADERRIS, Labillardière, Nov. Holl. pl. specim. I, 61, t. 86 (1804).
P. lanigera, Sims, Bot. Mag. t. 1823 (1816) ... ... — — T. V. N.S.W. — — B.fl.I,416
P. elliptica, Labillardière, Nov. Holl. pl. spec. I, t. 86 (1804) — — T. V. N.S.W. — — B.fl.I,417 M.fr.IX,139.
P. phyllyroides, Sieber in De Cand., prodr. II, 33 (1825) ... — — — N.S.W. Q. — B.fl.I,418
P. vaccinifolia, Reisseck & F. v. M. in Linnaea XXIX, 266 (1856) — — V. N.S.W. — — B.fl.I,419 M fr.III,71.
P. myrtilloides, Fenzl in Hueg. enum. 22 (1837) ... ... W.A. — — — — B.fl.I,419 M.fr.IX,139.
P. grandis, F. v. M., fragm. III, 68 (1862)... ... ... W.A. — — — — B.fl.I,417 M.fr.IX,139.
P. Calvertiana, F. v. M., fragm. IX, 138 (1875) ... ... — — — N.S.W. — — M.fr.IX,138.
P. lodifolia, Cunningham in Field, N.S.Wales 351 (1825) ... — — V. N.S.W. — — B.fl.I,419 M.fr.III,69.
P. apetala, Labillardière, Nov. Holl. pl. spec. I, 62, t. 87(1904) — S.A. T. V. N.S.W. — — B.fl.I,419 M.fr.III,69&73;IX,139.
P. cinerea, Bentham, Fl. Austr. I, 420 (1863) ... ... — — V. N.S.W. — — B.fl.I,420
P. prunifolia, Cunningham in Hueg. enum. 22 (1837) ... ... — — V. N.S.W. — — B.fl.I,420 M.fr.III,75;IX,139.
P. ligustrina, Sieber in De Cand., prodr. II, 34 (1825) ... — — — N.S.W. — — B.fl.I,420 M.fr.III,71;IX,139.
P. betulina, Cunningham in Bot. Mag. t. 3212 (1833) ... ... — — V. N.S.W. — — B.fl.I,421 M.fr.III,76;IX,139.
P. Forrestiana, F. v. M., fragm. IX, 139 (1875) ... ... W.A. — — — — — M.fr.IX,139.
P. obcordata, Fenzl in Hueg. enum. 23 (1837) ... ... W.A. S.A. — — — — B.fl.I,421 M.fr.III,73.
P. racemosa, Hooker, Journ. of Bot. I, 256 (1834) ... ... — S.A. T. V. N.S.W. — — B.fl.I,421 M.fr.III,75&163;IX,139.
P. subrepanda, Reisseck & F. v. M. in Linnaea XXIX, 267 (1856) —. — V. N.S.W. — — B.fl.I,422 M.fr.III,74.
P. elachophylla, F. v. M., fragm. II, 131 (1861) ... ... — — V. N.S.W. — — B.fl.I,422 M.fr.III,166;IX,139.
P. phylicifolia, Loddiges, Bot. Cabinet t. 120 (1818) ... ... — — T. V. — — B.fl.I,422 M.fr.III,73.
P. intangenda, F. v. M., fragm. X, 52 (1870) ... ... W.A. — — — — — M.fr.X,53.

CRYPTANDRA, Smith in Transact. Linn. Soc. IV, 217 (1798). (Trymalium, Spyridium, Stenanthemum, Wichuren, Stenodiscus.)
C. albicans, F. v. M.; Trym., Reisseck in pl. Preiss. II, 286 (1847) W.A. — — — — — B.fl.I,423
C. Billardieri, F. v. M.; Trym., Fenzl in Hueg. enum. 25 (1837) W.A. — — — — — B.fl.I,423
C. lodifolia, F. v. M.; Trymalium, Fenzl in Hueg. enum. 24 (1837) W.A. — — V. — — — M.fr.IX,135.
C. Daltoni; Trymalium, F. v. M., fragm. IX, 135 (1875) ... — — — V. — — — M.fr.IX,135.
C. angustifolia, F. v. M.; Trym., Reisseck in pl. Preiss. II,284(1847) W.A. — — — — — B.fl.I,424 M.fr.IX,135.
C. Wichurae, F.v.M.; Trym., Nees in Linnaea XXI, 281(1847) W.A. — — — — — B.fl.I,425 M.fr.IX,135.
C. ericifolia, Smith in Transact. Linn. Soc. X, 294, t. 18 (1811)... — — — N.S.W. Q. — B.fl.I,438 M.fr.IX,137.
C. propinqua, Cunningham in Hueg. enum. 23 (1837) ... — — — N.S.W. — — B.fl.I,442 M.fr.IX,137.
C. hispidula, Reisseck & F. v. M. in Linnaea XXIX, 204 (1856) — S.A. — — — — B.fl.I,439
C. spyridioides, F. v. M. in Linnaea III, 68 (1862) ... ... W.A. — — — — — B.fl.I,439 M.fr.IX,138.
C. scoparia, Reisseck in Lehm. pl. Preiss. II, 283 (1847) W.A. — — — — — B.fl.I,439
C. spinosecns, Sieber in De Cand., prodr. II, 36 (1825) ... — — — N.S.W. — — B.fl.I,439
C. amara, Smith in Transact. Linn. Soc. X, 295, t. 18 (1811) ... — S.A. T. V. N.S.W. Q. — B.fl.I,440 M.fr.III,66&67;IX,137.
C. lanceolata, F. v. M., fragm. III, 65 (1862) ... ... — — — N.S.W. — — B.fl.I,440 M.fr.IX,137.
C. tomentosa, Lindley in Mitch. Three Exped. II, 178 (1838) ... — S.A. — — N.S.W. — — B.fl.I,441 M.fr.III,67;IX,138.
C. nutans, Steudel in Lehm. pl. Preiss. I, 186 (1844) ... ... W.A. — — — — — B.fl.I,441 M.fr.IX,138.

C. glabriflora, Bentham, Fl. Austr. I, 441 (1863) ... ... W.A. — — — — — — B.fl.I,441
C. alpina, J. Hooker, Fl. Tasm. I, 75, t. 12 (1860) ... ... — T. — — — — B.fl,I,441
C. Leucopogon, Meisener in Lehm. pl. Preiss. II, 287 (1847) ... W.A. — — — — — — B.fl.I,442
C. parvifolia, Turczaninow in Bull. de Mosc. I, 459 (1858) ... W.A. — — — — — — B.fl.I,442
C. buxifolia, Fenzl in Hueg. enum. 23 (1837) ... ... — — — — N.S.W. — — B.fl.I,442 M.fr.IX,138.
C. pungens, Steudel in Lehm. pl. Preiss. I, 187 (1844) ... W.A. — — — — — — B.fl.I,443 M.fr.IX,138.
C. mutila, Nees in Lehm. pl. Preiss. II, 289 (1847) ... W.A. — — — — — — B.fl.I,443
C. longistaminea, F. v. M., fragm. III, 64 (1862)... ... ... — — — — N.S.W. — — B.fl.I,443 M.fr.IX,138.
C. arbutiflora, Fenzl in Hueg. enum. 26 (1837) ... ... W.A. — — — — — — B.fl.I,444 M.fr.IX,138.
C. miliaris, Reisseck in Lehm. pl. Preiss. II, 288 (1847)... ... W.A. — — — — — — B.fl.I,444
C. nudiflora, F. v. M., fragm. III, 64 (1862) ... ... W.A. — — — — — — B.fl.I,444 M.fr.III,64.
C. nitens, Hooker, Journ. of Bot. I, 257 (1834) ... ... — T. — — — — B.fl.I,434 M.fr.IX,137.
C. pomaderroides, Reisseck in Endl. nov. stirp. dec. 29 (1830)... W.A. — — — — — — B.fl.I,435 M.fr.IX,137.
C. pumila; Spyridium, F. v. M., fragm. IX, 137 (1875)... ... W.A. — — — — — — — M.fr.IX,137.
C. leucophracta, Schlechtendal, Linnaea XX, 640 (1847) — S.A. T. V. — — — B.fl.I,435 M.fr.III,77;IX,136.
C. coronata, Reisseck in Lehm. pl. Preiss. II, 288 (1847) ... W.A. — — — — — — B.fl.I,436 M.fr.IX,137.
C. humilis, F. v. M.; Stenanthemum. Benth., Fl. Austr. I,436(1863) W.A. — — — — — — B.fl.I,436
C. Waterhousii; Spyridium, F. v. M., fragm. III, 83 (1862) — S.A. — — — — — B.fl.I,436 M.fr.III,83.
C. tridentata, Steudel in Lehm. pl. Preiss. I, 186 (1844)... ... W.A. — — — — — — B fl.I,427 M.fr.IX,137.
C. divaricata, F. v. M.; Spyridium, Bentham, Fl. Austr. I,427(1863) W.A. — — — — — — B fl.I,427
C. obcordata, J. Hooker, Fl. Tasm. I, 71 (1860) ... ... — T. V. — — — B.fl.I,427 M.fr.III,80.
C. Lawrencii, J. Hooker, Fl. Tasm. I, 72 (1860) ... ... — T. — — — — B.fl.I,430 M.fr.IX,136.
C. Hookeri, F. v. M.; C. parvifolia, J. Hooker, Fl. Tasm. I,73(1860) — S.A. T. V. N.S.W. — B.fl.I,428 M.fr.III,79;IX,136.
C. spadicea, F. v. M.; Trymalium, Fenzl in Hueg. enum. 26 (1837) W.A. — — — — — — B.fl.I,428 M.fr.IX,137.
C. globulosa, F. v. M.; Pomad., G. Don in Loud. hort. Brit. 84(1830) W.A. — — — — — — B.fl.I,429 M.fr.IX,137.
C. obovata, J. Hooker, Fl. Tasm. I, 74 (1860) ... ... — S.A. T. — — — B.fl.I,429 M.fr.IX,136.
C. apathulata; Trymalium, F. v. M. in Trans.Vict.Inst.I,122(1855) — S.A. — — — — — B.fl.I,430 M.fr.IX,136.
C. cordata, Turczaninow in Bull. de Mosc. XXXI, 439 (1858) ... W.A. — — — — — — B.fl.I,430 M.fr.IV,25;IX,137.
C. rotundifolia; Spyridium, F. v. M., fragm. IV, 25 (1863) ... W.A. — — — — — — — M.fr.IV,25.
C. phlebophylla; Trymalium, F. v. M. in Linnaea XXIX, 272 (1856) — S.A. — — — — — B.fl.I,430
C. coactilifolia, F. v. M.; Spyr., Reiss. in Linnaea XXIX, 291 (1856) — S.A. — — — — — B.fl.I,431 M.fr.IX,136.
C. complicata; Spyridium, F. v. M., fragm. III, 78 (1862) ... — S.A. — — — — — B.fl.I,431 M.fr.III,78.
C. westringiifolia, F.v.M.; Trym., Reiss. in pl. Preiss. II, 284 (1847) W.A. — — — — — — B.fl.I,431
C. villosa, Turczaninow in Bull. de Mosc. XXXI, 458 (1858) ... W.A. — — — — — — B.fl.I,432
C. pauciflora, Turczaninow in Bull. de Mosc. XXXI, 458 (1858) W.A. — — — — — — B.fl.I,432
C. halmaturina; Trym., F. v. M. in Linnaea XXIX, 283 (1856) — S.A. — — — — — B.fl.I,432
C. bifida; Trymalium, F. v. M. in Linnaea XXIX, 282 (1856)... — S.A. — — — — — B.fl.I,432 M.fr.IX,136.
C. aubochreata; Trym., F. v. M. in Trans. Vict. Inst. I, 122 (1855) W.A. S.A. — V. N.S.W. — B.fl.I,432 M.fr.III,82;IX,136.
C. vexillifera, Hooker, Journ. Bot. I, 257 (1834) ... ... — W.A. S.A. T. V. N.S.W. — B.fl.I,433 M.fr.III,61;IX,136.
  COLLETIA, Commerçon in A. L. de Jussieu, gen. 380 (1789). (Discaria, Tetrapasma.)
C. pubescens, Brogniart in Annal. des sc. nat. X, 366 (1827) ... — — T. V. N.S.W. — B.fl.I,445 M.fr.III,83;IV,175.
  GOUANIA, N. J. Jacquin, stirp. Amer. hist. 263, t. 170 (1763).
G. Hillii, F. v. M., fragm. IX, 163 (1875) ... ... — — — — — Q. — M.fr.IX,163.
G. Australiana, F. v. M., fragm. IV, 144 (1864) ... ... — — — — — Q. — M.fr.IV,144.

ARALIACEAE.
Ventenat, Tabl. III (1790). (Umbelliferarum suborde.)
ASTROTRICHA, De Candolle, Mémoire sur la fam. des Ombellifères 25 (1829).
A. pterocarpa, Bentham, Fl. Austr. III, 379 (1866) ... ... — — — — — Q. — B.fl.III,379 M.fr.VII,148.
A. floccosa, De Candolle, Mém. des Ombell. 30, t. 5 (1829) ... — — — — N.S.W. Q. — B.fl.III,380 M.fr.VII,148.
A. longifolia, Bentham in Hueg. enum. 55 (1837) ... ... — — — — N.S.W. Q. — B.fl.III,380 M.fr.VII,148.
A. ledifolia, De Candolle, Mém. des Ombell. 30, t. 6 (1829) ... — — — V. N.S.W. — B.fl.III,380 M.fr.VII,148.
A. Hamptoni, F. v. M., fragm. VI, 123 (1868) ... ... W.A. — — — — — — M.fr.VI,125&250
  FOROSPERMUM, F. v. M., fragm. VII, 94 (1870).
P. Michicanum, F. v. M., fragm. VII, 95 (1870) ... ... — — — — — — — M.fr.VII,95.
  PANAX, Linné, spec. plant. edit. sec. 1513 (1763). (Nothopanax.)
P. Gunnii, J. Hooker in Lond. Journ. of Bot. V, 466 (1847) ... — T. — — — — B.fl.III,381
P. Murrayi, F. v. M., fragm. II, 106 (1861) ... ... — — N.S.W. Q. — B.fl.III,381 M.fr.VII,96.
P. mollis, Bentham, Fl. Austr. III, 382 (1866) ... ... — — — — — Q. — B.fl.III,382 M.fr.VII,96.
P. Macrophyllus, Bentham, Fl. Austr. III, 382 (1866) ... ... — — — — — Q. — B.fl.III,382
P. sambucifolius, Sieber in De Cand. prodr. III, 255 (1825) ... — T. V. N.S.W. Q. — B.fl.III,382 M.fr.VII,95.
P. cephalobotrys, F. v. M., fragm. II, 83 (1861) ... ... — — — N.S.W. Q. — B.fl.III,382 M.fr.II,83;VII,95.
P. elegans, C. Moore & F. v. M. in Trans. phil. Inst. Vict. II, 68(1858) — — — N.S.W. Q. — B.fl.III,383
P. Cissodendron, C. Moore & F. v. M., fragm. VII, 96 (1870) ... — — — N.S.W. Q. — B.fl.IX,96.
  MOTHERWELLIA, F. v. M., fragm. VII, 107 (1871).
M. haploscadia, F. v. M., fragm. VII, 107 (1871) ... ... — — — — — Q. — — M.fr.VII,107.
  CISSODENDRON, Seemann, Journ. of Bot. II, 303 (1864). (Hedera partly, Irvingia, Kissodendron.)
C. Australianum, Seemann, Journ. of Bot. II, 303 (1864) ... ... — — — — — Q. — B.fl.III,384 M.fr.IV,120;V,18.
  HEPTOPLEURUM, Gaertner, de fruct. II, 472, t. 178 (1791). (Paratropia.)
H. venulosum, Seemann, Journ. of Bot. III, 80 (1865) ... ... — — — — — Q. — B.fl.III,384
  BRASSAIA, Endlicher, nov. stirp. decad. I, 89 (1839).
B. actinophylla, Endlicher, nov. stirp. dec. I, 89 (1839) ... ... — — — — — Q. — B.fl.III,385
  MERYTA, R. & G. Forster, char. gen. 119, t. 60 (1776). (Bothryodendron.)
M. latifolia, Seemann in Bonplandia, 295 (1862) ... ... — — — — N.S.W. — — M.fr.IX,160.
M. angustifolia, Seemann in Bonplandia, 296 (1862) ... ... — — — — N.S.W. — — M.fr.IX,169.
  MACKINLAYA, F. v. M., fragm. IV, 120 (1864).
M. macrosciadea, F. v. M., fragm. IV, 120 (1864) ... ... — — — — — Q. — B.fl.III,383 M.fr.II,108&176;IV,120.

UMBELLIFERAE;

Morison, hist. pl. II, lib. 3, sect. 9 (1680).

HYDROCOTYLE, Tournefort, inst. 328, t. 173 (1700).

H. vulgaris, Linné, spec. plant. 234 (1753)...	...	...	—	S.A.	— V.	N.S.W.	Q.	— B.fl.III,339
H. peduncularis, R. Brown in annal. des sc. phys. VI, 62, t. 61 (1820)	—	T.	—	—	—	— B.fl.III,339
H. hirta, R. Brown in annal. des sc. phys. VI, 64 (1820)	...	W.A.	S.A.	T.	V.	N.S.W.	Q.	— B.fl.III,339
H. laxiflora, De Candolle, prodr. IV, 61 (1830)	...	...	—	S.A.	— V.	N.S.W.	Q.	— B.fl.III,340	M.fr.IV,179&180.
H. rhombifolia, F. v. M., fragm. VII, 147 (1871)...	...	...	—	—	— —	N.S.W.	—	—	M.fr.VII,147.
H. pedicellosa, F. v. M., fragm. IV, 162 (1864)	...	...	—	—	— —	N.S.W.	Q.	— B.fl.III,341	M.fr.VII,147.
H. tripartita, R. Brown in annal. des sc. phys. VI, 69, t. 61 (1820)	—	S.A.	T.	V.	N.S.W.	Q.	— B.fl.III,341	M.fr.VII,147.
H. plebeia, R. Brown in annal. des sc. phys. VI, 46, t. 60 (1820)	W.A.	—	—	—	—	— B.fl.III,341
H. pterocarpa, F. v. M. in Transact. Vict. Inst. 126 (1855)	—	—	T.	V.	—	—	— B.fl.III,342
H. geranifolia, F. v. M. in Transact. Vict. Inst. 126 (1855)	...	—	—	— V.	N.S.W.	—	— B.fl.III,342
H. medicaginoides, Turczaninow in Bull. Mosc. XXII, 27 (1849)	W.A.	S.A.	— V.	—	—	— B.fl.III,342	M.fr.IV,178.
H. muriculata, Turczaninow in Bull. Mosc. 28 (1849)	...	W.A.	—	—	—	— B.fl.III,342
H. callicarpa, Bunge in Lehm. pl. Preiss. I, 283 (1844)	...	W.A.	S.A.	T.	V.	N.S.W.	— B.fl.III,343
H. scutellifera, Bentham, Fl. Austr. III, 343 (1866)	...	W.A.	—	—	—	— B.fl.III,343
H. hispidula, Bunge in Lehm. pl. Preiss. I, 283 (1844)	...	W.A.	—	—	—	— B.fl.III,343
H. trachycarpa, F. v. M. in Schlecht. Linnaea XXV, 394 (1852)	—	S.A.	— —	N.S.W.	—	— B.fl.III,343
H. pilifera, Turczaninow in Bull. Mosc. XXII, 26 (1849)	...	W.A.	—	—	—	— B.fl.III,344
H. capillaris, F. v. M., fragm. IV, 178 (1864)	...	W.A.	S.A.	— V.	N.S.W.	— B.fl.III,344	M.fr.IV,178.
H. rugulosa, Turczaninow in Bull. Soc. Mosc. XXII, 27 (1849)...	W.A.	—	—	—	— N.A.	B.fl.III,344
H. grammatocarpa, F. v. M., fragm. II, 128 (1861)	...	W.A.	—	—	—	— —	B.fl.III,345
H. diantha, De Candolle, prodr. IV, 63 (1830)	...	W.A.	—	—	—	—	B.fl.III,345	M.fr.IV,179.
H. lemnoides, Bentham, Fl. Austr. III, 345 (1866)	...	W.A.	—	—	—	—	B.fl.III,345
H. alata, R. Brown in annal. des sc. phys. VI, 75, t. 61 (1820)..	W.A.	—	—	—	—	— B.fl.III,345
H. tetragonocarpa, Bunge in Lehm. pl. Preiss. I, 284 (1844)	...	W.A.	—	—	—	— B.fl.III,345
H. glochidiata, Bentham, Fl. Austr. III, 346 (1866)	...	W.A.	—	—	—	— B.fl.III,346	M.fr.II,129;VII,147.
H. verticillata, Turczaninow in Bull. Mosc. XXII, 28 (1849)	...	W.A.	—	—	—	— B.fl.III,346	M.fr.VII,147.
H. Asiatica, Linné, spec. plant. 234 (1753)...	...	...	W.A.	S.A.	T.	V.	N.S.W.	Q.	— B.fl.III,346

DIDISCUS, De Candolle in Bot. Mag. t. 2857 (1828). (Dimetopia, Pritzelia, Huegelia, Ceratia, Hemicarpus.)

D. pusillus, F. v. M., fragm. IX, 47 (1875)	...	...	W.A.	S.A.	— V.	N.S.W.	— B.fl.III,348	M.fr.IX,47.
D. cyanopetalus, F. v. M., fragm. IX, 46 (1875)	...	...	W.A.	S.A.	— V.	N.S.W.	— B.fl.III,348	M.fr.IX,47.
D. eriocarpus, F. v. M., fragm. IX, 46 (1875)	...	...	W.A.	S.A.	—	—	— B.fl.III,348	M.fr.IX,46.
D. villosus, F. v. M. in Proc. Roy. Soc. Tasm. IX, 238 (1857) ...	—	—	—	— N.A.	B.fl.III,349
D. coeruleus, De Candolle in Hook. Bot. Mag. 2875 (1828)	W.A.	—	—	—	— B.fl.III,349	M.fr.IX,47.
D. pilosus, Bentham in Hueg. enum. 54 (1837)	...	W.A.	S.A.	T.	V.	N.S.W.	Q.	— B.fl.III,349
D. glaucifolius, F. v. M. in Schlecht. Linnaea XXV, 395 (1852)	—	S.A.	— —	N.S.W.	—	— B.fl.III,350	M.fr.IX,47.
D. glandulosus, F. v. M. in Proc. Roy. Soc. Tasm. III, 238 (1857)	—	—	—	— N.A.	B.fl.III,350
D. incisus, Hooker, Bot. Magaz. 2875 (1828)	...	...	—	— —	N.S.W.	Q.	— B.fl.III,350
D. procumbens, F. v. M. in Proc. Roy. Soc. Tasm. III, 237 (1857)	—	—	— —	N.S.W.	Q.	— B.fl.III,350	M.fr.IX,47.
D. hemicarpus, F. v. M. in Trans. Bot. Soc. Edinb. VII, 491 (1857)	—	—	—	— N.A.	B.fl.III,351	M.fr.IX,47.
D. humilis, J. Hooker in icon. pl. t. 304 (1841)	...	...	W.A.	—	—	—	— B.fl.III,351

TRACHYMENE, Rudge in Transact. Linn. Soc. X, 300 (1810). (Siebera, Fischera, Platysace, Platycarpidium.)

T. compressa, Rudge in Transact. Linn. Soc. X, 300 (1810)	...	W.A.	—	—	—	— B.fl.III,352
T. juncea, Bunge in Lehm. pl. Preiss. I, 286 (1844)	...	W.A.	—	—	—	— B.fl.III,353
T. haplocladea, F. v. M.; Siebera, Benth., Fl. Austr. III,353 (1866)	W.A.	—	—	—	— B.fl.III,353
T. cirrosa, F. v. M.; Platysace, Bunge in pl. Preiss. I, 285 (1844)	—	S.A.	— V.	—	— B.fl.III,354
T. heterophylla, F. v. M., First Gen. Rep. 13 (1853)	...	W.A.	S.A.	—	—	— B.fl.III,354
T. tenuissima, F. v. M.; Siebera, Benth., Fl. Austr. III, 354 (1866)	W.A.	—	—	—	— B.fl.III,354
T. dissecta, F. v. M.; Siebera, Bentham, Fl. Austr. III, 354 (1866)	W.A.	—	—	—	— B.fl.III,355
T. commutata, Turczaninow in Bull. Mosc. XXII, 30 (1840)	...	W.A.	—	—	—	— B.fl.III,355
T. effusa, Turczaninow in Bull. Mosc. XXII, 31 (1849)	...	W.A.	—	—	—	— B.fl.III,355
T. deflexa, Turczaninow in Bull. Mosc. XXII, 31 (1849)...	...	W.A.	—	—	—	— B.fl.III,355
T. valida; Platycarpidium, F. v. M. in Kew. Misc. IX, 310 (1857)	—	—	—	—	Q.	— B.fl.III,355
T. cricoides, Sieber in De Candolle, prodr. IV, 73 (1830)	—	—	— V.	N.S.W.	Q.	— B.fl.III,356
T. linearis, Sprengel, spec. umbellif. 7 (1818)	...	...	—	— —	N.S.W.	— B.fl.III,356
T. Billardieri, F. v. M.; Siebera, Benth., Fl. Austr. III, 356 (1866)	—	— —	N.S.W.	—	— B.fl.III,356
T. Stephensonii, Turczaninow in Bull. do Mosc. XX, 170 (1847)	—	— —	N.S.W.	—	— B.fl.III,357

XANTHOSIA, Rudge in Transact. Linn. Soc. X, 361 (1810). (Leucolaena, Schoenolaena, Pentapeltis.)

X. juncea, Bentham, Fl. Austr. III, 359 (1866)	...	W.A.	—	—	—	— B.fl.III,359
X. tenuior, Bentham, Fl. Austr. III, 359 (1866)	...	...	W.A.	—	—	—	— B.fl.III,359
X. peltigera, Steudel, nomencl. bot. 791 (1841)	...	...	W.A.	—	—	—	— B.fl.III,359
X. hederifolia, Bentham, Fl. Austr. III, 359 (1866)	...	W.A.	—	—	—	— B.fl.III,359
X. candida, Steudel, nomencl. bot. 790 (1841)	...	...	W.A.	—	—	—	— B.fl.III,359
X. tridentata, De Candolle, prodr. IV, 75 (1830) ...	...	—	— —	N.S.W.	—	— B.fl.III,260
X. singuliflora, F. v. M., fragm. IV, 184 (1864)	...	...	—	—	—	— B.fl.III,260	M.fr.IV,184.
X. pilosa, Rudge in Transact. Linn. Soc. X. 301, t. 22 (1811) ...	—	—	T.	V.	N.S.W.	Q.	— B.fl.III,260
X. ciliata, Hooker, icon. pl. t. 726 (1848)	...	...	W.A.	—	—	—	— B.fl.III,260
X. pusilla, Bunge in Lehm. pl. Preiss. I, 291 (1844)	...	W.A.	S.A.	T.	V.	—	— B.fl.III,261
X. fruticulosa, Bentham, Fl. Austr. III, 361 (1866)	...	W.A.	—	—	—	— B.fl.III,261
X. Huegelii, Steudel, nomenclat. bot. 791 (1841)	...	...	W.A.	—	—	—	— B.fl.III,262
X. dissecta, J. Hooker in icon. plant. t. 302 (1841)	...	—	S.A.	T.	V.	N.S.W.	— B.fl.III,262
X. peduncularis, Bentham, Fl. Austr. III, 362 (1866)	...	W.A.	—	—	—	— B.fl.III,262
X. vestita, Bentham, Fl. Austr. III, 363 (1866)	...	...	W.A.	—	—	—	— B.fl.III,363
X. Atkinsoniana, F. v. M., fragm. II, 127 (1861) ...	...	—	— —	N.S.W.	—	— B.fl.III,363	M.fr.II,127.
X. rotundifolia, De Candolle, prodr. IV, 75 (1830)	...	W.A.	—	—	—	— B.fl.III,363

AZORELLA, Lamarck, Encycl. méthodiq. I, 344, t. 189 (1783).   (Fragosa, Pozoa, Microsciadium, Schizeilema, Oschatzia, Centella partly, Dichopetalum.)
A. Muelleri, Bentham, Fl. Austr. III, 364 (1866)... ... ... — — — V. — — — B.fl.III,364
A. cuneifolia, F. v. M. in Transact. phil. Inst. Vict. I, 103 (1855)   — — — V. — — — B.fl.III,365
A. saxifraga, Bentham, Fl. Austr. III, 365 (1866)... ... ... — — — T. — — — B.fl.III,365
A. dichopetala, Bentham, Fl. Austr. III, 365 (1866)   ... ... — — T. V. — — — B.fl.III,365
   HUANACA, Cavanilles, icon. VI, 18, t. 528 (1801).   (Diplaspis, Pozoopsis.)
H. hydrocotylea, Bentham & J. Hooker, gen. pl. I, 877 (1867)... — — — T. V. — — — B.fl.III,366
H. cordifolia, Bentham & J. Hooker, gen. pl. I, 877 (1867)   ... — — T. — — — B.fl.III,366
   ACTINOTUS, Labillardière, Nov. Holl. pl. spec. I, 67, t. 92 (1804).   (Hemiphues, Eriocalia, Holotome.)
A. Helianthi, Labillardière, Nov. Holl. pl. spec. I, 67, t. 92 (1804)   — — — V. N.S.W. Q. — B.fl.III,367
A. leucocephalus, Bentham in Huog. enum. 56 (1837)   ... ... W.A. — — — — — — B.fl.III,368
A. minor, De Candolle, prodr. IV, 83 (1830)   ... ... — — — — N.S.W. — — B.fl.III,368
A. Gibbonsii, F. v. M., fragm. VI, 23 (1867)   ... ... — — — — N.S.W. — — M.fr.VII,23.
A. omnifertilis, Bentham, Fl. Austr. III, 368 (1866)   ... W.A. — — — — — — B.fl.III,368
A. rhomboideus, Bentham, Fl. Austr. III, 368 (1866)   ... ... W.A. — — — — — — B.fl.III,368
A. bellidioides, Bentham, Fl. Austr. III, 369 (1866)   ... ... — — T. — — — — B.fl.III,369
A. glomeratus, Bentham, Fl. Austr. III, 369 (1866)   .. ... W.A. — — — — — — B.fl.III,369
   ERYNGIUM, Tournefort, inst. 327, t. 173 (1700) from Theophrastoe and Dioscorides).
E. rostratum, Cavanilles, icon. pl. VI, 35, t. 552 (1801)... ... W.A. S.A. — V. N.S.W. Q. — B.fl.III,370
E. vesiculosum, Labillardière, Nov. Holl. pl. spec. I, 73, t. 98 (1804)   — S.A. T. V. N.S.W. Q. — B.fl.III,370 M.fr.VII,148.
E. plantagineum, F. v. M. in Pap. Roy. Soc. Tasm. III, 235 (1857)   — S.A. — — N.S.W. Q. — B.fl.III,371 M.fr.VII,148.
E. expansum, F. v. M. in Pap. Roy. Soc. Tasm. III, 236 (1857)   — — — — N.S.W. Q. — B.fl.III,371 M.fr.VII,148.
   APIUM, Tournefort, inst. 305, t. 160 (1700).
A. prostratum, Labillardière, relat. du voy. I, 141 (1799)   ... W.A. S.A. T. V. N.S.W. Q. — B.fl.III,372 M.fr.VII,148.
A. leptophyllum, F. v. M. in Benth. Fl. Austr. III, 372 (1866)... — — — V. N.S.W. Q. — B.fl.III,372 M.fr.VII,148.
   SIUM, Tournefort, inst. 308, t. 162 (1700), from C. Bauhin (1623).
S. latifolium, C. Bauhin, pinax 154 (1623) ... ... — S.A. — V. N.S.W. Q. — —
S. angustifolium, Linné, spec. plant. ed. sec. 1672 (1763)   ... — — — — N.S.W. Q. — —
   SESELI, Rivinus in Rupp. Fl. Jen. 267 (1718).
S. Harveyanum, F. v. M. in Transact. phil. Soc. Vict. I, 104 (1855)   — — — V. — — — B.fl.III,373
S. algens, F. v. M. in Transact. phil. Soc. Vict. I, 104 (1855)   ... — — — V. — — — B.fl.III,373
   CRANTZIA, Nuttall, Gen. of North Amer. plants I, 177 (1818).   (Non Scopoli, 1777, Crantziola, F. v. M.)
C. lineata, Nuttall, Gen. of North Amer. pl. I, 178 (1818)   ... — S.A. T. V. N.S.W. Q. — B.fl.III,374
   OENANTHE, Tournefort, inst. 312, t. 166 (1700).
O. Javanica, De Candolle, prodr. IV, 138 (1830) ... ... — — — — — Q. — M.fr.V,182.
   ACIPHYLLA, R. & G. Forster, char. gen. 135, t. 68 (1776).   (Anisotome, Calosciadium, Gingidium partly, Ligustici subgenus.)
A. simplicifolia, F. v. M. in Benth. Fl. Austr. III, 375 (1866)   ... — — — V. N.S.W. — — B.fl.III,375
A. glacialis, F. v. M. in Benth. Fl. Austr. III, 375 (1866)   ... — — T. V. N.S.W. — — B.fl.III,375
A. procumbens, F. v. M. in Benth. Fl. Austr. III, 375 (1866)   ... — — T. — — — B.fl.III,375
   DAUCUS, Tournefort, inst. 307, t. 161 (1700), from l'Ecluse (1576).
D. brachiatus, Sieber in De Cand., prodr. IV, 214 (1830)   ... W.A. S.A. T. V. N.S.W. Q. — B.fl.III,376
   OREOMYRRHIS, Endlicher, gen. plantar. 787 (1839).
O. andicola, Endlicher, gen. plantar. 787 (1839)   ... ... — S.A. T. V. N.S.W. — — B.fl.III,377 M.fr.VIII,185.
O. pulvinifica, F. v. M., fragm. VIII, 185 (1874) ... ... — — — V. N.S.W. — — M.fr.VIII,185.

SYNPETALEAE PERIGYNAE.

F. v. M. in Woolls's plants of the neighb. of Sydney, 27 (1880).

AQUIFOLIACEAE.

De Candolle, Théor. élém. I, 217 (1813).
B. Arnhemensis, F. v. M., fragm. II, 119 (1861) ... ... — — — — — — N.A. B.fl.I,397 M.fr.II,119.
   ILEX, Linné, gen. plant. 33 (1737), from C. Bauhin (1623).
I. peduncularis, F. v. M., fragm. VII, 105 (1871)... ... ... — — — — — Q. — M.fr.VII,105.

OLACINEAE.

Mirbel in Bull. de la Soc. philomat. 377 (1813).
   XIMENIA, Plumier, nova plant. Amer. genera 6, t. 21 (1703).
X. Americana, Linné, spec. plant. 1103 (1753)   ... — — — — — Q. N.A. B.fl.I,391 M.fr.VI,3;IX,150.
   OLAX, Linné, amoen. acad. 387 (1747).   (Spermaxyrum, Lopadocalyx.)
O. phyllanthi, R. Brown, prodr. 358 (1810)   ... ... W.A. — — — — — — B.fl.I,392 M.fr.IX,151.
O. retusa, F. v. M. in Benth. Fl. Austr. I, 392 (1863)   ... — — — N.S.W. Q. — B.fl.I,392 M.fr.IX,151.
O. stricta, R. Brown, prodr. 358 (1810)   ... ... — — — N.S.W. Q. — B.fl.I,392 M.fr.IX,151.
O. Benthamiana, Miquel in Lehm. pl. Preiss. I, 228 (1844)   ... W.A. S.A. — — — — N.A. B.fl.I,393 M.fr.IX,151.
O. aphylla, R. Brown, prodr. 358 (1810)   ... ... — — — — — — N.A. B.fl.I,393 M.fr.IX,151.
   CANSJERA, A. L. de Jussieu, gen. pl. 448 (1789).
C. leptostachya, Benth. in Hook. Lond. Journ. of Bot. II, 231 (1843)   — — — — — Q. — B.fl.I,394 M.fr.VI,3.
   OPILIA, Roxburgh, pl. Corom. II, 31, t. 158 (1799).
O. amentacea, Roxburgh, pl. Corom. II, 31, t. 158 (1799)   ... — — — — — Q. N.A. B.fl.I,394 M.fr.VI,4;IX,150.
   PHLEBOCALYMMA, Griffith in Benth. & J. Hook. gen. pl. I, 333 (1862).   (Phlebocalymna).
P. lobospora, F. v. M., fragm. IX, 151 (1875)   ... ... — — — — — Q. — M.fr.IX,151.

PENNANTIA, R. & G. Forster, char. gen. 133, t. 67 (1776).
P. Cunninghamii, Miers in Ann. of nat. hist. sec. ser. IX, 491 (1852) — — — — N.S.W. Q. — B.fl.I,395 M.fr.VI,4;IX,150.
P. Endlicheri, Reissek in Schlecht. Linnaea XVI, 341, t. 13 (1843) — — — — N.S.W. — — — M.fr.IX,150.
APODYTES, E. Meyer in Hook. Journ. of Bot. III, 155 (1841).
A. brachystylis, F. v. M., fragm. IX, 140 (1875) ... ... — — — — — Q. — — M.fr.IX,149.
VILLARESIA, Ruiz & Pavon, flor. Peruv. et Chil. prodr. 35 (1794).
V. Moorei, F. v. M. in Benth. Fl. Austr. I, 396 (1863) ... ... — — — — N.S.W. — — B.fl.I,395
V. Smythii, F. v. M., fragm. V, 156 (1866) ... ... — — — — — Q. — — M.fr.IX,150.
GOMPHANDRA, Wallich, Catal. n. 3718 & 7204 (1832).
G. Australiana, F. v. M., fragm. VI, 3 (1877) ... ... — — — — — Q. — — M.fr.VI,3&253;IX,150.

ELAEAGNEAE.
R. Brown, prodr. I, 350 (1810.)
ELAEAGNUS, Tournefort, coroll. 53, t. 459 (1703) from Camerarius (1586).
E. latifolia, Linné, spec. plant. 121 (1753) ... ... — — — Q. — M.fr.VI,39 M.fr.IX,100.

SANTALACEAE.
R. Brown, prodr. I, 350 (1810).
THESIUM, Linné, gen. pl. 60 (1737).
T. australe, R. Brown, prodr. 353 (1810) ... ... — — — T. V. N.S.W. Q. — M.fr.VI,212 M.fr.IX,3.
SANTALUM, Linné, gen. ed. sec., 165 (1742), from C. Bauhin (1623). (Fusanus, Eucarya).
S. lanceolatum, R. Brown, prodr. 356 (1810) ... ... W.A. S.A. — — N.S.W. Q. N.A. B.fl.VI,214 M.fr.I,85;VIII,11;IX,3.
S. ovatum, R. Brown, prodr. 355 (1810) ... ... — — — — — — N.A. B.fl.VI,214
S. obtusifolium, R. Brown, prodr. 356 (1810) ... ... — — — V. N.S.W. — — B.fl.VI,215 M.fr.VIII,11.
S. acuminatum, A. de Caudolle, prodr. XIV, 684 (1856)... W.A. S.A. — V. N.S.W. — — B.fl.VI,216 M.fr.IX,169;XI,20.
S. persicarium, F. v. M. in Transact. Vict. Inst. 41 (1854) W.A. S.A. — V. N.S.W. — — B.fl.VI,216 M.fr.I,86;IX,3;XI,20.
S. cygnorum, Miquel in Lehm. pl. Preiss. I, 615 (1845) ... W.A. — — — — — — B.fl.VI,217 M.fr.VIII,11.
S. crassifolium, A. de Candolle, prodr. XIV, 685 (1856)... — — — — N.S.W. — — B.fl.VI,217
CHORETRUM, R. Brown, prodr. 353 (1810).
C. glomeratum, R. Brown, prodr. 354 (1810) ... ... W.A. S.A. — V. N.S.W. — — B.fl.VI,218
C. chrysanthum, F. v. M. in Transact. phil. Soc. Vict. I, 23 (1854) — S.A. — V. N.S.W. — — — M.fr.VIII,11;XI,138.
C. spicatum, F. v. M., fragm. I, 21 (1858) ... ... — S.A. — V. N.S.W. — — B.fl.VI,218 M.fr.VIII,11.
C. lateriflorum, R. Brown, prodr. 354 (1810) ... ... — — — V. N.S.W. — — B.fl.VI,219
C. Candollei, F. v. M. in Bentham's Fl. Austr. VII, 219 (1873) — — — — N.S.W. Q. — B.fl.VI,219
LEPTOMERIA, R. Brown, prodr. 353 (1810).
L. spinosa, A. de Candolle, prodr. XIV, 678 (1856) ... W.A. — — — — — B.fl.VI,220 M.fr.VIII,10.
L. Preissiana, A. de Candolle, prodr. XIV, 678 (1856) ... W.A. — — — — — B.fl.VI,221 M.fr.VIII,10.
L. pauciflora, R. Brown, prodr. 354 (1810)... ... ... W.A. — — — — — B.fl.VI,221 M.fr.VIII,10.
L. scrobiculata, R. Brown, prodr. 354 (1810) ... ... W.A. — — — — — B.fl.VI,221 M.fr.VIII,10.
L. acida, R. Brown, prodr. 353 (1810) ... ... — — — T. — N.S.W. Q. — B.fl.VI,222 M.fr.VIII,10.
L. Billardieri, R. Brown, prodr. 354 (1810) ... ... — — — T. — N.S.W. — — B.fl.VI,222 M.fr.VIII,10.
L. aphylla, R. Brown, prodr. 354 (1810) ... ... — — S.A. — V. N.S.W. — — B.fl.VI,222 M.fr.VIII,10.
L. glomerata, F. v. M. in J. Hook. Tasm. II, 370 (1860)... — — — T. — — — — B.fl.VI,223 M.fr.VIII,10.
L. squarrulosa, R. Brown, prodr. 354 (1810) ... ... W.A. — — — — — B.fl.VI,223 M.fr.VIII,11.
L. Cunninghamii, Miquel in Lehm. pl. Preiss. I, 611 (1845) W.A. — — — — — B.fl.VI,223 M.fr.VIII,10.
L. empetriformis, Miquel in Lehm. pl. Preiss. I, 610 (1845) W.A. — — — — — B.fl.VI,224
L. axillaris, R. Brown, prodr. 354 (1810) ... ... W.A. — — — — — B.fl.VI,224
L. laxa, Miquel in Lehm. pl. Preiss. I, 612 (1845)... ... .W.A. — — — — — B.fl.VI,224
L. obovata, Miquel in Lehm. pl. Preiss. I, 613 (1845) ... W.A. — — — — — B.fl.VI,224
OMPHACOMERIA, Endlicher, gen. pl. 326 (1838).
O. acerba, A. de Candolle, prodr. XIV, 681 (1856) ... — — — V. N.S.W. — — B.fl.VI,224
ANTHOBOLUS, R. Brown, prodr. 357 (1810).
A. filifolius, R. Brown, prodr. 357 (1810) ... ... — — — — — N.A. B.fl.VI,226 M.fr.IX,3.
A. triqueter, R. Brown, prodr. 357 (1810) ... ... — — — — Q. — B.fl.VI,226
A. foveolatus, F. v. M., fragm. I, 212 (1859) ... ... W.A. — — — — — B.fl.VI,226 M.fr.II,181;IX,3;XI,20.
A. leptomerioides, F. v. M., fragm. I, 21 (1858) ... ... — — — — Q. — B.fl.VI,227 M.fr.IX,3.
A. exocarpoides, F. v. M., fragm. IX, 3 (1875) ... ... — S.A. — — — — — M.fr.IX,3.
EXOCARPOS, Labillardière, Relat. du voy. à la rech. de La Pérouse I, 155 (1798). (Exocarpus.)
E. latifolia, R. Brown, prodr. 356 (1810) ... ... — — — N.S.W. Q. N.A. B.fl.VI,228 M.fr.IX,3.
E. odorata, A. de Candolle, prodr. XIV, 089 (1856) ... W.A. — — — — — B.fl.VI,228 M.fr.VIII,10.
E. cupressiformis, Labillardière, voy. I, 155, t. 14 (1798) W.A. S.A. T. V. N.S.W. Q. — B.fl.VI,229
E. spartea, R. Brown, prodr. 356 (1810) ... ... W.A. S.A. — V. N.S.W. Q. — B.fl.VI,230 M.fr.VIII,10;XI,20. [20.
E. aphylla, R. Brown, prodr. 357 (1810) ... ... W.A. S.A. — V. N.S.W. — — B.fl.VI,230 M.fr.VIII,10;IX,169;XI,
E. homaloclada, C. Moore & F. v. M., fragm. VIII, 9 (1871) — — — N.S.W. — — B.fl.VI,230 M.fr.VIII,9.
E. phyllanthoides, Endlicher, prodr. pl. ins. Norf. 46 (1833) — — — N.S.W. — — B.fl.VI,230 M.fr.IX,169.
E. stricta, R. Brown, prodr. 357 (1810) ... ... — S.A. T. V. N.S.W. — — B.fl.VI,230 M.fr.IX,169.
E. humifusa, R. Brown, prodr. 356 (1810) ... ... — — — T. V. N.S.W. — — B.fl.VI,231
E. nana, J. Hooker in Lond. Journ. of Bot. VI, 281 (1847) ... — — — T. V. — — — B.fl.VI,230

LORANTHACEAE.
A. L. de Jussieu, in Ann. du Mus. XII, 292 (1808).
VISCUM, Tournefort, inst. 609, t. 380 (1700), from Camerarius (1586).
V. orientale, Willdenow, spec. plant. IV, 737 (1805) ... ... — — — — Q. — B.fl.III,396
V. angulatum, Heyne in De Cand., prodr. IV, 283 (1830) ... ... — — — — Q. — B.fl.III,396
V. articulatum, N. L. Burmann, Fl. Ind. 311 (1768) ... ... — — — N.S.W. Q. — B.fl.III,396

NOTOTHIXOS, Oliver in Journ. Linn. Soc. VII, 104 (1865).
N. incanus, Oliver in Journ. Linn. Soc. VII, 104 (1865)... ... — — — — N.S.W. Q. — B.fl.III,307 M.fr.II,109,176;IV,173.
LORANTHUS, Linné, syst. nat. ed. sec. 22 (1740).
L. celastroides, Sieber in Roem. & Schult. syst. veg. VII, 163 (1829) — — V. N.S.W. Q. — B.fl.III,389 M.fr.VI,232.
L. Bidwillii, Bentham, Fl. Austr. III, 390 (1866) .. .. — — — Q — B.fl.III,390
L. myrtifolius, Cunningham in Benth. Fl. Austr. III, 390 (1866) — — — N.S.W. Q. — B.fl.III,390
L. longiflorus, Desroussaux in Lam. Encycl. méth. IV, 598 (1796) -- — — N.S.W. Q. N.A. B.fl.III,390
L. angustifolius, R. Brown in Benth. Fl. Austr. III, 390 (1866) — S.A. — — -- — B.fl.III,390
L. dictyophlebus, F. v. M., Rep. Burdek. Exped. 14 (1858) ... — — — N.S.W. Q. — B.fl.III,391
L. alyxifolius, F. v. M. in Bentham Fl. Austr. III, 391 (1866) .. — — — N.S.W. Q. — B.fl III,391
L. odontocalyx, F. v. M. in Bentham Fl. Austr. III, 391 (1866) — — — — Q. N.A. B.fl.III,391
L. linearifolius, Hooker in Mitchell, Trop. Austr. 102 (1848) ... W.A. S.A. — — — Q. — B.fl.III,391
L. Exocarpi, Behr in Schlecht. Linnaea XX, 624 (1847)... ... W.A. S.A. — V. N.S.W. Q. N.A. B.fl.III,392
L. acacioides, Cunningham in Benth. Fl. Austr. III, 392 (1866) — — — — — N.A. B.fl.III,392
L. signatus, F. v. M. in Bentham Fl. Austr. III, 302 (1866) ... — — — — Q. N.A. B.fl.III,392
L. maytenifolius, A. Gray, Bot.Wilk.Expl.Exped.I,739,t.99(1854) — — — N.S.W. — — B.fl.III,393
L. sanguineus, F. v. M., fragm. I, 177 (1859) ... ... — — — — — Q. N.A. B.fl.III,393 M.fr.IV,171.
L. bifurcatus, Bentham, Fl. Austr. III, 393 (1866) ... ... — — — — — N.A. B.fl.III,393
L. linophyllus, Fenzl in Huegel. enum. 56 (1837)... ... W.A. S.A. — V. N.S.W. Q. N.A. B.fl.III,393
L. pendulus, Sieber in De Candolle, prodr. IV, 294 (1830) W.A. S.A. — V. N.S.W. Q. N.A. B.fl.III,394
L. Quandang, Lindley in Mitch. Three Exped. II, 69 (1838) ... W.A. S.A. — V. N.S.W. Q. N.A. B.fl.III,394
L. grandibracteus, F. v. M., Rep. Burdek. Exped. 14 (1858) ... — S.A. — — — Q. N.A. B.fl.III,395

ATKINSONIA, F. v. Mueller, fragm. V, 34 (1865).
A. ligustrina, F. v. M., fragm. V, 34 (1865) ... ... — — -- — N.S.W. — — B.fl.III,388 M.fr.II,130;VI,252.

NUYTSIA, R. Brown in Journ. of the geogr. Soc. I, 17 (1831).
N. floribunda, R. Brown in Journ. geogr. Soc. I, 17 (1831) ... W.A. — — — — — B.fl.III,387 M.fr.VI,232.

PROTEACEAE.
A. L. de Jussieu, gen. pl. 78 (1789).

PETROPHILA, R. Brown in Transact. Linn. Soc. X, 67 (1809).
P. teretifolia, R. Brown in Transact. Linn. Soc. X, 68 (1809) ... W.A. — — — -- - B.fl.V,321 M.fr.VI,245.
P. longifolia, R. Brown, prot. nov. 5 (1830) ... ... W.A. — — — — B.fl.V,322 M.fr.VI,255.
P. media, R. Brown, prot. nov. 5 (1830) . ... W.A. — — -- — -1 B.fl.V,322 M.fr.VI,255.
P. acicularis, R. Brown in Transact. Linn. Soc. X, 69 (1800) ... W.A. — — — — B.fl.V,323 M.fr.VI,255.
P. megalostegia, F. v. M., fragm. X, 61 (1876) ... ... W.A. — — — — — M.fr.X,61.
P. linearis, R. Brown, prot. nov. 6 (1830) ... ... W.A. — — — — B.fl.V,323 M.fr.VI,243.
P. anceps, R. Brown, prot. nov. 5 (1830) ... ... W.A. — — — — — M.fr.VI,243;X,47.
P. heterophylla, Lindley, Bot. Regist. XXV, App. XXXV (1839) W.A. — — — — B.fl.V,324 M.fr.VI,244.
P. biloba, R. Brown, prot. nov. 7 (1830) ... ... W.A. — — — — B.fl.V,324 M.fr.VI,255.
P. propinqua, R. Brown, prot. nov. 7 (1830) ... ... W.A. — — — — B.fl.V,325
P. squamata, R. Brown in Transact. Linn. Soc. X, 70 (1809) ... W.A. — — — — B.fl.V,325 M.fr.VI,244.
P. colorata, Meissner in Lehm. pl. Preiss. II, 246 (1847) ... W.A. — — — — B.fl.V,326
P. striata, R. Brown, prot. nov. 6 (1830) ... ... W.A. — — — — B.fl.V,326 M.fr.VI,344.
P. divaricata, R. Brown, prot. nov. 7 (1830) ... ... W.A. — — — — B.fl.V,326 M.fr.VI,344.
P. Serruriae, R. Brown, prot. nov. 6 (1830) ... ... W.A. — — — — B.fl.V,327 M.fr.VI,245.
P. inconspicua, Meissner in Hook. Kew Misc. VII, 68 (1855) ... W.A. — — — — B.fl.V,327 M.fr.VI,243.
P. trifida, R. Brown in Transact. Linn. Soc. X, 70 (1800) ... W.A. — — — — B.fl.V,328 M.fr.VI,244.
P. carduacea, Meissner in Hook. Kew Misc. IV, 182 (1852) ... W.A. — — — — B.fl.V,328 M.fr.VI,244.
P. macrostachya, R. Brown, prot. nov. 7 (1830) ... ... W.A. — — — — B.fl.V,329 M.fr.VI,245.
P. diversifolia, R. Brown in Transact. Linn. Soc. X, 70 (1809)... W.A. — — — — B.fl.V,329 M.fr.VI,244.
P. biternata, Meissner in Hook. Kew Misc. VII, 69 (1855) ... W.A. — — — — B.fl.V,330 M.fr.VI,245.
P. plumosa, Meissner in Hook. Kew Misc. VII, 69 (1855) ... W.A. — — — — B.fl.V,330 M.fr.VI,244.
P. ericifolia, R. Brown, prot. nov. 5 (1830) ... ... W.A. — — — — B.fl.V,331 M.fr.VI,243.
P. chrysantha, Meissner in Hook. Kew Misc. VII, 69 (1855) ... W.A. — — — — B.fl.V,331 M.fr.VI,243.
P. pedunculata, R. Brown in Transact. Linn. Soc. X, 70 (1809) ... — — — N.S.W. Q. B.fl.V,332 M.fr.VI,242.
P. pulchella, R. Brown in Transact. Linn. Soc. X, 69 (1809) ... — — — N.S.W. Q. B.fl.V,332 M.fr.VI,242.
P. sessilis, Sieber in Roem. & Schult. mant. 262 (1822) ... — — — N.S.W. Q. B.fl.V,332
P. fastigiata, R. Brown in Transact. Linn. Soc. X, 70 (1800) ... W.A. — — — — B.fl.V,333 M.fr.VI,243.
P. seminuda, Lindley, Bot. Regist. XXV, App. XXXIV (1839) W.A. — — — — B.fl.V,333 M.fr.VI,243.
P. circinata, Kippist in Hook. Kew Misc. VII, 69 (1855) ... W.A. — — — — B.fl.V,333 M.fr.VI,243.
P. Drummondii, Meissner in Lehm. pl. Preiss. I, 490 (1845) ... W.A. — — — — B.fl.V,334 M.fr.VI,243.
P. crispata, R. Brown, prot. nov. 6 (1830)... ... ... W.A. — — — — B.fl.V,334
P. rigida, R. Brown in Transact. Linn. Soc. X, 69 (1809) ... — — — — — B.fl.V,334 M.fr.VI,242.
P. multisecta, F. v. M., fragm. VI, 242 (1868) ... ... — S.A. — — — B.fl.V,335 M.fr.VI,242.
P. conifera, Meissner in Hook. Kew Misc. VII, 67 (1855) ... W.A. — — — — B.fl.V,335 M.fr.VI,245.
P. scmifurcata, F. v. M. in Bentham Fl. Austr. III, 335 (1870) ... W.A. — — — — B.fl.V,335 M.fr.X,47.

ISOPOGON, R. Brown in Transact. Linn. Soc. X, 70 (1809).
I. latifolius, R. Brown, prot. nov. 8 (1830)... ... ... W.A. — — — — B.fl.V,338 M.fr.VI,237.
I. cuneatus, R. Brown in Transact. Linn. Soc. X, 73 (1809) ... W.A. — — — — B.fl.V,339 M.fr.VI,235.
I. linearis, Meissner in Hook. Kew Misc. VII, 69 (1855) ... W.A. — — — — M.fl.V,339 M.fr.VI,236.
I. polycephalus, R. Brown in Transact. Linn. Soc. X, 73 (1809) ... W.A. — — — — B.fl.V,339 M.fr.VI,236.
I. attenuatus, R. Brown in Transact. Linn. Soc. X, 73 (1809) ... W.A. — — — — B.fl.V,340 M.fr.VI,237.
I. sphaerocephalus, Lindl., Bot. Regist. XXV, App. XXXIV (1839)W.A. — — — — B.fl.V,340 M.fr.VI,237.
I. uncinatus, R. Brown, prot. nov. 8 (1830) ... ... W.A. — — — — B.fl.V,341 M.fr.VI,237.
I. buxifolius, R. Brown in Transact. Linn. Soc. X, 74 (1809) ... W.A. — — — — B.fl.V,341 M.fr.VI,237.
I. axillaris, R. Brown in Transact. Linn. Soc. X, 74 (1809) ... W.A. — — — — B.fl.V,341 M.fr.VI,237.
I. tridens, F. v. M., fragm. VI, 239 (1868)... ... ... W.A. — — — — B.fl.V,342 M.fr.VI,239.

J

66

Species	Distribution	B.fl. ref.	M.fr. ref.
I. Baxteri, R. Brown, prot. nov. 9 (1830)	W.A.	B.fl.V,342	M.fr.VI,240.
I. roseus, Lindley, Bot. Regist. Misc. n. 37 (1842)	W.A.	B.fl.V,343	M.fr.VI,240
I. adenanthoides, Meissner in Hook. Kew Misc. VII, 69 (1855)	W.A.	B.fl.V,343	M.fr.VI,241.
I. trilobus, R. Brown in Transact. Linn. Soc. X, 72 (1809)	W.A.	B.fl.V,343	M.fr.VI,239.
I. tripartitus, R. Brown, prot. nov. 8 (1830)	W.A.	B.fl.V,344	
I. longifolius, R. Brown in Transact. Linn. Soc. X, 73 (1809)	W.A.	B.fl.V,344	M.fr.VI,237.
I. Drummondii, Bentham, Fl. Austr. V, 344 (1870)	W.A.	B.fl.V,344	M.fr.VI,241.
I. heterophyllus, Meissner in Lehm. pl. Preiss. I, 504 (1845)	W.A.	B.fl.V,345	M.fr.VI,241.
I. villosus, Meissner in De Candolle, prodr. XIV, 277 ( )	.A.		M.fr.VI,241.
I. teretifolius, R. Brown in Transact. Linn. Soc. X, 71 (1809)	W.A.	B.fl.V,345	M.fr.VI,241.
I. anethifolius, R. Brown in J. Knight, proteac. 94 (1809)	W.A. N.S.W.	B.fl.V,346	M.fr.VI,238.
I. petiolaris, Cunningham in R. Br., prot. nov. 8 (1830)	N.S.W.	B.fl.V,346	
I. anemonifolius, R. Brown in J. Knight, prot. 93 (1809)	N.S.W. Q.	B.fl.V,347	M.fr.VI,238;X,90.
I. ceratophyllus, R. Brown in Transact. Linn. Soc. X, 72 (1809)	S.A. T. N.S.W.	B.fl.V,347	M.fr.VI,238.
I. asper, R. Brown, prot. nov. 8 (1830)	W.A. S.A. T. V.	B.fl.V,348	M.fr.VI,240.
I. crithmifolius, F. v. M., fragm. VI, 239 (1868)	W.A.	B.fl.V,348	M.fr.VI,239.
I. formosus, R. Brown in Transact. Linn. Soc. X, 72 (1809)	W.A.	B.fl.V,349	M.fr.VI,240.
I. divergens, R. Brown, prot. nov. 7 (1830)	W.A.	B.fl.V,349	M.fr.VI,241.
I. scabriusculus, Meissner in Hook. Kew Misc. IV, 182 (1852)	W.A.	B.fl.V,349	M.fr.VI,240.

ADENANTHOS, Labillardière, Nov. Holl. pl. specim. I, 28, t. 36 (1804).

Species	Distribution	B.fl. ref.	M.fr. ref.
A. Detmoldi, F. v. M., fragm. VIII, 140 (1874)	W.A.		M.fr.VIII,149.
A. barbigerus, Lindley, Bot. Regist. XXV, App. XXXVI (1839)	W.A.	B.fl.V,351	M.fr.VI,205;VIII,149
A. obovatus, Labillardière, Nov. Holl. pl. spec. I, 29, t. 37 (1804)	W.A.	B.fl.V,352	
A. cuneatus, Labillardière, Nov. Holl. pl. spec. I, 28, t. 36 (1804)	W.A.	B.fl.V,352	
A. Cunninghami, Meissner in Lehm. pl. Preiss. I, 513 (1845)	W.A.	B.fl.V,352	
A. pungens, Meissner in Lehm. pl. Preiss. I, 515 (1845)	W.A.	B.fl.V,353	
A. venosus, Meisner in Hook. Kew Misc. IV, 183 (1852)	W.A.	B.fl.V,353	
A. Dobsoni, F. v. M., fragm. VI, 204 (1868)	W.A.	B.fl.V,353	M.fr.VI,204.
A. linearis, Meissner in Hook. Kew Misc. IV, 183 (1852)	W.A.	B.fl.V,354	
A. sericeus, Labillardière, Nov. Holl. pl. specim. I, 29, t. 38 (1804)	W.A. S.A.	B.fl.V,354	
A. Meissneri, Lehmann, pl. Preiss. I, 515 (1845)	W.A.	B.fl.V,354	
A. filifolius, Bentham, Fl. Austr. V, 355 (1870)	W.A.	B.fl.V,355	
A. terminalis, R. Brown in Transact. Linn. Soc. X, 152 (1809)	S.A. V.	B.fl.V,355	
A. flavidiflorus, F. v. M., fragm. I, 157 (1859)	W.A.	B.fl.V,355	M.fr.I,157.
A. apiculatus, R. Brown, prot. nov. 9 (1830)	W.A.	B.fl.V,356	

SIMSIA, R. Brown in Transact. Linn. Soc. X, 152 (1809). (Stirlingia.)

Species	Distribution	B.fl. ref.	M.fr. ref.
S. simplex, F.v.M.; Stirl., Lindl. Bot. Reg. XXV, App. XXX (1839)	W.A.	B.fl.V,357	
S. abrotanoides, F.v.M.; Stirlingia, Meissn. in pl. Preiss. I, 517 (1845)	W.A.	B.fl.V,357	
S. teretifolia, F. v. M.; Stirlingia, Meissn. in pl. Preiss. 515 (1845)	W.A.	B.fl.V,357	
S. tenuifolia, R. Brown in Transact. Linn. Soc. X, 152 (1809)	W.A.	B.fl.V,358	
S. latifolia, R. Brown, prot. nov. 9 (1830)	W.A.	B.fl.V,358	

SYNAPHEA, R. Brown in Transact. Linn. Soc. X, 155 (1809).

Species	Distribution	B.fl. ref.	M.fr. ref.
S. polymorpha, R. Brown in Transact. Linn. Soc. X, 156 (1809)	W.A.	B.fl.V,360	
S. dilatata, R. Brown in Transact. Linn. Soc. X, 156 (1809)	W.A.	B.fl.V,360	
S. favosa, R. Brown in Transact. Linn. Soc. X, 156 (1809)	W.A.	B.fl.V,360	
S. Preissii, Moissner in Lehm. pl. Preiss. I, 529 (1845)	W.A.	B.fl.V,361	
S. acutiloba, Meissner in Lehm. pl. Preiss. I, 528 (1845)	W.A.	B.fl.V,361	
S. petiolaris, R. Brown in Transact. Linn. Soc. X, 156 (1809)	W.A.	B.fl.V,361	
S. decorticans, Lindley, Bot. Regist. XXV, App. XXXII (1839)	W.A.	B.fl.V,362	
S. pinnata, Lindley, Bot. Regist. XXV, App. XXXII (1839)	W.A.	B.fl.V,362	

CONOSPERMUM, Smith in Transact. Linn. Soc. IV, 213 (1798).

Species	Distribution	B.fl. ref.	M.fr. ref.
C. capitatum, R. Brown in Transact. Linn. Soc. X, 153 (1809)	W.A.	B.fl.V,363	
C. petiolare, R. Brown, prot. nov. 11 (1830)	W.A.	B.fl.V,365	
C. teretifolium, R. Brown, prot. nov. 11 (1830)	W.A.	B.fl.V,365	M.fr.VI,224.
C. flexuosum, R. Brown, prot. nov. 11 (1830)	W.A.	B.fl.V,366	M.fr.VI,224.
C. acerosum, Lindley, Bot. Regist. XXV, App. XXX (1839)	W.A.	B.fl.V,366	
C. amoenum, Meissner in Lehm. pl. Preiss. I, 522 (1845)	W.A.	B.fl.V,367	M.fr.I,157;VI,224.
C. nervosum, Meissner in Hook. Kew Misc. VII, (1855)	W.A.	B.fl.V,367	M.fr.VI,224.
C. diffusum, Bentham, Fl. Austr. V, 367 (1870)	W.A.	B.fl.V,367	
C. glumaceum, Lindley, Bot. Regist. XXV, App. XXX (1839)	W.A.	B.fl.V,367	
C. ephedroides, Kippist in Hook. Kew Misc. VII, 70 (1855)	W.A.	B.fl.V,368	
C. Toddii, F. v. M., fragm. X, 20 (1876)			M.fr.X,20,90.
C. polycephalum, Meissner in Lehm. pl. Preiss. II, 249 (1847)	W.A.	B.fl.V,368	
C. coeruleum, R. Brown in Transact. Linn. Soc. V, 154 (1809)	W.A.	B.fl.V,368	M.fr.VI,224.
C. debile, Kippist in Hook. Kew Misc. VII, 70 (1855)	W.A.	B.fl.V,369	
C. scaposum, Bentham, Fl. Austr. V, 369 (1870)	W.A.	B.fl.V,369	
C. Huegelii, R. Brown in Endlicher, nov. stirp. dec. 58 (1830)	W.A.	B.fl.V,369	
C. densiflorum, Lindley, Bot. Regist. XXV, App. XXXII (1839)	W.A.	B.fl.V,369	
C. Brownii, Meissner in Lehm. pl. Preiss. II, 248 (1847)	W.A.	B.fl.V,370	
C. longifolium, Smith, Exot. Bot. II, 45, t. 82 (1805)	N.S.W.	B.fl.V,370	
C. tenuifolium, R. Brown in Transact. Linn. Soc. X, 154 (1809)	N.S.W.	B.fl.V,371	M.fr.VI,223.
C. Mitchellii, Meissner in De Candolle. prodr. XIV, 320 (1856)	V.	B.fl.V,371	
C. sphacelatum, Hooker in Mitchell, Trop. Austr. 342 (1848)	Q. N.A.	B.fl.V,371	
C. patens, Schlechtendal, Linnaea XX, 587 (1847)	S.A. V. N.S.W.	B.fl.V,371	M.fr.VI,223.
C. taxifolium, Smith in Rees's Cycl. IX (1808)	N.S.W.	B.fl.V,372	
C. ericifolium, Smith in Rees's Cycl. IX (1808)	T. V. N.S.W. Q.	B.fl.V,372	M.fr.VI,224.
C. ellipticum, Smith in Rees's Cycl. IX (1808)	N.S.W.	B.fl.V,372	
C. distichum, R. Brown in Transact. Linn. Soc. X, 155 (1809)	W.A. N.S.W.	B.fl.V,373	M.fr.VI,224.
C. floribundum, Bentham, Fl. Austr. V, 373 (1870)	W.A.	B.fl.V,373	

MACADAMIA. F. v. M. in Transact. phil. Inst. Vict. II, 72 (1857). (Panopsis partly.)   [VII,59.
M. ternifolia, F. v. M. in Transact. phil. Inst. Vict. II, 72 (1857)   —   —   — — N.S.W. Q.   —   B.fl.V,406   M.fr.II,91;VI,101;224,

HELICIA, Loureiro, Fl. Cochinch. I, 83 (1790).
H. Youngiana, C. Moore & F. v. M., fragm. IV, 84 (1864)   ...   —   — — — N.S.W. Q.   --   B.fl.V,406   M.fr.V,186.
H. praealta, F. v. M., fragm III, 37 (1862)   ...   ...   —   — — N.S.W. Q.   —   B.fl.V,404   M.fr.IV,174;VI,224.
H. Australasica, F. v. M. in Hook. Kew Misc. IX, 27 (1857) ...   --   — — — — N.A.   B.fl.V,405   M.fr.II,91.
H. glabriflora, F. v. M., fragm. II. 91 (1861)   ...   ...   —   — — N.S.W. Q.   —   B.fl.V,405   M.fr.V,38,186.   [224.
H. ferruginea, F. v. M., fragm. III, 37 (1862)   ...   ...   —   — — N.S.W. Q.   —   B.fl.V,405   M.fr.IV,174;V,186;VI,

ROUPALA, Aublet, Hist. des pl. de la Guian. I, 83, t. 32 (1775). (Rhopala, Bleasdalea. Adenostephanus partly.)
R. Bleasdalei, F. v. M. in proceed. Roy. Soc. N.S. Wales, 28 (1881)   —   — — — — Q.   —   B.fl.V,417   M.fr.V,90.

XYLOMELUM, Smith in Transact. Linn. Soc. IV, 214 (1798).
X. pyriforme, Smith in J. Knight, prot. 105 (1809)   ...   ...   —   — — N.S.W. —   —   B.fl.V,406   M.fr.VI,220.
X. occidentale, R. Brown, prot. nov. 31 (1830)   ...   W.A.   — — — — —   B.fl.V,408
X. salicinum, Cunningham in R. Br. prot. 31 (1830)   ...   —   — — — Q.   --   B.fl.V,408   M.fr.IV,107,174;VI,220,
X. angustifolium, Kippist in Hook. Kew Misc. IV., 209 (1852)...   W.A.   — — — — —   B.fl.V,409   M.fr.VI,220.

LAMBERTIA, Smith in Transact. Linn. Soc. IV, 214 (1798).
L. uniflora, R. Brown in Transact. Linn. Soc. X, 188 (1809)   ..   W.A.   — — — — —   B.fl.V,413   M.fr.VI,248.
L. rariflora, Meissner in Lehm. pl. Preiss. II, 263 (1847)   W.A.   — — — — —   B.fl.V,414
L. inermis, R. Brown in Transact. Linn. Soc. X, 188 (1809)   W.A.   — — — — —   B.fl.V,414   M.fr.VI,248.
L. ericifolia, R. Brown, prot. nov. 30 (1830)   ...   W.A.   — — — — —   B fl.V,414   M.fr.VI,248.
L. multiflora, Lindley, Bot. Regist. XXV, App. XXXII (1839)   W.A.   — — — — —   B.fl.V,415
L. formosa, Smith in Transact. Linn. Soc. IV, 214, t. 20 (1799)   W.A.   — — — N.S.W. —   B.fl.V,415   M.fr.VII,133.
L. echinata, R. Brown in Transact. Linn. Soc. X, 189 (1809) ...   W.A.   — — — — —   B.fl.V,416   M.fr.VI,248.
L. ilicifolia, Hooker, icon. plant. t. 553 (1843)   ...   ...   W.A.   — — — — —   B.fl.V,416

ORITES, R. Brown in Transact. Linn. Soc. X, 189 (1809). (Oritina.)
O. excelsa, R. Brown, prot. nov. 32 (1832)...   ...   — — N.S.W. —   B.fl.V,411   M.fr.V,153;VI,223.
O. diversifolia, R. Brown in Transact. Linn. Soc. X, 190 (1809)   — — T. — — —   B.fl.V,411   M.fr.V,153.
O. Milligani, Meissner in Hook. Kew Misc. IV, 209 (1852)   -- — T. — — —   B.fl.V,411   [133.
O. lancifolia, F. v. M. in Transact. phil. Soc. Vict. I, 108 (1855)   — — V. — — —   B.fl.V,412   M.fr.V,153;VI,223;VII,
O. revoluta, R. Brown in Transact. Linn. Soc. X, 19 (1809)   ...   — — T. — — —   B.fl.V,412
O. acicularis, R. Brown, prot. nov. 32 (1830)   ...   ...   — — V. — — —   B.fl.V,412

STRANGEA, Meissner in Hooker's Kew Misc. VII, 66 (1855). (Molloya.)
S. linearis, Meissner in Hook. Kew Misc. VII, 66 (1855)   ...   — — — N.S.W. Q.   —   B.fl.V,453   M.fr.VI,223;VII,132.
S. cynanchocarpa, F. v. M., fragm. VII, 132 (1871)   ...   W.A.   — — — — —   M.fr.VII,132.

GREVILLEA, R. Brown in Transact. Linn. Soc. X, 168 (1809). (Anadenia, Lysanthe, Stylurus, Manglesia.)
G. Pinaster, Meissner in Hook. Kew Misc. VII, 76 (1855)   W.A.   — — — — —   B.fl.V,427   M.fr.VI,212.
G. obtusifolia, Meissner in Hook. Kew Misc. VII, 187 (1852)   W.A.   — — — — —   B.fl.V,427
G. sparsiflora, F. v. M., fragm. VI, 206 (1868)   ...   W.A.   — — — — —   B.fl.V,428   M.fr.VI,206.
G. macrostylis, F. v. M., fragm. I, 137 (1859)   ...   W.A.   - — — — —   B.fl.V,428   M.fr.VI,211.
G. tripartita, Meissner in Hook. Kew Misc. IV, 186 (1852)   W.A.   — — — — —   B.fl.V,428   M.fr.VI,213.
G. platypoda, F. v. M., fragm. VI, 205 (1868)   ...   W.A.   — — — — —   B.fl.V,428   M.fr.VI,205.
G. patentiloba, F. v. M., fragm. I, 137 (1859)   ...   W.A.   — — — — —   B.fl.V,429   M.fr.I,137.
G. pectinata, R. Brown, prot. nov. 23 (1830)   ...   W.A.   — — — — —   B.fl.V,429   M.fr.VI,212.
G. plurijuga, F. v. M., fragm. IV, 84 (1864)   ...   W.A.   — — — — —   B.fl.V,430   M.fr.VI,212.
G. nudiflora, Meissner in Hook. Kew Misc. IV, 186 (1852)   W.A.   — — — — —   B.fl.V,430   M.fr.I,35;VI,212.
G. stenomera, F. v. M., fragm. IV, 85 (1864)   ...   W.A.   — — — — —   B.fl.V,430   M.fr.IV,85.
G. Thelemanniana, Endlicher, nov. stirp. dec. 6 (1839)   ...   W.A.   — — — — —   B.fl.V,431   M.fr.VI,212.
G. concinna, R. Brown in Transact. Linn. Soc. X, 172 (1809)   ...   W.A.   — — — — —   B.fl.V,431   M.fr.VI,212.
G. Hookeriana, Meissner in Lehm. pl. Preiss. I, 546 (1845) ...   W.A.   — — — — —   B.fl.V,432   M.fr.VI,212.
G. Baxteri, R. Brown, prot. nov. 22 (1830)   ...   ...   W.A.   — — — — —   B.fl.V,432
G. orbotrya, F. v. M., fragm. X, 44 (1876)   ...   ...   W.A.   — — — — —   B.fl.X,44.
G. pterosperma, F. v. M. in Transact. phil. Soc. Vict. I, 22 (1854) W.A. S.A. — V. N.S.W. —   B.fl.V,432   M.fr.VI,210.
G. stenobotrya, F. v. M., fragm. IX, 3 (1875)   ...   — S.A. — — — —   M.fr.IX,3.
G. eriostachya, Lindley, Bot. Regist. XXV, App. XXXVI (1839)   — — — — — N.A.   B.fl.V,433   M.fr.VI,208&213.
G. thyrsoides, Meissner in Hooker's Kew Misc. VII, 77 (1855)   W.A.   — — — — —   B.fl.V,433
G. Chrysodendron, R. Brown in Transact. Linn. Soc. X, 176 (1809) — — — — Q. N.A.   B.fl.V,434   M.fr.VI,210.   [46.
G. Banksii, R. Brown in Transact. Linn. Soc. X, 176 (1809)   ...   — — — — Q. —   B.fl.V,434   M.fr.VI,210;VIII,150; X,
G. Caleyi, R. Brown, prot. nov. 22 (1830)   ...   ...   — — — N.S.W. —   B.fl.V,435
G. asplenifolia, R. Brown in J. Knight, prot. 120 (1809)...   — — — N.S.W. —   B.fl.V,435   M.fr.VI,212.
G. cirsiifolia, Meissner in Lehm. pl. Preiss. II, 253 (1847)   W.A.   — — — — —   B.fl.V,436
G. laurifolia, Sieber in Roem. & Schult. mant. 270 (1822)   '- — — N.S.W. —   B.fl.V,436   [X,46.
G. Barklyana, F. v. M., gen. report 14 & 18 (1861)   ...   — — V. — — —   B.fl.V,436   M.fr.VII,133;VIII,150;
G. repens, F. v. M. in Schlecht. Linnaea XXVI, 355 (1853)   ...   — — V. — — —   B.fl.V,437
G. Aquifolium, Lindley in Mitchell, Three Exped. II, 178 (1838)   — — V. — — —   B.fl.V,437
G. ilicifolia, R. Brown, prot. nov. 21 (1830)   ...   ...   — S.A. — — — —   B.fl.V,437   M.fr.VI,212.
G. Gaudichandii, R. Brown in Freyc. Voy. Bot. 443, t. 48 (1826)   — — — N.S.W. —   B.fl.V,438   M.fr.VI,212;VIII,150.
G. acanthifolia, Cunningham in Field, N.S.Wales, 328 (1825) ...   — — — N.S.W. —   B.fl.V,438   M.fr.VI,212.
G. bipinnatifida, R. Brown, prot. nov. 23 (1830)...   ...   W.A.   — — — — —   B.fl.V,439
G. armigera, Meissner in Hook. Kew Misc. IV, 186 (1852)   W.A.   — — — — —   B.fl.V,439
G. asparagoides, Meissner in Hook. Kew Misc. IV, 186 (1852) ..   W.A.   — — — — —   B.fl.V,439   M.fr.IX,123.
G. Treueriana, F. v. M., fragm. IX, 123 (1875)   ...   W.A. S.A.   — — — — —   M.fr.IX,123.
G. floribunda, R. Brown, prot. nov. 19 (1830)   ...   — — V. N.S.W. Q.   —   B.fl.V,440
G. cinerea, R. Brown in Transact. Linn. Soc. X, 173 (1809)   ...   — — — N.S.W. —   B.fl.V,440
G. alpina, Lindley in Mitchell, Three Exped. II, 179 (1838)   ...   — — V. — — —   B.fl.V,441
G. montana, R. Brown in Transact. Linn. Soc. X, 172 (1809)   ...   — — — N.S.W. —   B.fl.V,441
G. obtusiflora, R. Brown, prot. nov. 19 (1830)   ...   — — — N.S.W. —   B.fl.V,442
G. arenaria, R. Brown in Transact. Linn. Soc. X, 172 (1809)   ...   — — — N.S.W. —   B.fl.V,442
G. inncronulata, R. Brown in Transact. Linn. Soc. X, 173 (1809)   — — — N.S.W. —   B.fl.V,443

Species	W.A.	S.A.	V.	N.S.W.	Q.	N.A.	B.fl.	M.fr.
G. Baueri, R. Brown in Transact. Linn. Soc. X, 173 (1809)	—	—	—	N.S.W.	—	—	B.fl.V,443	
G. lanigera, Cunningham in R. Br., prot. nov. 20 (1830)	—	—	V.	N.S.W.	—	—	B.fl.V,444	
G. rosmarinifolia, Cunningham in Field, N.S.Wales, 328 (1825)	—	—	V.	N.S.W.	—	—	B.fl.V,445	
G. Goodii, R. Brown in Transact. Linn. Soc. X, 174 (1809)	—	—	—	—	Q.	N.A.	B.fl.V,446	M.fr.VI,211.
G. venusta, R. Brown in Transact. Linn. Soc. X, 175 (1809)	—	—	—	—	Q.	—	B.fl.V,446	
G. longistyla, Hooker in Mitchell, Trop. Austr. 343 (1848)	—	—	—	—	Q.	N.A.	B.fl.V,446	M.fr.VI,211.
G. juncifolia, Hooker in Mitchell, Trop. Austr. 341 (1848)	—	S.A.	—	N.S.W.	Q.	—	B.fl.V,447	M.fr.VI,209;VIII,150.
G. Wilsoni, Cunningham in Wilson's Voy. 273 (1835)	W.A.	—	—	—	—	—	B.fl.V,447	M.fr.VI,212;X,45.
G. eroctiloba, F. v. M., fragm. X, 44 (1876)	W.A.	—	—	—	—	—		M.fr.X,44.
G. lavandulacea, Schlechtendal, Linnaea XX, 586 (1847)	—	S.A.	V.	N.S.W.	—	—	B.fl.V,448	
G. insignis, Kippist in De Cand., prodr. XIV, 379 (1856)	W.A.	—	—	—	—	—	B.fl.V,448	M.fr.VIII,150;X,45.
G. Brownii, Meissner in Lehm. pl. Preiss. I, 537 (1845)	W.A.	—	—	—	—	—	B.fl.V,449	M.fr.VI,213.
G. fasciculata, R. Brown, prot. nov. 20 (1830)	W.A.	—	—	—	—	—	B.fl.V,449	
G. aspera, R. Brown in Transact. Linn. Soc. X, 172 (1809)	W.A.	S.A.	—	—	—	—	B.fl.V,450	M.fr.VI,211.
G. brachystylis, Meissner in Lehm. pl. Preiss. I, 538 (1845)	W.A.	—	—	—	—	—	B.fl.V,450	
G. saccata, Bentham, Fl. Austr. V, 450 (1870)	W.A.	—	—	—	—	—	B.fl.V,450	
G. Drummondii, Meissner in Lehn. pl. Preiss. I, 536 (1845)	W.A.	—	—	—	—	—	B.fl.V,451	
G. disjuncta, F. v. M., fragm. VI, 206 (1868)	W.A.	—	—	—	—	—	B.fl.V,451	M.fr.VI,206.
G. haplantha, F. v. M. in Bentham Fl. Austr. V, 451 (1870)	W.A.	—	—	—	—	—	B.fl.V,452	
G. pinifolia, Meissner in Hook. Kew Misc. 186 (1852)	W.A.	—	—	—	—	—	B.fl.V,452	
G. acuaria, F. v. M. in Bentham Fl. Austr. V, 452 (1870)	W.A.	—	—	—	—	—	B.fl.V,452	
G. singuliflora, F. v. M., fragm. VI, 92 (1868)	—	—	—	—	Q.	—	B.fl.V,452	M.fr.VI,92.
G. pauciflora, R. Brown in Transact. Linn. Soc. X, 171 (1809)	W.A.	S.A.	—	—	—	—	B.fl.V,453	M.fr.I,135;VI,206,208.
G. quercifolia, R. Brown, prot. nov. 23 (1830)	W.A.	—	—	—	—	—	B.fl.V,453	M.fr.VI,213.
G. angulata, R. Brown, prot. nov. 24 (1830)	—	S.A.	—	—	—	N.A.	B.fl.V,455	
G. Wickhami, Meissner in Hook. Kew Misc. 187 (1852)	—	S.A.	—	—	—	N.A.	B.fl.V,455	
G. agrifolia, Cunningham in R. Br., prot. nov. 24 (1830)	—	—	—	—	—	N.A.	B.fl.V,455	
G. Cunninghamii, R. Brown, prot. nov. 23 (1830)	—	—	—	—	—	N.A.	B.fl.V,456	
G. pungens, R. Brown in Transact. Linn. Soc. X, 175 (1809)	—	—	—	—	—	N.A.	B.fl.V,456	M.fr.VI,212;X,46.
G. dimidiata, F. v. M., fragm. III, 146 (1863)	—	—	—	—	—	N.A.	B.fl.V,457	M.fr.III,146.
G. heliosperma, R. Brown in Transact. Linn. Soc. X, 176 (1809)	—	—	—	—	—	N.A.	B.fl.V,457	M.fr.VI,209.
G. refracta, R. Brown in Transact. Linn. Soc. X, 176 (1809)	—	—	—	—	—	N.A.	B.fl.V,458	
G. Dryandri, R. Brown in Transact. Linn. Soc. X, 175 (1809)	—	—	—	—	Q.	N.A.	B.fl.V,458	M.fr.VI,209.
G. polystachya, R. Brown in Transact. Linn. Soc. X, 177 (1809)	—	—	—	—	Q.	N.A.	B.fl.V,459	M.fr.VI,210.
G. robusta, Cunningham in R. Br., prot. nov. 24 (1830)	—	—	—	N.S.W.	Q.	—	B.fl.V,459	M.fr.VI,210;VII.133.(XI 20.
G. annulifera, F. v. M., fragm. IV, 85 (1864)	W.A.	—	—	—	—	—	B.fl.V,460	M.fr.IV,85;VI,212;VII,132;
G. leucopteris, Meissner in Hook. Kew Misc. VII, 76 (1855)	W.A.	—	—	—	—	—	B.fl.V,460	M.fr.III,145;IV,85;176;
G. Leucadendron, Cunningham in R. Br., prot. nov. 25 (1830)	—	—	—	—	—	N.A.	B.fl.V,461	M.fr.I,136-7. [XI,20
G. pyramidalis, Cunningham in R. Br., prot. nov. 25 (1830)	—	—	—	—	—	N.A.	B.fl.V,462	
G. striata, R. Brown in Transact. Linn. Soc. X, 177 (1809)	—	S.A.	—	N.S.W.	Q.	N.A.	B.fl.V,462	
G. mimosoides, R. Brown in Transact. Linn. Soc. X, 177 (1809)	—	—	—	—	—	N.A.	B.fl.V,462	
G. Hilliana, F. v. M. in Transact. phil. Inst. Vict. II, 72 (1857)	—	—	—	N.S.W.	Q.	—	B.fl.V,463	
G. gibbosa, R. Brown in Transact. Linn. Soc. X, 177 (1809)	—	—	—	—	Q.	N.A.	B.fl.V,463	M.fr.VI,210.
G. buxifolia, R. Brown in Transact. Linn. Soc. X, 174 (1809)	—	—	—	N.S.W.	—	—	B.fl.V,464	
G. phylicoides, R. Brown in Transact. Linn. Soc. X, 174 (1809)	—	—	—	N.S.W.	—	—	B.fl.V,464	M.fr.VI,213.
G. sphacelata, R. Brown in Transact. Linn. Soc. X, 174 (1809)	—	—	—	N.S.W.	—	—	B.fl.V,465	
G. occidentalis, R. Brown in Transact. Linn. Soc. X, 173 (1809)	W.A.	—	—	—	—	—	B.fl.V,465	
G. acerosa, F. v. M., fragm. I, 136 (1859)	W.A.	—	—	—	—	—	B.fl.V,465	M.fr.I,136.
G. umbellulata, Meissner in Lehm. pl. Preiss. II, 252 (1847)	W.A.	—	—	—	—	—	B.fl.V,465	
G. oxystigma, Meissner in Lehm. pl. Preiss. I, 540 (1845)	W.A.	—	—	—	—	—	B.fl.V,466	
G. Candolleana, Meissner in Lehm. pl. Preiss. I, 541 (1845)	W.A.	—	—	—	—	—	B.fl.V,466	M.fr.VI,213.
G. scabra, Meissner in Lehm. pl. Preiss. I, 541 (1845)	W.A.	—	—	—	—	—	B.fl.V,466	
G. Miqueliana, F. v. M. in Transact. Vict. Inst. 132 (1855)	—	—	V.	—	—	—	B.fl.V,467	
G. Victoriae, F. v. M. in Transact. phil. Soc. Vict. I, 107 (1865)	—	—	V.	N.S.W.	—	—	B.fl.V,468	M.fr.VIII,142,150.
G. punicea, R. Brown in Transact. Linn. Soc. X, 160 (1809)	—	—	—	N.S.W.	—	—	B.fl.V,468	M.fr.XI,123.
G. oleoides, Sieber in Roem. & Schult. mant. 277 (1822)	—	—	—	N.S.W.	—	—	B.fl.V,468	
G. trinervis, R. Brown, prot. nov. 18 (1830)	—	—	—	N.S.W.	—	—	B.fl.V,469	M.fr.VI,211.
G. juniperina, R. Brown in Transact. Linn. Soc. X, 171 (1809)	—	—	—	N.S.W.	—	—	B.fl.V,469	
G. sericea, R. Brown in Transact. Linn. Soc. X, 170 (1809)	—	—	—	N.S.W.	—	—	B.fl.V,470	
G. capitellata, Meissner in Hook. Kew Misc. IV, 187 (1852)	—	—	—	N.S.W.	—	—	B.fl.V,470	
G. leiophylla, F. v. M. in Bentham, Fl. Austr. V, 471 (1870)	—	—	—	N.S.W.	Q.	—	B.fl.V,470	
G. linearis, R. Brown in Transact. Linn. Soc. X, 170 (1809)	—	—	—	N.S.W.	—	—	B.fl.V,471	
G. confertifolia, F. v. M. in Transact. phil. Soc. Vict. I, 22 (1854)	—	—	V.	—	—	—	B.fl.V,472	M.fr.VI,211.
G. parviflora, R. Brown in Transact. Linn. Soc. X, 171 (1809)	—	S.A.	V.	N.S.W.	—	—	B.fl.V,472	
G. australis, R. Brown in Transact. Linn. Soc. X, 171 (1809)	—	T.	V.	N.S.W.	—	—	B.fl.V,472	M.fr.VII,133.
G. commutata, F. v. M., fragm. VI, 207 (1868)	W.A.	—	—	—	—	—	B.fl.V,473	M.fr.VII,207.
G. pinnatisecta, F. v. M. in Bentham, Fl. Austr. V, 473 (1870)	W.A.	—	—	—	—	—	B.fl.V,473	
G. argyrophylla, Meissner in Hook. Kew Misc. VII, 75 (1855)	W.A.	—	—	—	—	—	B.fl.V,474	M.fr.VI,411.
G. brachystachys, Meissner in Lehn. pl. Preiss. II, 254 (1847)	W.A.	—	—	—	—	—	B.fl.V,474	M.fr.VI,211.
G. Endlicheriana, Meissner in Lehn. pl. Preiss. I, 546 (1845)	W.A.	—	—	—	—	—	B.fl.V,474	
G. manglesoides, Meissner in Lehn. pl. Preiss. I, 547 (1845)	W.A.	—	—	—	—	—	B.fl.V,475	
G. diversifolia, Meissner in Lehn. pl. Preiss. I, 547 (1845)	W.A.	—	—	—	—	—	B.fl.V,475	
G. filifolia, Meissner in Lehn. pl. Preiss. I, 547 (1845)	W.A.	—	—	—	—	—	B.fl.V,475	
G. hakeoides, Meissner in Lehn. pl. Preiss. II, 232 (1847)	W.A.	—	—	—	—	—	B.fl.V,476	
G. saretifolia, Meissner in Lehn. pl. Preiss. II, 255 (1847)	W.A.	—	—	—	—	—	B.fl.V,476	M.fr.X,46.
G. crynagioides, Bentham, Fl. Austr. V, 476 (1870)	W.A.	—	—	—	—	—	B.fl.V,476	
G. bracteosa, Meissner in Lehn. pl. Preiss. II, 254 (1847)	W.A.	—	—	—	—	—	B.fl.V,477	
G. crithmifolia, R. Brown, prot. nov. 23 (1830)	W.A.	—	—	—	—	—	B.fl.V,477	
G. trachytheca, F. v. M., fragm. VI, 207 (1868)	W.A.	—	—	—	—	—	B.fl.V,477	M.fr.VI,207.
G. triternata, R. Brown, prot. nov. 21 (1830)	—	—	V.	N.S.W.	—	—	B.fl.V,478	

Species	W.A.	S.A.	T.	V.	N.S.W.	Q.	N.A.	B.fl.	M.fr.
G. ramosissima, Meisner in De Candolle, prodr. XIV, 388					N.S.W.			B.fl.V,478	M.fr.VIII,150;X,46.
G. monticola, Meissner in Lehm. pl. Preiss. II, 259 (1847)	W.A.							B.fl.V,478	M.fr.VI,209.
G. Muelleri, Bentham, Fl. Austr. V, 479 (1870)	W.A.							B.fl.V,479	
G. trifida, Meissner in Lehm. pl. Preiss. I, 553 (1845)	W.A.							B.fl.V,479	M.fr.VI,209.
G. Synapheae, R. Brown, prot. nov. 23 (1830)	W.A.							B.fl.V,480	
G. flexuosa, Meisner in Lehm. pl. Preiss. I, 553 (1845)	W.A.							B.fl.V,480	
G. leptobotrya, Meisner in Lehm. pl. Preiss. II, 256 (1847)	W.A.							B.fl.V,480	M.fr.VI,209.
G. brevicuspis, Meisner in Lehm. pl. Preiss. II, 256 (1847)	W.A.							B.fl.V,481	
G. intricata, Meissner in Hook. Kew Misc. VII, 74 (1855)	W.A.							B.fl.V,481	
G. didymobotrya, Meissner in Hook. Kew Misc. IV, 186 (1852)	W.A.							B.fl.V,481	M.fr.I,136;VI,208.
G. polybotrya, Meissner in Hook. Kew Misc. IV, 185 (1852)	W.A.						N.A.	B.fl.V,482	M.fr.IV,120;VI,208.
G. nematophylla, F. v. M., fragm. I, 136 (1859)		S.A.			N.S.W.			B.fl.V,482	M.fr.I,136;VI,208.
G. paradoxa, F. v. M., fragm. VI, 246 (1868)	W.A.							B.fl.V,483	M.fr.VI,246.
G. petrophiloides, Meissner in Lehm. pl. Preiss. II, 257 (1847)	W.A.							B.fl.V,483	M.fr.VI,209;X,46.
G. tenuiflora, Meissner in Lehm. pl. Preiss. I, 554 (1845)	W.A.							B.fl.V,483	
G. pulchella, Meissner in Hook. Kew Misc. VII, 73 (1855)	W.A.							B.fl.V,484	M.fr.VI,209.
G. rudis, Meisner in Hook. Kew Misc. VII, 73 (1855)	W.A.							B.fl.V,484	
G. apiciloba, F. v. M., fragm. X, 45 (1876)	W.A.								M.fr.X,45.
G. Shuttleworthiana, Meisner in Lehm. pl. Preiss. II, 258 (1847)	W.A.							B.fl.V,484	
G. integrifolia, Meissner in De Candolle, prodr. XIV, 385 (1856)	W.A.							B.fl.V,485	
G. stenocarpa, F. v. M. in Bentham Fl. Austr. V, 485 (1870)	W.A.							B.fl.V,485	
G. acrobotrya, Meissner in Hooker's Kew Misc. VII, 74 (1855)	W.A.							B.fl.V,486	M.fr.VI,213.
G. glabrata, Meissner in Lehm. pl. Preiss. I, 549 (1845)	W.A.							B.fl.V,486	
G. ornithopoda, Meissner in Lehm. pl. Preiss. II, 256 (1847)	W.A.							B.fl.V,486	
G. paniculata, Meissner in Lehm. pl. Preiss. VII, 74 (1855)	W.A.							B.fl.V,487	
G. biternata, Meissner in Lehm. pl. Preiss. I, 549 (1845)	W.A.							B.fl.V,487	M.fr.VI,213.
G. triloba, Meissner in Hooker's Kew Misc. VII, 74 (1855)	W.A.							B.fl.V,487	M.fr.VI,213.
G. amplexans, F. v. M. in Bentham Fl. Austr. V, 488 (1870)	W.A.							B.fl.V,488	
G. vestita, Meissner in Lehm. pl. Preiss. I, 548 (1845)	W.A.							B.fl.V,488	M.fr.VI,213.
G. tridentifera, Meissner in Lehm. pl. Preiss. I, 547 (1845)	W.A.							B.fl.V,489	
G. crinacea, Meissner in Hooker's Kew Misc. VII, 74 (1855)	W.A.							B.fl.V,489	M.fr.X,45.
HAKEA, Schrader, sert. Hannov. I, fasc. 3, 27, t. 17 (1797). (Conchium.)									
H. chordophylla, F. v. M. in Hooker's Kew Misc. IX, 23 (1857)							N.A.	B.fl.V,495	M.fr.VI,190.
H. Cunninghamii, R. Brown, prot. nov. 26 (1830)							N.A.	B.fl.V,495	M.fr.VI,190.
H. lorea, R. Brown, prot. nov. 25 (1830)	W.A.	S.A.				Q.	N.A.	B.fl.V,495	M.fr.VI,189;VII,133.
H. Fraseri, R. Brown, prot. nov. 26 (1830)					N.S.W.			B.fl.V,496	
H. macrocarpa, Cunningham in R. Brown, prot. nov. 30 (1830)							N.A.	B.fl.V,496	M.fr.VI,190,210.
H. arborescens, R. Brown in Transact. Linn. Soc. X, 187 (1809)						Q.	N.A.	B.fl.V,497	M.fr.VI,190.
H. ctenophylla, Cunningham in De Cand. prodr. XIV, 417 (1856)							N.A.	B.fl.V,497	
H. trincura, F. v. M., fragm. III, 146 (1863)						Q.		B.fl.V,497	M.fr.VI,216.
H. cyclocarpa, Lindley, Bot. Regist. XXV, App. XXXVI (1839)	W.A.							B.fl.V,498	
H. crassifolia, Meissner in Lehm. pl. Preiss. I, 570 (1845)	W.A.							B.fl.V,498	M.fr.VI,216.
H. pandanocarpa, R. Brown, prot. nov. 29 (1830)	W.A.							B.fl.V,499	M.fr.VI,216.
H. Roei, Bentham, Fl. Austr. V, 499 (1870)	W.A.							B.fl.V,499	
H. adnata, R. Brown, prot. nov. 26 (1830)	W.A.							B.fl.V,499	
H. obliqua, R. Brown in Transact. Linn. Soc. X, 180 (1809)	W.A.							B.fl.V,500	M.fr.VI,218.
H. Hookeriana, Meissner in Hook. Kew Misc. IV, 208 (1852)	W.A.							B.fl.V,500	M.fr.VI,216.
H. incrassata, R. Brown, prot. nov. 29 (1839)	W.A.							B.fl.V,501	
H. flabellifolia, Meissner in Hooker's Kew Misc. VII, 116 (1855)	W.A.							B.fl.V,501	
H. Brownii, Meissner in Lehm. pl. Preiss. I, 569 (1845)	W.A.							B.fl.V,501	
H. Baxteri, R. Brown, prot. nov. 28 (1830)	W.A.							B.fl.V,501	
H. ceratophylla, R. Brown in Transact. Linn. Soc. X, 184 (1809)	W.A.							B.fl.V,502	M.fr.IV,49;VI,217.
H. lasiantha, R. Brown, prot. nov. 29 (1830)	W.A.							B.fl.V,502	M.fr.VI,215.
H. eriantha, R. Brown, prot. nov. 26 (1830)				V.	N.S.W.			B.fl.V,502	M.fr.VI,217.
H. megalosperma, Meissner in Hook. Kew Misc. VII, 117 (1855)	W.A.							B.fl.V,503	
H. clavata, Labillardière, Nov. Holl. pl. spec. I, 31, t. 41 (1804)	W.A.							B.fl.V,503	M.fr.VI,216.
H. orthorrhyncha, F. v. M., fragm. V, 214 (1868)	W.A.							B.fl.V,503	M.fr.VI,214.
H. Candolleana, Meissner in Lehm. pl. Preiss. I, 262 (1847)	W.A.							B.fl.V,504	M.fr.VI,215.
H. trifurcata, R. Brown in Transact. Linn. Soc. X, 183 (1809)	W.A.							B.fl.V,504	M.fr.VI,220.
H. crinacea, Meissner in Lehm. pl. Preiss. I, 559 (1845)	W.A.							B.fl.V,505	M.fr.VI,220.
H. platysperma, Hooker, icones pl. t. 433 (1842)	W.A.							B.fl.V,505	
H. brachyptera, Meissner in Hook. Kew Misc. IV, 208 (1852)	W.A.							B.fl.V,505	
H. Kippistiana, Meissner in Hooker's Kew Misc. VII, 115 (1855)	W.A.							B.fl.V,506	
H. Preissii, Meissner in Lehm. pl. Preiss. I, 557 (1845)	W.A.							B.fl.V,506	M.fr.VI,219.
H. pugioniformis, Cavanilles, Annal. hist. nat. I, 213, t. 11 (1800)			T.	V.	N.S.W.			B.fl.V,506	M.fr.VI,217.
H. Pampliniana, Kippist in Hook. Kew Misc. VII, 115 (1855)		S.A.		V.	N.S.W.	Q.		B.fl.V,507	
H. vittata, R. Brown in Transact. Linn. Soc. X, 182 (1809)		S.A.		V.				B.fl.V,507	
H. rostrata, F. v. M. in Schlecht. Linnaea XXVI, 259 (1853)		S.A.		V.				B.fl.V,508	M.fr.VI,218;VII,133.
H. rugosa, R. Brown in Transact. Linn. Soc. X, 179 (1809)		S.A.		V.				B.fl.V,508	M.fr.VI,218.
H. Epiglottis, Labillardière, Nov. pl. spec. I, 30, t. 40 (1804)			T.					B.fl.V,508	
H. amplexicaulis, R. Brown in Transact. Linn. Soc. X, 184 (1809)	W.A.							B.fl.V,509	M.fr.VI,217;VII,133.
H. glabella, R. Brown, prot. nov. 28 (1830)	W.A.							B.fl.V,509	
H. auriculata, Meissner in Hook. Kew Misc. VII, 116 (1855)	W.A.							B.fl.V,510	M.fr.VI,217.
H. cristata, R. Brown, prot. nov. 28 (1830)	W.A.							B.fl.V,510	
H. linearis, R. Brown in Transact. Linn. Soc. X, 183 (1809)	W.A.							B.fl.V,511	
H. stenocarpoides, F. v. M. in Bentham Fl. V, 511 (1870)	W.A.							B.fl.V,511	
H. ruscifolia, Labillardière, Nov. Holl. pl. spec. I, 30, t. 39 (1804)	W.A.							B.fl.V,511	
H. saligna, R. Brown in J. Knight, prot. 108 (1809)					N.S.W.	Q.		B.fl.V,512	M.fr.VI,217.
H. verrucosa, F. v. M., fragm. V, 25 (1865)	W.A.							B.fl.V,512	M.fr.V,25;VI,218.

Species	W.A.	S.A.	T.	V.	N.S.W.	Q.	N.A.	B.fl.V	M.fr
H. rhombalea, F. v. M., fragm. X, 90 (1876)							N.A.		M.fr.X,00.
H. purpurea, Hooker in Mitch. Trop. Austr. 348 (1848)... ...					N.S.W.	Q.		B.fl.V,513	
H. gibbosa, Cavanilles, Anal. hist. nat. I, 215 (1800) ...					N.S.W.			B.fl.V,513	M.fr.VI,218.
H. propinqua, Cunningham in Field N.S. Wales, 327 (1825) ...					N.S.W.			B.fl.V,513	M.fr.V,26.
H. nodosa, R. Brown in Transact. Linn. Soc. X, 179 (1809) ...			T.	V.				B.fl.V,514	M.fr.V,26;VI,217.
H. acicularis, R. Brown in Transact. Linn. Soc. X, 181 (1809)...			T.	V.	N.S.W.			B.fl.V,514	M.fr.VI,218.
H. leucoptera, R. Brown in Transact. Linn. Soc. X, 180 (1809)		S.A.		V.	N.S.W.	Q.		B.fl.V,515	M.fr.VI,219;VII,133.
H. cycloptera, R. Brown in Transact. Linn. Soc. X, 182 (1809)		S.A.						B.fl.V,515	M.fr.VI,217.
H. microcarpa, R. Brown in Transact. Linn. Soc. X, 182 (1809)			T.	V.	N.S.W.			B.fl.V,516	M.fr.VI,217.
H. recurva, Meissner in Hook. Kew Misc. IV, 207 (1852) ...	W.A.							B.fl.V,516	M.fr.VI,218.
H. circumalata, Meissner in Hook. Kew Misc. VII, 114 (1855)...	W.A.							B.fl.V,516	M.fr.VI,219.
H. commutata, F. v. M., fragm. V, 26 (1865)	W.A.							B.fl.V,517	M.fr.VI,219.
H. strumosa, Meissner in Hook. Kew Misc. IV, 208 (1852)	W.A.							B.fl.V,517	M.fr.VI,219.
H. multilineata, Meissner in Lehm. pl. Preiss. II, 261 (1847) ...	W.A.	S.A.						B.fl.V,518	M.fr.VI,215.
H. laurina, R. Brown, prot. nov. 29 (1830)... ...	W.A.							B.fl.V,518	M.fr.IV,130,210.
H. obtusa, Meissner in Hook. Kew Misc. IV, 209 (1852)	W.A.							B.fl.V,519	
H. cinerea, R. Brown in Transact. Linn. Soc. X, 186 (1809)	W.A.							B.fl.V,519	
H. corymbosa, R. Brown, prot. nov. 28 (1830) ...	W.A.							B.fl.V,519	M.fr.VI,216.
H. undulata, R. Brown in Transact. Linn. Soc. X, 185 (1809) ...	W.A.							B.fl.V,520	M.fr.VI,216.
H. petiolaris, Meissner in Lehm. pl. Preiss. I, 577 (1845)	W.A.							B.fl.V,520	
H. neurophylla, Meissner in Hook. Kew Misc. VII, 117 (1809)	W.A.							B.fl.V,521	
H. loranthifolia, Meissner in Lehm. pl. Preiss. I, 574 (1845) ...	W.A.							B.fl.V,521	
H. cucullata, R. Brown, prot. nov. 30 (1830)	W.A.							B.fl.V,521	M.fr.VI,216.
H. ferruginea, Sweet, Fl. Austral. t. 45 (1828) ...	W.A.							B.fl.V,522	M.fr.VI,216.
H. smilacifolia, Meissner in Lehm. pl. Preiss. I, 587 (1845)	W.A.							B.fl.V,522	M.fr.VI,217.
H. elliptica, R. Brown in Transact. Linn. Soc. X, 187 (1809) ...	W.A.							B.fl.V,523	M.fr.VI,216.
H. ambigua, Meissner in Lehm. pl. Preiss. II, 260 (1847) ...	W.A.							B.fl.V,523	
H. plurinervia, F. v. M. in Benth. Fl. Austr. V, 523 (1870)						Q.		B.fl.V,523	
H. dactyloides, Cavanilles, Anal. hist. nat. I, 215, t. 12 (1800)...					N.S.W.			B.fl.V,524	M.fr.VI,215.
H. olicina, R. Brown, prot. nov. 29 (1830)...		S.A.	T.	V.	N.S.W.			B.fl.V,524	
H. falcata, R. Brown, prot. nov. 29 (1830)...	W.A.							B.fl.V,524	
H. pycnoneura, Meissner in Hook. Kew Misc. VII, 117 (1855)	W.A.							B.fl.V,525	M.fr.VI,215.
H. stenocarpa, R. Brown, prot. nov. 29 (1830) ...	W.A.							B.fl.V,525	
H. marginata, R. Brown in Transact. Linn. Soc. X, 185 (1809)...	W.A.							B.fl.V,526	M.fr.VI,216.
H. myrtoides, Meissner in Lehm. pl. Preiss. I, 577 (1845)	W.A.							B.fl.V,526	
H. costata, Meissner in Lehm. pl. Preiss. I, 575 (1845) ...	W.A.							B.fl.V,526	
H. oleifolia, R. Brown in Transact. Linn. Soc. X, 183 (1809)	W.A.							B.fl.V,527	M.fr.VI,215.
H. florida, R. Brown in Transact. Linn. Soc. X, 183 (1809)	W.A.							B.fl.V,527	
H. varia, R. Brown in Transact. Linn. Soc. X, 183 (1809)	W.A.							B.fl.V,527	M.fr.VI,219.
H. anisata, R. Brown in Transact. Linn. Soc. X, 180 (1809)	W.A.							B.fl.V,528	M.fr.VI,219.
H. Meisneriana, Kippist in Hook. Kew Misc. VII, 114 (1809)	W.A.							B.fl.V,529	M.fr.VI,219.
H. subaulcata, Meissner in Lehm. pl. Preiss. I, 555 (1845)	W.A.							B.fl.V,529	M.fr.VI,219.
H. Lehmanniana, Meissner in Lehm. pl. Preiss. I, 557 (1845)	W.A.							B.fl.V,529	M.fr.VI,218.
H. flexilis, F. v. M. in Schlecht. Linnaea XXVI, 330 (1853)		S.A.		V.	N.S.W.			B.fl.V,530	M.fr.VI,216.
H. nitida, R. Brown in Transact. Linn. Soc. X, 184 (1809)	W.A.							B.fl.V,530	M.fr.V,72;VI,217.
H. Oldfieldii, Bentham, Fl. Austr. V, 530 (1870)...	W.A.							B.fl.V,530	
H. suaveolens, R. Brown in Transact. Linn. Soc. X, 182 (1809)	W.A.							B.fl.V,531	M.fr.VI,220.
H. lissocarpha, R. Brown, prot. nov. 27 (1830) ...	W.A.							B.fl.V,531	M.fr.VI,219.
H. bipinnatifida, R. Brown, prot. nov. 28 (1830)	W.A.							B.fl.V,532	

CANARVONIA, F. v. M., fragm. VI, 81 (1867).

C. araliifolia, F. v. M., fragm. VI, t. 55 (1867)						Q.		B.fl.V,410	M.fr.VI,81,247;VII,59.

BUCKINGHAMIA, F. v. M., fragm. VI, 246 (1868).

B. celsissima, F. v. M., fragm. VI, 248 (1868)						Q.		B.fl.V,532	M.fr.VI,248.

DARLINGIA, F. v. M., fragm. V, 152 (1866).

D. spectatissima, F. v. M., fragm. V, 152 (1866)						Q.		B.fl.V,533	M.fr.X,90.

CARDWELLIA, F. v. M., fragm. V, 23 (1865).

C. sublimis, F. v. M., fragm. V, 24 (1865)...						Q.		B.fl.V,533	M.fr.V,24.

STENOCARPUS, R. Brown in Transact. Linn. Soc. X, 201 (1809). (Cybele, Agnostus.)

S. sinuatus, Endlicher, gen. pl. suppl. IV, 88 (1847)					N.S.W.	Q.		B.fl.V,539	M.fr.VI,224. [VII,133.
S. salignus, R. Brown in Transact. Linn. Soc. X, 202 (1809)					N.S.W.	Q.		B.fl.V,539	M.fr.I,135,244;VI,224;
S. Cunninghamii, R. Brown, prot. nov. 34 (1830)...							N.A.	B.fl.V,540	

LOMATIA, R. Brown in Transact. Linn. Soc. X, 199 (1809). (Tricondylus.)

L. fraxinifolia, F. v. M. in Benth. Fl. Austr. V, 536 (1870) ...						Q.		B.fl.V,536	
L. ilicifolia, R. Brown in Transact. Linn. Soc. X, 200 (1809) ...				V.	N.S.W.			B.fl.V,536	M.fr.VII,133.
L. longifolia, R. Brown in Transact. Linn. Soc. X, 200 (1809) ...				V.	N.S.W.			B.fl.V,537	
L. silaifolia, R. Brown in Transact. Linn. Soc. X, 199 (1809) ...					N.S.W.	Q.		B.fl.V,537	M.fr.V,153;VI,101.
L. tinctoria, R. Brown in Transact. Linn. Soc. X, 200 (1809) ...			T.					B.fl.V,537	M.fr.VI,224.
L. polymorpha, R. Brown in Transact. Linn. Soc. X, 200 (1809) ...			T.					B.fl.V,538	

EMBOTHRIUM, R. & G. Forster, charact. gen. 15, t. 8 (1776).

E. Wickhami, F. v. M., fragm. VIII, 104 (1874)						Q.			M.fr.VIII,164;IX,194.

TELOPEA, R. Brown in Transact. Linn. Soc. X, 197 (1809).

T. speciosissima, R. Brown in Transact. Linn. Soc. X, 198 (1809)					N.S.W.			B.fl.V,534	M.fr.X,90.
T. oreades, F. v. M., Annual Report 18 (1861)				V.				B.fl.V,534	M.fr.II,170;VII,133.
T. truncata, R. Brown in Transact. Linn. Soc. X, 198 (1809) ...			T.					B.fl.V,535	M.fr.II,171.

BANKSIA, Linné, fil. suppl. 15 et 126 (1781).

B. pulchella, R. Brown in Transact. Linn. Soc. X, 202 (1809) ...	W.A.							B.fl.V,544	M.fr.VII,54.
B. Meisneri, Lehmann, plant. Preiss. I, 582 (1845)	W.A.							B.fl.V,545	

72

Species	W.A.	S.A.	T.	V.	N.S.W.	Q.	N.A.	B.fl.V	M.fr.
B. antans, R. Brown in Transact. Linn. Soc. X, 203 (1809)	W.A.	—	—	—	—	—	—	B.fl.V,545	M.fr.IV,108;VII,54.
B. sphaerocarpa, R. Brown in Transact. Linn. Soc. X, 203 (1809)	W.A.	—	—	—	—	—	—	B.fl.V,546	M.fr.VII,54.
B. tricuspis, Meissner in Hook. Kew Misc. VII, 118 (1855)	W.A.	—	—	—	—	—	—	B.fl.V,546	M.fr.VII.54.
B. occidentalis, R. Brown in Transact. Linn. Soc. X, 204 (1809)	W.A.	—	—	—	—	—	—	B.fl.V,546	M.fr.VII,54.
B. littoralis, R. Brown in Transact. Linn. Soc. X, 204 (1809)	W.A.	—	—	—	—	—	—	B.fl.V,547	M.fr.VII,55.
B. ericifolia, Linné, fil. suppl. 127 (1781)	—	—	—	—	N.S.W.	—	—	B.fl.V,547	M.fr.VII,54.
B. spinulosa, Smith, Specim. Bot. New Holl. 13, t. 4 (1793)	—	—	—	—	N.S.W.	—	—	B.fl.V,547	
B. collina, R. Brown in Transact. Linn. Soc. X, 204 (1809)	—	—	—	V.	N.S.W.	Q.	—	B.fl.V,548	M.fr.VII,54.
B. verticillata, R. Brown in Transact. Linn. Soc. X, 207 (1809)	W.A.	—	—	—	—	—	—	B.fl.V,548	
B. dryandroides, Baxter in Sweet, Fl. Austral. t. 56 (1828)	W.A.	—	—	—	—	—	—	B.fl.V,549	M.fr.VII,58.
B. Brownii, Baxter in R. Br. prot. nov. 37 (1830)	W.A.	—	—	—	—	—	—	B.fl.V,549	M.fr.VII,58.
B. attenuata, R. Brown in Transact. Linn. Soc. X, 209 (1809)	W.A.	—	—	—	—	—	—	B.fl.V,549	M.fr.VII,55.
B. media, R. Brown, prot. nov. 35 (1830)	W.A.	—	—	—	—	—	—	B.fl.V,550	M.fr.IV,100.
B. Solandri, R. Brown, prot. nov. 35 (1830)	W.A.	—	—	—	—	—	—	B.fl.V,550	M.fr.VII,58.
B. Goodii, R. Brown, prot. nov. 36 (1830)	W.A.	—	—	—	—	—	—	B.fl.V,550	M.fr.VII,58.
B. petiolaris, F. v. M., fragm. IV, 109 (1864)	W.A.	—	—	—	—	—	—	B.fl.V,551	M.fr.IV,109.
B. repens, Labillardière, Voy. I, 411, t. 23 (1798)	W.A.	—	—	—	—	—	—	B.fl.V,551	M.fr.IV,108,177;VII,58.
B. prostrata, R. Brown, prot. nov. 36 (1830)	W.A.	—	—	—	—	—	—	B.fl.V,551	M.fr.VII,57.
B. grandis, Willdenow, spec. plant. I, 535 (1797)	W.A.	—	—	—	—	—	—	B.fl.V,552	M.fr.VII,57.
B. quercifolia, R. Brown in Transact. Linn. Soc. X, 210 (1809)	W.A.	—	—	—	—	—	—	B.fl.V,552	M.fr.VII,57.
B. Baueri, R. Brown, prot. nov. 35 (1830)	W.A.	—	—	—	—	—	—	B.fl.V,552	M.fr.IV,107;VII,56.
B. marginata, Cavanilles, Anal. hist. nat. I, 227, t. 13 (1800)	—	S.A.	T.	V.	N.S.W.	—	—	B.fl.V,553	M.fr.VII,55,133;X,90.
B. integrifolia, Linné, fil. suppl. 127 (1781)	—	—	—	V.	N.S.W.	Q.	—	B.fl.V,554	M.fr.VII,55.
B. dentata, Linné, fil. suppl. 127 (1781)	—	—	—	—	—	Q.	N.A.	B.fl.V,555	M.fr.VII,57.
B. latifolia, R. Brown in Transact. Linn. Soc. X, 208 (1809)	—	—	—	—	N.S.W.	Q.	—	B.fl.V,555	M.fr.VII,56.
B. aemula, R. Brown in Transact. Linn. Soc. X, 210 (1809)	—	—	—	—	N.S.W.	Q.	—	B.fl.V,556	M.fr.VII,56.
B. ornata, F. v. M. in Schlecht. Linnaea XXVI, 352 (1853)	—	—	T.	V.	—	—	—	B.fl.V,556	M.fr.VII,56.
B. coccinea, R. Brown in Transact. Linn. Soc. X, 207 (1809)	W.A.	—	—	—	—	—	—	B.fl.V,557	M.fr.VII,56.
B. Sceptrum, Meissner in Hook. Kew Misc. VII, 120 (1855)	W.A.	—	—	—	—	—	—	B.fl.V,557	M.fr.VII,56.
B. Menziesii, R. Brown, prot. nov. 36 (1830)	W.A.	—	—	—	—	—	—	B.fl.V,558	
B. laevigata, Meissner in Hook. Kew Misc. IV, 210 (1852)	W.A.	—	—	—	—	—	—	B.fl.V,558	M.fr.VII,56.
B. Hookeriana, Meisner in Hook. Kew Misc. VII, 119 (1855)	W.A.	—	—	—	—	—	—	B.fl.V,558	
B. prionotes, Lindley, Bot. Regist. XXV, App. XXXIV (1839)	W.A.	—	—	—	—	—	—	B.fl.V,558	M.fr.VII,56.
B. Victoriae, Meissner in Hook. Kew Misc. VII, 119 (1855)	W.A.	—	—	—	—	—	—	B.fl.V,559	M.fr.VII,58.
B. speciosa, R. Brown in Transact. Linn. Soc. X, 210 (1809)	W.A.	—	—	—	—	—	—	R.fl.V,559	M.fr.VII,58.
B. Baxteri, R. Brown, prot. nov. 36 (1830)	W.A.	—	—	—	—	—	—	B.fl.V,560	M.fr.VII,58.
B. marcescens, R. Brown in Transact. Linn. Soc. X, 209 (1809)	W.A.	—	—	—	—	—	—	B.fl.V,560	M.fr.VII,56.
B. Lemanniana, Meissner in Hook. Kew Misc. 210 (1852)	W.A.	—	—	—	—	—	—	B.fl.V,560	M.fr.VII,57.
B. Caleyi, R. Brown, prot. nov. 35 (1830)	W.A.	—	—	—	—	—	—	B.fl.V,560	M.fr.VII,57.
B. Lindleyana, Meissner in Hook. Kew Misc. VII, 120 (1855)	W.A.	—	—	—	—	—	—	B.fl.V,561	M.fr.VII,58.
B. elegans, Meissner in Hook. Kew Misc. VII, 119 (1855)	W.A.	—	—	—	—	—	—	B.fl.V,561	
B. Candolleana, Meissner in Hook. Kew Misc. VII, 118 (1855)	W.A.	—	—	—	—	—	—	B.fl.V,561	M.fr.VII,58.
B. ilicifolia, R. Brown in Transact. Linn. Soc. X, 211 (1809)	W.A.	—	—	—	—	—	—	B.fl.V,561	M.fr.VII,58.

DRYANDRA, R. Brown in Transact. Linn. Soc. X, 211 (1809). (Josephia, Hemiclidia.)

Species	W.A.	S.A.	T.	V.	N.S.W.	Q.	N.A.	B.fl.V	M.fr.
D. quercifolia, Meissner in Hook. Kew Misc. IV, 210 (1852)	W.A.	—	—	—	—	—	—	B.fl.V,566	M.fr.VII,50.
D. praemorsa, Meissner in Lehm. pl. Preiss. II, 265 (1847)	W.A.	—	—	—	—	—	—	B.fl.V,566	M.fr.VII,50.
D. cuneata, R. Brown in Transact. Linn. Soc. X, 212 (1809)	W.A.	—	—	—	—	—	—	B.fl.V,566	M.fr.VII,50.
D. falcata, R. Brown in Transact. Linn. Soc. X, 213 (1809)	W.A.	—	—	—	—	—	—	B.fl.V,567	
D. armata, R. Brown in Transact. Linn. Soc. X, 212 (1809)	W.A.	—	—	—	—	—	—	B.fl.V,567	M.fr.VII,50.
D. longifolia, R. Brown in Transact. Linn. Soc. X, 215 (1809)	W.A.	—	—	—	—	—	—	B.fl.V,568	
D. Fraseri, R. Brown, prot. nov. 39 (1830)	W.A.	—	—	—	—	—	—	B.fl.V,568	M.fr.VII,52.
D. floribunda, R. Brown in Transact. Linn. Soc. X, 212 (1809)	W.A.	—	—	—	—	—	—	B.fl.V,569	M.fr.VI,92;VII,50.
D. carduacea, Lindley, Bot. Regist. XXV, App. XXXIII (1839)	W.A.	—	—	—	—	—	—	B.fl.V,569	M.fr.VII,51.
D. carlinoides, Meissner in Lehm. pl. Preiss. II, 267 (1847)	W.A.	—	—	—	—	—	—	B.fl.V,569	M.fr.VII,53.
D. polycephala, Bentham, Fl. Austr. V, 570 (1870)	W.A.	—	—	—	—	—	—	B.fl.V,570	
D. Kippistiana, Meissner in Hook. Kew Misc. VII, 122 (1855)	W.A.	—	—	—	—	—	—	B.fl.V,570	M.fr.VII,52.
D. squarrosa, R. Brown, prot. nov. 38 (1830)	W.A.	—	—	—	—	—	—	B.fl.V,571	M.fr.VII,52.
D. Serra, R. Brown, prot. nov. 38 (1830)	W.A.	—	—	—	—	—	—	B.fl.V,571	M.fr.VII,51.
D. concinna, R. Brown, prot. nov. 38 (1830)	W.A.	—	—	—	—	—	—	B.fl.V,572	M.fr.VII,51.
D. foliolata, R. Brown, prot. nov. 38 (1830)	W.A.	—	—	—	—	—	—	B.fl.V,572	
D. stuposa, Lindley, Bot. Regist. XXV, App. XXXIII (1839)	W.A.	—	—	—	—	—	—	B.fl.V,572	M.fr.VII,51.
D. nobilis, Lindley, Bot. Regist. XXV, App. XXXIII (1839)	W.A.	—	—	—	—	—	—	B.fl.V,573	
D. mucronulata, R. Brown, prot. nov. 38 (1830)	W.A.	—	—	—	—	—	—	B.fl.V,573	M.fr.VII,51.
D. formosa, R. Brown in Transact. Linn. Soc. X, 213, t. 3 (1800)	W.A.	—	—	—	—	—	—	B.fl.V,573	M.fr.VII,51.
D. Baxteri, R. Brown in Transact. Linn. Soc. X, 214 (1809)	W.A.	—	—	—	—	—	—	B.fl.V,574	M.fr.VII,51.
D. nivea, R. Brown in Transact. Linn. Soc. X, 214 (1809)	W.A.	—	—	—	—	—	—	B.fl.V,574	M.fr.VII,52.
D. Arctotidis, R. Brown, prot. nov. 39 (1830)	W.A.	—	—	—	—	—	—	B.fl.V,575	M.fr.VII,52.
D. nana, Meissner in Hook. Kew Misc. VII, 121 (1855)	W.A.	—	—	—	—	—	—	B.fl.V,575	M.fr.VII,52.
D. Preissii, Meissner in Lehm. pl. Preiss. I, 599 (1845)	W.A.	—	—	—	—	—	—	B.fl.V,575	M.fr.VII,54.
D. sclerophylla, Meissner in Hook. Kew Misc. VII, 123 (1855)	W.A.	—	—	—	—	—	—	B.fl.V,576	M.fr.VII,54.
D. pulchella, Meissner in Hook. Kew Misc. VII, 122 (1855)	W.A.	—	—	—	—	—	—	B.fl.V,576	M.fr.VII,52.
D. plumosa, R. Brown in Transact. Linn. Soc. X, 214 (1809)	W.A.	—	—	—	—	—	—	B.fl.V,576	
D. senecionifolia, R. Brown, prot. nov. 39 (1830)	W.A.	—	—	—	—	—	—	B.fl.V,577	M.fr.VII,51.
D. vestita, Kippist in Hook. Kew Misc. IV, 211 (1852)	W.A.	—	—	—	—	—	—	B.fl.V,577	M.fr.VII,52.
D. cirsioides, Meissner in Hook. Kew Misc. IV, 211 (1852)	W.A.	—	—	—	—	—	—	B.fl.V,577	M.fr.VII,52.
D. Hewardiana, Meissner in Hook. Kew Misc. IV, 210 (1852)	W.A.	—	—	—	—	—	—	B.fl.V,578	M.fr.VII,53.
D. patens, Bentham, Fl. Austr. V, 578 (1870)	W.A.	—	—	—	—	—	—	B.fl.V,578	
D. conferta, Bentham. Fl. Austr. V, 578 (1870)	W.A.	—	—	—	—	—	—	B.fl.V,578	
D. horrida, Meissner in Hook. Kew Misc. IV, 211 (1852)	W.A.	—	—	—	—	—	—	B.fl.V,579	M.fr.VII,52.

	W.A.	S.A.	T.	V.	N.S.W.	Q.	N.A.	B.fl.	M.fr.
D. serratuloides, Meissner in Hook. Kew Misc. VII, 123 (1855)	W.A.	—	—	—	—	—	—	B.fl.V,579	M.fr.VII,52.
D. comosa, Meissner in Hook. Kew Misc. IV, 211 (1852) ...	W.A.	—	—	—	—	—	—	B.fl.V,579	M.fr.VII,53.
D. Shuttleworthiana, Meissner in Hook. Kew Misc. VII, 122 (1855)	W.A.	—	—	—	—	—	—	B.fl.V,580	M.fr.VII,52.
D. speciosa, Meissner in Hook. Kew Misc. IV, 211 (1852) ...	W.A.	—	—	—	—	—	—	B.fl.V,580	
D. tridentata, Meissner in Hook. Kew Misc. VII, 120 (1855) ...	W.A.	—	—	—	—	—	··	B.fl.V,580	M.fr.VII,53.
D. tenuifolia, R. Brown in Transact. Linn. Soc. X, 215 (1809)...	W.A.	—	—	—	—	—	—	B.fl.V,581	M.fr.VII,53.
D. proteoides, Lindley, Bot. Regist. XXV, App. XXXIII (1839)	W.A.	—	—	—	—	—	—	B.fl.V,582	
D. runcinata, Meissner in Hook. Kew Misc. IV, 210 (1852) ...	W.A.	—	—	—	—	—	—	B.fl.V,582	
D. obtusa, R. Brown in Transact. Linn. Soc. X, 214 (1909) ...	W.A.	—	—	—	—	—	—	B.fl.V,582	M.fr.V,185.
D. bipinnatifida, R. Brown, prot. nov. 39 (1830)...	W.A.	—	—	—	—	—	—	B.fl.V,583	
D. pteridifolia, R. Brown in Transact. Linn. Soc. X, 215 (1809)	W.A.	—	—	—	—	—	—	B.fl.V,583	M.fr.VII,54.
D. oalophylla, R. Brown, prot. nov. 40 (1830) ...	W.A.	—	—	—	—	—	—	B.fl.V,583	M.fr.VII,54.

THYMELEAE.
A. L. de Jussieu, gen. plant. 76 (1789).

PIMELEA, Banks & Solander in Gaert. de fruct. I, 186 (1788). (Thecanthes, Gymnococca, Heterolaena, Calyptrostegia, Macrostegia;—Banksia, Forst., 1776.)

	W.A.	S.A.	T.	V.	N.S.W.	Q.	N.A.	B.fl.	M.fr.
P. punicea, R. Brown, prodr. 359 (1810)	—	—	—	—	—	—	N.A.	B.fl.VI,6	M.fr.V,74;VII,3.
P. concreta, F. v. M., fragm. V, 73 (1865)...	—	—	—	—	—	—	N.A.	B.fl.VI,6	M.fr.V,73.
P. cornucopiae, Vahl, enum. plant. I, 305 (1804) ...	—	—	—	—	—	Q.	—	B.fl.VI,6	M.fr.VII,3.
P. sanguinea, F. v. M., fragm. I, 84 (1858)...	—	—	—	—	—	Q.	—	B.fl.VI,7	M.fr.VII,3;XI,47.
P. alpina, F. v. M., second general report, 14 (1854)	—	—	—	V.	N.S.W.	—		B.fl.VI,7	M.fr.VII,4.
P. longifolia, Banks & Solander in Rees, Cycl. XXVI (1814) ...	—	—	—	—	N.S.W.	—		B.fl.VI,7	M.fr.VIII,9.
P. cinerea, R. Brown, prodr. 361 (1810) ...	—	—	T.	—	—	—		B.fl.VI,8	
P. Milligani, Meissner in De Candolle, prodr. XIV, 509 (1856)...	—	—	T.	—	—	—		B.fl.VI,8	
P. spectabilis, Lindley, Bot. Regist. t. 33 (1841) ...	W.A.	—	—	—	—	—		B.fl.VI,9	M.fr.VII,4.
P. Lehmanniana, Meissner in Lehm. pl. Preiss. I, 603 (1845)	W.A.	—	—	—	—	—		B.fl.VI,9	M.fr.VII,4.
P. hispida, R. Brown, prodr. 360 (1810) ...	W.A.	—	—	—	—	—		B.fl.VI,10	
P. rosea, R. Brown, prodr. 360 (1810) ...	W.A.	—	—	—	—	—		B.fl.VI,10	
P. ferruginea, Labillardière, Nov. Holl. pl. spec. I, 10, t. 5 (1804)	W.A.	—	—	—	—	—		B.fl.VI,10	M.fr.VII,4.
P. brachyphylla, Bentham, Fl. Austr. VI, 11 (1873) ...	W.A.	—	—	—	—	—		B.fl.VI,11	
P. sylvestris, R. Brown, prodr. 361 (1810)...	W.A.	—	—	—	—	—		B.fl.VI,11	M.fr.VII,2&5.
P. brevifolia, R. Brown, prodr. 360 (1810)...	W.A.	—	—	—	—	—		B.fl.VI,12	M.fr.VII,4.
P. Maxwelli, F. v. M. in Benth. Fl. Austr. VI, 12 (1873) ...	W.A.	—	—	—	—	—		B.fl.VI,12	
P. angustifolia, R. Brown, prodr. 360 (1810) ...	W.A.	—	—	—	—	—		B.fl.VI,13	M.fr.VII,3.
P. nervosa, Meissner in Lehm. pl. Preiss. II, 269 (1847)...	W.A.	—	—	—	—	—		B.fl.VI,13	M.fr.VII,3.
P. sulphurea, Meissner in Mohl & Schlecht. Bot. Zeitung, 396 (1848)	W.A.	—	—	—	—	—		B.fl.VI,14	M.fr.VII,5.
P. floribunda, Meissner in De Candolle, prodr. XIV, 505 (1856)	W.A.	—	—	—	—	—		B.fl.VI,14	M.fr.VII,5.
P. suaveolens, Meissner in Lehm. pl. Preiss. I, 603 (1845)	W.A.	—	—	—	—	—		B.fl.VI,14	M.fr.VII,2,4,5.
P. physodes, Hooker, icon. plant. t. 865 (1852) ...	W.A.	—	—	—	—	—		B.fl.VI,15	M.fr.VII,4.
P. glauca, R. Brown, prodr. 360 (1810)	—	S.A.	T.	V.	N.S.W.	Q.	—	B.fl.VI,15	M.fr.VII,4;XI,123.
P. colorans, Cunningham in De Cand. prodr. XIV, 499 (1856) ...	—	—	—	—	N.S.W.	Q.	—	B.fl.VI,16	
P. collina, R. Brown, prodr. 359 (1810)	—	—	—	V.	N.S.W.	Q.		B.fl.VI,16	
P. spathulata, Labillardière, Nov. Holl. pl. spec. I, 9, t. 4 (1804)	—	S.A.	T.	V.	N.S.W.	—		B.fl.VI,17	
P. linifolia, Smith, specif. of Bot. of N. Holl. 31, t. 11 (1793)...	—	S.A.	T.	V.	N.S.W.	Q.	—	B.fl.VI,17	
P. ligustrina, Labillardière, Nov. Holl. pl. spec. I, 9, t. 3 (1804)	—	S.A.	T.	V.	N.S.W.	—		B.fl.VI,18	M.fr.I,87;IV,163;VII,5.
P. humilis, R. Brown, prodr. 361 (1810) ...	—	S.A.	T.	V.	N.S.W.	—		B.fl.VI,19	M.fr.VII,4;XI,47.
P. sericea, R. Brown, prodr. 361 (1810) ...	—	—	T.	—	—	—		B.fl.VI,19	M.fr.VII,5.
P. nivea, Labillardière, Nov. Holl. pl. spec. I, 10, t. 6 (1804) ...	—	—	T.	—	—	—		B.fl.VI,20	M.fr.VII,5.
P. imbricata, R. Brown, prodr. 361 (1810)...	W.A.	—	—	—	—	—		B.fl.VI,20	M.fr.VII,5.
P. villifera, Meissner in Lehm. pl. Preiss. II, 271 (1847)	W.A.	—	—	—	—	—		B.fl.VI,21	M.fr.VII,8.
P. drupacea, Labillardière, Nov. Holl. pl. spec. I, 10, t. 7 (1804)	—	—	T.	V.	—	—		B.fl.VI,21	M.fr.VII,8.
P. haematostachya, F. v. M., fragm. I, 84 (1858)...	—	—	—	—	—	Q.	—	B.fl.VI,22	M.fr.I,84;VII,8;XI,47.
P. spicata, R. Brown, prodr. 362 (1810) ...	—	—	—	—	N.S.W.	—		B.fl.VI,22	M.fr.VII,7;VIII,9.
P. spiculigera, F. v. M. in Bentham Fl. Austr. VI, 23 (1873)	W.A.	—	—	—	—	—		B.fl.VI,23	M.fr.XI,46.
P. Forrestiana, F. v. M., fragm. XI, 46 (1879) ...	—	—	—	—	—	N.A.	—	M.fr.XI,46.	
P. filiformis, J. Hooker in Lond. Journ. VI, 280 (1847)...	—	—	T.	—	—	—		B.fl.VI,23	M.fr.VII,7.
P. latifolia, R. Brown, prodr. 362 (1810)	—	—	—	—	—	Q.	—	B.fl.VI,23	M.fr.IV,49;VII,7.
P. simplex, F. v. M. in Linnaea XXV, 443 (1852)	—	S.A.	—	V.	N.S.W.	—		B.fl.VI,24	M.fr.XI,47.
P. sericostachya, F. v. M., fragm. IV, 162 (1864)	—	—	—	—	N.S.W.	Q.	—	B.fl.VI,24	M.fr.VII,7.
P. trichostachya, Lindley in Mitch. Trop. Austr. 355 (1848)	—	—	—	—	N.S.W.	Q.	—	B.fl.VI,24	
P. leptostachya, Bentham, Fl. Austr. VI, 24 (1873)	—	—	—	—	—	Q.	—	B.fl.VI,24	
P. argentea, R. Brown, prodr. 362 (1810) ...	W.A.	—	—	—	—	—		B.fl.VI,25	M.fr.VII,7.
P. clavata, Labillardière, Nov. Holl. pl. spec. I, 11 (1804)	W.A.	—	—	—	—	—		B.fl.VI,25	M.fr.VI,160;VII,5.
P. axiflora, F. v. M., First general Report 17 (1853)	—	—	T.	V.	N.S.W.	—		B.fl.VI,26	M.fr.VII,5.
P. leptospermoides, F. v. M., fragm. VII, 2 (1870)	—	—	—	—	—	Q.	—	B.fl.VI,27	M.fr.VII,7.
P. microcephala, R. Brown, prodr. 361 (1810)	W.A.	S.A.	—	V.	N.S.W.	Q.	—	B.fl.VI,27	M.fr.XI,27,47.
P. pauciflora, R. Brown, prodr. 360 (1810)	—	—	T.	V.	N.S.W.	—		B.fl.VI,27	M.fr.VII,8;XI,47.
P. elaeobantha, F. v. M., First general Report 17 (1853)...	—	—	—	T.	—	—		B.fl.VI,28	M.fr.VII,6.
P. pygmaea, F. v. M. in Schlecht. Linnaea XXVI, 346 (1853)	—	—	T.	—	—	—		B.fl.VI,29	
P. serpyllifolia, R. Brown, prodr. 360 (1810)	W.A.	S.A.	T.	V.	N.S.W.	—		B.fl.VI,29	M.fr.VII,6.
P. flava, R. Brown, prodr. 361 (1810)	W.A.	S.A.	T.	V.	N.S.W.	—		B.fl.VI,29	M.fr.VII,6.
P. petrophila, F. v. M. in Schlecht. Linnaea XXV, 442 (1852)...	—	S.A.	—	—	N.S.W.	—		B.fl.VI,30	
P. Bowmanni, F. v. M. in Benth. Fl. Austr. IX, 30 (1873)	—	—	—	—	—	Q.	—	B.fl.VI,30	M.fr.VII,6.
P. ammocharis, F. v. M. in Hook. Kew Misc. IX, 24 (1857)	—	—	—	—	—	N.A.		B.fl.VI,30	M.fr.VII,6;XI,47.
P. curviflora, R. Brown, prodr. 362 (1810))	—	S.A.	T.	V.	N.S.W.	Q.	—	B.fl.VI,31	M.fr.VII,6;VIII,9.
P. hirsuta, Meissner in De Cand. prodr. XIV, 513 (1856)	—	—	—	—	N.S.W.	—		B.fl.VI,31	
P. altior, F. v. M., fragm. I, 84 (1858)	—	—	—	—	N.S.W.	Q.	—	B.fl.VI,32	M.fr.VII,7.
P. octophylla, R. Brown, prodr. 361 (1810)	—	S.A.	—	V.	N.S.W.	—		B.fl.VI,32	M.fr.VII,6.

K

P. petraea, Meissner in Schlecht. Linnaea XXVI, 347 (1853) ... — S.A. — — N.S.W. — — B.fl.VI,33
P. phylicoides, Meissner in Lehm. pl. Preiss. II, 271 (1847) ... — S.A. — V. N.S.W. — — B.fl.VI,33 M.fr.VII,6.
P. Eyrei, F. v. M., fragm. V, 109 (1865) ... ... W.A. — — — — — B.fl.VI,34 M.fr.VII,109.
P. longiflora, R. Brown, prodr. 361 (1810)... ... ... W.A. — — — — — B.fl.VI,34 M.fr.VII,5.
P. stricta, Meissner in Schlecht. Linnaea XXVI, 348 (1853) ... — S.A. T. V. — — — B.fl.VI,34 M.fr.VII,4.
P. Preissii, Meissner in Lehm. pl. Preiss. I, 601 (1845) .. ... W.A. — — — — — B.fl.VI,35 M.fr.VII,3.
P. Holroydi, F. v. M., fragm. VI, 159, t. 59 (1868) ... ... — — — — — N.A. B.fl.VI,35 M.fr.250.

DRAPETES, Lamarck, Journ. d'hist. nat. I, 119, t. 10 (1792). (Kelleria, Daphnobryon).
D. Tasmanica, J. Hooker in Kew Misc. V, 299, t. 7 (1853) — — — T. V. N.S.W. — — B.fl.VI,36 M.fr.VII,1.

WICKSTROEMIA, Endlicher, prodr. fl. insul. Norfolk. 47 (1833). (Daphne partly).
W. Indica, C. A. Meyer in Bull. de l'Acad. de Petersb. I, 357 (1843) — — — N.S.W. Q. N.A. B.fl.VI,37 M.fr.VII,1.

PHALERIA, Jack in Hook. Comp. to Bot. Mag. I, 156 (1835). (Drymispermum).
P. Blumei, Bentham, Fl. Austr. VI, 38 (1873) ... ... — — — — Q. — B.fl.VI,38
P. Neumanni, F. v. M., fragm. VIII, 9 (1873) ... ... — — — — Q. — B.fl.VI,38 M.fr.V,26.
P. Clerodendron, F. v. M., fragm. VIII, 9 (1873) ... ... — — — — — Q. — B.fl.VI,38 M.fr.VII,1.

CORNACEAE.
Humboldt, Bonpland & Kunth, nov. gen. Americ. III, 430 (1818).
STYLIDIUM, Loureiro, Fl. Cochinch. I, 220 (1790). (Marlea, Rhytidandra, Pseudalangium).
S. Vitiense, F.v.M.; Rhytid.,AsaGray,Bot.Wilk.Ex.I,303,t.26(1854) — — — — N.S.W. Q. — B.fl.III,386

RUBIACEAE.
A. L. de Jussieu, gen. plant. 196 (1789), from B. de Jussieu (1759).
SARCOCEPHALUS, Afzelius in Transact. hort. soc. Lond. V, 422, t. 18 (1824).
S. cordatus, Miquel, Fl. Ind. Batav. II, 133 (1856) ... ... — — — — Q. N.A. B.fl.III,402

OLDENLANDIA, Plumier, nov. plant. Americ. gen. 42, t. 36 (1703). (Hedyotis, Synaptantha).
O. corymbosa, Linné, sp. pl. 119 (1753) ... ... — — — — Q. — — M.fr.VII,45.
O. paniculata, Linné, spec. plant. edlt. sec. 1667 (1763) ... — — — — Q. — M.fr.VIII,47.
O. auricularia, F. v. M.; Hedyotis, Linné, spec. plant. 101 (1753) — — — — Q. — B.fl.III,404
O. polyclada; Hedyotis, F. v. M., fragm. VIII, 146 (1874) — — — — Q. — M.fr.VIII,146.
O. coerulescens; Hedyotis, F. v. M., fragm. VIII, 146 (1863) — — — — Q. — B.fl.III,404 M.fr.VIII,38.
O. spermacocoides; Hedyotis, F. v. M., fragm. VIII, 146 (1874) — — — — N.A. — M.fr.VIII,146.
O. Crouchiana; Hedyotis, F. v. M., fragm. X, 85 (1876) — — — — N.A. — M.fr.X,85.
O. mitrasacmoides; Hedyotis, F. v. M., fragm. IV, 37 (1863) ... — — — — N.A. B.fl.III,404 M.fr.IV,37.
O. galioides; Hedyotis, F. v. M., fragm. IV, 39 (1863) ... — — — — Q. N.A. B.fl.III,404 M.fr.IV,39.
O. scleranthoides; Hedyotis, F. v. M., fragm. IV, 39 (1863) ... — — — — N.A. B.fl.III,405 M.fr.IV,39.
O. elatinoides, F. v. M.; Hedyotis, Benth., Fl. Austr. III, 405 (1866) W.A. — — — — B.fl.III,405 M.fr.VIII,147.
O. tillacacea; Hedyotis, F. v. M., fragm. IV, 39 (1863) ... — S.A. — N.S.W. Q. — B.fl.III,405 M.fr.IV,39.
O. trachymenioides; Hedyotis, F. v. M., fragm. IV, 40 (1863)... — — — — Q. — B.fl.III,406 M.fr.IV,40.
O. pterospora; Hedyotis, F. v. M., fragm. IV, 40 (1863)... — — — — N.A. B.fl.III,407 M.fr.IV,40,177.

DENTELLA, R. et G. Forster, char. gen. 25, t. 13 (1776). (Lippaya).
D. repens, R. & G. Forster, char. gen. 25, t. 13 (1776) ... — S.A. — N.S.W. Q. N.A. B.fl.III,407 M.fr.IX,187.

OPHIORRHIZA, Linné, fl. Zeylan. 190 et 239 (1747).
O. Australiana, Bentham, Fl. Austr. III, 407 (1866) ... ... — — — — Q. — B.fl.III,408 M.fr.IX,187.

ABBOTTIA, F. v. M., fragm. IX, 181 (1875).
A. singularis, F. v. M., fragm. IX, 181 (1875) ... ... — — — — Q. — — M.fr.IX,181.

GARDENIA, Ellis in Phil. Transact. LI, 935, t. 23 (1761).
G. edulis, F. v. M., fragm. I, 54 (1858) ... ... — — — — N.A. B.fl.III,408 M.fr.I,54.
G. resinosa, F. v. M., fragm. I, 54 (1858) ... ... — — — — N.A. B.fl.III,408 M.fr.I,54.
G. pyriformis, Cunningham in Benth. Fl. Austr. III, 409 (1866) ... — — — — N.A. B.fl.III,409
G. megasperma, F. v. M., fragm. I, 54 (1858) ... ... — — — — N.A. B.fl.III,409 M.fr.I,54.
G. Macgillivraei, Bentham, Fl. Austr. III, 409 (1866) ... ... — — — — Q. — B.fl.III,409 M.fr.VII,46.
G. ochreata, F. v. M., fragm. I, 55 (1858) ... ... — — — — Q. — B.fl.III,409 M.fr.185;VII,46.
G. suffruticosa, R. Brown in Benth. Fl. Austr. III, 410 (1866)... — — — — N.A. B.fl.III,410
G. fucata, R. Brown in Benth. Fl. Austr. III, 410 (1866) ... — — — — N.A. B.fl.III,410
G. Jardinei, F. v. M. in Benth. Fl. Austr. III, 410 (1866) ... — — — — Q. — B.fl.III,410 M.fr.VII,46.

RANDIA, Houston in Linné, hort. Clifford. 485 (1737).
R. hirta, F. v. M., fragm. VII, 46 (1869) ... ... — — — — Q. — M.fr.IX,181.
R. chartacea, F. v. M., fragm. IX, 180 (1875) ... ... — — — N.S.W. Q. — B.fl.III,410 M.fr.VII,47.
R. Benthamiana, F. v. M., fragm. IX, 180 (1875) ... — — — N.S.W. Q. — M.fr.IX,180.
R. Moorei, F. v. M. in Benth. Fl. Austr. III, 411 (1866) ... — — — N.S.W. Q. — B.fl.III,411
R. stipularis, F. v. M., Papuan plants I, 69 (1876) ... ... — — — — Q. — M.fr.VII,47;IX,70,181.
R. Fitzalani, F. v. M., Rep. Burdek. Exped. 12 (1860) ... ... — — — — Q. — B.fl.III,411 M.fr.VII,47;IX,180.
R. sessilis, F. v. M., fragm. VII, 47 (1869) ... ... — — — — Q. N.A. M.fr.VII,47. [180.
R. densiflora, Bentham, Fl. Hongk. 155 (1861) ... ... — — — N.S.W. Q. N.A. B.fl.III,412 M.fr.II,132;III,166;IX,

DIPLOSPORA, De Candolle, prodr. IV, 477 (1830). (Discospermum.)
D. Australis, Bentham, Fl. Austr. III, 413 (1866) ... ... — — — — Q. — B.fl.III,413
D. ixoroides, F. v. M., fragm. IX, 182 (1875) ... ... — — — — Q. — B.fl.III,413 M.fr.IX,182.

IXORA, Linné, syst. nat. 8 (1735); Linné, gen. pl. 27 (1737). (Pavetta, Webera, Stylocorne.)
I. Pavetta, Roxburgh, Fl. Ind. I, 393, ed. Carey (1820) ... ... — — — — Q. — B.fl.III,414 M.fr.IX,182.
I. tomentosa, Roxburgh, Fl. Ind. I, 396, ed. Carey (1820) ... — — — — N.A. B.fl.III,414
I. coccinea, Linné, sp. pl. 110 (1753) ... ... — — — — N.A. B.fl.III,414
I. Timorensis, Decaisne in Nouv. Ann. du Mus. III, 419 (1834)... — — — — Q. N.A. B.fl.III,415 M.fr.V,19;IX,183.
I. Beckleri, Bentham, Fl. Austr. III, 415 (1866)... ... ... — — — N.S.W. Q. — B.fl.III,415 M.fr.IX,183.

L. pentamera, Bentham, Fl. Austr. III, 416 (1866) ... ... — — — — — — N.A. B.fl.III,416
I. Dallachyana; F. v. M., Webera, Benth., Fl. Austr. III, 412 (1866) — — — — — Q. — B.fl.III,412
I. expansilens; Webera, F. v. M., fragm. VI, 25 (1867) ... — — — — — — N.A. — M.fr.VI,25.
COFFEA, Linné, syst. nat. 8 (1735); Linné, gen. pl. 55 (1737), from Ray (1691).
C. Bengalensis, Roxburgh, hort. Bengal. 15 (1814) ... ... — — — Q. — — M.fr.VIII,147.
TIMONIUS, Rumphius, herb. Amboin. III, 216, t. 140 (1743). (Polyphragmon.)
T. Rumphii, De Candolle, prodr. IV, 461 (1830) ... ... — — — — Q. N.A. B.fl.III,417 M.fr.II,134;IX,187.
SCYPHIPHORA, K. F. Gaertner, de fruct. III, 91, t. 196 (1805). (Epithinia.)
S. hydrophylacea, K. F. Gaertner, de fruct. III, 91, t. 196 (1805) — — — — — Q. N.A. B.fl.III,418 M.fr.VII,46;IX,187.
GUETTARDA, Osbeck in Linné, spec. plant. 991 (1753). (Antirrhoea, Guettardella.)
G. speciosa, Linné, spec. plant. 991 (1753)... ... ... — — — — Q. N.A. B.fl.III,419 M.fr.IX,183.
G. tenuiflora, F. v. M. fragm. VII, 48 (1809) ... ... — — — — — Q. — B.fl.III,418 M.fr.IX,183.
G. putaminosa, F. v. M., fragm. IX, 183 (1875) ... ... — — — — — Q. — B.fl.III,419 M.fr.IV,92.
G. myrtoides, F. v. M., fragm. IX, 184 (1875) ... . — — — — — Q. — — M.fr.IX,184.
HODGKINSONIA, F. v. M., fragm. II, 132 (1861).
H. ovatiflora, F. v. M., fragm. II, 132 (1861) ... ... — — — — — N.S.W. Q. — B.fl.III,420 M.fr.III,166.
CANTHIUM, Lamarck, Encycl. method. I, 602 (1783). (Psydrax, Plectronia partly.)
C. eymosum, Persoon, synops. plant. I, 200 (1805) ... ... — — — Q. — — M.fr.IX,185.
C. latifolium, F. v. M. in Benth. Fl. Austr. III, 421 (1866) W.A. S.A. — — N.S.W. Q. — B.fl.III,421
C. attenuatum, R. Brown in Benth. Fl. Austr. III, 421 (1866)... ... — — N.A. B.fl.III,421 [135.
C. lucidum, Hooker & Arnott, Bot. of Beechey's voy. 65 (1841) — — N.S.W. Q. N.A. B.fl.III,421 M.fr.II,133; IX,185; XI,
C. oleifolium, Hooker in Mitch. Trop. Austr. 397 (1848)... ... — N.S.W. Q. — B.fl.III,422
C. buxifolium, Bentham, Fl. Austr. III, 422 (1866) ... — N.S.W. Q. — B.fl.III,422
C. vaccinifolium, F. v. M. in Trans. phil. Inst. Vict. III, 47 (1858) — — N.S.W. Q. — B.fl.III,422 M.fr.II,134;IX,186.
C. coprosmoides, F. v. M. in Trans. phil. Inst. Vict. III, 47 (1858) — — N.S.W. Q. — B.fl.III,422
MORINDA, Vaillant in act. Acad. Par. 202 (1722).
M. citrifolia, Linné, spec. plant. 176 (1753) ... ... — — — — Q. N.A. B.fl.III,422
M. acutifolia, F. v. M., fragm. IX, 179 (1875) ... ... — — — — — Q. — B.fl.III,421 M.fr.IX,179.
M. jasminoides, Cunningham in Hook. Bot. Mag. t. 3351 (1834) — — V. N.S.W. Q. — B.fl.III,423
M. umbellata, Linné, spec. plant. 176 (1753) ... ... — — — — — Q. — B.fl.III,423
M. reticulata, Bentham, Fl. Austr. III, 424 (1866) ... — — — — — Q. N.A. B.fl.III,423 M.fr.IX,180.
COELOSPERMUM, Blume, Bijdr. 904 (1826). (Pogonolobus.)
C. paniculatum, F. v. M., fragm. V, 19 (1865) ... ... — — N.S.W. — B.fl.III,425 M.fr.VII,46;IX,185.
C. reticulatum, Bentham, Fl. Austr. III, 425 (1866) ... ... — — — Q. N.A. B.fl.III,425 M.fr.I,56;IX,185.
HYDNOPHYTUM, Jack in Transact. Linn. Soc. XIV, 124 (1823).
H. formicarum, Jack in Transact. Linn. Soc. XIV, 124 (1823)... — — — — Q. N.A. — M.fr.VII,45.
MYRMECODIA, Jack in Transact. Linn. Soc. XIV, 122 (1823).
M. echinata, Gaudichaud, Bot. voy. Bonito 472, t. 96 (1844) ... — — — — Q. N.A. — M.fr.VII,45.
LASIANTHUS, Jack in Transact. Linn. Soc. XIV, 125 (1823).
L. strigosus, Wight in Calc. Journ. nat. hist. VI, 512 (1846) ... — — — — — Q. — B.fl.III,426
PSYCHOTRIA, Linné, syst. ed. decim. 929 (1759). (Uragoga, 1737, Grumilia.)
P. nesophila, F. v. M., fragm. II, 135 (1861) ... ... — — — — Q. N.A. B.fl.III,427 M.fr.IX,184.
P. Dallachyana, Bentham, Fl. Austr. III, 427 (1866) ... ... — — — — Q. — B.fl.III,427
P. nematopoda, F. v. M., fragm. VII, 48 (1809) ... ... — — — Q. — — M.fr.IX,184.
P. loniceroides, Sieber in De Cand., prodr. IV, 523 (1830) ... ... — — N.S.W. — B.fl.III,427 M.fr.IX,184.
P. daphnoides, Cunningham in Hook. Bot. Mag. 2228 (1833) ... ... — — N.S.W. — B.fl.III,428 M.fr.IX,184.
P. poliostemma, Bentham, Fl. Austr. III, 427 (1866) ... ... — — — Q. — B.fl.III,428
P. Fitzalani, Bentham, Fl. Austr. III, 427 (1866) ... ... — — — Q. — B.fl.III,428
P. Carronis, Moore & F. v. M., fragm. VII, 49 (1809) ... ... — — N.S.W. — — M.fr.IX,70,185.
GEOPHILA, D. Don, prodr. fl. Nepal. 136 (1825). (Cephaelis partly.)
G. reniformis, D. Don, prodr. fl. Nepal. 136 (1825) ... ... — — — Q. — — M.fr.VII,45;VIII,147.
COPROSMA, R. et G. Forster, char. gen. 137, t. 69 (1776). (Nertera, Marquisia.)
C. Nertera, F. v. M., fragm. IX, 186 (1875) ... ... — — T. V. N.S.W. — — M.fr.IX,186.
C. reptans, F. v. M., fragm. IX, 186 (1875) ... ... — — V. — — — M.fr.IX,186.
C. pumila, J. Hooker in Lond. Journ. VI, 465 (1847) ... ... — T. V. — — B.fl.III,430 M.fr.IX,186.
C. nitida, J. Hooker in Lond. Journ. VI, 465 (1847) ... ... — T. V. — B.C.III,430 M.fr.IX,186.
C. Billardieri, J. Hooker in Lond. Journ. VI, 465 (1847) ... — T. V. N.S.W. — B.fl.III,430
C. hirtella, Labillardière, Nov. Holl. pl. spec. J, 70, t. 95 (1804) S.A. T. V. N.S.W. — B.fl.III,429 M.fr.IX,186.
C. Baueri, Endlicher, iconogr. gen. pl. XI, t. 111 (1838) ... ... — N.S.W. — — M.fr.IX,69.
C. putida, C. Moore & F. v. M., fragm. VII, 45 (1809) ... ... — N.S.W. — — M.fr.IX,69.
C. lanceolaris, F. v. M., fragm. IX, 70 (1875) ... ... — N.S.W. — — M.fr.IX,70.
OPERCULARIA, J. Gaertner, de fruct. I, 111, t. 24 (1788).
O. vaginata, Labillardière, Nov. Holl. pl. spec. I, 34, t. 46 (1804) W.A. — — — — B.fl.III,433
O. spermacocea, Labillardi., Nov. Holl. pl. spec. I, 35, t. 47 (1804) W.A. — — — — B.fl.III,433
O. scabrida, Schlechtendal, Linnaea XX, 604 (1847) ... ... — S.A. — N.S.W. — B.fl.III,433
O. hirsuta, F. v. M. in Benth. Fl. Austr. III, 434 (1866) W.A. — — — — B.fl.III,434
O. aspera, Gaertner, de fruct. I, 112, t. 24 (1788)... ... — V. N.S.W. Q. — B.fl.III,434
O. hispida, Sprengel, syst. veg. I, 385 (1825) ... ... — — N.S.W. Q. — B.fl.III,434
O. diphylla, Gaertner, de fruct. I, 113 (1788) ... ... — V. N.S.W. — B.fl.III,434
O. ovata, J. Hooker in Lond. Journ. VI, 465 (1847) ... ... — S.A. T. V. N.S.W. — B.fl.III,435
O. varia, J. Hooker in Lond. Journ. VI, 466 (1847) ... ... — S.A. T. V. N.S.W. — B.fl.III,435
O. rubioides, A. L. de Jussieu in Annal. du Mus. IV, 428 (1804) W.A. — — — — B.fl.III,435
O. volubilis, R. Brown in Benth. Fl. Austr. III, 435 (1866) W.A. — — — — B.fl.III,435
O. hispidula, Endlicher in Hueg. enum. 58 (1837) ... W.A. — — — — B.fl.III,436
O. echinocephala, Bentham, Fl. Austr. III, 436 (1866) ... W.A. — — — — B.fl.III,436
O. apiciflora, Labillardière, Nov. Holl. pl. spec. I, 35, t. 46 (1804) W.A. — — — — B.fl.III,436

POMAX, Solander in Gaertn., de fruct. I, 112 (1788).
P. umbellata, Solander in Gaertn., de fruct. I, 112 (1788) ... — S.A. — V. N.S.W. Q. — B.fl.III,437 M.fr.IX,187.
ELEUTHRANTHES, F. v. M., fragm. IV, 92 (1864).
E. opercularina, F. v. M., fragm. IV, 92 (1864) ... ... W.A. — — — — — — B.fl.III,437 M.fr.IV,92.
KNOXIA, Linné, Fl. Zeyl. 189 (1747).
K. corymbosa, Willdenow, spec. pl. I, 582 (1798) .. .. — — — — — Q. N.A. B.fl.III,438 M.fr.IX,187.
SPERMACOCE, Dillenius, hort. Elth. 369, t. 277 (1732).
S. brachystoma, R. Brown in Benth. Fl. Austr. III, 439 (1866) — — — — — Q. N.A. B.fl.III,430 M.fr.IV,41.
S. pogostoma, Bentham, Fl. Austr. III, 440 (1866) ... ... — — — — — — N.A. B.fl.III,440
S. leptoloba, Bentham, Fl. Austr. III, 440 (1866) ... ... — — — — — — N.A. B.fl.III,440
S. marginata, Bentham, Fl. Austr. III, 440 (1866) ... ... — — — — — Q. N.A. B.fl.III,440
S. multicaulis, Bentham, Fl. Austr. III, 440 (1866) ... .. — — — — N.S.W. Q. N.A. B.fl.III,440
S. exserta, Bentham, Fl. Austr. III, 441 (1866) ... ... — — — — — — — N.A. B.fl.III,441
S. membranacea, R. Brown in Benth. Fl. Austr. III, 441 (1866) — — — — — Q. — B.fl.III,441
S. debilis, Bentham, Fl. Austr. III, 441 (1866) ... ... — — — — — — N.A. B.fl.III,441 M.fr.IV,43.
S. Insperta, F. v. M., fragm. IV, 43 (1863) ... ... — — — — — — N.A. B.fl.III,441 M.fr.IV,43.
S. stenophylla, F. v. M., fragm. IV, 43 (1863) ... ... — — — — — — N.A. B.fl.III,442 M.fr.IV,41.
S. laevigata, F. v. M., fragm. IV, 41 (1863) ... ... — — — — — — N.A. B.fl.III,442
S. breviflora, F. v. M. in Benth. Fl. Austr. III, 442 (1866) ... — — — — — — N.A. B.fl.III,442
S. auriculata, F. v. M., fragm. IV, 42 (1863) ... ... — — — — — — N.A. B.fl.III,442 M.fr.IV,42.
S. suffruticosa, R. Brown in Benth. Fl. Austr. III, 443 (1866) ... — — — — — — N.A. B.fl.III,443
ASPERULA, Dodoens, pemptad. 355 (1583).
A. geminifolia, F. v. M., fragm. V, 147 (1865) ... ... — S.A. — V. N.S.W. Q. — B.fl.III,443 M.fr.IX,188;XI,27.
A. oligantha, F. v. M. in Neerl. Kruidk. Arch. IV, 111 et 112 (1859) — S.A. T. V. N.S.W. Q. — B.fl.III,444 M.fr.IX,187;XI,27.
GALIUM, Dodoens, pemptad. 335 (1583), from Dioscorides.
G. umbrosum, Solander in G. Forst. prodr. 89 (1786) ... ... — S.A. T. V. N.S.W. — — B.fl.III,446 M.fr.IX,188.
G. australe, De Candolle, prodr. IV, 608 (1830) ... ... — S.A. T. V. N.S.W. Q. — B.fl.III,446

CAPRIFOLIACEAE.
Adanson, Familles des plantes II, 133 (1763).

SAMBUCUS, Tournefort, inst. 606, t. 306 (1700), from Dodoens (1583). (Tripetelus.)
S. xanthocarpa, F. v. M. in Transact. phil. Inst. Vict. I, 42 (1855) — — — V. N.S.W. Q. — B.fl.III,398
S. Gaudichaudiana, De Candolle, prodr. IV, 322 (1830) ... — — — — V. N.S.W. Q. — B.fl.III,398

PASSIFLOREAE.
A. L. de Jussieu in Ann. du Mus. VI, 102 (1805).

PASSIFLORA, Plukenet, phytograph. 202 et 282, t. 104, 210, 211, 212 (1692). (Disemma.)
P. Herbertiana, Lindley, Bot. Regist. t. 737 (1823) ... ... — — — N.S.W. Q. — B.fl.III,311 M.fr.IX,68.
P. cinnabarina, Lindley in Gard. Chron. 724 (1855) ... — — — V. N.S.W. — — — M.fr.IX,68.
P. aurantia, G. Forster, fl. ins. Austr. prodr. 62 (1786) ... — — — — N.S.W. Q. — B.fl.III,312 M.fr.IX,68.
P. brachystephanea, F. v. M. in Benth. Fl. Austr. III, 312 (1866) — — — — — Q. — B.fl.III,312 M.fr.I.56.
MODECCA, Lamarck, Encycl. méthod. IV, 208 (1797).
M. australis, R. Brown in De Cand. prodr. III, 337 (1828) ... — — — — — Q. N.A. B.fl.III,313 M.fr.IX,69.

CUCURBITACEAE.
Haller, enum. stirp. Helv. praef. 34 (1742).

TRICHOSANTHES, Linné, gen. plant. 295 (1737).
T. pentaphylla, F. v. M. in Benth. Fl. Austr. III, 314 (1866) ... — — — — — Q. — B.fl.III,314
T. cucumerina, Linné, spec. plant. 1008 (1753) ... . — — — — — — N.A. B.fl.III,314
T. palmata, Roxburgh, Fl. Ind. III, 704 (1832) ... ... — — — — N.S.W. Q. — B.fl.III,315 M.fr.VI,187.
T. Hearnii, F. v. N. in Benth. Fl. Austr. III, 315 (1866) ... — — — — — — Q. — B.fl.III,315
LAGENARIA, Seringe in Mém. Soc. Genèv. III, 25, t. 2 (1825). ·
L. vulgaris, Seringe in De Cand. prodr. III, 279 (1828) ... — — — — — — — B.fl.III,310 M.fr.VII,62.
LUFFA, Tournefort in act. Acad. Paris. 84, t. 2 (1706), from Vesling (1638).
L. Aegyptiaca, Miller, Gardener's Dictionary (1731) ... ... — — — — — — Q.' — B.fl.III,316 M.fr.III,107.
L. graveolens, Roxburgh, Fl. Ind. III, 716 (1832) ... ... — — — — — — N.A. B.fl.III,316 M.fr.III,106;IV,173.
L. foetida, Cavanilles, icon. I, 7, t. 9—10 (1791) ... ... — — — — — — N.A. — M.fr.XI,138.
ZANONIA, Linné, coroll. 19 (1737). (Alsomitra.)
Z. Capricornica, F. v. M., fragm. VI, 61 (1870) ... ... — — — — — — Q. — M.fr.VII,61.
Z. Hookeri, F. v. M., fragm. VI, 188 (1869) ... ... — — — — — Q. — M.fr.V,181;VII,62.
Z. Stephousiana, F. v. M., fragm. VIII, 181 (1874) ... ... — — — — — Q. — M.fr.VIII,181.
CUCUMIS, Tournefort, inst. 104, t. 31 (1700), from Plinius.
C. trigonus, Roxburgh, Fl. Ind. III, 274 (1841). ... ... — — — N.S.W. Q. N.A. B.fl.III,317 M.fr.VI,187.
C. Chate, Linné, syst. nat. edit. decim. 5 (1759) ... ... — S.A. — N.S.W. Q. N.A. — M.fr.VI,186.
BENINCASA, Savi in Bibl. Ital. IX, 158 (1818).
B. vacua; Cucurbita vacua, F. v. M., fragm. VI, 186 (1868) ... — — — — — Q. — M.fr.VI,186.
MOMORDICA, Tournefort, Institut. rei herbar. 103, t. 29 et 30 (1700).
M. Balsamina, Linné, sp. pl. 1009 (1753) ... ... — — — — N.S.W. Q. — B.fl.III,318 M.fr.VI,187.
M. Charantia, Linné, sp. pl. 1009 (1753) ... ... — — — — — Q. — ·
BRYONOPSIS, Arnott in Hook. Journ. of Bot. III, 274 (1841). (Bryonia partly.)
B. laciniosa, Naudin in Annal. des sc. nat. sér. cinq. VI, 30 (1866) — — — — N.S.W. Q. N.A. B.fl.III,310
B. Pancheri, Naudin in Annal. des sc. nat. sér. cinq. V, 30 (1866) — — — — N.S.W. —

MELOTHRIA, Linné, coroll. 1 (1737). (Zehneria, Mukia.)
M. Cunninghamii, Bentham, Fl. Austr. III, 320 (1866) ... ... — — — — N.S.W. Q. N.A. B.fl.III,320 M.fr.VI,186;VII,02.
M. Baueriana, F. v. M , fragm. VI, 188 (1868) ... ... — — — N.S.W. — — .M.fr.VI,188.
M. Muelleri, Bentham, Fl. Austr. III, 320 (1866)... ... ... — S.A. — V. N.S.W. — — R.fl.III,320
M. Maderaspatana, Cogniaux in De Cand. mon. phan. III,623(1881) W.A. S.A. — — N.S.W. Q. N.A. B.fl.III,321 M.fr.VI,187.

SICYOS, Linné, syst. nat. 9 (1735); Linné, gen. plant. 297 (1737).
S. angulata, Linné, spec. plant. 1013 (1753) ... ... — T. V. N.S.W. Q. — B.fl.III,322 M.fr.VI,187;VII,62.

COMPOSITAE.

Vaillant in act. Acad. Paris. 143 (1718).

CENTRATHERUM, Cassini in Bull. de la Soc. philom. (1917).
C. muticum, Lessing in Schlecht. Linnaea II, 320 (1829) ... — — — — N.S.W. Q. — B.fl.III,460

PLEUROCARPAEA, Bentham, Fl. Austr. III, 460 (1866).
P. denticulata, Bentham, Fl. Austr. III, 460 (1866) ... ... — — — — — N.A. D.fl.III,460

VERNONIA, Schreber, gen. plant. II, 541 (1791).
V. cinerea, Lessing in Schlecht. Linnaea II, 291 (1829) ... ... — — — — N.S.W. Q. N.A. B.fl.III,459
V. vagans, De Candolle, prodr. V, 32 (1843) ... ... — — — — — Q. — — M.fr.VI,234.

ELEPHANTOPUS, Vaillant in act. Acad. Par. 309 (1710).
E. scaber, Linné, spec. plant. 814 (1753) ... ... — — — — — Q. N.A. B.fl.III,461

ADENOSTEMMA, R. et G. Forster, char. gen. 89, t. 45 (1776).
A. viscosum, R. & G. Forster, char. gen. 89, t. 45 (1776) ... — S.A. — V. N.S.W. Q. — B.fl.III,462

AGERATUM, Linné, gen. plant. 247 (1737).
A. conyzoides, Linné, spec. plant. 830 (1753) ... ... — — — — — Q. — B.fl.III,462

EUPATORIUM, Tournefort, inst. 455, t. 259 (1700), from C. Bauhin (1623).
E. cannabinum, Linné, spec. pl. 838 (1753)... ... ... — — — N.S.W. Q. — B.fl.III,462 M.fr.V,62.

LAGENOPHORA, Cassini in Bull. de la Soc. philom. 34 (1818). (Ixauchenus, Solenogyne, Emphysopus.)
L. Billardieri, Cassini in Diction. XXV, 111 (1822) ... W.A. S.A. T. V. N.S.W. Q. — B.fl.III,507
L. Huegelii, Bentham in Hueg. enum. 59 (1837) ... ... W.A. S.A. T. V. — — — B.fl.III,507
L. Solenogyne, F. v. M., fragm. V, 62 (1865) ... ... — — — N.S.W. Q. — B.fl.III,508 M.fr.V,62.
L. Emphysopus, J. Hooker, Fl. Tasman. I, 189 (1860) ... — T. V. N.S.W. — — D.fl.III,508

BRACHYCOME, Cassini in Bull. de la Soc. philom. 199 (1816). (Bellidis subgenus, Pacquerina, Brachystephium, Steiroglossa, Silphiospermum.)
B. diversifolia, Fischer & Meyer, ind. sem. hort. Petrop. 31 (1835) — S.A. T. V. N.S.W. — B.fl.III,511
B. segmentosa, Moore & F. v. M., fragm. VIII, 144 (1873) — — — N.S.W. — — M.fr.VIII,144.
B. melanocarpa, Sonder & F. v. M. in Schl. Linn. XXV, 476 (1852) — S.A. — V. N.S.W. — B.fl.III,511
B. radicans, Steetz in Lehm. pl. Preiss. 1, 420 (1845) — — T. V. N.S.W. — B.fl.III,512
B. goniocarpa, Sonder & F. v. M. in Schl. Linn. XXV, 474 (1852) W.A. S.A. — V. N.S.W. — B.fl.III,512
B. pachyptera, Turczaninow in Bull. Mosc. XXIV, 175 (1851) W.A. S.A. — V. N.S.W. — B.fl.III,512
B. iberidifolia, Bentham in Hueg. enum. 59 (1837) ... W.A. — — — — — B.fl.III,512 M.fr.XI,123.
B. pusilla, Steetz in Lehm. pl. Preiss. I, 427 (1845) ... W.A. — — — — — B.fl.III,513
B. microcarpa, F. v. M., fragm. I, 50 (1858) ... ... — — — N.S.W. Q. — B.fl.III,513 M.fr.I,50.
B. Stuartii, Bentham, Fl. Austr. III, 513 (1866) ... ... — — — N.S.W. — — B.fl.III,513
R. scapigera, De Candolle, prodr. VII, 277 (1838)... ... — — — N.S.W. — — B.fl.III,513
B. Muelleri, Sonder in Schlecht. Linnaea XXV, 475 (1852) — S.A. — — — — B.fl.III,514
B. graminea, F. v. M., fragm. I, 49 (1858)... ... — S.A. T. V. N.S.W. — B.fl.III,514 M.fr.VIII,142.
B. angustifolia, Cunningham in De Cand , prodr. V, 306 (1836) — — T. V. N.S.W. — B.fl.III,514 M.fr.I,49.
B. linearifolia, De Candolle, prodr. V, 306 (1836)... ... — — — N.S.W. — — B.fl.III,515
B. basaltica, F. v. M., fragm. I, 50 (1858) ... ... — S.A. — N.S.W. Q. — B.fl.III,515 M.fr.I,50.
B. trachycarpa, F. v. M. in Schlecht. Linnaea XXV, 339 (1852) — S.A. — V. N.S.W. — B.fl.III,516
B. exilis, Sonder in Linnaea XXV, 473 (1852) ... — S.A. — V. N.S.W. — B.fl.III,516
B. ptychocarpa, F. v. M. in Transact. phil. Soc. Vict. I, 43 (1854) — — — N.S.W. — — B.fl.III,516
B. debilis, Sonder in Linnaea XXV, 477 (1852) ... ... — S.A. — V. N.S.W. — B.fl.III,516
B. decipiens, J. Hooker in Lond. Journ. VI, 114 (1847) ... — S.A. T. V. N.S.W. — B.fl.III,516
B. cardiocarpa, F. v. M. in Benth. Fl. Austr. III, 516 (1866) ... — S.A. — V. N.S.W. — B.fl.III,517
B. nivalis, F. v. M. in Transact. phil. Soc. Vict. I, 43 (1854) ... — — — V. N.S.W. — B.fl.III,517
B. scapiformis, De Candolle, prodr. V, 306 (1836) ... ... — — T. V. N.S.W. — B.fl.III,517
B. stricta, De Candolle, prodr. V, 305 (1836) ... ... — — T. V. N.S.W. — B.fl.III,518
R. heterodonta, De Candolle, prodr. V, 305 (1836) ... ... — — — N.S.W. — — B.fl.III,518
B. Billardieri, Bentham, Fl. Austr. III, 519 (1866) ... ... — W.A. — — — — B.fl.III,518
B. ciliaris, Lessing, syn. comp. 172 (1832) .. .. W.A. S.A. T. V. N.S.W. — B.fl.III,519
B. calocarpa, F. v. M. in Schlecht. Linnaea XXV, 309 (1852) ... — S.A. — V. N.S.W. — B.fl.III,519 M.fr.XI,27.
B. chrysoglossa, F. v. M. in Transact. phil. Soc. Vict. I, 44 (1854) — S.A. — — — — B.fl.III,519
B. marginata, Bentham in Hueg. enum. 60 (1837) ... ... — — — — Q. — B.fl.III,519
B. Sieberi, De Candolle, prodr. V, 306 (1836) ... ... — — — N.S.W. — — B.fl.III,520
B. discolor, C. Stuart in Benth. Fl. Austr. III, 520 (1866) ... — — T. — — — B.fl.III,520
B. multifida, De Candolle, prodr. V, 306 (1836) ... ... — — — N.S.W. — — B.fl.III,520 M.fr.VIII,145.
B. latisquamea, F. v. M., fragm. XI, 16 (1878) ... ... W.A. — — — — — M.fr.XI,16.
B. glandulosa, Bentham, Fl. Austr. III, 520 (1866) ... ... W.A. — — N.S.W. — — D.fl.III,520
B. collina, Bentham, Fl. Austr. III, 521 (1866) ... ... W.A. S.A. — N.S.W. — B.fl.III,521

ERODIOPHYLLUM, F. v. M., fragm. IX, 120 (1875).
E. Elderi, F. v. M., fragm. IX, 120 (1875)... ... ... — S.A. — — — — M.fr.IX,119.

MINURIA, De Candolle, prodr. V, 298 (1836). (Therogeron, Elachothamnus, Kippistia.)
M. leptophylla, De Candolle, prodr. V, 298 (1836). ... W.A. S.A. — V. N.S.W. Q. — B.fl.III,498 M.fr.IX,119;X,56.
M. Cunninghamii, Bentham, Fl. Austr. III, 498 (1866) ... ... — S.A. — V. N.S.W. — — B.fl.III,498

M. Candollei, F. v. M., fragm. IX, 119 (1875) ... ... — S.A. — V. N.S.W. Q. N.A. B.fl.III,499 M.fr.X,56.
M. suaedifolia, F. v. M. in Bentham, Fl. Austr. III, 499 (1866) — S.A. — V. N.S.W. — — B.fl.III,499 M.fr.X,56.

CALOTIS, R. Brown in Bot. Regist. t. 504 (1820). (Huenefeldia, Goniopogon, Cheiroloma.)
C. dentex, R. Brown in Bot. Regist. 504 (1820) ... ... — — — N.S.W. Q. — B.fl.III,501
C. cuneifolia, R. Brown in Bot. Regist. t. 504 (1820) ... ... — S.A. — V. N.S.W. Q. — B.fl.III,501
C. glandulosa, F. v. M. in Transact. Vict. Inst. 129 (1855) ... — — V. N.S.W. — — B.fl.III,502
C. cymbacantha, F. v. M. in Schlecht. Linnaea XXV, 400 (1852) — S.A. — V. N.S.W. — — B.fl.III,502
C. erinacea, Steetz in Lehm. pl. Preiss. I, 424 (1845) ... ... W.A. S.A. — V. N.S.W. — — B.fl.III,502
C. scabiosifolia, Sonder & F. v. M. in Linnaea XXV, 471 (1852) — S.A. — V. N.S.W. — — B.fl.III,503 M.fr.XI,27.
C. scapigera, Hooker in Mitch. Trop. Austr. 75 (1848) .. ... — S.A. — V. N.S.W. Q. N.A. B.fl.III,503 M.fr.XI,27.
C. anthemoides, F. v. M. in Transact. phil. Soc. Vict. I, 44 (1854) — — — V. — — B.fl.III,504
C. lappulacea, Bentham in Hueg. enum. 60 (1837) ... ... W.A. S.A. — V. N.S.W. Q. — B.fl.III,504
C. microcephala, Bentham, Fl. Austr. III, 504 (1866) ... — S.A. — V. N.S.W. — — B.fl.III,504
C. breviseta, Bentham in Hueg. enum. 60 (1837) ... ... — — — — — N.A. B.fl.III,505
C. plumulifera, F. v. M. in Transact. Vict. Inst. III, 57 (1859) W.A. S.A. — V. N.S.W. — — B.fl.III,505
C. porphyroglossa, F. v. M. in Benth. Fl. Austr. III, 505 (1866) — S.A. — — N.S.W. — N.A. B.fl.III,505
C. pterosperma, R. Brown in Benth. Fl. Austr. III, 505 (1866)... — — — — — N.A. B.fl.III,505
C. hispidula, F. v. M. in Transact. Vict. Inst. 130 (1855) ... W.A. S.A. — V. N.S.W. — — B.fl.III,506
C. Kempci, F. v. M. in Transact. Roy. Soc. S.Austr. IV, 112 (1881)— S.A. — — — — — M.fr.XI,139.

ASTER, Tournefort, inst. 481, t. 274 (1700), from Dioscorides. (Olearia, Celmisia, Eurybia, Steetzia.)
A. megalophyllus, F. v. M., fragm. V. 70 (1865)... ... — — — V. N.S.W. — B.fl.III,468 M.fr.V,70.
A. chrysophyllus, Cunningham in De Cand. prodr. V, 266 (1836) — — — — N.S.W. — B.fl.III,408
A. alpicola, F. v. M., fragm. V, 70 (1865) ... ... — — — V. N.S.W. — B.fl.III,468 M.fr.V,70.
A. rosmarinifolius, Cunningham in De Cand. prodr. V, 208 (1836) — — — — N.S.W. — B.fl.III,468
A. viscosus, Labillardière, Nov. Holl. pl. spec. II, 53, t. 203 (1806) — T. V. — — — B.fl.III,468 M.fr.V,71.
A. Sonderi, F. v. M., fragm. V, 83 (1865)... ... — S.A. — — — — B.fl.III,469 M.fr.V,83.
A. pannosus, F. v. M., fragm. V, 83 (1865) ... ... — S.A. — — — — B.fl.III,469 M.fr.V,83.
A. oliganthemus; Olearia, F. v. M. in Benth. Fl. Austr. III, 499 (1866) — — — N.S.W. — — B.fl.III,469
A. argophyllus, Labillard., Nov. Holl. pl. spec. II, 52, t. 201 (1806) — T. V. N.S.W. — — B.fl.III,470 M.fr.V,68.
A. cydoniifolius, Cunningham in De Cand. prodr. V, 267 (1836) — — — N.S.W. — — B.fl.III,470
A. myrsinoides, Labillard., Nov. Holl. pl. spec. II, 53, t. 202 (1806) S.A. T. V. N.S.W. — — B.fl.III,471 M.fr.V,69.
A. persoonioides, Cunningham in De Cand. prodr. V, 267 (1836) — T. — — — — B.fl.III,471 M.fr.V,69.
A. obcordatus, F. v. M., fragm. V, 69 (1865) ... — T. V. — — — B.fl.III,471 M.fr.V,69.
A. pinifolius, F. v. M., fragm. V, 71 (1865) ... — T. — — — — B.fl.III,472 M.fr.V,71.
A. lodifolius, Cunningham in De Cand. prodr. V, 269 (1836) ... — T. — — — — B.fl.III,472
A. dentatus, Andrews, Bot. Reposit. t. 61 (1799)... — — — N.S.W. — — B.fl.III,472 M.fr.V,82.
A. stellulatus, Labillard., Nov. Holl. pl. spec. II, 50, t. 106 (1806) — S.A. T. V. N.S.W. Q. — B.fl.III,473 M.fr.V,81.
A. asterotrichus, F. v. M., fragm. V, 79 (1865) ... ... — S.A. — V. N.S.W. — — B.fl.III,473 M.fr.I,111.
A. gravis, F. v. M., fragm. V, 82 (1865) ... ... — — — N.S.W. — — B.fl.III,474 M.fr.V,82.
A. Nernstii, F. v. M., fragm. V, 81 (1865)... ... ... — — — N.S.W. Q. — B.fl.III,474 M.fr.V,81.
A. hygrophilus, Cunningham in De Cand. prodr. V, 269 (1836) ... — — — — Q. — B.fl.III,474
A. Siemssenii, F. v. M., fragm. V, 71 (1865) ... ... — — — N.S.W. — — B.fl.III,475 M.fr.I,50.
A. tubuliflorus, F. v. M., fragm. V, 65 (1865) ... ... — S.A. — V. N.S.W. — — B.fl.III,475 M.fr.V,65.
A. axillaris, F. v. M., fragm. V, 64 (1865)... ... ... W.A. S.A. T. V. N.S.W. — N.A. B.fl.III,475 M.fr.V,64.
A. revolutus; Olearia, F. v. M. in Benth. Fl. Austr. III, 476 (1866) W.A. — — — — — B.fl.III,476
A. oxilifolius, Labillard., Nov. Holl.plant. spec. II, 51, t. 199 (1806) — S.A. T. V. N.S.W. — — B.fl.III,476 M.fr.V,69.
A. fiorulentus, F. v. M., fragm. V, 82 (1865) ... ... — S.A. — — N.S.W. — — B.fl.III,477 M.fr.V,82.
A. microphyllus, Persoon, synops. II, 442 (1807) ... ... — S.A. T. V. N.S.W. — — B.fl.III,478 M.fr.V,70.
A. Mitchelli, F. v. M., fragm. V, 78 (1865) ... ... — S.A. — V. N.S.W. Q. — B.fl.III,478 M.fr.V,78.
A. exiguifolius, F. v. M., fragm. V, 67 (1865) ... ... W.A. — — — — — B.fl.III,479 M.fr.V,67.
A. Cassiniae, F. v. M., fragm. V, 68 (1865) ... ... W.A. — — — — — B.fl.III,479 M.fr.V,68.
A. cyanodiscus, F. v. M., fragm. V, 82 (1865) ... ... — — — N.S.W. — — B.fl.III,479 M.fr.V,82.
A. pimeleoides, Cunningham in De Cand. prodr. V, 268 (1836)... — S.A. — V. N.S.W. — — B.fl.III,479
A. iodochrous, F. v. M., fragm. V, 81 (1865) ... ... — — — V. N.S.W. — — B.fl.III,480 M.fr.II,110.
A. adenolasius, F. v. M., fragm. V, 67 (1865) ... ... W.A. — — — — — B.fl.III,480 M.fr.V,67.
A. caloocephalus, F. v. M., fragm. V, 70 (1865) ... ... W.A. — — — — — B.fl.III,480 M.fr.V,70.
A. calcareus; Olearia, F. v. M. in Benth. Fl. Austr. III, 481 (1866) — S.A. — — — — B.fl.III,481
A. magniflorus, F. v. M., fragm. V, 50 (1865) ... ... — S.A. — — — — B.fl.III,481 M.fr.V,50.
A. Muelleri; Eurybia, Sonder in Schlecht. Linn. XXV, 459 (1852) W.A. S.A. — — — — B.fl.III,481
A. Stuartii, F. v. M., fragm. V, 76 (1865) ... ... — S.A. — — — — B.fl.III,481 M.fr.I,202.
A. occurrens, Cunningham in De Cand. prodr. V, 269 (1836) ... — S.A. — — — — B.fl.III,481
A. glutescens, F. v. M., fragm. V, 77 (1865) ... ... — S.A. T. V. N.S.W. — — B.fl.III,482 M.fr.V,77.
A. orarius, F. v. M., fragm. V, 78 (1865) ... ... W.A. — — — — — B.fl.III,482 M.fr.V,67.
A. vernicosus, F. v. M., fragm. V, 67 (1865) ... ... W.A. — — — — — B.fl.III,482 M.fr.V,67.
A. teretifolius, F. v. M., fragm. V, 77 (1865) ... ... — S.A. — V. N.S.W. — — B.fl.III,483 M.fr.V,77.
A. Turczaninowii, F. v. M., fragm. V, 67 (1865) ... ... W.A. — — — — — B.fl.III,483 M.fr.V,67.
A. ellipticus, Cunningham in De Cand. prodr. V, 271 (1836) W.A. — — — — — B.fl.III,483 M.fr.I,16.
A. glandulosus, Labillard., Nov. Holl. pl. spec. II, 50, t. 197 (1806) S.A. T. V. N.S.W. — — B.fl.III,484
A. heleophilus, F. v. M., fragm. V, 66 (1865) ... ... W.A. — — — — — B.fl.III,484 M.fr.V,66.
A. muricatus, F. v. M., fragm. V, 66 (1865) ... ... W.A. — — — — — B.fl.III,484 M.fr.V,66.
A. Benthami, F.v.M.; Oleariastricta, Benth. Fl. Austr.III, 485(1866) — — — V. — — B.fl.III,485
A. Steetzii, F. v. M., fragm. V, 66 (1865) ... ... W.A. — — — — — B.fl.III,485 M.fr.V,66.
A. Cunninghami, F. v. M.; Eurybia, De Cand. prodr. V, 269 (1836) — — — N.S.W. — — B.fl.III,485 M.fr.V,66.
A. paucidentatus, F. v. M., fragm. V, 66 (1865) ... ... W.A. — — — — — B.fl.III,485 M.fr.V,66.
A. megalodontus, F. v. M., fragm. VIII, 244 (1874) ... W.A. S.A. — — — — B.fl.III,486 M.fr.VIII,244.
A. adenophorus, F. v. M., fragm. V, 78 (1865) ... — — — V. N.S.W. — — B.fl.III,486 M.fr.I,111.
A. homolepis, F. v. M., fragm. V, 65 (1865) ... W.A. — — — — — B.fl.III,486 M.fr.V,65.
A. xerophilus, F. v. M., fragm. V, 76 (1865) ... ... — — — — Q. — B.fl.III,486 M.fr.I,51,76&86.
A. Ferresii, F. v. M., fragm. V, 75 (1865) ... ... — S.A. — — — N.A. B.fl.III,487 M.fr.III,18.

A. exul, Lindley, Bot. Regist. XXV, App. XXIV (1830) ... W.A. S.A. — V. N.S.W. — — B.fl.III,487 M.fr.V,75.
A. argutus, R. Brown in Benth. Fl. Austr. III, 488 (1866) — — — — — — N.A. B.fl.III,488
A. Huegelii, F. v. M., fragm. V, 79 (1865)... ... ... W.A. S.A. T. V. N.S.W. — — B.fl.III,489 M.fr.V,79.
A. Balli, F. v. M., fragm. VIII, 143 (1874)... ... ... — — — — N.S.W. — — M.fr.VIII,143.
A. Mooneyi, F. v. M., fragm. VIII, 144 (1874) ... ... — — — — N.S.W. — — M.fr.VIII,144.
A. Celmisia, F. v. M., fragm. V, 84 (1865)... ... . — — — T. V. N.S.W. — — B.fl.III,489 M.fr.V,84.

VITTADINIA, Ach. Richard, Voy. d'Astrol. bot. 250 (1832). (Eurybiopsis, Microgyne.)
V. brachycomoides, F. v. M., fragm. V, 86 (1865) ... ... — — — — — — Q. N.A. B.fl.III,490 M.fr.V,86.
V. australis, A. Richard, Voy. d'Astrol. 251 (1832) ... ... W.A. S.A. T. V. N.S.W. Q. — B.fl.III,490 M.fr.XI,20.
V. scabra, De Candolle, prodr. V, 281 (1836) ... ... — — — — N.S.W. Q. — B.fl.III,491
V. macrorrhiza, A. Gray in Proc. Am. Acad. V, 118 (1863) ... ... — — — — — — N.A. B.fl.III,492

PODOCOMA, Cassini in Bull. de la Soc. philom. 137 (1817). (Ixiochlamys.)
P. cuneifolia, R. Brown, App. to Sturt, Centr. Austr. 17 (1849) — S.A. — — N.S.W. — N.A B.fl.III,493

ERIGERON, Linné, hort. Cliffort. 407 (1737). (Haplopappus partly.)
E. pappochromus, Labillard., Nov. Holl. pl. spec. II, 47, t. 193 (1806) — — T. V. N.S.W. — — B.fl.III,494
E. ambiguus, F. v. M. in Transact. phil. Inst. Vict. III, 58 (1858) — — — — — — Q. N.A. B.fl.III,494
E. sessilifolius, F. v. M., fragm. X, 100 (1851) ... ... — — — — — — — N.A. — M.fr.X,137.
E. minurioides, Bentham, Fl. Austr. III, 495 (1866) ... ... — — — — — — — — — M.fr.VIII,144.
E. conyzoides, F.v. M. in Transact. phil. Soc. Vict. I, 105 (1855) ... — — — V. N.S.W. — — B.fl.III,495

CONYZA, Lamarck, Encycl. méthod. II, 87, 89 (1786).
C. viscidula, Wallich, numer. list 3006 (1828) ... ... — — — — N.S.W. Q. — B.fl.III,496
C. Aegyptiaca, Aiton, hort. Kew. II, 183 (1789)... ... ... — — — — — — Q. N.A. B.fl.III,497

BLUMEA, De Candolle in Guillemin, Arch. de Bot. II, 514 (1833).
B. glandulosa, De Candolle in Wight, contrib. 14 (1834)... ... — — — — — Q. N.A B.fl.III,525
B. amplectens, De Candolle in Wight, contrib. 13 (1834) ... ... — — — — — — B.fl.III,525
B. integrifolia, De Candolle, prodr. V, 433 (1836)... ... ... — — — — — N.A. B.fl.III,525
B. diffusa, R. Brown in Benth. Fl. Austr. III, 525 (1866) ... ... — — — — — N.A. B.fl.III,525
B. hieracifolia, De Candolle in Wight, contrib. 15 (1834)... ... — — — — — Q. N.A B.fl.III,526
B. Cunninghamii, De Candolle, prodr. V, 435 (1836) ... ... — — — — — N.A. B.fl.III,526
B. acutata, De Cand. in Ann. du Mus. d'hist. nat. Par. III, 409(1834) — — — — Q. — — M.fr.IX,160.
B. lacera, De Candolle in Wight, contrib. 14 (1834) ... ... — — — — — Q. N.A B.fl.III,526

PLUCHEA, Cassini in Bull. de la Soc. philom. 31 (1817). (Spiropodium, Eyrea.)
P. Indica, Lessing in Schlecht. Linnaea VI, 150 (1831) ... ... — — — — — Q. N.A. B.fl.III,527 M.fr.X,101.
P. tetranthera, F. v. M., Rep. Babb. Exped. 12 (1860) ... ... — — — — — Q. N.A B.fl.III,529 M.fr.I,34.
P. baccharoides, F. v. M. in Benth. Fl. Austr. III, 528 (1866)... ... — — — — — Q. — B.fl.III,528
P. Eyrea, F. v. M., Rep. Babb. Exped. 11, 12 (1860) ... ... W.A. S.A. — — — Q. N.A. B.fl.III,528
P. squarrosa, Bentham, Fl. Austr. III, 529 (1866)... ... ... W.A. — — — — — B.fl.III,529
P. dentex, R. Brown in Benth. Fl. Austr. III, 529 (1866) ... ... — — — — — Q. — B.fl.III,529

PTERIGERON, De Candolle, prodr. V, 293 (1836). (Streptoglossa, Oliganthemum.)
P. decurrens, De Candolle, prodr. V, 293 (1836) ... ... — — — — — N.A. B.fl.III,531
P. liatroides, Bentham, Fl. Austr. III, 532 (1866)... ... ... W.A. S.A. — — — — B.fl.III,532
P. odorus, Bentham, Fl. Austr. III, 532 (1866) ... ... — — — — — Q. N.A. B.fl.III,532
P. macrocephalus, Bentham, Fl. Austr. III, 532 (1866) ... ... — — — — — N.A. B.fl.III,532
P. microglossus, Bentham, Fl. Austr. III, 532 (1866) ... ... — — — — — N.A. B.fl.III,532
P. dentatifolius, F. v. M., fragm. IX, 110 (1875) .. .. — S.A. — — — — — M.fr.X,119.
P. adscendens, Bentham, Fl. Austr. III, 533 (1866) ... ... — — — — — Q. N.A. B.fl.III,533
P. filifolius, Bentham, Fl. Austr. III, 533 (1866) ... ... — — — — — N.A. B.fl.III,533

THESPIDIUM, F. v. M. in Journ. of Landsborough's Exped. app. (1862).
T. basiflorum, F. v. M. in Journ. of Landsb. Exped. App. (1862) ... — — — — — N.A. B.fl.III,534 M.fr.XI,101.

COLEOCOMA, F. v. M. in Hook. Kew Misc. IX, 19 (1857).
C. Centaurea, F. v. M. in Hook. Kew Misc. IX, 19 (1857) ... ... — — — — — N.A. B.fl.III,533 M.fr.X,101.

EPALTES, Cassini in Bull. de la Soc. philom. 139 (1818). (Sphaeromorphaea partly, Ethuliopsis.)
E. Cunninghamii, Bentham, Fl. Austr. III, 530 (1866) ... ... — S.A. — V. N.S.W. — B.fl.III,530
E. australis, Lessing in Linnaea V, 148 (1831) ... ... — S.A. — V. N.S.W. Q. N.A. B.fl.III,530 M.fr.X,101.
E. pleiochaeta, F. v. M., fragm. X, 100 (1877) ... ... — — — — N.S.W. — — M.fr.X,100.
E. Harrisii, F. v. M., fragm. XI, 101 (1881) ... ... — — — — — Q. — M.fr.XI,101.

SPHAERANTHOS, Vaillant in act. Acad. Par. 289 (1719). (Sphaeranthus.)
S. hirtus, Willdenow, spec. plant. III, 2395 (1800) ... ... — — — — — Q. N.A. B.fl.III,521 M.fr.III,138.
S. microcephalus, Willdenow, spec. plant. III, 2395 (1800) ... ... — — — — — — N.A. B.fl.III,521 M.fr.III,138.

PTEROCAULON, Elliot, Sketch of the Bot. of S. Carolina and Georgia II, 323 (1824.). (Monenteles.)
P. verbascifolina, Bentham & J. Hooker, gen. pl. II, 204 (1873) — — — — — Q. N.A. B.fl.III,523
P. Billardieri, F. v. M., Papuan plants, 43 (1876)... ... ... — S.A. — — N.S.W. Q. N.A. B.fl.III,523 M.fr.XI,101.
P. sphacelatus, Bentham & J. Hooker, gen. plant. II, 94 (1873) — S.A. — — — Q. N.A. B.fl.III,523
P. glandulosus, Bentham & J. Hooker, gen. plant. II, 94 (1873) — — — — — Q. N.A. B.fl.III,524
P. sphaeranthoides, Bentham & J. Hooker, gen. plant. II, 94 (1873) — — — — — Q. N.A. B.fl.III,524

STUARTINA, Sonder in Schlechtendal's Linnaea XXV, 521 (1852).
S. Muelleri, Sonder in Schlecht. Linnaea XXV, 522 (1852) ... — S.A. — V. N.S.W. — — B.fl.III,557 M.fr.VIII,145;XI,27.

GNAPHALIUM, Linné, gen. plant. 250 (1737), from J. & C. Bauhin (1619). (Euchiton.)
G. luteo-album, Linné, spec. plant. 851 (1753) ... ... W.A. S.A. T. V. N.S.W. Q. N.A. B.fl.III,653 M.fr.V,150.
G. Japonicum, Thunberg, Fl. Japon. 311 (1784) ... ... W.A. S.A. T. V. — — B.fl.III,653 M.fr.V,150;XI,27.
G. alpigenum, F. v. M., second gen. Report 12 (1854) ... ... — — T. V. — — B.fl.III,654 M.fr.V,150.
G. purpureum, Linné, spec. plant. 854 (1753) ... ... — — — N.S.W. — — B.fl.III,655
G. Indicum, Linné, spec. plant. 852 (1753)... ... ... — — — N.S.W. Q. N.A. B.fl.III,655 M.fr.V,140.
G. indutum, J. Hooker in Lond. Journ. of Bot. VI, 121 (1847)... W.A. S.A. T. V. N.S.W. — — B.fl.III,655 M.fr.V,150.
G. Traversii, J. Hooker, Handb. N. Zeal. Fl. 154 ... ... — — — V. N.S.W. — — B.fl.III,656

ANTENNARIA, Gaertner, de fruct. II, 410, t. 167 (1791).   (Raoulia partly.)
A. Planchoni, F.v.M.; Gnaphalium, Hook. Fl. Tasm. I, 217, t. 62 (1860)   —   —   —   —   —   —   —
A. unicops, F. v. M. in Transact. phil. Soc. Vict. t. 105 (1854)..   —   —   —   V.   N.S.W.   —   —   B.fl.III,652   M.fr.VIII,145.

LEONTOPODIUM, R. Brown in Transact. Linn. Soc. XII, 124 (1817).   (Raoulia partly).
L. Catipes, F. v. M. in Papers of the Roy. Soc. of Tasm. 44 (1882)   —   T.   V.   N.S.W.   —   —   B.fl.III,651
L. Meredithae; Antennaria, F.v.M. in Pap. Roy. Soc.of Tasm. 15 (1870)   —   —   T.   —   —   —   —

PTERYGOPAPPUS, J. Hooker in Lond. Journ. of Bot. VII, 120 (1847).
P. Lawrencii, J. Hooker in Lond. Journ. VII, 120 (1847)   ...   —   —   T.   —   —   —   —   B.fl.III,656

PODOTHECA, Cassini in Diction. XXIII, 561 (1822).   (Podospermum, Phaenopoda, Lophoclinium.)
P. gnaphalioides, Graham in Bot. Mag. t. 3920 (1842)   ...   W.A.   —   —   —   —   —   —   B.fl.III,001
P. angustifolia, Lessing, synops. gen. compos. 273 (1832)   ..   W.A.   S.A.   T.   V.   N.S.W.   —   —   B.fl.III,001   M.fr.XI,20.
P. pygmaea, A. Gray in Hook. Kew Misc. IV, 227 (1852)   ..   W.A.   —   —   —   —   —   —   B.fl.III,002
P. chrysantha, Bentham, Fl. Austr. III, 602 (1866)   ...   W.A.   —   —   —   —   —   —   B.fl.III,602
P. fuscescens, Bentham, Fl. Austr. III, 602 (1866)   ...   W.A.   —   —   —   —   —   —   B.fl.III,602

IXIOLAENA, Bentham in Hueg. enum. 66 (1837).
I. brevicompta, F. v. M., fragm. I, 53 (1858)   ...   —   —   —   —   N.S.W.   Q.   —   B.fl.III,597   M.fr.I,53.
I. leptolepis, Bentham, Fl. Austr. III, 597 (1866)...   ...   ...   —   S.A.   —   V.   N.S.W.   —   N.A   B.fl.III,597
I. supina, F. v. M. in Transact. Vict. Inst. 37 (1855)   ...   —   S.A.   T.   —   —   —   —   B.fl.III,598
I. tomentosa, Sonder & F. v. M. in Linnaea XXV, 504 (1852)   ···   —   S.A.   —   V.   N.S.W.   —   —   B.fl.III,598   M.fr.I,53;XI,27.
I. viscosa, Bentham in Hueg. enum. 66 (1837)   ...   W.A.   —   —   —   —   —   —   B.fl.III,598

PODOLEPIS, Labillardière, Nov. Holl. pl. specim. II, 57, t. 28 (1806).   (Scalia, Panaetia, Scaliopsis, Siemssenia, Stylolepis, Rutidochlamys.)
P. rhytidochlamys, F. v. M., fragm. IV, 79 (1864)   ...   —   —   —   —   N.S.W.   Q.   —   B.fl.III,603   M.fr.IV,79.
P. longipedata, Cunningham in De Cand. prodr. VI, 163 (1837)...   —   —   —   V.   N.S.W.   Q.   —   B.fl.III,604   M.fr.I,112.
P. acuminata, R. Brown in Ait. hort. Kew. ed. sec. V, 82 (1813)   —   S.A.   T.   V.   N.S.W.   Q.   —   B.fl.III,604   M.fr.XI,27.
P. canescens, Cunningham in De Cand. prodr. VI, 163 (1837)   ·.   —   S.A.   —   V.   N.S.W.   Q.   —   B.fl.III,605
P. aristata, Bentham in Hueg. enum. 66 (1837)   ...   W.A.   —   —   —   —   —   —   B.fl.III,605
P. pallida, Turczaninow in Bull. de Mosc. XXIV, 78 (1851)   ...   W.A.   —   —   —   —   —   N.A   B.fl.III,605
P. nutans, Steetz in Lehm. pl. Preiss. I, 464 (1845)   ...   W.A.   —   —   —   —   —   —   B.fl.III,605
P. gracilis, Graham in Edinb. new phil. journ. V, 379 (1828)   ...   W.A.   —   —   —   —   —   —   B.fl.III,006
P. rugata, Labillardière, Nov. Holl. pl. spec. II, 57, t. 208 (1806)   W.A.   S.A.   —   V.   —   —   —   B.fl.III,000
P. Lessoni, Bentham, Fl. Austr. III, 606 (1866)   ...   W.A.   S.A.   —   V.   N.S.W.   —   —   B.fl.III,006   M.fr.XI,27.
P. Siemsscnia, F. v. M. in Benth. Fl. Austr. III, 607 (1866)   ...   W.A.   —   —   —   —   —   —   B.fl.III,607
P. microcephala, Bentham, Fl. Austr. 607 (1866)...   ...   W.A.   —   —   —   —   —   —   B.fl.III,607

ATHRIXIA, Ker in Bot. Regist. VIII, 681 (1823).   (Astorides, Chrysodiscus, Trichostegia.)
A. australis, Steetz in Lehm. pl. Preiss. I, 482 (1845)   ...   ...   W.A.   —   —   —   —   —   —   B.fl.III,599
A. gracilis, Bentham, Fl. Austr. III, 599 (1866)   ...   ...   W.A.   —   —   —   —   —   —   B.fl.III,599
A. multiceps, Bentham, Fl. Austr. III, 599 (1806)   ...   W.A.   —   —   —   —   —   —   B.fl.III,599
A. stricta, Bentham, Fl. Austr. III, 600 (1866)   ...   W.A.   —   —   —   —   —   —   B.fl.III,600
A. tenella, Bentham, Fl. Austr. III, 600 (1866)   ...   W.A.   S.A.   —   V.   N.S.W.   —   —   B.fl.III,600
A. chaetopoda, F. v. M., fragm. X, 56 (1876)   ...   ...   —   —   —   —   —   —   —   M.fr.X,56.

LEPTORRHYNCHOS, Lessing, syn. composit. 273 (1832).   (Rhytidanthe.)
L. squamatus, Lessing, syn. comp. 273 (1832)   ...   ...   —   S.A.   T.   V.   N.S.W.   —   —   B.fl.III,608   M.fr.I,52.
L. panactioides, Bentham, Fl. Austr. III, 609 (1866)   ...   —   S.A.   —   V.   N.S.W.   —   —   B.fl.III,609
L. tonuifolius, F. v. M., fragm. I, 52 (1858)   ...   —   S.A.   —   V.   —   —   —   B.fl.III,609
L. pulchellus, F. v. M. in Linnaea XXV, 500 (1852)   ...   —   S.A.   —   V.   N.S.W.   —   —   B.fl.III,610   M.fr.I,53;XI,27.
L. Bellyi, F. v. M., fragm. X, 101 (1877)...   ...   ...   —   —   —   —   —   Q.   —   M.fr.X,101.
L. elongatus, De Candolle, prodr. VI, 160 (1837)   ...   W.A.   S.A.   T.   V.   N.S.W.   —   —   B.fl.III,610   M.fr.XI,27,28.
L. medius, Cunningham in De Cand. prodr. VI, 160 (1837)   W.A.   S.A.   —   V.   N.S.W.   —   —   —   M.fr.XI,95.
L. Waitzia, Sonder in Schlecht. Linnaea XXV, 501 (1852)   ...   —   S.A.   —   V.   N.S.W.   —   —   B.fl.III,610
L. nitidulus, De Candolle, prodr. VI, 160 (1837)   ...   —   —   T.   V.   N.S.W.   —   —   B.fl.III,611

WAITZIA, Wendland, Collect. pl. II, 13, t. 42 (1808).   (Viraya, Morna, Pterochaeta.)
W. corymbosa, Wendland, Coll. pl. II, 13, t. 42 (1808)...   ..   W.A.   S.A.   —   V.   N.S.W.   —   —   B.fl.III,035
W. aurea, Steetz in Lehm. pl. Preiss. I, 452 (1845)   ...   W.A.   —   —   —   —   —   —   B.fl.III,636
W. nivea, Bentham, Fl. Austr. III, 636 (1866)   ...   ...   W.A.   —   —   —   —   —   —   B.fl.III,636
W. Steetziana, Lehmann, pl. Preiss. I, 454 (1845)   ...   W.A.   —   —   —   —   —   —   B.fl.III,036
W. podolepis, Steetz in Lehm. pl. Preiss. I, 450 (1845)   ...   W.A.   —   —   —   —   —   —   B.fl.III,637
W. paniculata, F. v. M. in Benth. Fl. Austr. III, 637 (1866)   ...   W.A.   —   —   —   —   —   —   B.fl.III,637

HELIPTERUM, De Candolle, prodr. VI, 211 (1837).   (Argyrocome, 1822, Pteropogon, Rodanthe, Xyridanthe, Anisolepis, Hyalosperma, Triptilodiscus, Acroclinium, Monencyanthes, Dimorpholepis, Duttonia, Cassiniola.)
H. Manglesii, F. v. M., fragm. V, 200 (1866)   ...   ...   —   —   —   —   —   —   B.fl.III,040   M.fr.X,108.
H. Margaritae, F. v. M., fragm. XI, 48 (1878)   ...   ...   —   —   —   —   —   N.A.   —   M.fr.XI,48.
H. roseum, Bentham, Fl. Austr. III, 640 (1886)   ...   W.A.   S.A.   —   —   —   —   —   B.fl.III,640   M.fr.X,108.
H. anthemoides, De Candolle, prodr. VI, 216 (1837)   ...   —   S.A.   T.   V.   N.S.W.   Q.   —   B.fl.III,641   M.fr.X,108.
H. polygalifolium, De Candolle, prodr. VI, 216 (1837)   ...   —   S.A.   T.   V.   N.S.W.   Q.   —   B.fl.III,641   M.fr.X,107;108.
H. rubellum, Bentham, Fl. Austr. III, 641 (1866)   ...   W.A.   —   —   —   —   —   —   B.fl.III,641   M.fr.X,108.
H. chlorocephalum, Bentham, Fl. Austr. 641 (1866)   ...   W.A.   —   —   —   —   —   —   B.fl.III,641
H. floribundum, De Candolle, prodr. VI, 217 (1837)   ...   W.A.   —   —   —   —   —   —   B.fl.III,642   M.fr.X,108.
H. Pyrethrum, Bentham, Fl. Austr. III, 642 (1866)   ...   W.A.   —   —   —   —   —   —   B.fl.III,642   M.fr.X,10.
H. heteranthum, Turczan. in Bull. de Mosc. XXIV, 198 (1851)   W.A.   S.A.   —   V.   N.S.W.   Q.   N.A.   B.fl.III,642   M.fr.III,137;X,108.
H. Kendallii, F. v. M., fragm. VIII, 108 (1874)   ...   W.A.   —   —   —   —   —   —   B.fl.III,642   M.fr.VIII,169;X,107.
H. stipitatum, F. v. M. in Bentham, Fl. Austr. III, 643 (1866)   —   S.A.   —   —   —   —   —   B.fl.III,643   M.fr.III,133;X,100.
H. incanum, De Candolle, prodr. VI, 215 (1837)..   ...   —   S.A.   T.   V.   N.S.W.   Q.   —   B.fl.III,643   [110.
H. Cotula, De Candolle, prodr. VI, 215 (1837)   ...   W.A.   S.A.   T.   V.   N.S.W.   Q.   —   B.fl.III,644   M.fr.II,157;III,134;X,
H. hyalospermum, F. v. M. in Benth., Fl. Austr. III, 644 (1866)   W.A.   S.A.   —   V.   N.S.W.   Q.   —   B.fl.III,644
H. condensatum, F. v. M., fragm. III, 136 (1863)   W.A.   S.A.   —   —   —   —   —   B.fl.III,644   M.fr.III,136.
H. polyphyllum, F. v. M., fragm. I, 35 (1858)   ...   ..   W.A.   —   —   —   —   N.A.   B.fl.III,645   M.fr.III,136.
H. polyphyllum, F. v. M., fragm. I, 35 (1858)   ...   —   —   —   —   N.S.W.   Q.   —   B.fl.III,645   M.fr.I,35.

H. Humboldtianum, De Candolle, prodr. VI, 216 (1866) ... W.A. — — — — — N.A. B.fl.III,645 M.fr.III,135;X,109.
H. Haigii, F. v. M., fragm. X, 107 (1877) ... ... W.A. S.A. — — — — — — M.fr.X,107. [20.
H. tenellum, Turczaninow in Bull. de Mosc. XXIV, 198 (1851) W.A. S.A. — — — — — B.fl.III,646 M.fr.III,135;X,109;XI,
H. strictum, Bentham, Fl. Austr. III, 646 (1866) ... ... W.A. S.A. — — N.S.W. — — B.fl.III,646 M.fr.X,100.
H. corymbiflorum, Schlechtendal, Linnaea XXI, 448 (1848) ... — S.A. — V. N.S.W. Q. — B.fl.III,647 M.fr.X,100.
H. pygmaeum, Bentham, Fl. Austr. III, 647 (1866) ... ... W.A. S.A. — V. N.S.W. — — B.fl.III,647 M.fr.X,109.
H. apicatum, F. v. M. in Bentham, Fl. Austr. III, 647 (1866)... W.A. — — — — — — B.fl.III,647 M.fr.X.109.
H. Charsleyae, F. v. M., fragm. VIII, 168 (1874)... ... ... W.A. S.A. — — — — — — M.fr.VIII,168;XI,134.
H. moschatum, Bentham, Fl. Austr. III, 648 (1866) ... ... — S.A. — V. N.S.W. Q. — B.fl.III,648 M.fr.X,107.
H. Tietkensii, F. v. M., fragm. VIII, 227 (1874)... ... ... — S.A. — — — — — M.fr.VIII,227.
H. pterochaetum, Bentham, Fl. Austr. III, 648 (1866) ... ... — S.A. — — N.S.W. — — B.fl.III,648 M.fr.X,109.
H. polycephalum, Bentham, Fl. Austr. III. 649 (1866) ... ... W.A. — — — — — — B.fl.III,649 M.fr.III,139;X,109.
H. corymbosum, Bentham, Fl. Austr. III, 649 (1866) ... ... W.A. — — — — — — B.fl.III,649 M.fr.V,200.
H. laeve, Bentham, Fl. Austr. III, 649 (1866) ... ... W.A. — — — — — — B.fl.III,649 M.fr.V,200.
H. exiguum, F. v. M. in Transact. Vict. Inst. 39 (1855)... ... W.A. S.A. T. V. N.S.W. — — B.fl.III,649 M.fr.X,109.
H. Dimorpholepis, Bentham, Fl. Austr. III, 650 (1866)... ... W.A. S.A. — V. N.S.W. — — B.fl.III,650 M.fr.X,100;XI,27.

HELICHRYSUM, Vaillant in Act. Acad. Par. 290 (1719), from Theophrastos and Dioscorides. (Helichrysum, Schoenia, Petalolepis, Faustula, Ozothamnus, Swammerdamia, Lawrencella, Argyrophanes, Chrysocephalum, Conanthodium, Xanthochrysum, Argyroglottis, Acanthocladium, Raoulia partly.)

H. Cassinianum, Gaudichaud in Freyc. Voy. Bot. 466, t. 87 (1826) W.A. S.A. — — — — — B.fl.III,616
H. Lawrencella, F. v. M. in Benth. Fl. Austr. III, 616 (1866)... W.A. S.A. — — — — — B.fl.III,616
H. subullifolium, F. v. M., fragm. III, 134 (1863).. ... — W.A. — — — — — B.fl.III,616 M.fr.III,134.
H. filifolium, F. v. M., fragm. III, 134 (1863) ... ... — W.A. — — — — — B.fl.III,617 M.fr.III,134.
H. semifertile, F. v. M., Rep. Babb. Exped. 14 (1858) ... ... — S.A. — N.S.W. — — B.fl.III,617
H. Baxteri, Cunningham in De Cand. prodr. VI, 193 (1837) ... — S.A. V. — N.S.W. — — B.fl.III,617
H. Ayersii, F. v. M., fragm. VIII, 167 (1874) ... ... — S.A. — — — — — M.fr.IX,195.
H. rutidolepis, De Candolle, prodr. VI, 194 (1837) ... ... — S.A. — V. N.S.W. — — B.fl.III,618
H. scorpioides, Labillardière, Nov. Holl. pl. spec. II, 45, t. 191 (1806) — S.A. T. V. N.S.W. Q. — B.fl.III,618 M.fr.VIII,167.
H. oligochaetum, F. v. M., fragm. VI, 235 (1868)... ... ... W.A. — — — — — N.A. — M.fr.VI,235.
H. Tepperi, F. v. M. in Wing's South. Science Record, II, 1 (1882) — S.A. — — — — — —
H. obtusifolium, Sonder & F. v. M. in Linnaea XXV, 515 (1852) W.A. S.A. — V. N.S.W. — — B.fl.III,619
H. Calvertianum, F. v. M., fragm. X, 108 (1877)... ... ... — — — N.S.W. — — — M.fr.XI,130.
H. Spiceri, F. v. M., fragm. XI, 47 (1878) ... ... — T. — — — — M.fr.XI,47.
H. dealbatum, Labillardière, Nov. Holl. pl. spec. II, 45, t. 190 (1806) — — T. — — — — B.fl.III,619
H. pumilum, J. Hooker, Fl. Tasman. I, 213, t. 60 (1860) ... — — T. — — — — B.fl.III,619
H. lucidum, Henckel, adumbr. pl. hort. Hal. 5 (1806) ... ... W.A. S.A. T. V. N.S.W. Q. N.A. B.fl.III,620 M.fr.XI,48.
H. elatum, Cunningham in De Cand. prodr. VI, 193 (1837) ... — — — V. N.S.W. Q. — B.fl.III,621
H. adenophorum, F. v. M. in Transact. Vict. Inst. 38 (1855) ... — S.A. — — — — — B.fl.III,622
H. leucopsidium, De Candolle, prodr. VI, 193 (1837) ... ... W.A. S.A. T. V. N.S.W. — — B.fl.III,622 M.fr.VI,255.
H. Blandowskianum, Steetz in Schlecht. Linnaea XXV, 512 (1852) — S.A. — V. N.S.W. — — B.fl.III,622
H. oxylepis, F. v. M., fragm. I, 35 (1859) ... ... — — — N.S.W. Q. — B.fl.III,623 M.fr.I,35.
H. collinum, De Candolle, prodr. VI, 190 (1837) ... ... — — — N.S.W. Q. — B.fl.III,623
H. rupicola, De Candolle, prodr. VI, 190 (1837) ... ... — — — — Q. — B.fl.III,623
H. podolepideum, F. v. M., Rep. Babb. Exped. 13 (1858) ... — S.A. — V. N.S.W. — — B.fl.III,624
H. Gilesii, F. v. M., fragm. X, 85 (1876) ... ... W.A. — — — — — N.A. — M.fr.X,86.
H. ambiguum, Turczaninow in Bull. de Mosc. 195 (1851) ... W.A. S.A. — — N.S.W. — — B.fl.III,600
H. apiculatum, De Candolle, prodr. VI, 195 (1837) ... ... W.A. S.A. T. V. N.S.W. Q. N.A. B.fl.III,624
H. semipapposum, De Candolle, prodr. VI, 195 (1837) ... ... W.A. S.A. T. V. N.S.W. — — B.fl.III,625
H. Dockeri, F. v. M. in Benth. Fl. Austr. III, 626 (1866) ... — S.A. — — N.S.W. — — B.fl.III,625 M.fr.II,150.
H. Agyroglottis, Bentham, Fl. Austr. III, 626 (1866) ... ... W.A. — — — — — B.fl.III,626
H. ramosum, De Candolle, prodr. VI, 181 (1837) ... ... W.A. — — — — — B.fl.III,626
H. cordatum, De Candolle, prodr. VI, 180 (1837)... ... ... W.A. — — — — — B.fl.III,627
H. obovatum, De Candolle, prodr. VI, 180 (1837)... ... ... — — — N.S.W. — — B.fl.III,627 M.fr.II,89.
H. Bidwillii, Beutham, Fl. Austr. III, 627 (1866)... ... ... — — — N.S.W. Q. — B.fl.III,627
H. Beckleri, F. v. M. in Benth. Fl. Austr. III, 627 (1866) ... — — — N.S.W. Q. — B.fl.III,627 M.fr.I,183.
H. diotophyllum, F. v. M., fragm. V, 150 (1865) ... ... — — — N.S.W. Q. — B.fl.III,628 M.fr.VIII,46.
H. Thomsoni, F. v. M., fragm. VIII, 45 (1873) ... ... — S.A. — — — — — M.fr.VIII,45.
H. dicosmifolium, Don in Steud. Nom. Bot. ed. sec. 738 (1841) ... — — — N.S.W. Q. — B.fl.III,628
H. retusum, F. v. M., fragm. VIII, 46 (1873) ... ... — S.A. — V. N.S.W. — — M.fr.VIII,46.
H. decurrens, F. v. M., fragm. VIII, 46 (1873) ... ... — S.A. — V. N.S.W. — — M.fr.VIII,46.
H. Cunninghamii, Bentham, Fl. Austr. III, 629 (1866) ... ... — — — N.S.W. — — B.fl.III,629
H. reticulatum, Lessing in Steud. Nom. Bot. ed. sec. 740 (1841) — — — — — — B.fl.III,629
H. clavereum, F. v. M. in Benth. Fl. Austr. III, 629 (1866) ... S.A. — T. — N.S.W. — — B.fl.III,629 M.fr.V,199.
H. bracteolatum, Bentham, Fl. Austr. III, 630 (1866) ... ... — — T. — — — — B.fl.III,630
H. Kempei, F. v. M. in Melb. Chemist, Jan. (1882)]... ... ... S.A. — — — — — —
H. cassinioides, Bentham, Fl. Austr. III, 630 (1866) ... ... — — — — — Q. B.fl.III,630 M.fr.V,199.
H. Gunnii, F. v. M. in Benth. Fl. Austr. III, 630 (1866) ... ... — — T. — — — — B.fl.III,630 M.fr.V,199.
H. rosmarinifolium, Lessing, synops. gen. comp. 274 (1832) ... — — T. — — — — B.fl.III,631 M.fr.V,199.
H. ledifolium, F. v. M. in Benth. Fl. Austr. III, 630 (1866) ... — — T. — — — — B.fl.III,630 M.fr.V,199.
H. ferrugineum, Lessing, synops. gen. comp. 274 (1832) ... ... S.A. — T. V. N.S.W. — — B.fl.III,631
H. antennarium, F. v. M. in Benth. Fl. Austr. III, 632 (1866)... — — T. — — — — B.fl.III,632 M.fr.V,199.
H. obcordatum, F. v. M. in Benth. Fl. Austr. III, 632 (1866) ... — — T. — — — — B.fl.III,632 M.fr.V,199.
H. Backhousii, F. v. M. in Benth. Fl. Austr. III, 632 (1866) ... — — T. — — — — B.fl.III,632 M.fr.V,199.
H. cuneifolium, F. v. M. in Benth. Fl. Austr. III, 632 (1866) ... — — T. — — — — B.fl.III,632 M.fr.V,199.
H. baccharoides, F. v. M. in Benth. Fl. Austr. III, 633 (1866)... — — — V. — — — B.fl.III,633 M.fr.V,199.
H. lepidophyllum, F. v. M. in Benth. Fl. Austr. III, 633 (1866) W.A. — — — — — B.fl.III,633 M.fr.V,199.
H. scutellifolium, Bentham, Fl. Austr. III, 633 (1866) ... ... — — T. — — — — B.fl.III,633 M.fr.V,199.
H. pholidotum, F. v. M. in Benth. Fl. Austr. III, 634 (1866) ... S.A. — T. — — — — B.fl.III,634 M.fr.II,131;V,109.
H. lycopodioides, Bentham, Fl. Austr. III, 634 (1866) ... ... — — T. — — — — B.fl.III,634 M.fr.V,199.
H. selaginoides, F. v. M. in Benth. Fl. Austr. III, 634 (1866) ... — — T. — — — — B.fl.III,634 M.fr.V,199.

CASSINIA, R. Brown in Transact. Linn. Soc. XII, 126 (1817).   (Non zoologorum, Apolochlamys, Achromolaena, Chromochiton.)
C. leptocephala, F. v. M., fragm. III, 138 (1863) .. .. — — — — N.S.W. — — B.fl.III,585   M.fr.III,138.
C. compacta, F. v. M., fragm. I, 18 (1858)... ... ... — — — — N.S.W. Q. — B.fl.III,585   M.fr.I,18.
C. denticulata, R. Brown in Transact. Linn. Soc. XII, 127 (1817) — — — — N.S.W. — — B.fl.III,586
C. longifolia, R. Brown in Transact. Linn. Soc. XII, 127 (1817) — — — T. V. N.S.W. Q. — B.fl.III,586
C. aurea, R. Brown in Transact. Linn. Soc. XII, 127 (1817) ... — — — — N.S.W. — — B.fl.III,586
C. aculeata, R. Brown in Transact. Linn. Soc. XII, 127 (1817)... — — S.A. T. V. N.S.W. — — B.fl.III,586
C. laevis, R. Brown in Transact. Linn. Soc. XII, 128 (1817) ... — S.A. — — N.S.W. Q. — B.fl III,586
C. tenuifolia, Bentham, Fl. Austr. III, 585 (1866) ... ... — — — — N.S.W. — — B.fl.III,587
C. quinquefaria, R. Brown in Transact. Linn. Soc. XII, 128 (1817) ... — — — — N.S.W. Q. — B.fl.III,587
C. arcuata, R. Brown in Transact. Linn. Soc. XII, 128 (1817) ... W.A. S.A. — V. N.S.W. — — B.fl.III,587
C. subtropica, F. v. M., fragm. I, 17 (1858) ... ... — — — — N.S.W. Q. — B.fl.III,588   M.fr.I,17.
C. Theodori, F. v. M., fragm. V, 148 (1866) ... ... — — — — N.S.W. — — B.fl.III,588   M.fr.V,148.
C. spectabilis, R. Brown in Transact. Linn. Soc. XII, 128 (1817) — S.A. T. V. — — B.fl.III,588

HUMEA, Smith, Exot. bot. I, t. 1 (1804).   (Haeckeria, Calomeria.)
H. elegans, Smith, Exot. bot. I, t. 1 (1804) ... ... — — — V. N.S.W. — — B.fl.III,589   M.fr.I,17.
H. cassiniacea, F. v. M., fragm. I, 17 (1858) ... ... — S.A. — — — — B.fl.III,589   M.fr.I,17.
H. punctulata, F. v. M., fragm. III, 137 (1863) ... ... — S.A. — — — — B.fl.III,589   M.fr.III,137.
H. ozothamnoides, F. v. M., fragm. I, 17 (1858) ... ... — — — V. N.S.W. — — B.fl III,590   M.fr.VIII,142.
H. squamata, F. v. M., fragm. XI, 86 (1860) ... ... — S.A. — V. — — — M.fr.XI,86.

ACOMIS, F. v. M., fragm. II, 89 (1860).
A. Rutidosis, F. v. M., fragm. II, 89 (1860) ... ... — — — — N.S.W. — — B.fl.III,591   M.fr.II,89.
A. macra, F. v. M., fragm. IV, 145 (1864)... ... ... — — — — — Q. — B.fl.III,591   M.fr.IV,145.

RUTIDOSIS, De Candolle, prodr. VI, 158 (1837).   (Pumilo, Actinopappus, Lepidocoma.)
R. leiolepis, F. v. M. in Transact. Vict. Inst. 131 (1855) ... — — — — — — — B.fl.III,593
R. leptorrhynchoides, F. v. M., fragm. V, 148 (1866) ... — — — V. N.S.W. — — B.fl.III,593   M.fr.V,148.
R. Brownii, Bentham, Fl. Austr. III, 594 (1866) ... ... — — — — — Q. N.A. B.fl.III,594
R. helichrysoides, De Candolle, prodr. VI, 159 (1837) ... — S.A. — V. N.S.W. — — B.fl.III,594
R. leucantha, F. v. M., fragm. I, 35 (1858) ... ... — — — — — Q. — B.fl.III,594   M.fr.I,35.
R. Murchisonii, F. v. M., fragm. I, 34 (1858) ... ... — — — — — Q. — B.fl.III,594   M.fr.I,34.
R. Pumilo, Bentham, Fl. Austr. III, 595 (1866) ... ... W.A. S.A. T. V. N.S.W. — — B.fl.III,595

PITHOCARPA, Lindley, Bot. Regist. XXV, App. XXIII (1839).
P. corymbosa, Lindley, Bot. Regist. XXV, App. XXIII (1839) W.A. — — — — — — B.fl.III,590

AMMOBIUM, R. Brown in Bot. Magaz. t. 2459 (1824).
A. alatum, R. Brown in Bot. Magaz. t. 2459 (1824) ... — — — — N.S.W. Q. — B.fl.III,583
A. craspedioides, Bentham, Fl. Austr. III, 584 (1866) ... — — — — N.S.W. — — B.fl.III,584

IXODIA, R. Brown in Ait. hort. Kew. sec. ed. IV, 17 (1812).
I. achilleoides, R. Brown in Ait. hort. Kew. sec. ed. IV, 17 (1812) — S.A. — V. — — — B.fl.III,583

MILLOTIA, Cassini in Ann. sc. nat. XVII, 416 (1829).
M. tenuifolia, Cassini in Ann. sc. nat. XVII, 416 (1820)... — W.A. S.A. T. V. N.S.W. — — B.fl.III,596
M. Greevesii, F. v. M., fragm. III, 18, t. 19 (1863) ... — S.A. — — N.S.W. — — B.fl.III,596   M.fr.III,18.
M. Kempel, F. v. M. in Wing's South. Sc. Record II, 1 (1882)... — S.A. — — — — —

TOXANTHUS, Turczaninow in Bull. de Mosc. XXIV, 177 (1851).   (Anthocrastes, Scyphocoronis.)
T. perpusillus, Turczaninow in Bull. Mosc. XXIV, 177 (1851) ... W.A. S.A. — V. N.S.W. — — B.fl.III,592   M.fr.XI,49.
T. Muelleri, Bentham, Fl. Austr. III, 592 (1866) ... W.A. S.A. — — — — — B.fl.III,592   M.fr.XI,49.
T. major, Turczaninow in Bull. de Mosc. XXIV, 64 (1851) W.A. — — — — — — B.fl.III,592   M.fr.XI,49.

QUINETIA, Cassini in Diction. sc. nat. LX, 570 (1830).
Q. Urvillei, Cassini in Diction. LX, 579 (1830) ... ... W.A. S.A. — — — — — B.fl.III,593

PHACELOTHRIX, F. v. M., fragm. XI, 49 (1878).
P. cladochaeta, F. v. M., fragm. XI, 49 (1878) ... ... — — — — — Q. — — M.fr.V,199;XI,49.

ERIOCHLAMYS, Sonder and F. v. M. in Schlecht. Linnaea XXV, 488 (1852).
E. Behrii, Sonder & F. v. M. in Linnaea XXV, 488 (1852) ... — S.A. — V. — — B.fl.III,591

MYRIOCEPHALUS, Bentham in Hueg. enum. 61 (1837).   (Hyalolepis, Antheidosorus, Gilberta, Lamprochlaena, Klachopappus,
Polycalymma; some readily to be restituted.)
M. rhizocephalus, Bentham, Fl. Austr. III, 557 (1866)... ... W.A. S.A. — V. N.S.W. — — B.fl.III,557   M.fr.III,155.
M. nudus, A. Gray in Hook. Kew Misc. III, 174 (1851) ... ... W.A. — — — — — — B.fl.III,558   M.fr.III,157.
M. appendiculatus, Bentham in Hueg. enum. 61 (1837) ... W.A. — — — — — — B.fl.III,558   [145.
M. Rudallii, F. v. M., fragm. III, 157 (1863) ... ... — W.A. — — — — — B.fl.III,558   M.fr.III,157;VI,235;VII,
M. gracilis, Bentham, Fl. Austr. III, 559 (1866) ... ... — W.A. — — — — — B.fl.III,559   M.fr.VIII,169.
M. helichrysoides, A. Gray in Hook. Kew Misc. III, 175 (1851) W.A. — — — — — — B.fl.III,559
M. suffruticosus, Bentham, Fl. Austr. III, 559 (1866) ... W.A. — — — — — — B.fl.III,559
M. Stuartii, Bentham, Fl. Austr. III, 560 (1866) ... ... — S.A. — V. N.S.W. Q. — B.fl.III,560
M. Guerinae, F. v. M., fragm. VIII, 169 (1874) ... ... W.A. — — — — — — • M.fr.VIII,169.

ANGIANTHUS, Wendland, Collect. pl. II, 31 (1808).   (Siloxerus, Styloncerus, Ogcerostylus, Cylindrosorus, Phyllocalymma, Skirro-
phorus, Chrysocoryne, Eriocladium, Pogonolepis, Piptostemma, Epitriche, Gamozygis, Cephalosorus partly, Hyalochlamys,
Dithyrostegia, Pleuropappus; some readily to be restituted.)
A. humifusus, Bentham, Fl. Austr. III, 563 (1866) ... W.A. — — — — — — B.fl.III,563
A. tomentosus, Wendland, Collect. pl. II, 31, t. 48 (1809) W.A. S.A. — V. N.S.W. — N.A. B.fl.III,563
A. pleuropappus, Bentham, Fl. Austr. III, 563 (1866) ... — S.A. — — — — — B.fl.III,563
A. brachypappus, F. v. M. in Trans. phil. Soc. Vict. I, 44 (1854) — S.A. — V. N.S.W. — — B.fl.III,563
A. myosuroides, Bentham, Fl. Austr. III, 563 (1866) ... W.A. — — — — — — B.fl.III,563
A. tenellus, Bentham, Fl. Austr. III, 564 (1866) ... ... W.A. S.A. — — — — — B.fl.III,564
A. pusillus, Bentham, Fl. Austr. III, 564 (1866) ... ... W.A. S.A. — V. N.S.W. — — B.fl.III,564
A. Milnei, Bentham, Fl. Austr. III, 564 (1866) ... ... — S.A. — V. N.S.W. — — B.fl.III,564
A. Cunninghamii, Bentham, Fl. Austr. III, 565 (1866) ... W.A. — — — — — — B.fl.III,565
A. phyllocephalus, Bentham, Fl. Austr. III, 565 (1866) ... W.A. — — — — — — B.fl.III,565   M.fr.III,159.

A. micropoides, Bentham, Fl. Austr. III, 565 (1866) ... ... W.A. — — — — — — — B.fl.III,565
A. microcephalus, Bentham, Fl. Austr. III, 566 (1866) ... ... W.A. — — — — — — — B.fl.III,566 M.fr.III,158.
A. platycephalus, Bentham, Fl. Austr. III, 566 (1866) ... ... W.A. — — — — — — — B.fl.III,566
A. Drummondii, Bentham, Fl. Austr. III, 566 (1866) ... ... W.A. — — — — — — — B.fl.III,566
A. Preissianus, Bentham, Fl. Austr. III, 566 (1866) ... W.A. S.A. T. V. N.S.W. — B.fl.III,566
A. pygmaeus, Bentham, Fl. Austr. III, 567 (1866) ... ... W.A. — — — — — — — B.fl.III,567
A. globifer, Bentham, Fl. Austr. III, 567 (1866) ... ... W.A. — — — — — — — B.fl.III,567
A. demissus, Bentham, Fl. Austr. III, 567 (1866)... ... ... W.A. — — — — — — — B.fl.III,567
A. strictus, Bentham, Fl. Austr. III, 568 (1866) ... ... W.A. S.A. — V. N.S.W. — B.fl.III,568
A. plumiger, Bentham, Fl. Austr. III, 568 (1866)... ... ... W.A. — — — — — — — B.fl.III,568
A. amplexicaulis, Bentham, Fl. Austr. III, 568 (1866) ... W.A. — — — — — — — B.fl.III,568

GNEPHOSIS, Cassini in Bull. de la Soc. philom. 45 (1820). (Cephalosorus partly, Nematopus, Crossolepis, Leptotriche, Trichanthodium, Cyathopappus; some readily to be restituted.)
G. Burkittii, Bentham, Fl. Austr. III, 570 (1866) ... — S.A. — — — — — B.fl.III,570
G. eriocarpa, Bentham, Fl. Austr. III, 570 (1866) ... ... — S.A. — N.S.W. — — B.fl.III,570
G. macrocephala, Turcz. in Bull. Soc. Mosc. XXIV, 190 (1851)... W.A. — — — — — — B.fl.III,570
G. skirrophora, Bentham, Fl. Austr. III, 570 (1866) ... W.A. S.A. — V. N.S.W. — — B.fl.III,570
G. cyathopappa, Bentham, Fl. Austr. III, 571 (1866) ... — S.A. — N.S.W. — N.A. B.fl.III,571
G. codonopappa, F. v. M. in Giles, geogr. trav. 217 (1875) — S.A. — — — — — M.fr.IX,2.
G. leptoclada, Bentham, Fl. Austr. III, 571 (1866) ... W.A. — — — — — — B.fl.III,571 M.fr.III,158.
G. arachnoidea, Turcz. in Bull. Soc. Mosc. XXIV, 189 (1851) ... W.A. S.A. — — — — — B.fl.III,571
G. tenuissima, Cassini in Bull. de la Soc. philom. 43 (1820) .. W.A. — — — — — — B.fl.III,572
G. acicularis, Bentham, Fl. Austr. III, 572 (1866) ... ... W.A. — — — — — — B.fl.III,572
G. pygmaea, Bentham, Fl. Austr. III, 572 (1866)... ... ... W.A. — — — — — — B.fl.III,572
G. brevifolia, Bentham, Fl. Austr. III, 572 (1866) ... ... W.A. — — — — — — B.fl.III,572
G. eriocephala, Bentham, Fl. Austr. III, 573 (1866) ... W.A. — — — — — — B.fl.III,573

DECAZESIA, F. v. M., fragm. XI, 71 (1879).
D. hecatocephala, F. v. M., fragm. XI, 72 (1879)... ... ... — — — — — N.A. — M.fr.XI,72.

CALOCEPHALUS, R. Brown in Transact. Linn. Soc. XII, 106 (1817). (Leucophyta, Pachysurus, Blennospora, Achrysum.)
C. Drummondii, Bentham, Fl. Austr. III, 574 (1866) ... ... W.A. — — — — — — B.fl.III,574
C. Brownii, F. v. M., Rep. Babb. Exped. 13 (1858) ... W.A. S.A. T. V. N.S.W. — B.fl.III,574
C. Sonderi, F. v. M., Rep. Babb. Exped. 13 (1858) ... — S.A. T. V. N.S.W. — B.fl.III,775
C. lacteus, Lessing, syn. gen. comp. 271 (1832) ... ... W.A. S.A. T. V. N.S.W. — B.fl.III,575
C. citreus, Lessing, syn. gen. comp. 271 (1832) ... — S.A. T. V. N.S.W. — B.fl.III,575
C. angianthoides, Bentham, Fl. Austr. III, 575 (1866) ... W.A. — T. — — — — B.fl.III,575
C. Francisii, Bentham, Fl. Austr. III, 576 (1866)... ... W.A. — — — — — — B.fl.III,576
C. platycephalus, Bentham, Fl. Austr. 115, 576 (1866) ... — S.A. — N.S.W. — B.fl.III,576
C. multiflorus, Bentham, Fl. Austr. III, 576 (1866) ... W.A. — — — — — — B.fl.III,576
C. aeruoides, Bentham, Fl. Austr. III, 576 (1866) ... W.A. — — — — — — B.fl.III,576

CEPHALIPTERUM, A. Gray in Hook. Kew Misc. IV, 271 (1852).
C. Drummondii, A. Gray in Hook. Kew Misc. IV, 272 (1852) ... W.A. S.A. — — — — B.fl.III,577

GNAPHALODES, A. Gray in Hook. Kew Misc. IV, 228 (1852).
G. uliginosum, A. Gray in Hook. Kew Misc. IV, 228 (1852) ... W.A. S.A. — V. N.S.W. Q. — B.fl.III,578 M.fr.XI,27.
G. condensatum, A. Gray in Hook. Kew Misc. IV, 228 (1852) ... W.A. — — — — — — B.fl.III,578
G. filifolium, Bentham, Fl. Austr. III, 578 (1866) ... ... W.A. — — — — — — B.fl.III,578

CRASPEDIA, G. Forster, florul. insul. Austr. prodr. 306 (1786). (Richea, Pycnosorus.)
C. Richea, Cassini in Diction. XI, 353 (1818) ... ... W.A. S.A. T. V. N.S.W. Q. — B.fl.III,579
C. pleiocephala, F. v. M. in Schlecht. Linnaea XXV, 404 (1852) — S.A. — V. N.S.W. — B.fl.III,580 M.fr.XI,27.
C. chrysantha, Bentham, Fl. Austr. III, 580 (1866) ... — S.A. — V. N.S.W. Q. — B.fl.III,580
C. globosa, Bentham, Fl. Austr. III, 580 (1866) ... ... — S.A. — V. N.S.W. — B.fl.III,580

CHTHONOCEPHALUS, Steetz in Lehm. pl. Preiss. I, 444 (1845). (Chamaesphaerion, Gyrostephium, Lachnothalamus.)
C. tomentellus, Bentham, Fl. Austr. III, 581 (1866) ... ... — — — — — — — B.fl.III,581
C. Pseudevax, Steetz in Lehm. pl. Preiss. I, 445 (1845) ... W.A. S.A. — V. N.S.W. — B.fl.III,582
C. pygmaeus, Bentham, Fl. Austr. III, 582 (1866) ... W.A. — — — — — — B.fl.III,582

NABLONIUM, Cassini in Diction. XXXIV, 101 (1825).
N. calyceroides, Cassini, Dict. XXXIV, 101 (1825) ... ... — — — — — T. — — B.fl.III,545

CHRYSOGONUM, Linné, hort. Cliffort. 424 (1737). (Moonia, Pentalepis.)
C. trichodermoides, Bentham & J. Hooker, gen. pl. II, 350 (1873) — — — — — — N.A. B.fl.III,540 M.fr.VIII,145.
C. ecliptoides, Bentham & J. Hooker, gen. pl. II, 350 (1873) ... — — — — — — N.A. B.fl.III,540 M.fr.VIII,145.
C. procumbens, Bentham & J. Hooker, gen. pl. II, 350 (1873) ... — — — — — — N.A. B.fl.III,540 M.fr.VIII,145.

SIEGESBECKIA, Linné, hort. Cliffort. 412 (1737).
S. orientalis, Linné, spec. plant. 900 (1753) ... ... W.A. S.A. — V. N.S.W. Q. — B.fl.III,535

ENHYDRA, Loureiro, Flor. Cochinch. II, 510 (1790). (Enydra.)
E. paludosa, De Candolle, prodr. V, 637 (1836) ... ... — — — — N.S.W. Q. — B.fl.III,546 M.fr.III,130.

ECLIPTA, Linné, mantissa altera, 157 (1771).
E. alba, Hasskarl, plant. Javan. rar. 528 (1848) ... ... — — — — N.S.W. Q. — B.fl.III,536 M.fr.VIII,145.
E. latifolia, Linné, fil. suppl. 378 (1781) ... ... — — — — — Q. N.A. — M.fr.VIII,145.
E. platyglossa, F. v. M., fragm. II, 135 (1861) ... ... — S.A. — V. N.S.W. Q. N.A. B.fl.III,536 M.fr.XI,27.

WEDELIA, Jacquin, stirp. Amer. hist. 217, t. 130 (1763). (Wollastonia.)
W. calendulacea, Lessing, synops. gen. comp. 222 (1832) ... — — — — — Q. — B.fl.III,537
W. urticifolia, De Candolle in Wight, Contrib. Bot. Ind. 18 (1834) — — — — — — N.A. B.fl.III,538
W. spilanthoides, F. v. M., fragm. V, 64 (1865) ... ... — — — — N.S.W. Q. — B.fl.III,538 M.fr.V,64.
W. verbesinoides, F. v. M. in Benth. Fl. Austr. III, 538 (1866) — — — — — — N.A. B.fl.III,538
W. biflora, De Candolle in Wight, Contrib. Bot. Ind. 18 (1834) — — — — N.S.W. Q. N.A. B.fl.III,539
W. asperrima, Bentham, Fl. Austr. III, 539 (1866) ... ... — — — — — — N.A. B.fl.III,539

84

SPILANTHES, Jacquin, stirp. Amer. hist. 214, t. 126 (1763).
S. grandiflora, Turczaninow in Bull. de Mosc. XXIV, 185 (1851) — — — — N.S.W. Q. N.A. B.fl.III,541 M.fr.V,63.
S. anactina, F. v. M., fragm. V, 63 (1865) ... ... — — — — N.A. B.fl.III,541 M.fr.V,63.

BIDENS, Dillenius, hort. Eltham. 51 and 52, t. 43 et 44 (1732), from Cæsalpini (1583). (Possibly immigrated.)
B. tripartitus, Linné, spec. plant. 831 (1753) ... ... — — — V. N.S.W. — — B.fl.III,543
B. pilosus, Linné, spec. plant. 832 (1753) ... ... — — — — N.S.W. Q. — B.fl.III,543
B. bipinnatus, Linné, spec. plant. 832 (1753) ... ... — — — — N.S.W. Q. N.A. B.fl.III,543

GLOSSOGYNE, Cassini in Diction. LI, 475 (1827). (Diodontium.)
G. tenuifolia, Cassini in Dict. LI, 475 (1827) ... ... — S.A. — — N.S.W. Q. N.A. B.fl.III,544 M.fr.I,51.
G. retroflexa, F. v. M., fragm. I, 51 (1858) ... ... — — — — — — Q. — B.fl.III,544
G. filifolia, F. v. M. in Benth. Fl. Austr. III, 544 (1866) ... — — — — — N.A. B.fl.III,544

FLAVERIA, A. I,. de Jussieu, gen. plant. 186 (1789).
F. Australasica, Hooker in Mitch. Trop. Austr. 118 (1848) ... — — — — — Q. N.A. B.fl.III,546 M.fr.I,183.

SOLIVA, Ruiz et Pavon, prodr. 113, t. 24 (1794). (Gymnostyles; possibly immigrated.)
S. anthemifolia, R. Brown in Transact. Linn. Soc. XII, 102 (1817) — — — — N.S.W. Q. — B.fl.III,552

COTULA, Linné, syst. nat. 9 (1735); Linné, gen. pl. 256 (1737). (Gymnogyne, Strongylospermum, Pleiogyne, Symphyomera, Stenosperma, Leptinella.)
C. filifolia, Thunberg, prodr. pl. Capens. 161 (1800) ... ... W.A. S.A. T. V. N.S.W. — — B.fl.III,548 M.fr.VI,114.
C. coronopifolia, Linné, sp. pl. 892 (1753) ... ... W.A. S.A. T. V. N.S.W. — — B.fl.III,549 M.fr.VI,114.
C. Gymnogyne, F. v. M. in Benth. Fl. Austr. III, 549 (1866) ... W.A. — — V. — — — B.fl.III,549
C. australis, J. Hooker, Fl. N. Zeal. I, 128 (1853) ... ... W.A. S.A. T. V. N.S.W. Q. — B.fl.III,550
C. alpina, J. Hooker, Fl. Tasman. I, 192, t. 51 (1860) ... — — T. V. N.S.W. — — B.fl.III,550
C. Drummondii, Bentham, Fl. Austr. III, 550 (1866) ... W.A. — — — — — — B.fl.III,550
C. reptans, Bentham, Fl. Austr. III, 551 (1866) ... ... — S.A. — T. V. N.S.W. — — B.fl.III,551
C. Filicula, J. Hooker in Benth. Fl. Austr. III, 551 (1866) ... — — T. V. N.S.W. — — B.fl.III,551

CENTIPEDA, Loureiro, Fl. Cochinch. III, 492 (1790). (Myriogyne, Sphæromorphaea partly.)
C. racemosa, F.v.M.; Myriogyne, Hook.in Mitch.Tr.Austr.353(1848) — — — N.S.W. Q. N.A. B.fl.III,553
C. orbicularis, Loureiro, Fl. Cochinch. III, 492 (1790) ... W.A. S.A. T. V. N.S.W. Q. N.A. B.fl.III,553 M.fr.VIII,142.
C. Cunninghami, F. v. M., fragm. VIII, 143 (1874) ... W.A. S.A. — V. N.S.W. — — —
C. thespidioides, F. v. M., fragm. VIII, 143 (1874) ... — S.A. — V. N.S.W. — — —

ABROTANELLA, Cassini in Dict. XXXVI, 27 (1825). (Scleroleima, Trineuron.)
A. fosterioides, J. Hooker, Hand. N. Zeal. Fl. 139 (1864) ... — — T. — — — — B.fl.III,554
A. nivigena, F. v. M. in Benth. Fl. Austr. III, 554 (1866) ... — — — V. — — — B.fl.III,554
A. scapigera, F. v. M. in Benth. Fl. Austr. III, 554 (1866) ... — — T. — — — — B.fl.III,554

ELACHANTHUS, F. v. M. in Schlecht. Linnaea XXV, 410 (1852).
E. pusillus, F. v. M. in Schlecht. Linnaea XXV, 411 (1852) ... — S.A. — V. N.S.W. — — B.fl.III,555

CERATOGYNE, Turczaninow in Bull. de Mosc. XXIV, 68 (1851). (Diotosperma.)
C. obionoides, Turczaninow in Bull. de Mosc. XXIV, 68 (1851) W.A. — — — — — — B.fl.III,555

ISOETOPSIS, Turczaninow in Bull. de Mosc. XXIV, 175 (1851).
I. graminifolia, Turcz. in Bull. de Mosc. XXIV, 175, t. 3 (1851) W.A. S.A. — V. N.S.W. — — B.fl.III,556

EMILIA, Cassini in Bulletin de la Soc. philomat. 68 (1817).
E. purpurea, Cassini in Diction. des sc. nat. XXXIV, 393 (1825) — — — — — Q. — — M.fr.XII.

GYNURA, Cassini in Diction. des sc. nat. XXXIV, 391 (1825).
G. Pseudo-China, De Candolle, prodr. VI, 299 (1837) ... ... — — — — N.S.W. Q. — B.fl.III,661

SENECIO, Tournefort, inst. 456, t. 260 (1700), from Plinius. (Bedfordia, Centropappus.)
S. Gregorii, F. v. M., pl. of Greg. Search-Exped. for Leichh. 7 (1859) — S.A. — V. N.S.W. Q. — B.fl.III,661
S. platylepis, De Candolle, prodr. VI, 371 (1837)... ... — — — V. N.S.W. Q. — B.fl.III,662 M.fr.XI,27.
S. papillosus, F. v. M. in Transact. phil. Inst. Vict. II, 69 (1857) ... — — — T. — — — B.fl.III,662
S. primulifolius, F. v. M. in Transact. phil. Inst. Vict. II, 69 (1857) — — — T. — — — B.fl.III,662
S. pectinatus, De Candolle, prodr. VI, 372 (1837)... ... — — T. V. N.S.W. — — B.fl.III,662
S. spathulatus, A. Richard, sert. Astrol. 125 (1833) ... ... — S.A. T. V. N.S.W. — — B.fl.III,663 M.fr.VI,255.
S. megaglossus, F. v. M. in Linnaea XXV, 419 (1852) ... ... — S.A. — — — — — B.fl.III,663
S. magnificus, F. v. M. in Linnaea XXV, 418 (1852) ... ... — S.A. — V. N.S.W. — — B.fl.III,663
S. insularis, Bentham, Fl. Austr. III, 666 (1866) ... ... — — — — N.S.W. — — B.fl.III,664
S. Centropappus, F. v. M., annual report 26 (1858) ... ... — — T. — — — — B.fl.III,664
S. macranthus, A. Richard, sert. Astrol. 126 (1833) ... ... — — — — N.S.W. — — B.fl.III,664
S. Daltoni, F. v. M., fragm. VI, 27 (1867) ... ... — — — — N.S.W. — — M.fr.VI,27.
S. lautus, Solander in G. Forst. prodr. 91 (1786) ... ... W.A. S.A. T. V. N.S.W. — — B.fl.III,665 M.fr.XI,20.
S. capillifolius, J. Hooker in Lond. Journ. VI, 123 (1847) ... — — — — N.S.W. — — B.fl.III,667
S. vagus, F. v. M. in Transact. phil. Soc. Vict. I, 46 (1854) ... — — — V. N.S.W. — — B.fl.III,667
S. amygdalifolius, F. v. M. ,fragm. I, 232 (1859) ... ... — — — — N.S.W. Q. — B.fl.III,668 M.fr.I,232.
S. velleioides, Cunningham in De Cand. prodr. VI, 374 (1837) ... — — T. V. N.S.W. — — B.fl.III,668
S. australis, A. Richard, sert. Astrol. 131, t. 39 (1833) ... ... — S.A. T. V. N.S.W. — — B.fl.III,668
S. Behrianus, Sonder and F. v. M. in Linnaea XXV, 527 (1852) ... — S.A. — V. N.S.W. — — B.fl.III,669 M.fr.II,15.
S. leucoglossus, F. v. M., fragm. II, 15 (1861) ... ... W.A. — — — — — — B.fl.III,669 M.fr.II,27.
S. brachyglossus, F. v. M. in Schlecht. Linnaea XXV, 525 (1852) W.A. S.A. — V. N.S.W. — — B.fl.III,670
S. Georgianus, De Candolle, prodr. VI, 371 (1837) ... ... W.A. S.A. — V. N.S.W. — — B.fl.III,670
S. Gilberti, Turczaninow in Bull. Soc. Mosc. XXIV, 208 (1851) W.A. — — — — — — B.fl.III,670
S. ramosissimus, De Candolle, prodr. VI, 371 (1837) ... ... W.A. — — — — — — B.fl.III,671
S. odoratus, Horneman, hort. reg. bot. Hafn. II, 309 (1815) ... — S.A. T. V. N.S.W. — — B.fl.III,671
S. Cunninghami, De Candolle, prodr. VI, 371 (1837) ... ... — — — — N.S.W. Q. — B.fl.III,671
S. ancthifolius, Cunningham in De Cand. prodr. VI, 371 (1837).. — — — — N.S.W. Q. — B.fl.III,672
S. Bedfordii, F. v. M. annual report 26 (1858) ... ... — — T. V. N.S.W. — — B.fl.III,673
S. Billardieri, F. v. M. annual report 26 (1858) ... ... — — T. — — — — B.fl.III,673

ERECHTITES, Rafinesque, Fl. Ludov. 65 (1817).
E. prenanthoides, De Candolle, prodr. VI, 296 (1837) ... ... — T. V. N.S.W. — — B.fl.III,658
E. Atkinsoniae, F. v. M., fragm. V, 88 (1866) ... ... — — — N.S.W. Q. — R.fl.III,658 M.fr.V,88.
E. arguta, De Candolle, prodr. VI, 296 (1837) ... ... W.A. S.A. T. V. N.S.W. Q. — R.fl.III,650
E. mixta, De Candolle, prodr. VI, 297 (1837) ... ... — — — N.S.W. — — B.fl.III,659
E. quadridentata, De Candolle, prodr. VI, 295 (1837) ... ... W.A. S.A. T. V. N.S.W. Q. — B.fl.III,660
E. hispidula, De Candolle, prodr. VI, 296 (1837) ... ... W.A. S.A. T. V. N.S.W. — — R.fl.III,660

CYMBONOTUS, Cassini in Diction. XXXV, 397 (1825).
C. Lawsonianus, Gaudichaud in Freyc. Voy. Bot. 462, t. 86 (1826) W.A. S.A. T. V. N.S.W. Q. — B.fl.III,674 M.fr.XI,27.

SAUSSUREA, De Candolle in Ann. du Mus. XVI, 156 (1810). (Aplotaxis, Haplotaxis.)
S. carthamoides, Bentham, Fl. Hongk. 168 (1861) ... ... — — — N.S.W. Q. — B.fl.III,456

CENTAUREA, Linné, hort. Cliffort. 420 (1737). (Leuzea partly.)
C. Australis, Bentham & J. Hooker, gen. plant. II, 479 (1873)... — — — V. N.S.W. Q. — B.fl.III,457

TRICHOCLINE, Cassini in Bull. de la Soc. philom. 13 (1817). (Amblysperma.)
T. scapigera, Bentham & J. Hooker, gen. plant. II, 497 (1873)... W.A — — — — — B.fl.III,676

MICROSERIS, D. Don in Edinb. phil. Mag. XI, 388 (1832). (Phyllopappus, Scozonera partly.)
M. Forsteri, J. Hooker, Fl. N. Zeal. I, 151 (1853) ... ... W.A. S.A. T. V. N.S.W. — — B.fl.III,676 M.fr.XI,27.

CREPIS, Linné, gen. pl. 240 (1737). (Youngia.)
C. Japonica, Bentham, Fl. Hongk. 194 (1861) ... ... — — — — N.S.W. Q. — B.fl.III,679

CAMPANULACEAE.
A. L. de Jussieu, gen. plant. 163 (1789), from B. de Jussieu (1759).

LOBELIA, Linné, Fl. Lappon. 227 (1737). (Rapuntium, Pratia, Grammatotheca, Holostigma.)
L. heterophylla, Labillard., Nov. Holl. pl. spec. I, 52, t. 74 (1804) W.A. S.A. — — — — — B.fl.IV,124 M.fr.IV,184;X,41.
L. Browniana, Roemer & Schultes, syst. veg. V, 71 (1819) ... — S.A. T. V. N.S.W. Q. — —
L. simplicicaulis, R. Brown, prodr. 564 (1810) ... ... — S.A. T. V. N.S.W. — — B.fl.IV,124 M.fr.IV,184;X,41.
L. microsperma, F. v. M., fragm. X, 41 (1876) ... ... W.A. S.A. T. V. N.S.W. Q. — — M.fr.X,41.
L. dentata, Cavanilles, icon. pl. VI, 14, t. 522 (1801) ... — — — N.S.W. Q. — B.fl.IV,125 M.fr.IV,183;X,42.
L. gracilis, Andrews, Bot. Reposit. t. 340 (1805) ... ... — — — N.S.W. Q. — B.fl.IV,125 M.fr.IV,183;X,42.
L. rhytidosperma, Bentham, Fl. Austr. IV, 126 (1869) ... W.A. — — — — — B.fl.IV,126 M.fr.X,40.
L. tenuior, R. Brown, prodr. 564 (1810) ... ... W.A. — — — — — B.fl.IV,126 M.fr.IV,183;X,42.
L. rhombifolia, De Vriese in Lehm. pl. Preiss. I, 397 (1845) ... W.A. S.A. — V. — — — B.fl.IV,127 M.fr.X,42.
L. parvifolia, R. Brown, prodr. 564 (1810) ... ... — — — N.S.W. — — B.fl.IV,127 M.fr.IV,183.
L. trigonocaulis, F. v. M., fragm. I, 18 (1858) ... ... — — — N.S.W. Q. — B.fl.IV,127 M.fr.II,20;IV,171;X,43.
L. Bergiana, Chamisso in Schlecht. Linnaea VIII, 217 (1835) W.A. — — — — — B.fl.IV,128 M.fr.IV,184;VI,114;X,43
L. anceps, Thunberg, prodr. plant. cap. 40 (1794) ... W.A. S.A. T. V. N.S.W. Q. — B.fl.IV,128 M.fr.IV,183.
L. surrepens, J. Hooker, Fl. Tasman. I, 237, t. 66 (1860) ... — — — — T. — — B.fl.IV,129 M.fr.IV,183.
L. membranacea, R. Brown, prodr. 563 (1810) ... ... — — — — N.S.W. — — B.fl.IV,129 M.fr.IV,171;X,42.
L. stenophylla, Bentham, Fl. Austr. III, 130 (1869) ... — — — — — — Q. N.A. B.fl.IV,130 M.fr.X,42.
L. humistrata, F. v. M. in Benth. Fl. Austr. III, 130 (1869) ... — — — — — — Q. N.A. B.fl.IV,130 M.fr.IV,171;X,42.
L. dioica, R. Brown, prodr. 565 (1810) ... ... — — — — — — Q. N.A. B.fl.IV,130 M.fr.IV,183;X,43.
L. purpurascens, R. Brown, prodr. 563 (1810) ... ... — S.A. — V. N.S.W. Q. — B.fl.IV,131 M.fr.IV,183;X,42.
L. pratioides, Bentham, Fl. Austr. IV, 131 (1869) ... — S.A. T. V. N.S.W. — — B.fl.IV,131 M.fr.X,43.
L. irrigua, R. Brown, prodr. 563 (1810) ... ... — — — — N.S.W. — — B.fl.IV,132
L. gelida, F. v. M., fragm. IV, 183 (1864)... ... ... — — — V. N.S.W. — — B.fl.IV,132 M.fr.IV,183.
L. platycalyx, F. v. M., fragm. IV, 183 (1864) ... ... — — T. V. — — — B.fl.IV,133 M.fr.X,43.
L. concolor, R. Brown, prodr. 563 (1810) ... ... — S.A. — V. N.S.W. — — B.fl.IV,133 M.fr.X,43.
L. Benthami, F.v.M.; Pratia puberula, Benth. Fl.Austr.IV,133 (1869) — — — V. N.S.W. — — B.fl.IV,133
L. pedunculata, R. Brown, prodr. 563 (1810) ... ... — S.A. T. V. N.S.W. Q. — B.fl.IV,133 M.fr.X,42.

ISOTOMA, R. Brown, prodr. 505 (1810). (Lobelia partly, Enchysia partly, Laurencia partly.)
I. Brownii, G. Don, gen. syst. III, 716 (1834) ... ... — — — V. N.S.W. Q. — B.fl.IV,134 M.fr.X,40.
I. axillaris, Lindley, Bot. Regist. t. 964 (1826) ... ... — — — V. N.S.W. Q. — B.fl.IV,135 M.fr.X,39.
I. petraea, F. v. M. in Linnaea XXV, 420 (1852) ... ..., W.A. S.A. — — — — — B.fl.IV,135 M.fr.X,39.
I. pusilla, Bentham in Hueg. enum. 75 (1837) ... ... W.A. — — — — — B.fl.IV,135
I. Guilfverdi, F. v. M., fragm. X, 39 (1876) ... ... — — — — — — N.A. — M.fr.X,39.
I. scapigera, G. Don, gen. syst. III, 716 (1834) ... ... W.A. S.A. — — — — — B.fl.IV,136 M.fr.IV,183;X,39.
I. fluviatilis, F. v. M. in Bentham, Fl. Austr. IV, 136 (1869) ... — S.A. T. V. N.S.W. Q. — B.fl.IV,136 M.fr.X,39.

WAHLENBERGIA, Schrader, cat. hort. bot. Goetting. (1814).
W. gracilis, A. De Candolle, Monogr. comp. 142 (1830) ... ... W.A. S.A. T. V. N.S.W. Q. N.A. B.fl.IV,137

CANDOLLEACEAE.
F. v. M., fragm. VIII, 41 (1873). (Stylideae, R. Brown, prodr. 565, 1810.)

CANDOLLEA, Labillardière in Ann. du Mus. VI, 451 (1805). (Stylidium, Ventenata, Forsteropsis.)
C. (St.) carnosa, Bentham in Hueg. enum. 71 (1837) ... W.A. — — — — — B.fl.IV,6
C. (St.) pilosa, Labillardière, Nov. Holl. pl. spec. II, 63, t. 213 (1806)W.A. — — — — — B.fl.IV,7
C. (St.) reduplicata, R. Brown, prodr. 568 (1810) ... ... W.A. — — — — — B.fl.IV,7
C. (St.) scabrida, Lindley, Bot. Regist. XXV, App. XXVIII (1830)W.A. — — — — — B.fl.IV,8
C. (St.) hirsuta, R. Brown, prodr. 568 (1810) ... ... W.A. — — — — — B.fl.IV,8
C. (St.) crossocephala, F. v. M., fragm. VI, 5 (1866) ... ... W.A. — — — — — B.fl.IV,8 M.fr.VI,5,254;X,58.
C. (St.) juncea, R. Brown, prodr. 569 (1810) ... ... W.A. — — — — — B.fl.IV,9 M.fr.X,58.
C. (St.) repens, R. Brown, prodr. 571 (1810) ... ... W.A. — — — — — B.fl.IV,9
C. (St.) graminifolia, Swartz, Mag. der nat. Ges. Berl. 40 (1807) — S.A. T. V. N.S.W. Q. — B.fl.IV,10 M.fr.VI,6;X,58.
C. (St.) linearis, Swartz, Mag. der nat. Ges. Berl. 50 (1807) ... — — — N.S.W. — — B.fl.IV,11
C. (St.) elongata, Bentham, Fl. Austr. IV, 11 (1869) ... ... W.A. — — — — — B.fl.IV,11 M.fr.X,86.
C. (St.) spinulosa, R. Brown, prodr. 569 (1815) ... ... W.A. — — — — — B.fl.IV,11

C. (St.) limbata, F. v. M., fragm. X, 57 (1876) ... ... W.A. — — — — — — — M.fr.X,57.
C. (St.) caespitosa, R. Brown, prodr. 569 (1810) ... ... W.A. — — — — — — B.fl.IV,11 M.fr.X,57.
C. (St.) squamellosa, De Candolle, prodr. VII, 782 (1838) ... W.A. — — — — — — M.fr.X,56.
C. (St.) violacea, R. Rrown, prodr. 569 (1810) ... ... W.A. — — — — — — B.fl.IV,12
C. (St.) lutea, R. Brown, prodr. 570 (1810)... ... ... W.A. — — — — — — B.fl.IV,12
C. (St.) filifera, R. Brown, prodr. 569 (1810) ... ... W.A. — — — — — — B.fl.IV,12 M.fr.X,58,86.
C. (St.) ciliata, Lindley, Bot. Regist. XXV, App. XXVIII (1839) W.A. — — — — — — B.fl.IV,13
C. (St.) sobolifera, F. v. M. in transact. Vict. Inst. 131 (1855) ... — — — V. — — — B.fl.IV,13
C. (St.) Floodii, F. v. M., fragm. I, 149 (1859) ... ... — — — — — Q. N.A. B.fl.IV,13 M.fr.VI,6;X,58.
C. (St.) disperma, F. v. M., fragm. IV, 93 (1864) ... ... W.A. — — — — — — B.fl.IV,14 M.fr.IV,93.
C. (St.) calcarata, R. Brown, prodr. 570 (1810) ... ... W.A. S.A. — V. — — — B.fl.IV,14 M.fr.VI,6.
C. (St.) perpusilla, J. Hooker in Lond. Journ. of Bot. VI, 266 (1847)W.A. S.A. T. V. — — — B.fl.IV,15 M.fr.VI,78;X,58.
C. (St.) eriorrhiza, R. Brown, prodr. 569 (1810) ... .. — — — — — Q. — B.fl.IV,15 M.fr.I,147;VI,6;X,58.
C. (St.) debilis, F. v. M., fragm. I, 149 (1859) ... ... — — — — N.S.W. Q. — B.fl.IV,15 M.fr.VI,6;X,58.
C. (St.) floribunda, R. Brown, prodr. 569 (1810) ... .. — — S.A. — — — N.A. B.fl.IV,15 M.fr.I,148;VI,6;X,58.
C. (St.) leptorrhiza, F. v. M., fragm. I, 148 (1859) ... — — — — — — N.A. B.fl.IV,16 M.fr.I,148.
C. (St.) assimilis, R. Brown, prodr. 569 (1810) ... ... W.A. — — — — — — B.fl.IV,16 M.fr.X,57.
C. (St.) rupestris, Sonder in Lehm. pl. Preiss. I, 375 (1845) W.A. — — — — — — B.fl.IV,16
C. (St.) spathulata, R. Brown, prodr. 569 (1810) ... ... W.A. — — — — — — B.fl.IV,17
C. (St.) Barleei, F. v. M., fragm. VI, 5, t. 60 (1866) ... W.A. — — — — — — B.fl.IV,17 M.fr.VI,5,249.
C. (St.) lineata, Sonder in Lehm. pl. Preiss. I, 376 (1845) ... W.A. — — — — — — B.fl.IV,17
C. (St.) glauca, Labillard., Nov. Holl. pl. spec. II, 64, t. 214 (1806) W.A. — — — — — — B.fl.IV,18
C. (St.) amoena, R. Brown, prodr. 570 (1810) ... ... W.A. — — — — — — B.fl.IV,18
C. (St.) striata, Lindley, Bot. Regist. XXV, App. XXVIII (1839) W.A. — — — — — — B.fl.IV,18
C. (St.) diversifolia, R. Brown, prodr. 570 (1810)... ... W.A. — — — — — — B.fl.IV,19 M.fr.X,58.
C. (St.) articulata, R. Brown, prodr. 570 (1810) ... ... W.A. — — — — — — B.fl.IV,19
C. (St.) Brunoniana, Bentham in Hueg. enum. 72 (1837) ... W.A. — — — — — — B.fl.IV,20
C. (St.) diuroides, Lindley, Bot. Regist. XXV, App. XXIX (1839) W.A. — — — — — — B.fl.IV,20 M.fr.VI,78.
C. (St.) scandens, R. Brown, prodr. 570 (1810) ... ... W.A. — — — — — — B.fl.IV,20 M.fr.IV,94.
C. (St.) verticillata, F. v. M., fragm. IV, 94 (1864) ... W.A. — — — — — — B.fl.IV,20
C. (St.) glandulosa, Salisbury, Parad. Lond. t. 77 (1807) ... W.A. — — — — — — B.fl.IV,21
C. (St.) laricifolia, Richard in Pers. syn. II, 210 (1807) ... — — — — N.S.W. Q. — B.fl.IV,21 M.fr.III,122.
C. (St.) Preissii, F. v. M., fragm. III, 122 (1863) ... ... W.A. — — — — — — B.fl.IV,21
C. (St.) imbricata, Bentham in Hueg. enum. 73 (1837) .. W.A. — — — — — — B.fl.IV,22
C. (St.) adpressa, Bentham, Fl. Austr. IV, 22 (1869) ... W.A. — — — — — — B.fl.IV,22 M.fr.VI,58.
C. (St.) utricularioides, Bentham in Hueg. enum. 73 (1837) ... W.A. S.A. T. V. N.S.W. — — B.fl.IV,22
C. (St.) pygmaea, R. Brown, prodr. 571 (1810) ... ... W.A. — — — — — — B.fl.IV,23
C. (St.) longitubea, Bentham in Hueg. enum. 73 (1837) ... W.A. — — — — — — B.fl.IV,23
C. (St.) diffusa, R. Brown, prodr. 571 (1810) ... ... — — — — — Q. — B.fl.IV,23
C. (St.) fissiloba, F. v. M., fragm. I, 154 (1859) ... ... — — — — — N.A. B.fl.IV,24 M.fr.I,154.
C. (St.) trichopoda, F. v. M., fragm. X, 86 (1877) ... ... — — — — — Q. — — M.fr.X,86.
C. (St.) alsinoides, R. Brown, prodr. 572 (1810) ... ... — — — — — Q. N.A. B.fl.IV,24 M.fr.I,151;VI,6;X,58.
C. (St.) tenerrima, F. v. M., fragm. I, 150 (1859)... ... — — — — — — N.A. B.fl.IV,24 M.fr.I,150.
C. (St.) brachyphylla, Sonder in Lehm. pl. Preiss. I, 386 (1845) W.A. — — — — — — B.fl.IV,24
C. (St.) capillaris, R. Brown, prodr. 570 (1810) ... ... W.A. — — — — — — B.fl.IV,25
C. (St.) rotundifolia, R. Brown, prodr. 571 (1810) ... ... — — — — — Q. N.A. B.fl.IV,25 M.fr.I,151.
C. (St.) schizantha, F. v. M., fragm. I, 152 (1859) ... ... — — — — — N.A. B.fl.IV,25 M.fr.I,152.
C. (St.) lobuliflora, F. v. M., fragm. I, 153 (1859) ... ... — — — — — N.A. B.fl.IV,25 M.fr.I,153.
C. (St.) uliginosa, Swartz in Mag. nat. Fr. Berl. 62, t. 2 (1807) — — — — — Q. — B.fl.IV,26
C. (St.) pulchella, Sonder in Lehm. pl. Preiss. I, 381 (1845) ... W.A. — — — — — — B.fl.IV,26
C. (St.) petiolaris, Sonder in Lehm. pl. Preiss. I, 382 (1845) ... W.A. — — — — — — B.fl.IV,27
C. (St.) emarginata, Sonder in Lehm. pl. Preiss. I, 383 (1845) ... W.A. — — — — — — B.fl.IV,27
C. (St.) corymbosa, R. Brown, prodr. 571 (1810) ... ... W.A. — — — — — — B.fl.IV,27
C. (St.) lepida, F. v. M. in Benth. Fl. Austr. IV, 27 (1869) ... W.A. — — — — — — B.fl.IV,27
C. (St.) streptocarpa, Sonder in Lehm. pl. Preiss. I, 385 (1845) W.A. — — — — — — B.fl.IV,28 M.fr.VI,6.
C. (St.) uniflora, Sonder in Lehm. pl. Preiss. I, 381 (1845) ... W.A. — — — — — — B.fl.IV,28 M.fr.VI,91.
C. (St.) podunculata, R. Brown, prodr. 571 (1810) ... ... — — — — — Q. N.A. B.fl.IV,28 M.fr.I,152.
C. (St.) pachyrrhiza, F. v. M., fragm. I, 152 (1859) ... ... — — — — — N.A. B.fl.IV,28 M.fr.I,153.
C. (St.) muscicola, F. v. M., fragm. I, 153 (1859)... ... — — — — — N.A. B.fl.IV,29 M.fr.I,153.
C. (St.) crassifolia, R. Brown, prodr. 571 (1810) ... ... W.A. — — — — — — B.fl.IV,29
C. (St.) pycnostachya, Lindley, Bot. Regist. XXV, App.XXIX (1839) W.A. — — — — — — B.fl.IV,30
C. (St.) pubigera, Sonder in Lehm. pl. Preiss. I, 383 (1845) ... W.A. — — — — — — B.fl.IV,30
C. (St.) canaliculata, Lindley, Bot. Regist. XXV, App.XXIX (1830) W.A. — — — — — — B.fl.IV,30
C. (St.) leptophylla, De Candolle, prodr. VII, 785 (1838) ... W.A. — — — — — — B.fl.IV,30
C. (St.) dichotoma, De Candolle, prodr. VII, 783 (1838)... ... W.A. — — — — — — B.fl.IV,30
C. (St.) bulbifera, Bentham in Hueg. enum. 73 (1837) ... ... W.A. — — — — — — B.fl.IV,31
C. (St.) breviscapea, R. Brown, prodr. 572 (1810)... ... W.A. — — — — — — B.fl.IV,31
C. (St.) eglandulosa, F. v. M., fragm. I, 150 (1859) ... ... — — — — N.S.W. Q. — B.fl.IV,31 M.fr.VI,6;X,58.
C. (St.) fasciculata, R. Brown, prodr. 572 (1810) ... ... W.A. — — — — — — B.fl.IV,32
C. (St.) falcata, R. Brown, prodr. 572 (1810) ... ... W.A. — — — — — — B.fl.IV,33
C. (St.) rhynchocarpa, Sonder in Lehm. pl. Preiss. I, 389 (1845) W.A. — — — — — — B.fl.IV,33 M.fr.X,58.
C. (St.) adnata, R. Brown, prodr. 572 (1810) ... ... W.A. — — — — — — B.fl.IV,33

LEEWENHOEKIA, R. Brown, prodr. 572 (1810). (Levenhookia, Colcostylis.)
L. pusilla, R. Brown, prodr. 573 (1810) ... ... W.A. — — — — — — B.fl.IV,34 M.fr.X,58.
L. dubia, Sonder in Lehm. pl. Preiss. I, 392 (1845) ... ... W.A. S.A. T. V. N.S.W. — — B.fl.IV,34 M.fr.III,121;X,58.
L. Sonderi, F. v. M., fragm. I, 18 (1858) ... ... — — — V. — — — B.fl.IV,35
L. pauciflora, Bentham in Hueg. enum. 74 (1837)... ... W.A. — — — — — — B.fl.IV,35 M.fr.VI,77.
L. leptantha, Bentham, Fl. Austr. IV, 35 (1869) ... ... W.A. — — — — — — B.fl.IV,35
L. stipitata, F. v. M., fragm. IV, 94 (1864) ... ... — — — — — — — B.fl.IV,36 M.fr.IV,94.
L. Preissii, F. v. M., fragm. IV, 94 (1864)... ... ... W.A. — — — — — — B.fl.IV,36 M.fr.IV,94.

PHYLLACHNE, R. & G. Forster, char. gen. 115, t. 58 (1776).   (Forsters, Helophyllum, Oreostylidium.)
P. bellidifolia, F. v. M., fragm. VIII, 39 (1873) ...   .   —   — T. —   —   —   —   —   M.fr.X,58.
DONATIA, R. et G. Forster, char. gen. 9, t. 5 (1776).
D. Novae Zealandiae, J. Hooker, Fl. N. Zeal. I, 81, t. 20 (1853)   —   — T. —   —   —   —   B.fl.II,450   M.fr.VIII,46;X,58.

GOODENIACEAE.

R. Brown, prodr. 573 (1810).

BRUNONIA, Smith in Transact. Linn. Soc. X, 366 (1809).
B. australis, Smith in Transact. Linn. Soc. X, 367, t. 28 (1809)   W.A. S.A. T. V. N.S.W. Q.   —   B.fl.IV,121   M.fr.X,111.

DAMPIERA, R. Brown, prodr. 587 (1810). (Linschotenia.)
D. luteifolia, F. v. M., fragm. X, 11 (1876)   ...   .   W.A.   —   —   —   —   —   —   M.fr.X,11.
D. Linschotenii, F. v. M., fragm. VI, 28 (1866) ...   ...   —   —   —   —   —   Q.   —   B.fl.IV,108   M.fr.VI,28.
D. spicigera, Bentham, Fl. Austr. IV, 109 (1869)...   ...   W.A.   —   —   —   —   —   —   B.fl.IV,109
D. candicans, F. v. M., fragm. X, 86 (1876)   ...   ...   W.A. S.A.   —   —   —   —   —   M.fr.X,86.
D. teres, Lindley, Bot. Regist. XXV, App. XXVII (1839)   ...   W.A.   —   —   —   —   —   B.fl.IV,109
D. trigona, De Vriese in Lehm. pl. Preiss. I, 401 (1845)...   W.A.   —   —   —   —   —   B.fl.IV,109   M.fr.II,17;VIII,245;X,11
D. prostrata, De Vriese, Goodenov. 83 (1854)   ...   ...   W.A.   —   —   —   —   —   B.fl.IV,110
D. alata, Lindley, Bot. Regist. XXV, App. XXVII (1839)   ...   W.A.   —   —   —   —   —   B.fl.IV,110   M.fr.X,11.
D. coronata, Lindley, Bot. Regist. XXV, App. XXVII (1839)...   W.A.   —   —   —   —   —   B.fl.IV,110
D. carinata, Bentham, Fl. Austr. IV, 111 (1869)   ...   W.A.   —   —   —   —   —   B.fl.IV,111
D. sacculata, F. v. M. in Benth. Fl. Austr. IV, 111 (1869)   W.A.   —   —   —   —   —   B.fl.IV,111
D. incana, R. Brown, prodr. 588 (1810)   ...   ...   W.A.   —   —   —   —   —   B.fl.IV,111   M.fr.II,17.
D. hederacea, R. Brown, prodr. 588 (1810)...   ...   ...   W.A.   —   —   —   —   —   B.fl.IV,112   M.fr.X,12.
D. ferruginea, R. Brown, prodr. 588 (1810)   ...   ...   —   —   —   — V. N.S.W.   —   Q.   B.fl.IV,112   M.fr.VI,29.
D. Brownii, F. v. M., fragm. VI, 29 (1866)   ...   ...   —   —   —   — V. N.S.W.   —   B.fl.IV,112   M.fr.VI,29.
D. lanceolata, Cunningham in De Cand., prodr. VII, 503 (1839)   —   S.A.   — V. N.S.W.   —   B.fl.IV,113   M.fr.VI,29.
D. altissima, F. v. M. in Benth. Fl. Austr. IV, 113 (1869)   W.A.   —   —   —   —   —   B.fl.IV,113
D. marifolia, Bentham, Fl. Austr. IV, 114 (1869)...   ...   —   S.A.   — V. N.S.W.   —   B.fl.IV,114
D. rosmarinifolia, Schlechtendal, Linnaea XX, 603 (1847)   ...   —   S.A.   — V. N.S.W.   —   B.fl.IV,114   M.fr.VI,20;X,12.
D. lavandulacea, Lindley, Bot. Regist. XXV, App. XXVII (1839)   W.A.   —   —   —   —   —   B.fl.IV,114
D. juncea, Bentham, Fl. Austr. IV, 115 (1869)   ...   ...   W.A.   —   —   —   —   —   B.fl.IV,115
D. oligophylla, Bentham, Fl. Austr. IV, 115 (1869)   ...   W.A.   —   —   —   —   —   B.fl.IV,115
D. loranthifolia, F. v. M. in Benth. Fl. Austr. IV, 115 (1869)   W.A.   —   —   —   —   —   B.fl.IV,115
D. stricta, R. Brown, prodr. 589 (1810)   ...   .   —   S.A. T. V. N.S.W.   —   B.fl.IV,116   M.fr.X,12;XI,121.
D. Scottiana, F. v. M., fragm. XI, 120 (1881)   ...   ...   —   —   —   — N.S.W.   —   —   M.fr.XI,120.
D. leptoclada, Bentham, Fl. Austr. IV, 116 (1869)   ...   W.A.   —   —   —   —   —   B.fl.IV,116
D. fasciculata, R. Brown, prodr. 588 (1810)   ...   ...   W.A.   —   —   —   —   —   B.fl.IV,116   M.fr.XI,121.
D. eubspicata, Bentham, Fl. Austr. IV, 117 (1869)   ...   W.A.   —   —   —   —   —   B.fl.IV,117
D. triloba, Lindley, Bot. Regist. XXV, App. XXVII (1839)   ...   W.A.   —   —   —   —   —   B.fl.IV,117
D. linearis, R. Brown, prodr. 588 (1810) ...   .   W.A.   —   —   —   —   —   B.fl.IV,117
D. cuneata, R. Brown, prodr. 588 (1810) ...   .   W.A.   —   —   —   —   —   B.fl.IV,118
D. sericantha, F. v. M. in Benth. Fl. Austr. IV, 118 (1869)   W.A.   —   —   —   —   —   B.fl.IV,118
D. parvifolia, R. Brown, prodr. 589 (1810)...   ...   ...   W.A.   —   —   —   —   —   B.fl.IV,118
D. glabrescens, Bentham, Fl. Austr. IV, 119 (1869)   ...   W.A.   —   —   —   —   —   B.fl.IV,119
D. adpressa, Cunningham in De Cand., prodr. VII, 503 (1839)...   W.A.   —   —   — N.S.W. Q.   —   B.fl.IV,119
D. diversifolia, De Vriese in Lehm. pl. Preiss. I, 403 (1845)   ...   W.A.   —   —   —   —   —   B.fl.IV,120   M.fr.VI,29;X,12.
D. oriocephala, De Vriese, Goodenov. 118, t. 21 (1854)   ...   W.A.   —   —   —   —   —   —   M.fr.X,12.
D. Wellsiana, F. v. M., fragm. X, 12 (1876)   ...   ...   W.A.   —   —   —   —   —   —   M.fr.X,12.

DIASPASIS, R. Brown, prodr. 586 (1810).
D. filifolia, R. Brown, prodr. 587 (1810)   ...   .   W.A.   —   —   —   —   —   B.fl.IV,104   M.fr.I,206;II,181;VIII,58

LESCHENAULTIA, R. Brown, prodr. 581 (1810). (Latouria.)
L. formosa, R. Brown, prodr. 581 (1810)   ...   .   W.A.   —   —   —   —   —   B.fl.IV,40   M.fr.VI,226.
L. oblata, Sweet, flor. Australas. 26, 46, t. 46 (1829)   ...   W.A.   —   —   —   —   —   —   M.fr.VI,26.
L. chlorantha, F. v. M., fragm. II, 20 (1860)   ...   ...   W.A.   —   —   —   —   —   B.fl.IV,40   M.fr.II,20.
L. linarioides, De Candolle, prodr. VII, 519 (1818)   ...   W.A.   —   —   —   —   —   B.fl.IV,41
L. tubiflora, R. Brown, prodr. 581 (1810) ...   .   W.A.   —   —   —   —   —   B.fl.IV,41
L. superba, F. v. M., fragm. VI, 10 (1866)   ...   ...   W.A.   —   —   —   —   —   B.fl.IV,41   M.fr.VI,10.
L. striata, F. v. M., fragm. VIII, 245 (1874)   ...   ...   —   S.A.   —   —   —   —   —   M.fr.VIII,245.
L. acutiloba, Bentham, Fl. Austr. IV, 41 (1869)   ...   W.A.   —   —   —   —   —   B.fl.IV,41
L. laricina, Lindley, Bot. Regist. XXV, App. XXVII (1839)   ...   W.A.   —   —   —   —   —   B.fl.IV,41
L. hirsuta, F. v. M., fragm. VI, 9 (1866)   ...   .   W.A.   —   —   —   —   —   B.fl.IV,42   M.fr.VI,9.
L. longiloba, F. v. M., fragm. VI, 10 (1866)   ...   ...   W.A.   —   —   —   —   —   B.fl.IV,42   M.fr.VI,10.
L. expansa, R. Brown, prodr. 581 (1810)   ...   ...   W.A.   —   —   —   —   —   B.fl.IV,42
L. floribunda, Bentham in Hueg. enum. 70 (1837)...   ...   W.A.   —   —   —   —   —   B.fl.IV,43
L. heteromera, Bentham, Fl. Austr. IV, 43 (1869)   ...   W.A.   —   —   —   —   —   B.fl.IV,43
L. divaricata, F. v. M., fragm. III, 33, 167 (1863)   ...   —   S.A.   —   —   —   —   B.fl.IV,43   M.fr.VIII,246.
L. filiformis, R. Brown, prodr. 581 (1810) ...   .   —   —   —   —   Q. N.A. B.fl.IV,44   M.fr.VI,9,249.
L. agrostophylla, F. v. M., fragm. VI, 8, pl. XLVII (1866) ...   —   —   —   —   — N.A. B.fl.IV,44   M.fr.VI,8;249.

ANTHOTIUM, R. Brown, prodr. 582 (1810).
A. humile, R. Brown, prodr. 582 (1810)   ...   .   W.A.   —   —   —   —   —   B.fl.IV,44   M.fr.VIII,58.
A. rubriflorum, F. v. M. in Benth. Fl. Austr. IV, 45 (1869)   W.A.   —   —   —   —   —   B.fl.IV,45

CATOSPERMA, Bentham in Hooker, icon. pl. t. 1023 (1868).
C. Muelleri, Bentham in Hook. icon. pl. t. 1028 (1868) ...   —   —   —   —   —   — N.A. B.fl.IV,83   M.fr.I,121;VIII,58.

SCAEVOLA, Linné, mantiss. II, 145 (1771). (Pogonetes, Crossotoma, Temminckia, Kamphusia, Molkenboeria, Merkusia, Verreauxia.)
S. Koenigii, Vahl, symb. bot. III, 36 (1794)   ...   ...   —   —   —   —   —   Q. N.A. B fl.IV,86   M.fr.XI,76.
S. enantophylla, F. v. M., fragm. VIII, 58 (1873)...   ...   —   —   —   —   —   Q.   —   M.fr.VI,225.

	W.A.	S.A.	V.	N.S.W.	Q.	N.A.		
S. spinescens, R. Brown, prodr. 586 (1810)...	W.A.	S.A.	— V.	N.S.W.	Q.	N.A.	B.fl.IV,87	M.fr.XI,20,27,76.
S. Groeneri, F. v. M., fragm. VI, 15 (1866)	W.A.	S.A.	— —	—	—	—	B.fl.IV,88	M.fr.VIII,58.
S. tomentosa, Gaudichaud in Freyc. Voy. Bot. 460, t. 81 (1826)	W.A.	—	— —	—	—	—	B.fl.IV,88	
S. atriplicina, F. v. M., fragm. II, 18 (1860)	W.A.	—	— —	—	—	—	B.fl.IV,88	M.fr.III,181.
S. striata, R. Brown, prodr. 586 (1810)	W.A.	—	— —	—	—	—	B.fl.IV,89	
S. phlebopetala, F. v. M., fragm. II, 18 (1860)	W.A.	—	— —	—	—	—	B.fl.IV,89	M.fr.II,VI,16;XI,76.
S. pilosa, Bentham in Hueg. enum. 69 (1837)	W.A.	—	— —	—	—	—	B.fl.IV,89	
S. hispida, Cavanilles, icon. et descr. pl. VI, 7, t. 510 (1801)	—	—	— V.	N.S.W.	Q.	—	B.fl.IV,90	M.fr.XI,76.
S. apterantha, F. v. M., fragm. I, 121 (1859)	—	—	— V.	—	—	—	B.fl.IV,90	M.fr.I,121.
S. Hookeri, F. v. M. in Benth. Fl. Austr. IV, 90 (1869)...	—	—	— T. V.	N.S.W.	—	—	B.fl.IV,90	M.fr.XI,76.
S. parvifolia, F. v. M. in Benth. Fl. Austr. IV, 91 (1869)	W.A.	S.A.	— —	—	—	Q.	N.A.	B.fl.IV,91	M.fr.VIII,58;X,59;XI,76.
S. oxyclona, F. v. M., fragm. X, 58 (1876)...	—	—	— —	—	—	—	—	M.fr.X,58.
S. restiacea, Bentham, Fl. Austr. IV, 91 (1869)	W.A.	—	— —	—	—	—	B.fl.IV,91	[76.
S. depauperata, R. Brown, App. Sturt's Exped. 20 (1849)	—	S.A.	— —	—	—	—	B.fl.IV,91	M.fr.III,33;VIII,58;XI,
S. tortuosa, Bentham, Fl. Austr. IV, 91 (1869)	W.A.	—	— —	—	—	—	B.fl.IV,91	
S. Cunninghamii, De Candolle, prodr. VII, 508 (1838)	—	—	— —	—	—	N.A.	B.fl.IV,92	
S. collaris, F. v. M., Rep. Babb. Exped. 15 (1858)	—	S.A.	— —	—	—	—	B.fl.IV,92	M.fr.VIII,58;XI,76.
S. angulata, R. Brown, prodr. 586 (1810)	—	—	— —	—	—	N.A.	B.fl.IV,92	
S. nitida, R. Brown, prodr. 584 (1810)	W.A.	—	— —	—	—	—	B.fl.IV,93	
S. globuliflora, Lahillardière, Nov. Holl. pl. spec. I, 55, t. 78 (1804)	W.A.	—	— —	—	—	—	B.fl.IV,93	
S. porocarya, F. v. M., fragm. II, 19 (1860)	—	—	— —	—	—	—	B.fl.IV,94	M.fr.II,19;XI,76.
S. attenuata, R. Brown, prodr. 583 (1810)	W.A.	—	— —	—	—	—	B.fl.IV,94	
S. glandulifera, De Candolle, prodr. VII, 510 (1838)	W.A.	—	— —	—	—	—	B.fl.IV,94	
S. anchusifolia, Bentham in Hueg. enum. 68 (1837)	—	—	— —	—	—	—	B.fl.IV,94	
S. holosericea, De Vriese in Lehm. pl. Preiss. I, 408 (1845)	W.A.	—	— —	—	—	—	B.fl.IV,95	
S. suaveolens, R. Brown, prodr. 585 (1810)	—	S.A.	— V.	N.S.W.	Q.	—	B.fl.IV,95	
S. revoluta, R. Brown, prodr. 586 (1810)	—	—	— —	—	—	N.A.	B.fl.IV,96	
S. ovalifolia, R. Brown, prodr. 584 (1810)	—	S.A.	— —	N.S.W.	Q.	N.A.	B.fl.IV,96	M.fr.VIII,58;XI,76.
S. crassifolia, Labillardière, Nov. Holl. pl. spec. I, 56, t. 79 (1804)	W.A.	S.A.	— V.	—	—	—	B.fl.IV,96	M.fr.VIII,245;X1,20,76.
S. macrostachya, Bentham, Fl. Austr. IV, 97 (1869)	—	—	— —	—	—	N.A.	B.fl.IV,97	M.fr.VIII,58;XI,76.
S. longifolia, De Vriese in Lehm. pl. Preiss. I, 410 (1845)	W.A.	—	— —	—	—	—	B.fl.IV,97	
S. lanceolata, Bentham in Hueg. enum. 69 (1837)...	W.A.	—	— —	—	—	—	B.fl.IV,97	M.fr.I,207;XI,76.
S. thesioides, Bentham in Hueg. enum. 68 (1837)...	W.A.	—	— —	—	—	—	B.fl.IV,98	M.fr.XI,77.
S. macrophylla, Bentham, Fl. Austr. IV, 98 (1869)	W.A.	—	— —	—	—	—	B.fl.IV,98	
S. platyphylla, Lindley, Bot. Regist. XXV, App. XXVI (1839)	W.A.	—	— —	—	—	—	B.fl.IV,98	M.fr.XI,77.
S. auriculata, Bentham, Fl. Austr. IV, 99 (1869)...	W.A.	—	— —	—	—	—	B.fl.IV,99	M.fr.XI,77.
S. aemula, R. Brown, prodr. 584 (1810)	W.A.	S.A.	T. V.	N.S.W.	—	—	B.fl.IV,99	M.fr.XI,77.
S. humilis, R. Brown, prodr. 585 (1810)	—	S.A.	— —	—	—	N.A.	B.fl.IV,100	
S. amblyanthera, F. v. M., fragm. I, 121 (1859) ...	—	—	— —	—	—	—	B.fl.IV,100	M.fr.I,121.
S. microphylla, Bentham, Fl. Austr. IV, 100 (1869)	W.A.	—	— —	—	—	—	B.fl.IV,100	M.fr.XI,77.
S. cuneiformis, Labillard., Nov. Holl. pl. spec. I, 56, t. 80 (1804)	W.A.	—	— —	—	—	—	B.fl.IV,101	M.fr.XI,77.
S. microcarpa, Cavanilles, icon. et descr. pl. VI, 6, t. 509 (1801)	—	S.A.	T. V.	N.S.W.	Q.	—	B.fl.IV,101	M.fr.VI,16;XI,77.
S. linearis, R. Brown, prodr. 586 (1810)	—	S.A.	— —	—	—	—	B.fl.IV,102	
S. Oldfieldii, F. v. M., fragm. II, 19 (1860)	W.A.	—	— —	—	—	—	B.fl.IV,102	M.fr.II,19;XI,77.
S. paludosa, R. Brown, prodr. 586 (1810)	—	—	— —	—	—	—	B.fl.IV,102	M.fr.XI,77.
S. sericophylla, F. v. M. in Benth. Fl. Austr. IV, 102 (1869)	W.A.	—	— —	—	—	—	B.fl.IV,102	M.fr.XI,77.
S. canescens, Bentham in Hueg. enum. 69 (1837)	W.A.	—	— —	—	—	—	B.fl.IV,103	M.fr.XI,77.
S. humifusa, De Vriese in Lehm. pl. Preiss. I, 410 (1845)	W.A.	—	— —	—	—	—	B.fl.IV,103	M.fr.XI,77.
S. fasciculata, Bentham in Hueg. enum. 68 (1837)	W.A.	—	— —	—	—	—	B.fl.IV,104	M.fr.XI,77.
S. stenophylla, Bentham, Fl. Austr. IV, 104 (1869)	W.A.	—	— —	—	—	—	B.fl.IV,104	M.fr.I,113;XI,77.
S. Reinwardtii, De Vriese in Lehm. pl. Preiss. I, 409 (1845)	W.A.	—	— —	—	—	—	B.fl.IV,105	
S. Vorreauxii, F. v. M., Bot. Teachings, 65 (1877)	—	—	— —	—	—	—	B.fl.IV,105	
SELLIERA, Cavanilles in Anal. Cienz. Nat. I, 41, t. 5 (1799).								
S. radicans, Cavanilles, icon. et descr. pl. V, 49, t. 474 (1799)	—	S.A.	T. V.	N.S.W.	—	—	B.fl.IV,82	M.fr.VIII,58.
S. exigua, F. v. M., fragm. III, 142 (1862)...	W.A.	—	— —	—	—	—	B.fl.IV,83	M.fr.III,142.
CALOGYNE, R. Brown, prodr. 579 (1810). (Distylis.)								
C. pilosa, R. Brown, prodr. 579 (1810)	—	—	— —	—	Q.	N.A.	B.fl.IV,81	M.fr.VI,7,28.
C. Berardiana, F. v. M., fragm. VI, 7 (1867)	—	—	— —	—	—	N.A.	B.fl.IV,81	M.fr.VI,7;VIII,58.
C. heteroptera, F. v. Mueller, fragm. X, 43 (1876)	—	—	— —	—	Q.	N.A.	B.fl.X,43.	
C. purpurea, F. v. M., fragm. VIII, 57 (1873)	—	—	— —	—	—	N.A.	—	M.fr.VIII,57.
GOODENIA, Smith in Transact. Linn. Soc. II, 347 (1794). (Goodenoughia, Tetraphylax, Picrophyta, Stekhovia, Aillya.)								
G. phylicoides, F. v. M., fragm. I, 206 (1859)	W.A.	—	— —	—	—	—	B.fl.IV,55	M.fr.I,206.
G. viscida, R. Brown, prodr. 578 (1810)	—	—	— —	—	—	—	B.fl.IV,55	M.fr.I,114;III,163;XI,50.
G. xanthotricha, De Vriese, Gooden. 155 (1854)	W.A.	—	— —	—	—	—	B.fl.IV,56	M.fr.VI,13.
G. (ex Picrophyta R. Brown, prodr. 578 (1810)	—	—	— —	—	—	—	B.fl.IV,56	M.fr.I,114;III,163;XI,50.
G. (St.) brevicarpa, R. Brown, prodr. 578 (1810)	W.A.	—	— —	—	—	—	B.fl.IV,57	M.fr.III,141;V,13.
G. (St.) eglandulosa, F. v. M., III, 20, t. 17 (1862)	—	—	— —	—	—	N.A.	B.fl.IV,57	M.fr.VIII,245;X,13.
G. (St.) fasciculata, R. Brown, pr. 67, t. 30 (1854)	W.A.	—	— —	—	—	—	B.fl.IV,58	
G. (St.) falcata, R. Brown, prodr. 57 (1810)	—	—	— —	—	—	—	B.fl.IV,58	
G. (St.) rhynchocarpa, Sonder in Lehm.	—	—	— —	N.S.W.	Q.	—	B.fl.IV,58	M.fr.I,114.
G. (St.) adnata, R. Brown, prodr. 572 (1810) II, 349 (1794)	—	—	— —	N.S.W.	Q.	—	B.fl.IV,59	M.fr.XI,50.
LEEWENIIOEKIA, R. Brown, prodr. 572	—	—	— —	N.S.W.	Q.	—	B.fl.IV,59	M.fr.V,14.
L. pusilla, R. Brown, prodr. 573 (1810) [94]	—	—	— —	N.S.W.	—	—	B.fl.IV,59	M.fr.X,110.
L. dubia, Sonder in Lehm. pl. Preiss. I, 392 (1845) II, 70 (1850)	—	—	— —	—	—	—	B.fl.IV,60	M.fr.VI,15.
L. Sonderi, F. v. M., fragm. I, 18 (1858)	W.A.	—	— —	—	—	—	B.fl.IV,60	M.fr.I,119.
L. pauciflora, Bentham in Hueg. enum. 67 (1837)...	—	S.A.	— V.	N.S.W.	—	—	B.fl.IV,61	M.fr.I,205.
L. leptantha, Bentham, Fl. Austr. IV, 35 (1869) ...	W.A.	—	— —	—	—	—	B.fl.IV,61	
L. stipitata, F. v. M., fragm. IV, 94 (1864)	—	—	— T. V.	N.S.W.	—	—	B.fl.IV,61	M.fr.II,110,176;VI,15.
L. Preissii, F. v. M., fragm. IV, 94 (1864)	W.A.	—	— —	—	—	—	B.fl.IV,62	
	—	—	— —	—	Q.	—	B.fl.IV,62	M.fr.IV,145;VI,15.

	W.A.	S.A.	T.	V.	N.S.W.	Q.			
G. geniculata, R. Brown, prodr. 577 (1810)	W.A.	S.A.	T.	V.	N.S.W.	Q.	—	B.fl.IV,02	M.fr.VI,14.
G. hederacea, Smith in Transact. Linn. Soc. II, 349 (1794)	—	—	—	V.	N.S.W.	Q.	—	B.fl.IV,63	M.fr.VI,15,110.
G. xanthosperma, F. v. M., fragm. X, 12 (1876)	W.A.	—	—	—	—	—	—		M.fr.X,12
G. hirsuta, F. v. M., fragm. III, 35 (1802)	—	S.A.	—	—	—	—	N.A.	B.fl.IV,64	M.fr.III,35.
G. heterophylla, Smith in Transact. Linn. Soc. II, 349 (1794)	—	—	—	—	N.S.W.	Q.	—	B.fl.IV,04	
G. glabra, R. Brown, prodr. 577 (1810)	—	—	—	—	N.S.W.	Q.	—	B.fl.IV,64	M.fr.VI,14.
G. strongylophylla, F. v. M., fragm. VI, 12 (1867)	—	—	—	—	—	Q.	—	B.fl.IV,65	M.fr.VI,12.
G. rotundifolia, R. Brown, prodr. 576 (1810)	—	—	—	—	N.S.W.	Q.	—	B.fl.IV,65	M.fr.VI,15.
G. azurea, F. v. M., fragm. I, 117 (1859)	—	—	—	—	—	—	N.A.	B.fl.IV,66	M.fr.I,117.
G. scaevolina, F. v. M., fragm. I, 118 (1859)	—	—	—	—	—	—	N.A.	B.fl.IV,66	M.fr.I,118.
G. Stobbeiana, F. v. M., fragm. XI, 49 (1879)	—	—	—	—	—	—	N.A.		M.fr.XI,49,137.
G. incana, R. Brown, prodr. 578 (1810)	W.A.	—	—	—	—	—	—	B.fl.IV,66	M.fr.I,155;III,141. [50.
G. leptoclada, Bentham, Fl. Austr. IV, 67 (1869)	W.A.	—	—	—	—	—	—	B.fl.IV,67	M.fr.VI,227;VIII,186;XI,
G. Eatoniana, F. v. M., fragm. VIII, 186 (1874)	W.A.	—	—	—	—	—	—		M.fr.XI,51.
G. coerulea, R. Brown, prodr. 578 (1810)	—	—	—	—	—	—	—	B.fl.IV,67	M.fr.III,140;XI,20.
G. trichophylla, De Vriese in Benth. Fl. Austr. IV, 67 (1809)	W.A.	—	—	—	—	—	—	B.fl.IV,67	
G. Hassallii, F. v. M., fragm. VI, 10, t. 51 (1867)	W.A.	—	—	—	—	—	—	B.fl.IV,68	M.fr.XI,50.
G. pterygosperma, R. Brown, prodr. 578 (1810)	W.A.	—	—	—	—	—	—	B.fl.IV,68	M.fr.I,155;III,163.
G. Vilmoriniae, F. v. M., fragm. III, 19, t. 16 (1862)	—	S.A.	—	—	—	—	N.A.	B.fl.IV,68	M.fr.III,19;VIII,245.
G. Bonneyana, F. v. M. in Benth. Fl. Austr. IV, 69 (1869)	W.A.	—	—	—	—	—	—	B.fl.IV,69	M.fr.VI,226,230.
G. calcarata, F. v. M., fragm. VI, 14 (1867)	—	S.A.	—	—	N.S.W.	—	—	B.fl.IV,09	M.fr.VIII,245.
G. albiflora, Schlechtendal, Linnaea XX, 599 (1847)	—	S.A.	—	—	—	—	—	B.fl.IV,70	
G. Nicholsoni, F. v. M., fragm. I, 203, t. 4 (1859)	—	S.A.	—	—	—	—	—	B.fl.IV,09	M.fr.I,203.
G. Macmillani, F. v. M., fragm. I, 119, t. 5 (1859)	—	—	—	V.	—	—	—	B.fl.IV,69	M.fr.I,119.
G. grandiflora, Sims, Bot. Mag. t. 890 (1806)	—	—	—	—	N.S.W.	Q.	—	B.fl.IV,69	M.fr.I,204;VI,14.
G. Chamberaii, F. v. M., fragm. I, 204 (1859)	—	S.A.	—	—	—	—	—	B.fl.IV,70	M.fr.I,204.
G. Strangfordii, F. v. M., fragm. VI, 11, t. 52 (1867)	—	—	—	—	—	Q.	N.A.	B.fl.IV,70	
G. Mitchellii, Bentham, Fl. Austr. IV, 71 (1869)	—	S.A.	—	—	—	Q.	—	B.fl.IV,70	M.fr.XI,75.
G. Mueckeana, F. v. M., fragm. VIII, 56 (1873)	—	S.A.	—	—	—	—	—		M.fr.IX,194;XI,133.
G. melanoptera, F. v. M., fragm. I, 115 (1859)	—	—	—	—	—	—	N.A.		M.fr.I,115.
G. heterochila, F. v. M., fragm. III, 142 (1863)	—	S.A.	—	—	—	—	N.A.	B.fl.IV,71	M.fr.VIII,245.
G. sepalosa, F. v. M. in Benth. Fl. Austr. IV, 72 (1869)	—	—	—	—	—	—	N.A.	B.fl.IV,72	
G. hispida, R. Brown, prodr. 577 (1810)	—	—	—	—	—	—	N.A.	B.fl.IV,72	M.fr.VI,12.
G. auriculata, Bentham, Fl. Austr. IV, 72 (1869)	—	—	—	—	—	—	N.A.	B.fl.IV,73	
G. Armstrongiana, De Vriese, Goodenov. 138, t. 24 (1854)	—	—	—	—	—	—	N.A.	B.fl.IV,73	M.fr.I,205;VII,40.
G. corynocarpa, F. v. M., fragm. II, 16 (1860)	W.A.	—	—	—	—	—	—	B.fl.IV,73	M.fr.II,16;VI,15.
G. mollissima, F. v. M. in Benth. Fl. Austr. IV, 73 (1869)	—	—	—	—	—	Q.	—	B.fl.IV,73	
G. cycloptera, R. Brown in App. Sturt's Exped. 20 (1849)	—	S.A.	—	V.	N.S.W.	—	—	B.fl.IV,74	M.fr.VI,14;XI,51.
G. tenella, R. Brown, prodr. 577 (1810)	W.A.	—	—	—	—	—	—	B.fl.IV,74	M.fr.XI,51. [X,110.
G. elongata, Labillardiere, Nov. Holl. pl. spec. I, 52, t. 75 (1804)	—	S.A.	T.	V.	N.S.W.	—	—	B.fl.IV,74	M.fr.VI,14;VIII,142,245;
G. pinnatifida, Schlechtendal, Linnaea XXI, 440 (1848)	—	S.A.	—	V.	N.S.W.	—	—	B.fl.IV,75	M.fr.VI,14;XI,51.
G. coronopifolia, R. Brown, prodr. 577 (1810)	—	—	—	—	—	—	N.A.	B.fl.IV,75	
G. heteromera, F. v. M., fragm. I, 115 (1859)	—	S.A.	—	V.	N.S.W.	—	—	B.fl.IV,76	M.fr.VI,14;XI,27.
G. concinna, Bentham, Fl. Austr. IV, 76 (1869)	W.A.	—	—	—	—	—	—	B.fl.IV,76	
G. glauca, F. v. M. in Transact. Vict. Inst. 40 (1855)	W.A.	S.A.	—	V.	N.S.W.	Q.	N.A.	B.fl.IV,76	M.fr.VI,14;XI,27.
G. filiformis, R. Brown, prodr. 578 (1810)	W.A.	—	—	—	—	—	—	B.fl.IV,77	M.fr.VI,14;VIII,245.
G. Armitiana, F. v. M., fragm. X, 110 (1877)	—	—	—	—	—	Q.	N.A.		M.fr.X,110.
G. microptera, F. v. M., fragm. III, 34 (1862)	—	S.A.	—	—	—	—	N.A.	B.fl.IV,77	M.fr.VI,227;XI,75.
G. bicolor, F. v. M. in Benth. Fl. Austr. IV, 80 (1800)	—	—	—	—	—	—	N.A.	B.fl.IV,80	[245;X,110.
G. humilis, R. Brown, prodr. 575 (1810)	S.A.	T.	V.	N.S.W.	—	—	—	B.fl.IV,79	M.fr.VII,227;VIII,142,
G. paniculata, Smith in Transact. Linn. Soc. II, 348 (1794)	—	—	—	V.	N.S.W.	Q.	—	B.fl.IV,78	M.fr.I,110.
G. purpurascens, R. Brown, prodr. 578 (1810)	—	—	—	—	—	Q.	N.A.	B.fl.IV,79	M.fr.I,117.
G. gracilis, R. Brown, prodr. 575 (1810)	—	—	—	V.	N.S.W.	Q.	N.A.	B.fl.IV,79	M.fr.XI,51.
G. lamprosperma, F. v. M., fragm. I, 116 (1859)	—	—	—	—	—	—	N.A.	B.fl.IV,79	M.fr.I,116.
G. minutiflora, F. v. M., fragm. VIII, 244 (1874)	—	—	—	—	—	—	N.A.		M.fr.VIII,244.
G. claytoniacea, F. v. M. in Benth. Fl. Austr. IV, 79 (1869)	W.A.	—	—	—	—	—	—	B.fl.IV,79	M.fr.II,111.
G. Punilio, R. Brown, prodr. 579 (1810)	—	—	—	—	—	Q.	N.A.	B.fl.IV,80	M.fr.XI,111.

VELLEYA, Smith in Transact. Linn. Soc. IV, 217 (1798). (Velleia, Euthales.)

	W.A.	S.A.	T.	V.	N.S.W.	Q.			
V. Daviesii, F. v. M., fragm. X, 10 (1876)	W.A.	—	—	—	—	—	—		M.fr.X,10.
V. panduriformis, Cunningham in Benth. Fl. Austr. IV, 46 (1869)	—	—	—	—	—	—	N.A.	B.fl.IV,40	M.fr.XI,73.
V. connata, F. v. M. in Transact. phil. Soc. Vict. I, 18 (1855)	—	S.A.	—	V.	N.S.W.	—	N.A.	B.fl.IV,40	M.fr.VI,227;VIII,246.
V. perfoliata, R. Brown, prodr. 581 (1810)	—	—	—	—	N.S.W.	—	—	B.fl.IV,46	
V. discophora, F. v. M., fragm. X, 10 (1876)	W.A.	—	—	—	—	—	—		M.fr.X,10.
V. trinervis, Labillard., Nov. Holl. pl. spec. I, 54, t. 77 (1804)	W.A.	—	—	—	—	—	—	B.fl.IV,47	M.fr.XI,75.
V. macrophylla, Bentham, Fl. Austr. IV, 47 (1869)	—	—	—	—	—	—	N.A.	B.fl.IV,47	M.fr.XI,75.
V. paradoxa, R. Brown, prodr. 580 (1810)	—	S.A.	T.	V.	N.S.W.	Q.	—	B.fl.IV,48	M.fr.VI,11;XI,75.
V. cyanopetamica, F. v. M., fragm. VI, 7 (1867)	W.A.	—	—	—	—	—	—	B.fl.IV,48	M.fr.VI,249;X,10.
V. lyrata, R. Brown, prodr. 580 (1810)	—	—	—	—	N.S.W.	—	—	B.fl.IV,49	
V. macrocalyx, De Vriese, Goodenov. 176, t. 34 (1854)	—	—	—	—	N.S.W.	Q.	—	B.fl.IV,49	M.fr.X,75.
V. spathulata, R. Brown, prodr. 580 (1810)	—	—	—	—	N.S.W.	Q.	—	B.fl.IV,50	M.fr.XI,75.
V. pubescens, R. Brown, prodr. 581 (1810)	—	—	—	—	N.S.W.	Q.	—	B.fl.IV,49	M.fr.VI,11;XI,75.
V. montana, J. Hooker in Lond. Journ. of Bot. VI, 205 (1847)	—	—	—	T.	V.	N.S.W.	—	B.fl.IV,50	

SYNPETALEAE HYPOGYNAE.

F. v. M. in Woolls, pl. of the neighb. of Sydney 34 (1880).

GENTIANEAE.

Necker in Act. Acad. Theod. Pal. II, 477 (1770), from B. de Jussieu (1759).

LIMNANTHEMUM, Gmelin, Nov. Comm. Acad. Petrop. XIV, 257 (1770). (Villarsia, Liparophyllum.)

L. Indicum, Thwaites, enum. pl. Ceyl. 205 (1862)	...	—	—	—	--	N.S.W.	Q. N.A.	B.fl.IV,378	M.fr.VI,136.
L. Moonii, Thwaites, enum. pl. Ceyl. 205 (1862) ...	...	—	—	—	—	—	Q. —		M.fr.IX,163.
L. minimum, F. v. M., fragm. I, 40 (1858)...	...	...	—	—	—	—	— N.A.	B.fl.IV,379	M.fr.VI,137. [165.
L. crenatum, F. v. M. in Transact. phil. Soc. Vict. I, 17 (1854)	—	S.A.	— V. N.S.W.	Q. N.A.	B.fl.IV,379	M.fr.IV,127;VI,137;IX,
L. geminatum, Grisebach, gen. et spec. Gent. 346 (1839)	...	—	— V. N.S.W.	Q. N.A.	B.fl.IV,380	M.fr.VI,137,254.
L. hydrocharoides, F. v. M. in Benth. Fl. Austr. IV, 380 (1869)	—	—	—	—	Q. —	B.fl.IV,380	M.fr.VI,139;IX,165.
L. exiliflorum, F. v. M., fragm. V, 46 (1865)	...	...	...	—	—	—	Q. —	B.fl.IV,381	M.fr.VI,137.
L. exiguum, F. v. M., fragm. I, 40 (1858)...	...	...	—	T. —	—	—	—	B.fl.IV,381	M.fr.VI,137.
L. Gunnii, J. Hooker, Fl. Tasman. II, 368 (1860)...	...	...	—	T. —	—	—	—	B.fl.IV,382	M.fr.VI,137;IX,164.
L. capitatum, F. v. M., fragm. IX, 164 (1875) ...	...	W.A.	—	—	—	—	—	B.fl.IV,375	M.fr.VI,142.
L. calthifolium, F. v. M., fragm. IX, 164 (1875)...	...	W.A.	—	—	—	—	—	B.fl.IV,374	M.fr.VI,140.
L. congestiflorum; Villarsia, F. v. M., fragm. VI, 141 (1868)	...	W.A.	—	—	—	—	—	B.fl.IV,375	M.fl.VI,141.
L. latifolium, F. v. M., fragm. IX, 164 (1875)	...	...	W.A.	—	—	—	—	—	B.fl.IV,375	M.fr.VI,141.
L. exaltatum, F. v. M., fragm. IX, 165 (1875)	...	...	—	S.A.	T. V. N.S.W.	Q.	—	B.fl.IV,376	M.fr.II,21;VI,139.
L. parnassifolium, F. v. M., fragm. IX, 165 (1875)	...	W.A.	—	—	—	—	—	B.fl.IV,376	M.fr.VI,140.
L. violifolium; Villarsia, F. v. M., fragm. VI, 138 (1868)	...	W.A.	—	—	—	—	—	B.fl.IV,377	M.fr.VI,138.
L. lasiospermum, F. v. M., fragm. IX, 165 (1875)	...	W.A.	—	—	—	—	—	B.fl.IV,377	M.fr.VI,137.
L. albiflorum; Villarsia, F. v. M., fragm. II, 21 (1860) ...	...	W.A.	—	—	—	—	—	B.fl.IV,377	M.fr.VI,138.

SEBAEA, Solander in R. Brown, prodr. 451 (1810).

S. ovata, R. Brown, prodr. 452 (1810)	...	...	...	...	W.A.	S.A.	T. V. N.S.W. Q.	—	B.fl.IV,371	M.fr.VII,130;IX,165.
S. albidiflora, F. v. M. in Transact. phil. Soc. Vict. I, 46 (1854)	—	S.A.	T. V.	—	—	—	B.fl.IV,371	M.fr.VI,130.

ERYTHRAEA, Reneaulme, specim. hist. plant. 77, t. 76 (1611). [20.

E. australis, R. Brown, prodr. 451 (1810) ...	...	W.A.	S.A.	T. V. N.S.W.	Q. N.A.	B.fl.IV,371	M.fr.VI,136;IX,165;XI,

CANSCORA, Lamarck, Eucycl. method. I, 602 (1783). (Orthostemon.)

C. diffusa, R. Brown, prodr. 451 (1810)	...	...	...	—	—	—	Q. N.A.	B.fl.IV,372	M.fr.VI,136;IX,165.

GENTIANA, Tournefort, inst. 80, t. 40 (1700), from Dioscorides.

G. saxosa, Forster in Svensk. Kongl. Vet. Ac. Handl. 183, t. 5 (1777)	—	S.A.	T. V. N.S.W.	—	—	B.fl.IV,373	M.fr.VI,136,255;IX,165.

LOGANIACEAE.

R. Brown in App. Flind. Voy. 564 (1814).

STRYCHNOS, Linné, syst. nat. 8 (1735); Linné, spec. plant. 189 (1753).

S. lucida, R. Brown, prodr. 469 (1810) ...	...	...	—	—	—	—	Q. N.A.	B.fl.IV,309	
S. psilosperma, F. v. M., fragm. IV, 44 (1864)	...	...	—	—	—	—	Q. —	B.fl.IV,309	M.fr.VI,131.

FAGRAEA, Thunberg in Kongl. Svensk. Vetensk. Acad. Handl. 125 (1782).

F. racemosa, Jack in Roxb. Fl. Ind. ed. Carey. II, 35 (1824)	..	—	—	—	—	— N.A.	B.fl.IV,367	M.fr.II,137.
F. Muelleri, Bentham, Fl. Austr. IV, 368 (1869)...	...	..	—	—	—	—	Q. —	B.fl.IV,368	

MITREOLA, Linné, hort. Cliffort. 492 (1737).

M. oldenlandioides, Wallich, numeric. list 4350 (1828)	...	...	—	—	—	—	— N.A.	B.fl.IV,349	M.fr.VI,131.

MITRASACME, Labillardière, Nov. Holl. pl. spec. I, 36, t. 49 (1804). (Mitragyne.)

M. Archeri, J. Hooker, Fl. Tasman. II, 368 (1860)	...	—	T. —	—	—	B.fl.IV,351	
M. montana, J. Hooker, Fl. Tasman. I, 274, t. 88 (1859)	...	—	T. V.	—	—	B.fl.IV,351	M.fr.VI,131.
M. serpillifolia, R. Brown, prodr. 454 (1810)	...	—	T. V. N.S.W.	—	B.fl.IV,352	
M. paludosa, R. Brown, prodr. 453 (1810)	...	...	—	—	— N.S.W.	Q.	B.fl.IV,352	M.fr.I,134.
M. pilosa, Labillardière, Nov. Holl. pl. spec. I, 36, t. 49 (1804)	—	—	T. N.S.W.	—	B.fl.IV,352	M.fr.VI,131.
M. alsinoides, R. Brown, prodr. 453 (1810)...	...	...	—	—	— N.S.W.	Q.	B.fl.IV,353	
M. polymorpha, R. Brown, prodr. 452 (1810)	...	...	—	—	— V. N.S.W.	Q.	B.fl.IV,353	M.fr.VI,131.
M. longiflora, F. v. M. in Benth. Fl. Austr. IV, 353 (1869)	...	—	—	—	— N.A.	B.fl.IV,353	
M. elata, R. Brown, prodr. 453 (1810)	...	...	—	—	—	Q. N.A.	B.fl.IV,354	M.fr.I,132.
M. exserta, F. v. M., fragm. I, 131 (1859)...	...	...	—	—	— N.A.	B.fl.IV,354	M.fr.I,131.
M. tenuiflora, Benth. Fl. Austr. IV, 354 (1869)	...	—	—	— N.A.	B.fl.IV,354	
M. ambigua, R. Brown, prodr. 454 (1810) ...	...	—	—	—	Q. —	B.fl.IV,354	
M. nudicaulis, Reinwardt in Blume, Bijdr. 849 (1826)	...	—	—	— N.A.	B.fl.IV,355	
M. connata, R. Brown, prodr. 454 (1810)	...	...	—	—	—	Q. N.A.	B.fl.IV,355	M.fr.I,131.
M. laevis, Bentham in Journ. Linn. Soc. I, 93 (1857)	...	—	—	— N.S.W.	Q. N.A.	B.fl.IV,356	M.fr.I,133.
M. stellata, R. Brown, prodr. 454 (1810)	...	...	—	—	—	Q. N.A.	B.fl.IV,356	
M. Cunninghamii, Bentham, Fl. Austr. IV, 357 (1869) ...	—	—	—	Q. —	B.fl.IV,357	
M. pygmaea, R. Brown, prodr. 453 (1810) ...	...	—	—	—	Q. —	B.fl.IV,357	
M. lutea, F. v. M., fragm. I, 133 (1859) ...	...	—	—	— N.A.	B.fl.IV,357	M.fr.VI,170.
M. multicaulis, R. Brown, prodr. 453 (1810)	...	...	—	—	—	Q. N.A.	B.fl.IV,357	
M. laricifolia, R. Brown, prodr. 453 (1810)...	...	...	—	—	—	Q. N.A.	B.fl.IV,358	M.fr.VI,131.
M. prolifera, R. Brown, prodr. 454 (1810) ...	...	—	—	—	Q. N.A.	B.fl.IV,358	
M. gentianea, F. v. M., fragm. I, 130 (1859)	...	—	—	— N.A.	B.fl.IV,358	M.fr.I,130.
M. phascoides, R. Brown, prodr. 454 (1810) ...	...	—	—	—	Q. —	B.fl.IV,359	
M. paradoxa, R. Brown, prodr. 454 (1810)...	...	...	W.A.	S.A.	T. V. N.S.W.	—	B.fl.IV,359	M.fr.VI,131.
M. distylis, F. v. M. in Transact. phil. Soc. Vict. I, 20 (1854)...	—	T. V.	—	—	B.fl.IV,359	M.fr.VI,131.

GENIOSTOMA, R. & G. Forster, char. gen. 23, t. 12 (1776).

G. Australianum, F. v. M., fragm. V, 19 (1865) ...	...	—	—	—	—	Q. —	B.fl.IV,367	M.fr.V,19.
G. petiolosum, Moore & F. v. M., fragm. VII, 28 (1869)...	...	—	—	—	— N.S.W.	—	—	M.fr.IX,193.

91

LOGANIA, R. Brown, prodr. 454 (1810).   (Euosma 1808, restorable.)
L. longifolia, R. Brown, prodr. 456 (1810) ... ... W.A. S.A. — — — — — B.fl.IV,361
L. latifolia, R. Brown, prodr. 455 (1810) .. .. W.A. — — — — — B.fl.IV,361 M.fr.VI,132.
L. crassifolia, R. Brown, prodr. 455 (1810)... ... ... W.A. S.A. — — — — — B.fl.IV,362
L. ovata, R. Brown, prodr. 455 (1810) ... ... W.A. S.A. — V. — — B.fl.IV,362
L. buxifolia, F. v. M., fragm. VI, 132 (1868) ... ... W.A. — — — — — B.fl.IV,362 M.fr.IV,132.
L. stenophylla, F. v. M., fragm. I, 128 (1859) ... ... W.A. — — — — — B.fl.IV,362 M.fr.I,128.
L. micrantha, Bentham in Journ. Linn. Soc. I, 94 (1857) ... W.A. — — — — — B.fl.IV,363 M.fr.VI,132.
L. linifolia, Schlechtendal, Linnaea XX, 605 (1847) ... — S.A. — V. N.S.W. — — B.fl.IV,363 M.fr.VI,131.
L. fasciculata, R. Brown, prodr. 456 (1810) ... ... W.A. — — — — — B.fl.IV,363 M.fr.VI,131.
L. floribunda, R. Brown, prodr. 456 (1810) ... ... — — V. N.S.W. — B.fl.IV,364 M.fr.VI,131.
L. cordifolia, Hooker in Mitch. Trop. Austr. 341 (1848)... ... ... — — — — Q. — B.fl.IV,364
L. nuda, F. v. M., fragm. I, 129 (1859) ... ... W.A. S.A. — V. N.S.W. — B.fl.IV,365 M.fr.III,163; VI,131.
L. spermacocea, F. v. M., fragm. VI, 134 (1868)... ... ... W.A. — — — — — B.fl.IV,365 M.fr.VI,134.
L. callosa, F. v. M., fragm. VI, 134 (1868) ... ... W.A. — — — — — B.fl.IV,365 M.fr.VI,134.
L. campanulata, R. Brown, prodr. 456 (1810) ... ... W.A. — — — — — B.fl.IV,365 M.fr.VI,133.
L. serpillifolia, R. Brown, prodr. 456 (1810) ... ... W.A. — — — — — B.fl.IV,366 M.fr.VI,133.
L. pusilla, R. Brown, prodr. 456 (1810) ... ... — — N.S.W. Q. — B.fl.IV,366 M.fr.VI,131.

PLANTAGINEAE.
A. L. de Jussieu, gen. plant. 89 (1789).

PLANTAGO, Tournefort, inst. 126, t. 48 (1700), from l'Ecluse (1576).
P. varia, R. Brown, prodr. 424 (1810) ... ... W.A. S.A. T. V. N.S.W. Q. — B.fl.V,139 M.fr.XI,20.
P. stellaris, F. v. M., fragm. II, 23 (1860) ... ... — — V. N.S.W. — — B.fl.V,141 M.fr.VIII,148.
P. Brownii, Rapin in Mém. de la Soc. Linn. Par. VI, 484 (1827) — — T. — — — B.fl.V,141 M.fr.II,23.
P. Gunnii, J. Hooker in Lond. Journ. V, 446, t. 13 (1846) ... — — T. V. N.S.W. — B.fl.V,142 M.fr.VIII,148; XI,134.

PRIMULACEAE.
Ventenat, Tabl. du régn. végét. II, 285 (1799).

ANAGALLIS, Tournefort, inst. 142, t. 59 (1700), from Hippocrates and Dioscorides.   (Centunculus, Micropyxis.)
A. pumila, Swartz, Fl. Ind. occ. I, 345 (1797) ... ... — — — Q. N.A. B.fl.IV,270
LYSIMACHIA, Tournefort, inst. 141, t. 59 (1700), from Dioscorides.
L. salicifolia, F. v. M. in Benth. Fl. Austr. IV, 269 (1869) ... — — V. N.S.W. — B.fl.IV,269
L. Japonica, Thunberg, F. Japon. 83 (1784) ... ... — — N.S.W. — B.fl.IV,269
SAMOLUS, Tournefort, inst. 143, t. 60 (1700).   (Sheffieldia.)
S. Valerandi, Linné, spec. plant. 171 (1753) ... ... — V. N.S.W. Q. B.fl.IV,270
S. repens, Persoon, synops. plant. I, 171 (1805) ... ... W.A. S.A. T. V. N.S.W. Q. N.A. B.fl.IV,271 M.fr.XI,20.
S. platyphyllus, F. v. M. inedit. ... ... — S.A. — — N.A. —

MYRSINEAE.
R. Brown, prodr. 532 (1810).   (Primulacearum subordo.)

MAESA, Forskael, Fl. Aegypt. Arab. 66 (1775).   (Baeobotrys.)
M. dependens, F. v. M., fragm. V, 107 (1865) ... ... — — — — Q. — B.fl.IV,273 M.fr.V,107.
M. haplobotrys, F. v. M., fragm. V, 161 (1866) ... ... — — — — Q. — B.fl.IV,273 M.fr.V,161.
SAMARA, Linné, mantiss. II, 144 (1771).   (Choripetalum.)
S. Australiana, F. v. M., fragm. VI, 164 (1868) ... ... — — — N.S.W. Q. — B.fl.IV,274 M.fr.IV,36.
MYRSINE, Linné, syst. nat. 8 (1735); Linné, gen. plant. 54 (1737).
M. urceolata, R. Brown, prodr. 534 (1810)... ... ... — — — — Q. — B.fl.IV,274
M. campanulata, F. v. M., fragm. VI, 235 (1868)... ... ... — — — — — — M.fr.VI,235.   [48.
M. crassifolia, R. Brown, prodr. 534 (1810) ... ... — — — N.S.W. Q. — B.fl.IV,275 M.fr.IV,81; VI,164; VIII,
M. platystigma, F. v. M., fragm. VIII, 48 (1873) ... ... — — N.S.W. — — M.fr.VIII,48.
M. variabilis, R. Brown, prodr. 534 (1810)... ... ... — — V. N.S.W. Q. — B.fl.IV,275
M. achradifolia, F. v. M., fragm. VI, 164 (1868) ... ... — — — — Q. — B.fl.IV,275 M.fr.VI,164.
ARDISIA, Swartz, nov. gen. et spec. 3 et 48 (1788).
A. Pseudo-Jamboaa, F. v. M., fragm. IV, 81 (1864) ... ... — — N.S.W. Q. — B.fl.IV,276 M.fr.IV,81.
A. brevipedata, F. v. M., fragm. VI, 103 (1868) ... ... — — N.S.W. Q. — B.fl.IV,276 M.fr.VI,163.
AEGICERAS, Gaertner, de fruct. I, 216, t. 46 (1788).
A. majus, Gaertner, de fruct. I, 216, t. 46 (1789) ... ... — — N.S.W. Q. N.A. B.fl.IV,277

SAPOTACEAE.
A. L. de Jussieu, gen. 151 (1789), from B. de Jussieu (1759).

NIEMEYERA, F. v. M., fragm. VII, 114 (1870).   (Chrysophyllum partly.)
N. prunifera, F.v. M., fragm. VII, 114 (1870) ... ... — N.S.W. Q. — B.fl.IV,278 M.fr.VI,26.
AMORPHOSPERMUM, F. v. M., fragm. VII, 112 (1870).
A. antilogum, F. v. M., fragm. VII, 113 (1870) ... ... — — — N.S.W. Q. — M.fr.X,119.
LUCUMA, Molina, Saggio, 186 (1782).   (Sersalisia.)
L. sericea, Bentham & J. Hooker, gen. pl. II, 654 (1876) ... — — — — Q. N.A. B.fl.IV,279 M.fr.VII,112.
L. Galactoxylon, Bentham & J. Hooker, gen. pl. II, 654 (1876) — — — — Q. — B.fl.IV,279
SIDEROXYLON, Dillenius, hort. Eltham. 357, t. 265 (1732).   (Achras partly, Sapota partly, Sersalisia partly).
S. Arnhemicum, Bentham & J. Hooker, gen. pl. II, 655 (1876)... — — — — N.A. B.fl.IV,280
S. Pohlmanianum, Bentham & J. Hooker, gen. pl. II, 655 (1876) — — — — Q. — B.fl.IV,281 M.fr.V,184.
S. xerocarpum, Bentham & J. Hooker, gen. pl. II, 655 (1876) .. — — — — Q. — B.fl.IV,281
S. chartaceum, Bentham & J. Hooker, gen. pl. II, 655 (1876) ... — — — — Q. — B.fl.IV,281

S. Richardi, F. v. M.; Sersalisia laurifolia, A. Rich. sert. Astrol.
84 t. 13 (1834) ... ... — — — — N.S.W. Q. — B.fl.IV,282
S. australe, Bentham & J. Hooker, gen. pl. II, 655 (1876) ... — — — — N.S.W. Q. — B.fl.IV,282
S. Brownii, F.v.M.; Sersalisia obovata, R. Brown, prodr. 530 (1810) — — — — — Q. — B.fl.IV,283
S. myrsinoides, Bentham & J. Hooker, gen. pl. II, 655 (1876) ... — — — — N.S.W. Q. N.A. B.fl.IV,283
S euphlebium; Achras, F. v. M., fragm. VII (1870) ... .— — — — — Q. — — M.fr.VII,110.
S. Brownlessianum, F. v. M., fragm. VIII, 111 (1870) ... — — — — — Q. — — M.fr.VII,111.
S. Howeanum; Achras, F. v. M., fragm. IX, 72 (1875) ... ... — — — — N.S.W. — — — M.fr.IX,72.
S. costatum; Achras, Endlicher, prodr. fl. ins. Norf. 40 (1833) ... — — — — N.S.W. — — — M.fr.IX,72.
    HORMOGYNE, A. De Candolle, prodr. VIII, 176 (1844).
H. cotinifolia, A. De Candolle, prodr. VIII, 176 (1844) ... ... — — — — N.S.W. Q. — B.fl.IV,284 M.fr.VII,111.
    MIMUSOPS, Linné, Fl. Zeil. 57 (1747).
M. parvifolia, R. Brown, prodr. 531 (1810)... ... ... — — — — — -- Q. N.A. B.fl.IV,284 M.fr.VI,178;VII,114.
M. Browniana, Bentham, Fl. Austr. IV, 285 (1869) ... ... — — — — — Q. — B.fl.IV,285

EBENACEAE.
Ventenat, Tabl. II, 443 (1799).

    DIOSPYROS, Linné, gen. plant. 143 (1737). (Cargillia.)
D. cordifolia, Roxburgh, pl. Corom. I, 38, t. 50 (1795) ... — — — — — — N.A. B.fl.IV,286
D. hebecarpa, Cunningham in Benth. pl. Austr. IV, 286 (1869) — — — — — Q. N.A. B.fl.IV,286
D. maritima, Blume, Bijdrag. 669 (1825) ... ... — — — — — Q. N.A. B.fl.IV,287 M.fr.V,163.
D. manacca, F. v. M., Docum. intercol. Exhib. 35 (1866) — — — — N.S.W. Q. — B.fl.IV,288 M.fr.V,162.
D. Cargillia, F. v. M., Docum. intercol. Exhib. 35 (1866) — — — — N.S.W. Q. — B.fl.IV,289 M.fr.V,162.
D. pentamera, F. v. M. & Woolls, Doc. intercol. Exhib. 35 (1866) — — — — N.S.W. Q. — B.fl.IV,288 M.fr.V,163.
    MABA, R. & G. Forster, char. gen. 121, t. 61 (1776).
M. laurina, R. Brown, prodr. 527 (1810) ... ... — — — — — Q. — B.fl.IV,280
M. rufa, Labillardière, sert. Austr. Caled. 33, t. 36 (1824) — — — — — Q. — B.fl.IV,280 M.fr.V,164;VI,253.
M. hemicycloides, F. v. M. in Benth. Fl. Austr. IV, 290 (1869) — — — — — Q. — B.fl.IV,290
M. fasciculosa, F. v. M., fragm. V, 163 (1866) ... ... — — — — N.S.W. Q. — B.fl.IV,290 M.fr.V,163;X,74.
M. compacta, R. Brown, prodr. 528 (1810) ... ... — — — — — Q. N.A. B.fl.IV,290
M. reticulata, R. Brown, prodr. 528 (1810)... ... ... — — — — — Q. N.A. B.fl.IV,291
M. geminata, R. Brown, prodr. 527 (1810) ... ... — — — — — Q. N.A. B.fl.IV,291
M. humilis, R. Brown, prodr. 527 (1810) ... ... — — — — — Q. N.A. B.fl.IV,291 M.fr.X,74.
M. buxifolia, Persoon, synops. pl. II, 602 (1807) ... ... — — — — — -- N.A. — M.fr.X,74.

STYRACEAE.
L. C. Richard, Analyse du fruit. 48 (1808).

    SYMPLOCOS, N. J. Jacquin, enum. plant. Carib. 24 (1760).
S. Stawellii, F. v. M., fragm. V, 60 (1865)... ... — — — — N.S.W. Q. — B.fl.IV,292 M.fr.V,60.
S. Thwaitesii, F. v. M., fragm. III, 22 (1862) ... ... — — — — N.S.W. Q. — B.fl.IV,293 M.fr.V,211.

JASMINEAE.
A. L. de Jussieu, gen. 104 (1789), from Neoker (1770).

    JASMINUM, Tournefort, inst. 368, t. 368 (1700), from l'Ecluse (1611). [43.
J. didymum, G. Forster, prodr. 3 (1786) ... ... — — — — N.S.W. Q. N.A. B.fl.IV,294 M.fr.IV,150;VI,67;VIII,
J. racemosum, F. v. M., fragm. I, 19 (1858) ... ... — — — — — Q. — B.fl.IV,295 M.fr.I,19;VI,67;VIII,43.
J. lineare, R. Brown, prodr. 521 (1810) ... ... W.A. S.A. — V. N.S.W. Q. N.A. B.fl.IV,295 M.fr.VI,67;VIII,43.
J. simplicifolium, G. Forster, prodr. 3 (1786) ... ... — — — — N.S.W. Q. N.A. B.fl.IV,296 M.fr.VI,67;VIII,43.
J. aemulum, R. Brown, prodr. 521 (1810) ... ... — — — — — Q. N.A. B.fl.IV,296 M.fr.VI,86. [43.
J. calcareum, F. v. M., fragm. I, 212 (1859) ... ... W.A. S.A. — — — N.A. B.fl.IV,297 M.fr.I,212;VI,87,VIII,
J. suavissimum, Lindley in Mitch. Trop. Austr. 355 (1848) ... — — — — N.S.W. Q. — B.fl.IV,297 M.fr.I,183;VI,87.
    OLEA, Tournefort, inst. 598, t. 370 (1700), from Plinius.
O. paniculata, R. Brown, prodr. 523 (1810) ... ... — — — — N.S.W. Q. — B.fl.IV,297 M.fr.VIII,43.
O. apetala, Vahl, symbol. III, 3 (1704) ... ... — — — — N.S.W. — — — M.fr.VIII,43;IX,169.
    LIGUSTRUM, Tournefort, inst. 596, t. 367 (1700), from C. Bauhin (1623).
L. Australianum, F. v. M., fragm. V, 20 (1865) ... ... — — — — — Q. — B.fl.IV,298 M.fr.V,20.
    NOTELAEA, Ventenat, Choix des pl. dans le jard. de Cels, t. 25 (1803).
N. ovata, R. Brown, prodr. 524 (1810) ... ... — — — — N.S.W. Q. — B.fl.IV,299 M.fr.VIII,43.
N. longifolia, Ventenat, Choix t. 25 (1803)... ... ... — — — V. N.S.W. Q. — B.fl.IV,299 M.fr.VIII,43.
N. punctata, R. Brown, prodr. 524 (1810) ... ... — — — — N.S.W. Q. — B.fl.IV,300 M.fr.VIII,43.
N. microcarpa, R. Brown, prodr. 524 (1810) ... ... — — — — N.S.W. Q. — B.fl.IV,300 M.fr.VIII,43.
N. ligustrina, Ventenat, Choix t. 25 (1803) ... ... — — — T. V. N.S.W. — — B.fl.IV,300 M.fr.VIII,43,142.
N. linearis, Bentham, Fl. Austr. IV, 300 (1869) ... ... — — — — N.S.W. — B.fl.IV,300
    MAYEPEA, Aublet, Hist. des pl. de la Guian. I, 81, t. 31 (1775). (Chionanthus partly, Linociera partly, Ceranthes.)
M. ramiflora, F.v.M.; Chion. Roxb. Fl. Ind. I, 106 ed. Carey (1820) — — — — — Q. — B.fl.IV,301 M.fr.V,83.
M. picrophloia; Chionanthus, F. v. M., fragm. III, 139, t. 24 (1863) — — — — — Q. — B.fl.IV,301 M.fr.IV,176.
M. axillaris, F. v. M.; Chionanthus, R. Brown, prodr. 523 (1810) — — — — — Q. N.A. B.fl.IV,301
M. quadristamines, F. v. M., fragm. X, 89 (1876)... ... — — — — N.S.W. — — M.fr.VIII,41.

APOCYNEAE.
A. L. de Jussieu, gen. plant. 143 (1789).

    CHILOCARPUS, Blume, catal. hort. Buitenz. (1823).
C. australis, F. v. M., fragm. II, 90 (1861)... ... ... — — — — N.S.W. Q. — B.fl.IV,303 M.fr.II,90;VI,118.
    MELODINUS, R. & G. Forster, char. gen. 37, t. 10 (1775).
M. acutiflorus, F. v. M. in Transact. phil. Soc. Vict. II, 71 (1857) — — — — — Q. — B.fl.IV,304
M. Guilfoylei, F. v. M., fragm. VI, 118 (1868) ... ... — — — — — Q. — B.fl.IV,304 M.fr.VI,118.
M. Baueri, Endlicher, prodr. pl. Norfolk. 59 (1833) ... ... — — — — N.S.W. — — M.fr.IX,160.

CARISSA, Linné, mantiss, pl. 7 (1767).
C. laxiflora, Bentham, Fl. Austr. IV, 305 (1869) ...   — — — — —   Q.   —   B.fl.IV,305
C. Brownii, F. v. M., fragm. IV, 45 (1863)... ...   —   S.A.   — —   N.S.W.   Q.   N.A.   B.fl.IV,305   M.fr.IV,45.

CERBERA, Linné, gen. plant. 62 (1737).
C. Odollam, Gaertner, de fruct. 193, t. 124 (1791)...   ...   .   —   Q.   N.A.   B.fl.IV,306   M.fr.VI,117.

ALYXIA, Banks in R. Brown, prodr. 469 (1810).   (Gynopogon.)
A. buxifolia, R. Brown, prodr. 470 (1810) ...   .   W.A.   S.A.   T.   V.   N.S.W.   —   —   B.fl.IV,307   M.fr.VI,117.
A. ruscifolia, R. Brown, prodr. 470 (1810)...   ...   ...   —   —   —   N.S.W.   Q.   —   B.fl.IV,308   M.fr.VI,117.
A. ilicifolia, F. v. M., fragm. IV, 149 (1864)   ...   ...   —   —   —   —   —   Q.   —   B.fl.IV,308   M.fr.IV,149.
A. squamulosa, C. Moore & F. v. M., fragm. VIII, 47 (1873) ...   —   —   —   — N.S.W.   —   —   —   M.fr.VI,117;VIII,47.
A. obtusifolia, R. Brown, prodr. 470 (1810)   ...   .   —   —   —   —   Q.   —   B.fl.IV,308
A. Gynopogon, Roemer & Schultes, syst. veg. IV, 440 (1810) ...   —   —   —   — N.S.W.   —   —   —   M.fr.VIII,47.
A. spicata, R. Brown, prodr. 470 (1810)   ...   .   —   —   —   —   Q.   N.A.   B.fl.IV,306   M.fr.VI,117.
A. Lindii, F. v. M., fragm. VIII, 46 (1873)   ...   .   —   —   —   — N.S.W.   —   —   —   M.fr.VIII,46.
A. thyrsiflora, Bentham, Fl. Austr. IV, 309 (1869)   ...   ...   —   —   —   —   —   Q.   —   B.fl.IV,309
A. Thozetti, F. v. M., fragm. X, 103 (1877)   ...   .   —   —   —   —   —   Q.   —   M.fr.X,103.

OCHROSIA, A. L. de Jussieu, gen. 143 (1780).   (Lactaria.)
O. elliptica, Labillardière, sert. Austr. Caled. 27, t. 30 (1824) ...   —   —   —   —   —   Q.   —   B.fl.IV,310   M.fr.VI,118.
O. Moorei, F. v. M. in Benth. Fl. Austr. IV, 310 (1869)...   ...   —   —   —   — N.S.W.   Q.   —   B.fl.IV,310   M.fr.VI,118.
O. Kilneri, F. v. M., fragm. VII, 129 (1871)   ...   .   —   —   —   —   —   Q.   —   M.fr.VIII,129.

NOTONERIUM, Bentham in B. & J. H. gen. II, 698 (1876).
N. Gossei, Bentham in B. & J. H. gen. 698 (1876)   ...   ...   —   S.A.   — —   —   —   —   —   M.fr.X,103.

TABERNAEMONTANA, Plumier, nov. pl. Amer. gen. 16, t. 30 (1703).
T. orientalis, R. Brown, prodr. 468 (1810) ...   .   —   —   —   —   N.S.W.   Q.   N.A.   B.fl.IV,311   M.fr.VI,117.
T. pubescens, R. Brown, prodr. 468 (1810)...   ...   .   —   —   —   —   Q.   N.A.   B.fl.IV,311   M.fr.VI,118.

VINCA, Rivinus in Rupp. Fl. Jenens. 27 (1818), from Plinius.   (Perhaps immigrated.)
V. rosea, Linné, spec. plant. ed. sec. 305 (1762)   ...   ...   —   —   —   —   —   Q.   —   —   M.fr.XI,123.

ALSTONIA, R. Brown in Mem. Wern. Soc. I, 75 (1809).
A. scholaris, R. Brown in Mem. Wern. Soc. I, 75 (1809)   ...   —   —   —   —   —   Q.   —   B.fl.IV,312   M.fr.VI,117.
A. verticillosa, F. v. M., fragm. VI, 116 (1868)   ...   ...   —   —   —   —   —   Q.   N.A.   B.fl.IV,313   M.fr.VI,116.
A. ophioxyloides, F. v. M., fragm. I, 57 (1858) ...   ...   —   —   —   —   —   —   N.A.   B.fl.IV,313   M.fr.VI,117.
A. villosa, Blume, Bijdr. 1038 (1826)   ...   .   —   —   —   —   —   —   N.A.   B.fl.IV,313   M.fr.VI,117.
A. linearis, Bentham, Fl. Austr. IV, 314 (1869) ...   ...   —   —   —   —   —   —   N.A.   B.fl.IV,314
A. constricta, F. v. M., fragm. I, 57 (1858)   ...   .   —   —   —   — N.S.W.   Q.   —   B.fl.IV,314   M.fr.III,163;IV,170;VI,{117.

ICHNOCARPUS, R. Brown in Mem. Wern. Soc. I, 75 (1809).
I. frutescens, R. Brown in Mem. Wern. Soc. I, 75 (1809)   ...   ...   —   —   —   —   —   Q.   N.A.   B.fl.IV,315   M.fr.VI,118.

WRIGHTIA, R. Brown in Mem. Wern. Soc. I, 75 (1809).   (Balfouria.)
W. pubescens, R. Brown, prodr. 467 (1810)   ...   ...   —   —   —   —   —   N.A.   B.fl.IV,316
W. saligna, F. v. M. in Benth. Fl. Austr. IV, 316 (1869)   ...   —   —   —   —   —   Q.   N.A.   B.fl.IV,316   M.fr.V,151;VI,118.
W. Cunninghamii, Bentham, Fl. Austr. IV, 317 (1869) ...   ...   W.A.   —   —   —   —   —   —   B.fl.IV,317

PARSONSIA, R. Brown in Mem. Wern. Soc. I, 64 (1809).
P. lanceolata, R. Brown, prodr. 466 (1810)   ...   .   —   —   — N.S.W.   Q.   —   B.fl.IV,318   M.fr.VI,117.
P. velutina, R. Brown, prodr. 466 (1810)   ...   .   —   —   — N.S.W.   Q.   N.A.   B.fl.IV,318   M.fr.VI,130.
P. Leichhardtii, F. v. M., fragm. VI, 128 (1868) ...   ...   —   —   —   —   —   Q.   —   B.fl.IV,319   M.fr.VI,128.
P. ventricosa, F. v. M. in Transact. phil. Inst. Vict. II, 71 (1857) ...   —   —   — N.S.W.   —   —   B.fl.IV,319   M.fr.VI,130.

LYONSIA, R. Brown in Mem. Wern. Soc. I, 66 (1809).
L. lilacina, F. v. M. in Bentham Fl. Austr. IV, 321 (1869)   ...   —   —   —   — N.S.W.   Q.   —   B.fl.IV,321   M.fr.VI,127.
L. induplicata, F. v. M. in Benth. Fl. Austr. IV, 321 (1869)   ...   —   —   —   — N.S.W.   —   —   B.fl.IV,321   M.fr.VI,129.
L. straminea, R. Brown, prodr. 466 (1810)...   ...   ...   —   —   — T.   V.   N.S.W.   —   —   B.fl.IV,321   M.fr.VI,129.
L. reticulata, F. v. M., Rep. Burdek. Exped. 16 (1860)...   ...   —   —   — N.S.W.   Q.   —   B.fl.IV,321   M.fr.VI,129.
L. diaphanophleba, F. v. M. in Benth. Fl. Austr. IV, 322 (1860)   W.A.   —   —   —   —   —   —   B.fl.IV,322   M.fr.II,158;VI,130.
L. Langiana, F. v. M., fragm. VI, 128 (1868)   ...   ...   —   —   —   —   —   —   B.fl.IV,322   M.fr.VI,128.
L. largiflorens, F. v. M. in Benth. Fl. Austr. IV, 322 (1860) ...   —   —   — N.S.W.   Q.   —   B.fl.IV,322
L. latifolia, Bentham, Fl. Austr. IV, 323 (1869)   ...   ...   —   —   —   —   —   —   B.fl.IV,323
L. oblongifolia, Bentham, Fl. Austr. IV, 323 (1869)   ...   ...   —   —   —   —   —   —   B.fl.IV,323
L. eucalyptifolia, F. v. M. in Benth. Fl. Austr. IV, 323 (1860)...   —   —   —   —   —   —   B.fl.IV,323   M.fr.II,159;VI,130

ASCLEPIADEAE.
N. J. Jacquin, Misc. Austr. I, 35 (1778).

GYMNANTHERA, R. Brown in Mem. Wern. Soc. I, 58 (1809).
G. nitida, R. Brown, prodr. 464 (1810)   ...   .   —   —   —   —   Q.   N.A.   B.fl.IV,326   M.fr.V,160.

SECAMONE, R. Brown in Mem. Wern. Soc. I, 55 (1809).
S. elliptica, R. Brown, prodr. 464 (1810)   ...   ...   —   —   — N.S.W.   Q.   N.A.   B.fl.IV,327   M.fr.V,161.
S. ovata, R. Brown, prodr. 464 (1810)   ...   .   —   —   —   —   —   Q.   —   B.fl.IV,327

VINCETOXICUM, Dodoens, stirp. hist. pemptades, 704 (1583).   (Oxystelma partly, Cynoctonum.)
V. ovatum, Bentham, Fl. Austr. IV, 330 (1869) ...   ...   —   —   —   —   —   Q.   —   B.fl.IV,330
V. elegans, Bentham, Fl. Austr. IV, 330 (1869) ...   ...   —   —   — N.S.W.   Q.   N.A.   B.fl.IV,330   M.fr.V,160.
V. carnosum, Bentham, Fl. Austr. IV, 331 (1869)   ...   ...   —   —   — N.S.W.   Q.   N.A.   B.fl.IV,331
V. leptolepis, Bentham, Fl. Austr. IV, 331 (1869)   ...   ...   —   —   —   —   —   Q.   —   B.fl.IV,331

CYNANCHUM, Linné, gen. plant. 63 (1737).
C. erubescens, R. Brown, prodr. 463 (1810)   ...   .   —   —   —   —   —   Q.   —   B.fl.IV,332
C. floribundum, R. Brown, prodr. 463 (1810)   ...   ...   —   S.A.   — —   —   Q.   N.A.   B.fl.IV,332   M.fr.V,160;
C. pedunculatum, R. Brown, prodr. 463 (1810)   ...   ...   —   —   —   —   — N.A.   B.fl.IV,333   M.fr.V,160.
C. puberulum, F. v. M. in Benth. Fl. Austr. IV, 333 (1869)   ...   —   —   —   —   — N.A.   B.fl.IV,333

SARCOSTEMMA, R. Brown in Mem. Wern. Soc. I, 50 (1809).
S. Australe, R. Brown, prodr. 463 (1810) ... ... W.A. S.A. — — N.S.W. Q. N.A. B.fl.IV,328 M.fr.V,160;XI,80.

DAEMIA, R. Brown in Mem. Wern. Soc. I, 50 (1809). (Pentatropis, Rhyncharrhena.)
D. linearis, F. v. M.; Pentatropis, Decaisne in De Cand., prodr.
   VIII, 536 (1833) ... ... W.A. — — — — — B.fl.IV,329
D. atropurpurea; Ryncharrh., F. v. M., fr. I, 128 (1859) ... — — — — Q. — B.fl.IV,329 M.fr.I,128.
D. quinquepartita; Ryncharrh., F. v. M., fr. I, 128 (1859) ... — — — V. N.S.W. — — B.fl.IV,329 M.fr.I,128;XI,80.
D. Kempeana, F. v. M. in Wing's S. Sc. Record, Aug. (1882) ... — S.A. — — — — —

GYMNEMA, R. Brown in Mem. Wern. Soc. I, 33 (1809). (Bidaria.)
G. sylvestre, R. Brown in Mem. Wern. Soc. I, 33 (1809) ... — — — — — N.A. B.fl.IV,343
G. pleiadenium, F. v. M., fragm, XI, 77 (1879) ... ... — — — — — Q. — M.fr.XI,77.
G. Muelleri, Bentham, Fl. Austr. IV, 343 (1869).. ... ... — — — — — N.A. B.fl.IV,343
G. brevifolium, Bentham, Fl. Austr. IV, 343 (1869) ... ... — — — — — Q. — B.fl.IV,343
G. trinerve, R. Brown, prodr. 462 (1810) ... ... — — — — — N.A. B.fl.IV,343
G. stenophyllum, A. Gray in Proc. Am. Ac. of Sc. V, 335 (1864) — — — — — N.A. B.fl.IV,344

GONGRONEMA, Decaisne in De Candolle, prodr. VIII, 624 (1844).
G. micradenia, Bentham & J. Hooker, gen. plant. II, 770 (1876) — — — — — Q. — B.fl.IV,344 M.fr.XI,78.

TYLOPHORA, R. Brown in Mem. Wern. Soc. I, 28 (1809).
T. erecta, F. v. M. in Benth. Fl. Austr. IV, 334 (1869) ... ... — — — — — Q. — B.fl.IV,334 M.fr.XI,79.
T. macrophylla, Bentham, Fl. Austr. IV, 334 (1869) ... ... — — — — — N.A. B.fl.IV,334
T. grandiflora, R. Brown, prodr. 460 (1810) ... ... — — — — N.S.W. Q. — B.fl.IV,334 M.fr.XI,80.
T. scribunda, Bentham, Fl. Austr. IV, 335 (1869) ... ... — — — — N.S.W. Q. — B.fl.IV,335
T. calcarata, Bentham, Fl. Austr. IV, 335 (1869)... ... ... — — — — — Q. — B.fl.IV,335
T. barbata, R. Brown, prodr. 460 (1810) ... ... — — — V. N.S.W. — B.fl.IV,335 M.fr.V,160;XI,79.
T. Woollsii, Bentham, Fl. Austr. IV, 335 (1869) ... ... — — — — N.S.W. — B.fl.IV,335
T. paniculata, R. Brown, prodr. 460 (1810) ... ... — — — — N.S.W. — B.fl.IV,336
T. enervis, F. v. M., fragm. IX, 70 (1875)... ... ... — — — — N.S.W. — — M.fr.IX,70.
T. flexuosa, R. Brown, prodr. 460 (1810) ... ... — — — — N.A. B.fl.IV,336 M.fr.XI,79.
T. biglandulosa, A. Gray in Proc. of Am. acad. sc. vol. V (1864) — — — — N.S.W. — — M.fr.IX,169.

MARSDENIA, R. Brown in Mem. Wern. Soc. I, 28 (1809). (Leichhardtia.)
M. cinerascens, R. Brown, prodr. 461 (1810) ... ... — — — — — N.A. B.fl.IV,337
M. rhyncholepis, F. v. M., fragm. XI, 78 (1879) ... ... — — — — — Q. — M.fr.XI,78.
M. flavescens, Cunningham in Bot. Mag., t. 3289 (1833) ... — — — — N.S.W. — B.fl.IV,337 M.fr.V,160.
M. cymulosa, Bentham, Fl. Austr. IV, 338 (1869) ... ... — — — — — Q. — B.fl.IV,338
M. velutina, R. Brown, prodr. 461 (1810) ... ... — — — — — N.A. B.fl.IV,338
M. Hullsii, F. v. M. in Benth. Fl. Austr. IV, 338 (1869) ... — — — — — N.A. B.fl.IV,338
M. araujacea, F. v. M., fragm. VI, 135 (1868) ... ... — — — — — Q. — B.fl.IV,339 M.fr.VI,135.
M. tubulosa, F. v. M., fragm. IX, 71 (1875) ... ... — — — — N.S.W. — — M.fr.IX,71.
M. rostrata, R. Brown, prodr. 461 (1810) ... ... — — — V. N.S.W. Q. — B.fl.IV,339 M.fr.V,160;VI,135.
M. Fraseri, Bentham, Fl. Austr. IV, 339 (1869) ... ... — — — — N.S.W. — B.fl.IV,339
M. longiloba, Bentham, Fl. Austr. IV, 340 (1869) ... ... — — — — N.S.W. Q. — B.fl.IV,340 M.fr.XI,79.
M. suaveolens, R. Brown, prodr. 461 (1810) ... ... — — — — N.S.W. — B.fl.IV,340
M. leptophylla, F. v. M. in Benth. Fl. Austr. IV, 340 (1869) ... ... W.A. S.A. — V. N.S.W. Q. N.A. B.fl.IV,341 M.fr.V,160.
M. Leichhardtiana, F. v. M., fragm. V, 160 (1866) ... ... W.A. S.A. — V. N.S.W. Q. N.A. B.fl.IV,341 M.fr.V,160.
M. viridiflora, R. Brown, prodr. 461 (1810) ... ... — — — — N.S.W. Q. N.A. B.fl.IV,341 M.fr.V,160.
M. coronata, Bentham, Fl. Austr. IV, 341 (1810)... ... ... — — — — — Q. — B.fl.IV,341
M. microlepis, Bentham, Fl. Austr. IV, 342 (1869) ... ... — — — — — Q. — B.fl.IV,342

THOZETIA, F. v. M. in Benth. Fl. Austr. IV, 347 (1869).
T. racemosa, F. v. M. in Benth. Fl. Austr. IV, 347 (1869) ... — — — — — Q. — B.fl.IV,347

HOYA, R. Brown in Mem. Wern. Soc. I, 26 (1809).
H. carnosa, R. Brown in Mem. Wern. Soc. I, 26 (1809)... ... — — — — — Q. — B.fl.IV,346
H. australis, R. Brown in Transact. hort. Soc. VII, 28 (1826) ... — — — — N.S.W. Q. — B.fl.IV,346 M.fr.XI,80.
H. Nicholsoniae, F. v. M., fragm. V, 159 (1866) ... ... — — — — — — — B.fl.IV,347 M.fr.V,159;IX,70,190.

DISCHIDIA, R. Brown in Mem. Wern. Soc. I, 32 (1809).
D. nummularia, R. Brown, prodr. 461 (1810) ... ... — — — — — Q. — B.fl.IV,345 M.fr.V,161.
D. Timorensis, Decaisne in nouv. annal. du Mus. III, 377, t. 17 (1835) — — — — — Q. N.A. — —

MICROSTEMMA, R. Brown in Mem. Wern. Soc. I, 25 (1809).
M. tuberosum, R. Brown, prodr. 450 (1810) ... ... — — — — — Q. N.A. B.fl.IV,345 M.fr.I,58.
M. glabriflorum, F. v. M., fragm. I, 58 (1858) ... ... — — — — — N.A. B.fl.IV,345 M.fr.I,58.

CEROPEGIA, Linné, gen. plant. 65 (1737).
C. Cunningham, Decaisne in De Cand., prodr. VIII, 643 (1844) — — — — — Q. — B.fl.IV,348 M.fr.V,159.

CONVOLVULACEAE.

A. L. de Jussieu, gen. plant. 132 (1789), from B. de Jussieu (1759).

ERYCIBE, Roxburgh, pl. Corom. III, 31, t. 159 (1798).
E. paniculata, Roxburgh, pl. Corom. III, 31, t. 159 (1798) ... — — — — — Q. — B.fl.IV,411 M.fr.VI,100.

LEPISTEMON, Blume, Bijdrag. 722 (1825).
L. urceolatus, F. v. M.; L. Fitzalani, F. v. M., fragm. X, 111 (1877)— — — — — Q. — B.fl.IV,427 M.fr.X,111.

IPOMOEA, Linné, syst. nat. 8 (1735); Linné, gen. plant. 47 (1737). (Pharbitis, Batatas, Calonyction, Aniseia, Skimmera.)
I. paniculata, R. Brown, prodr. 486 (1810) ... ... — — — — — Q. N.A. B.fl.IV,414 M.fr.VI,98;X,111.
I. Davenporti, F. v. M., fragm. VI, 97 (1868) ... ... — N.A. — — — Q. — B.fl.IV,413 M.fr.VIII,17;X,111.
I. bona nox, Linné, spec. plant. edit. secund. 228 (1762) ... — — — — N.S.W. Q. — — M.fr.IX,74;X,113;X,73.
I. palmata, Forskael, Fl. Aegypt. Arabic. 43 (1775) ... ... — — — N.S.W. Q. — B.fl.IV,413 M.fr.V,98;X,111.
I. quinata, R. Brown, prodr. 486 (1810) ... ... — — — — Q. N.A. B.fl.IV,415 M.fr.VI,98;X,112.
I. diversifolia, R. Brown, prodr. 487 (1810) ... ... — — — — Q. N.A. B.fl.IV,416 M.fr.X,112.

I. dissecta, Willdenow, phytogr. 5, t. 2 (1794) ... ... — — — — — Q. N.A. B.fl.IV,416 M.fr.VI,98;X,112.
I. sinuata, Ortega, decad. nov. pl. hort. Matrit. VII, 84 (1799) — — — — — Q. — — M.fr.VIII,17;X,113.
I. hederacea, N. J. Jacquin, Collectan, I, 124 (1786) ... ... — — — — — Q. N.A. B.fl.IV,416 M.fr.VI,99;X,112.
I. congesta, R. Brown, prodr. 485 (1810) ... ... — — — — — Q. — B.fl.IV,417 M.fr.VI,99;X,112.
I. purpurea, Roth, Catalect. bot. I, 36 (1797) ... ... — — — — — Q. — B.fl.IV,417 M.fr.VI,99;X,112.
I. peltata, Choisy, Convolvulac. orient. 70 (1834).. ... ... — — — — N.S.W. Q. — R.fl.IV,418
I. Calobra, Hill & F. v. M., fragm. XI, 73 (1879) ... ... — — — — — Q. — — M.fr.XI,73,137.
I. alata, R. Brown, prodr. 484 (1810) ... ... — — — — — — N.A. B.fl.IV,418 M.fr.X.112.
I. Turpethum, R. Brown, prodr. 485 (1810) ... ... — — — — — Q. — D.fl.IV,418 M.fr.VI,98.
I. longiflora, R. Brown, prodr. 484 (1810) ... ... — — — — — Q. N.A. B.fl.IV,418 M.fr.VI,98;IX,74; X,112.
I. costata, F. v. M. in Benth, Fl. Austr. IV, 419 (1869)... W.A S.A. — — — — — N.A. B.fl.IV,419 M.fr.X,112.
I. Pes Caprae, Roth, nov. spec. plant. 109· ... ... W.A. — — — N.S.W. Q. N.A. B.fl.IV,419 M.fr.VI,97;X,112.
I. carnosa, R. Brown, prodr. 485 (1810) ... ... — — — — — — N.A. B.fl.IV,420
I. reptans, Poiret, Encycl. méth. suppl. III, 460 (1813).. W.A. — — — — Q. N.A. B.fl.IV,420 M.fr.VI,97.
I. graminea, R. Brown, prodr. 485 (1810) ... ... — — — — — Q. N.A. B.fl.IV,421
I. velutina, R. Brown, prodr. 485 (1810) ... ... — — — — — — N.A. B.fl.IV,421
I. abrupta, R. Brown, prodr. 485 (1810) ... ... — — — — — Q. N.A. B.fl.IV,421 M.fr.X,112.
I. denticulata, Choisy in De Cand. prodr. IX, 370 (1845) ... — — — — — Q. N.A. B.fl.IV,421 M.fr.I,185;X,112.
I. gracilis, R. Brown, prodr. 484 (1810) ... ... — — — — — Q. N.A. B.fl.IV,422 M.fr.I,185;X,112.
I. sepiaria, Koenig in Roxb. Fl. Indic. ed. Carey II, 90 (1824) .. — — — N.S.W. Q. — B.fl.IV,422
I. Muelleri, Bentham, Fl. Austr. IV, 423 (1869) ... ... W.A. — — — — N.A. B.fl.IV,423 M.fr.X,112.
I. cynosa, Roemer & Schultes, syst. veg. IV, 241 (1819) ... ... — — — — — Q. — B.fl.IV,423
I. linifolia, Blume, Bijdrag. 721 (1925) ... ... — — — — — — Q. — B.fl.IV,423
I. chryseides, Ker in Bot. Regist. t. 270 (1818) ... ... — — — — — Q. — B.fl.IV,423 M.fr.VI,98.
I. flava, F. v. M. in Benth. Fl. Austr. IV, 424 (1869) ... — — — — — — N.A. B.fl.IV,424
I. obscura, Ker in Bot. Regist. t. 239 (1817) ... ... — — — — — Q. — B.fl.IV,424
I. incisa, R. Brown, prodr. 486 (1810) ... ... — — — — — — N.A. B.fl.IV,424 M.fr.I,186.
I. uniflora, Roemer & Schultes, syst. IV, 247 (1819) ... ... — — — — — Q. — B.fl.IV,425 M.fr.VI,98.
I. angustifolia, N. J. Jacquin, Collect. II, 367 (1788) ... — — — — — Q. N.A. B.fl.IV,425 M.fr.VI,97;X,112.
I. plebeja, R. Brown, prodr. 484 (1810) ... ... — — — — — Q. — B.fl.IV,426 M.fr.VI,99;X,112.
I. eriocarpa, R. Brown, prodr. 484 (1810) ... ... — — — — — Q. N.A. B.fl.IV,426 M.fr.VI,98;X,112.
I. heterophylla, R. Brown, prodr. 487 (1810) ... ... — S.A. — — — Q. N.A. B.fl.IV,426 M.fr.VI,98;X,112.
I. erecta, R. Brown, prodr. 487 (1810) ... ... — — — — — — N.A. B.fl.IV,427 M.fr.VI,98.

CONVOLVULUS, Tournefort, inst. 82, t. 17 (1700). (Calystegia.)
C. erubescens, Sims, Bot. Magaz. t. 1067 (1808) ... ... W.A S.A T. V. N.S.W. Q. — M.fr.VI,99;X,113.
C. erubiflorus, Vahl, symbol. bot. III, 29 (1794) ... ... — — — — — V. N.S.W. Q. N.A. B.fl.IV,430 M.fr.VI,99;X,113.
C. marginatus, Poiret, Encycl. méth. suppl. II (1811) ... ... — — — V. N.S.W. Q. — B.fl.IV,430 M.fr.VI,99;X,113.
C. sepium, Linné, spec. plant. 153 (1753) ... ... W.A. S.A. T. V. N.S.W. — B.fl.IV,430 M.fr.VI,90;X,113.

POLYMERIA, R. Brown, prodr. 488 (1810).
P. marginata, Bentham, Fl. Austr. IV, 432 (1869) ... ... — — — Q. — B.fl.IV,432
P. longifolia, Lindley in Mitch. Trop. Austr. 398 (1848)... ... — S.A. — — N.S.W. Q. — B.fl.IV,432 M.fr.X,114.
P. angusta, F. v. M., fragm. VI, 100 (1868) ... ... — — — — — N.A. B.fl.IV,432 M.fr.VI,100.
P. distigma, Bentham, Fl. Austr. IV, 433 (1869) ... ... — — — — — N.A. B.fl.IV,433
P. occidentalis, F. v. M., inedit. ... ... W.A. — — — — — B.fl.IV,433
P. calycina, R. Brown, prodr. 488 (1810) ... ... — — — N.S.W. Q. — B.fl.IV,433 M.fr.VI,100.
P. ambigua, R. Brown, prodr. 488 (1810) ... ... — — — — — Q. N.A. B.fl.IV,433

PORANA, Burmann, Fl. Ind. 51, t. 21 (1768). (Duperreya.)
P. sericea, F. v. M., fragm. VI, 100 (1868) ... ... W.A. — — — — — N.A. B.fl.IV,434 M.fr.II,22.

BREWERIA, R. Brown, prodr. 487 (1810).
B. linearis, R. Brown, prodr. 488 (1810) ... ... — — — — — — N.A. B.fl.IV,435
B. media, R. Brown, prodr. 488 (1810) ... ... — S.A. — N.S.W. Q. N.A. B.fl.IV,436 M.fr.X,113.
B. brevifolia, Bentham, Fl. Austr. IV, 436 (1869) ... ... — — — — — N.A. B.fl.IV,436
B. rosea, F. v M., fragm. I, 233 (1858) ... ... W.A. S.A. — — — N.A. B.fl.IV,436 M.fr.VI,100;X,113.
B. pannosa, R. Brown, prodr. 488 (1810) ... ... — — — — — Q. N.A. B.fl.IV,436 M.fr.VI,100;X,113.

EVOLVULUS, Linné, spec. plant. ed. sec. 391 (1763).
E. linifolius, Linné, spec. plant. ed. sec. 392 (1762) ... ... W.A. S.A. — — N.S.W. Q. N.A. B.fl.IV,437 M.fr.VI,100;X,113.

DICHONDRA, R. & G. Forster, char. gen. 39, t. 20 (1776).
D. repens, R. & G. Forster, char. gen. 39, t. 20 (1776) ... ... W.A. S.A. T. V. N.S.W. Q. N.A. B.fl.IV,438 M.fr.VI,100;X,113.

CRESSA, Linné, Amoen. Acad. I, 121 (1747).
C. Cretica, Linné, spec. plant. 223 (1753) ... ... W.A. S.A. — V. N.S.W. Q. N.A. B.fl.IV,437 M.fr.VI,100.

WILSONIA, R. Brown, prodr. 490 (1810).
W. humilis, R. Brown, prodr. 490 (1810) ... ... W.A. S.A. T. V. N.S.W. — — B.fl.IV,439 M.fr.VI,101;X,113.
W. rotundifolia, Hooker, icon. plant. t. 410 (1842) ... ... W.A. S.A. — V. N.S.W. — — B.fl.IV,440 M.fr.VI,101.
W. Backhousi, J. Hooker in Lond. Journ. VI, 275 (1842) ... W.A. S.A. T. V. N.S.W. — — B.fl.IV,440 M.fr.VI,101;VIII,17.

CUSCUTA, Tournefort, inst. app. 652, t. 422 (1700), from C. Bauhin (1623).
C. Chinensis, Lamarck, Encycl. méth. II, 229 (1786) ... — — — — — Q. — B.fl.IV,441 M.fr.X,113.
C. Australis, R. Brown, prodr. 491 (1810) .. .. — S.A. — — N.S.W. Q — B.fl.IV,441 M.fr.I,100;X,113.
C. Tasmanica, Engelmann in Transact. Acad. St. Louis I, 512 (1860) — — — T. V. N.S.W. — — B fl.IV,441

SOLANACEAE.
Haller, enum. stirp. Helv. praef. 34 (1742).

PHYSALIS, Linné, syst. nat. 8 (1735); Linné, gen. plant. 51 (1737).
P. minima, Linné, spec. plant. 183 (1753) ... ... — — — N.S.W. Q. N.A. B.fl.IV,406 M.fr.VI,144.

SOLANUM, Tournefort, inst. 148, t. 62 (1700), from Celsus.
S. nigrum, Linné, spec. plant. 186 (1753) ... ... W.A. S.A. T. V. N.S.W. Q. N.A. B.fl.IV,446 M.fr.VI,145.
S. vescum, F. v. M. in Transact. Vict. Inst. 69 (1855) ... ... — — T. V. N.S.W. — — —

S. aviculare, G. Forster, prodr. 18 (1786) ... ... — S.A. T. V. N.S.W. Q. — B.fl.IV,447 M.fr.VI,144.
S. simile, F. v. M. in Transact. phil. Soc. Vict. I, 19 (1854) ... W.A. S.A. — V. N.S.W. — — B.fl.IV,448 M.fr.I,123;VI,144.
S. fasciculatum, F. v. M., fragm. I, 123 (1859) ... ... W.A. S.A. — — — — — M.fr.VI,144.
S. Bauerianum, Endlicher, prodr. fl. Norfolk, 54 (1833)... ... — — — — N.S.W. — — — M.fr.IX,74.
S. viride, R. Brown, prodr. 445 (1810) ... ... — — — — — Q. — B.fl.IV,449 M.fr.VI,145.
S. tetrandrum, R. Brown, prodr. 445 (1810) ... . — — — — — — N.A. B.fl.IV,449
S. Shanesii, F. v. M., fragm. VI, 144 (1868) ... ... — — — — — Q. — B.fl.IV,448 M.fr.VI,144.
S. verbascifolium, Linné, spec. plant. 184 (1753)... ... ... — — — — N.S.W. Q. — B.fl.IV,449 M.fr.VI,145.
S. discolor, R. Brown, prodr. 445 (1810) ... ... — — — — N.S.W. Q. N.A. B.fl.IV,450 M.fr.II,166;VI,147.
S. stelligerum, Smith, Exot. Bot. II, 57, t. 88 (1805) ... ... — — — — N.S.W. Q. — B.fl.IV,450 M.fr.VI,147.
S. parvifolium, R. Brown, prodr. 446 (1810) ... ... — — — — N.S.W. Q. — B.fl.IV,451 M.fr.II,164;VI,147.
S. ferocissimum, Lindley in Mitch. Three Exped. II, 58 (1838)... — S.A. — — N.S.W. Q. — B.fl.IV,451
S. sporadotrichum, F. v. M. in Melb. Chemist, Oct. (1862) ... — — — — — Q. — — —
S. defensum, F. v. M., fragm. V, 103 (1866) ... ... — — — — — — — B.fl.IV,451 M.fr.V,193.
S. violaceum, R. Brown, prodr. 445 (1810)... ... ... — — — — N.S.W. Q. — B.fl.IV,452
S. tetrathecum, F. v. M., fragm. II, 165 (1861) ... ... — — — — N.S.W. Q. — B.fl.IV,453 M.fr.II,165.
S. elachophyllum, F. v. M., fragm. II, 164 (1861)... ... ... — — — — — — Q. — B.fl.IV,453 M.fr.II,164.
S. orbiculatum, Dunal in Poir. Encycl. méth. XIII, 762 (1817)... W.A. S.A. — — — — — B.fl.IV,453 M.fr.VI,145.
S. oligacanthum, F. v. M. in Transact. phil. Soc. Vict. I, 19 (1854) — S.A. — — N.S.W. — — B.fl.IV,454 M.fr.VI,145.
S. esuriale, Lindley in Mitch. Three Exped. II, 43 (1838) ... — S.A. — V. N.S.W. Q. N.A. B.fl.IV,454
S. chenopodinum, F. v. M., fragm. II, 165 (1861) ... ... — S.A. — — N.S.W. — N.A. B.fl.IV,454 M.fr.VI,146.
S. Sturtianum, F. v. M. in Transact. phil. Soc. Vict. I, 19 (1854) — S.A. — — N.S.W. — N.A. B.fl.IV,454 M.fr.VI,146.
S. furfuraceum, R. Brown, prodr. 446 (1810) ... ... — — — — — — — B.fl.IV,455
S. dianthophorum, Dunal, hist. des Solan. 183 (1813) ... ... — — — — — — Q. — B.fl.IV,455 M.fr.VI,145.
S. Dallachyi, Bentham, Fl. Austr. IV, 456 (1869) ... ... — — — — — — Q. — B.fl.IV,456 M.fr.VI,145.
S. densivestitum, F. v. M. in Benth. Fl. Austr. IV, 456 (1869) — — — — N.S.W. Q. — B.fl.IV,456
S. semiarmatum, F. v. M., fragm. II, 161 (1861) ... ... — — — — — — Q. — B.fl.IV,456 M.fr.II,161.
S. Oldfieldii, F. v. M., fragm. II, 161 (1861) ... ... W.A. — — — — — — B.fl.IV,457 M.fr.VI,146.
S. semiarmatum, F. v. M., fragm. II, 163 (1861)... ... ... — — — — N.S.W. Q. — B.fl.IV,458
S. armatum, R. Brown, prodr. 446 (1810) ... ... — — — V. N.S.W. Q. — B.fl.IV,458
S. Hystrix, R. Brown, prodr. 446 (1810) ... ... — S.A. — — — — — B.fl.IV,458
S. cataphractum, Cunningham in Benth. Fl. Austr. IV, 459 (1869) — — — — — — N.A. B.fl.IV,459
S. pungetium, R. Brown, prodr. 446 (1810) ... ... — — — V. N.S.W. Q. — B.fl.IV,459 M.fr.VI,146.
S. eremophilum, F. v. M. in Schlecht. Linnaea XXV, 432 (1852) — S.A. — — N.S.W. — Q. — B.fl.IV,459 M.fr.VI,146.
S. campanulatum, R. Brown, prodr. 446 (1810) ... ... — — — — N.S.W. — Q. — B.fl.IV,460 M.fr.VI,146.
S. adenophorum, F. v. M., fragm. II, 162 (1861)... ... ... — — — — — — Q. — B.fl.IV,460 M.fr.II,162.
S. cinereum, R. Brown, prodr. 446 (1810) ... ... — — — — N.S.W. Q. — B.fl.IV,460
S. lacunarium, F. v. M. in Transact. phil. Soc. Vict. I, 18 (1854) — S.A. — V. N.S.W. — — B.fl.IV,461 M.fr.VI,146.
S. petrophilum, F. v. M. in Linnaea XXV, 433 (1852) ... — S.A. — N.S.W. — — B.fl.IV,461 M.fr.VI,146.
S. diversiflorum, F. v. M., fragm. VI, 145 (1868)... ... ... — — — — — — N.A. B.fl.IV,461 M.fr.VI,146.
S. carduiforme, F. v. M., fragm. II, 163 (1861) ... ... — — — — — — N.A. B.fl.IV,462 M.fr.II,163.
S. melanospermum, F. v. M., fragm. II, 163 (1861) ... ... — — — — — — N.A. B.fl.IV,462 M.fr.II,163.
S. horridum, Dunal, Solan. synops. 28 (1816) ... ... — — — — — — N.A. B.fl.IV,463
S. echinatum, R. Brown, prodr. 447 (1810) ... ... — — — — — — N.A. B.fl.IV,463 M.fr.VI,145.
S. lasiophyllum, Dunal in Poir. Encycl. méth. XIII, 764 (1817) W.A. — — — — N.A. B.fl.IV,463 M.fr.VI,145.
S. ellipticum, R. Brown, prodr. 446 (1810) ... ... W.A. S.A. — — N.S.W. Q. N.A. B.fl.IV,464 M.fr.II,161.
S. quadriloculatum, F. v. M., fragm. II, 161 (1861) ... ... — — — — — — N.A. B.fl.IV,464 M.fr.VI,146.
S. phlomoides, Cunningham in Benth. Fl. Austr. IV, 464 (1869) — — — — — — N.A. B.fl.IV,465
S. Cunninghamii, Bentham, Fl. Austr. IV, 465 (1869) ... — — — — — — N.A. B.fl.IV,465

LYCIUM, Linné, gen. plant. 57 (1737), from Celsus.
L. australe, F. v. M. in Transact. phil. Soc. Vict. I, 20 (1854)... W.A. S.A. — V. N.S.W. — — B.fl.IV,467 M.fr.I,83;243;II,179.

DATURA, Linné, gen. plant. 48 (1737).
D. Leichhardtii, F. v. M. in Transact. phil. Soc. Vict. I, 20 (1854) — S.A. — — — Q. N.A. B.fl.IV,468 M.fr.VI,144.

NICOTIANA, Tournefort, inst. 117, t. 41 (1700), from C. Bauhin (1623).
N. suaveolens, Lehmann, gener. Nicot. hist. 43 (1818) ... ... — — — V. N.S.W. Q. N.A. B.fl.IV,469 M.fr.VI,144.

ANTHOTROCHE, Endlicher, nov. stirp. mus. Vindob. dec. 6 (1839).
A. pannosa, Endlicher, nov. stirp. mus. Vind. dec. 7 (1839) ... W.A. — — — — — B.fl.IV,467 M.fr.VI,142.
A. Blackii, F. v. M., fragm. VIII, 232 (1874) ... ... W.A. S.A. — — — — M.fr.VIII,232.
A. Walcottii, F. v. M., fragm. I, 123 (1859) ... ... W.A. — — — — — B.fl.IV,468 M.fr.I,123.

ANTHOCERCIS, Labillardière, Nov. Holl. pl. spec. II, 19 (1806). (Cyphanthera, Eadesia.)
A. viscosa, R. Brown, prodr. 448 (1810) ... ... W.A. — — — — — B.fl.IV,475 M.fr.VI,143.
A. litorea, Labillard., Nov. Holl. pl. spec. II, 19, t. 158 (1806) W.A. — — — — — B.fl.IV,476 M.fr.VI,143;XI,123.
A. gracilis, Bentham in De Cand., prodr. X, 192 (1846)... ... W.A. — — — — — B.fl.IV,476 M.fr.VI,143.
A. genistoides, Miers, Illustr. of plants II, App. 27, t. 83 (1857) W.A. — — — — — B.fl.IV,476 M.fr.I,122;VI,143.
A. anisantha, Endlicher in Ann. Wien. Mus. II, 201 (1838) ... W.A. S.A. — — — — B.fl.IV,477 M.fr.VI,143.
A. intricata, F. v. M., fragm. I, 211 (1859) ... ... W.A. — — — — — B.fl.IV,477 M.fr.I,211.
A. arborea, F. v. M., fragm. I, 212 (1859)... ... ... W.A. — — — — — B.fl.IV,477 M.fr.I,212;VI,143.
A. angustifolia, F. v. M. in Transact. phil. Soc. I, 21 (1854) — S.A. — — — — B.fl.IV,478
A. fasciculata, F. v. M., fragm. I, 122 (1859) ... ... W.A. — — — — — B.fl.IV,478 M.fr.I,122;VI,143.
A. microphylla, F. v. M., fragm. I, 179 (1859) ... ... W.A. — — — — — B.fl.IV,478 M.fr.I,179;VI,143.
A. myosotidea, F. v. M. in Transact. phil. Soc. Vict. I, 20 (1854) — S.A. — V. N.S.W. — — B.fl.IV,479 M.fr.VI,143.
A. scabrella, Bentham in De Cand., prodr. X, 192 (1846) ... — — — N.S.W. — — B.fl.IV,479
A. albicans, Cunningham in Field's N.S.W. 335, t. 2 (1825) ... — — — N.S.W. — N.A. B.fl.IV,479 M.fr.VI,143.
A. Odgersii, F. v. M., fragm. X, 19 (1876)... ... ... — — — — — — B.fl.IV,479 M.fr.X,19.
A. Tasmanica, J. Hooker, Fl. Tasman. I, 289, t. 92 (1860) ... — — T. — — — B.fl.IV,479 M.fr.VI,143.
A. Eadesii, F. v. M., fragm. II, 139 (1861) ... ... — S.A. — V. N.S.W. — — B.fl.IV,480 M.fr.VI,142.
A. racemosa, F. v. M., fragm. I, 211 (1859) ... ... W.A. — — — — — B.fl.IV,480 M.fr.I,211.

DUBOISIA, R. Brown, prodr. fl. Nov. Holl. 448 (1810).
D. myoporoides, R. Brown, prodr. 448 (1810) ... ... — — — — N.S.W. Q. — B.fl.IV,474 M.fr.VI,144.
D. Hopwoodii, F. v. M., fragm. VIII, 232 (1874) ... ... W.A. S.A. — — — N.S.W. Q. — B.fl.IV,480 M.fr.II,139;VI,143;X,20.
D. Leichhardti, F. v. M. in Wing's S. Sc. Record, II, 222 (1892) — — — — N.S.W. — — B.fl.IV,480 M.fr.VI,142.

SCROPHULARINAE.
Mirbel, Elém. de Botan. II, 879 (1815).

MIMULUS, Linné in Act. Upsal. 82 (1741). (Uvedalia.)
M. Uvedaliae, Bentham in De Cand. X, 369 (1846) ... ... — — — — — Q. N.A. B.fl.IV,482 M.fr.VI,103;IX,167.
M. debilis, F. v. M. in Transact. phil. Inst. Vict. I, 62 (1857) ... — — — — — — N.A. — M.fr.VI,103.
M. gracilis, R. Brown, prodr. 439 (1810) ... ... — — — — — Q. N.A. B.fl.IV,482 M.fr.VI,103;IX,167.
M. repens, R. Brown, prodr. 439 (1810) ... ... W.A. S.A. T. V. N.S.W. Q. — B.fl.IV,482 M.fr.VI,103;IX,166.
M. prostratus, Bentham in De Cand. X, 373 (1846) ... ... — S.A. — V. N.S.W. Q. — B.fl.IV,483 M.fr.IX,167.

MAZUS, Loureiro, Fl. Cochinch. II, 385 (1790).
M. Pumilio, R. Brown, prodr. 439 (1810) ... ... — S.A. T. V. N.S.W. Q. — B.fl.IV,484 M.fr.VI,104;IX,167.

ADENOSMA, R. Brown, prodr. fl. Nov. Holl. 442 (1810). (Pterostigma.)
A. coerulea, R. Brown, prodr. 443 (1810) ... ... — — — — — — Q. B.fl.IV,484
A. Muelleri, Bentham, Fl. Austr. IV, 485 (1869)... ... — — — — — — N.A. B.fl.IV,485

STEMODIA, Linné, syst. ed. doc. 1118 (1759). (Morgania, Limnophila.)
S. lythrifolia, F. v. M. in Benth. Fl. Austr. IV, 486 (1869) ... — — — — — — N.A. B.fl.IV,486
S. grossa, Bentham, Fl. Austr. IV, 486 (1869) ... ... — — — — — — N.A. B.fl.IV,486
S. viscosa, Roxburgh, pl. of Coromand. II, 33, t. 163 (1798) ... W.A. S.A. — — — — N.A. B.fl.IV,486
S. linophylla, F. v. M., fragm. X, 88 (1876) ... ... W.A. — — — — — M.fr.X,88.
S. debilis, Bentham, Fl. Austr. IV, 487 (1869) ... ... — — — — — — N.A. B.fl.IV,487
S. pedicellaris, F. v. M., fragm. VIII, 231 (1874)... ... ... W.A. S.A. — — — — — M.fr.VIII,231.
S. Morgania, F. v. M., fragm. X, 89 (1876) ... ... W.A. S.A. — V. N.S.W. Q. N.A. B.fl.IV,488 M.fr.VI,104.
S. sessiliflora, F. v. M., Limnophila, Blume, Bijdr. 750 (1826) ... — — — — — — N.A. —
S. gratioloides, F. v. M., fragm. X, 89 (1876) ... ... — — — — — Q. N.A. B.fl.IV,489 M.fr.X,89.
S. punctata, F. v. M.; Limnophila, Blume, Bijdr. 750 (1826) ... — — — — — Q. — B.fl.IV,490
S. hirsuta, Heyne in Wallich, numer. list. 3930 (1828) ... ... — — — — — — N.A. B.fl.IV,490
S. tenuiflora, Bentham, Scrophul. Indic. 23 (1835) ... ... — — — — — — N.A. B.fl.IV,490

BRAMIA, Lamarck, Encycl. méthod. I, 459 (1783). (Herpestis.)
B. florihunda, F. v. M., fragm. IX, 167 (1875) ... ... — — — — — Q. N.A. B.fl.IV,491 M.fr.VI,104.
B. Indica, Lamarck, Encycl. méth. I, 459 (1783) ... ... — — — — N.S.W. Q. N.A. B.fl.IV,491 M.fr.IX,167.

GRATIOLA, Ruppius, Fl. Jenens. 241 (1718), from Dodoens (1556).
G. pedunculata, R. Brown, prodr. 435 (1810) ... ... W.A. S.A. — V. N.S.W. Q. — B.fl.IV,492 M.fr.VI,108;IX,167.
G. Peruviana, Linné, spec. plant. 17 (1753).. ... ... W.A. S.A. T. V. N.S.W. Q. N.A. B.fl.IV,493 M.fr.VI,103;IX,167.
G. nana, Bentham in De Cand. prodr. X, 404 (1846) ... ... — T. V. N.S.W. — — B.fl.IV,493 M.fr.IX,167.

DOPATRIUM, Hamilton in Benth. Scroph. Ind. 4 (1835).
D. junceum, Hamilton in Benth. Scroph. Ind. 4 (1745) ... ... — — — — — Q. — B.fl.IV,494

ARTANEMA, D. Don in Sweet, Fl. gard. t. 234 (1829).
A. fimbriatum, D. Don in Sweet, Fl. Gard. t. 234 (1829) ... — — — — N.S.W. Q. — B.fl.IV,495 M.fr.VI,104;IX,167.

LINDERNIA, Allioni, rarior. Pedemont. stirp. specim. t. 5 (1755). (Vandellia, Bonnaya, Ilysanthes, Torenia partly, Tittmannia partly.)
L. crustacea, F. v. M.; Vandellia, Bentham, Scroph. Indic. 35 (1835) — — — — N.S.W. Q. N.A. —
L. pubescens, F.v.M.; Vandellia, Benth. in De Cand. pr. X, 415(1846) — — — — — N.A. B.fl.IV,496 M.fr.VI,102.
L. alsinoides, F. v. M.; Vandellia, Benth. in De Cand. ... ... W.A. — — — N.S.W. Q. N.A. B.fl.IV,496 M.fr.VI,102.
L. scapigera, R. Brown, prodr. 441 (1810)... ... ... — — — — — N.A. B.fl.IV,497 M.fr.VI,102.
L. subulata, R. Brown, prodr. 441 (1810) ... ... — — — — — Q. N.A. B.fl.IV,497 M.fr.VI,102.
L. lobelioides ; Vandellia, F. v. M. in Transact. phil. Inst. Vict.
   III, 61 (1858) ... ... — — — — — N.A. B.fl.IV,497 M.fr.VI,102.
L. clausa, F. v. M., fragm. VI, 102 (1868) ... ... — — — — — N.A. B.fl.IV,498 M.fr.VI,102.
L. veronicifolia, F. v. M., fragm. VI, 101 (1868) ... ... — — — — — Q. B.fl.IV,498 M.fr.VI,101.
L. serrata, F.v.M.; Gratiola, Roxburgh. fl. Ind. ed. Car. I, 140 (1820) — — — — — Q. — B.fl.IV,498

HEMIARRHENA, Bentham, Fl. Austr. IV, 518 (1869).
H. plantaginea, Bentham in Hook. icon. pl. t. 1059 (1869) ... — — — — — — N.A. B.fl.IV,518 M.fr.VI,102.

PEPLIDIUM, Delile, Fl. Aegypt. 148 (1813).
P. humifusum, Delile, Fl. Aegypt. 148 (1813) ... ... — — — — N.S.W. N.A. B.fl.IV,500 M.fr.VI,105.
P. Muelleri, Bentham, Fl. Austr. IV, 500 (1869) ... ... — — — — — Q. N.A. B.fl.IV,500

MICROCARPAEA, R. Brown, prodr. 435 (1810).
M. muscosa, R. Brown, prodr. 435 (1810) ... ... — — — — — Q. N.A. B.fl.IV,501 M.fr.VI,104.

GLOSSOSTIGMA, Arnott in nov. act. acad. Caes. Leop. Carol. XVIII, 355 (1836). (Tricholoma.)
G. spathulatum, Arnott in nov. act. Leop. XVIII, 355 (1836) ... — — — — — Q. — B.fl.IV,501
G. Drummondii, Bentham in De Cand. prodr. X, 426 (1846) ... W.A. — — — — — — B.fl.IV,502
G. elatinoides, Bentham in J. Hook. Fl. New Zeal. I, 190 (1853) ... — S.A. T. V. N.S.W. — — B.fl.IV,502

LIMOSELLA, Lindern, opusc. plant. Argentorat. 156, t. 5 (1728).
L. aquatica, Linné, spec. plant. 631 (1753)... ... ... W.A. S.A. T. V. N.S.W. — — B.fl.IV,502 M.fr.VI,103;IX,165.
L. Curdieana, F. v. M., fragm. IX, 166 (1875) ... ... — S.A. — V. N.S.W. — — M.fr.IX,166.

SCOPARIA, Linné, syst. nat. ed. sext. 87 (1748).
S. dulcis, Linné, spec. plant. 116 (1753) ... ... — — — — — Q. N.A. B.fl.IV,504 M.fr.VI,104.

OURISIA, Commerçon in A. L. de Jussieu, gen. pl. 100 (1789).
O. integrifolia, R. Brown, prodr. 439 (1810) ... ... — — — T. — — — B.fl.IV,512 M.fr.VI,105;IX,168.

VERONICA, Tournefort, inst. 143, t. 60 (1700), from Fuchs (1543). (Pygmaea.)
V. densifolia, F. v. M., fragm. II, 137, t. 63 (1861) ... ... — — — V. N.S.W. — B.fl.IV,505 M.fr.VI,103.
V. formosa, R. Brown, prodr. 434 (1810) ... ... — — — T. — — B.fl.IV,506 M.fr.VI,102;IX,167.

N

V. decorosa, F. v. M. in Schlechtend., Linnaea XXV, 430 (1852)   — S.A. — —   —   —   B.fl.IV,506
V. perfoliata, R. Brown, prodr. 434 (1810) ..   ..   —   — — V. N.S.W. —   —   B.fl.IV,507   M.fr.VI,102;IX,167.
V. Derwentia, Littlejohn in Andr. Bot. Reposit. t. 531 (1808) ...   — S.A. T. V. N.S.W. Q.   —   B.fl.IV,507   M.fr.VI,102;IX,168.
V. arenaria, Cunningham in De Cand. prodr. X, 463 (1846)   ...   —   —   — N.S.W.   —   B.fl.IV,507   M.fr.VI,102;IX,167.
V. nivea, Lindley, Bot. Regist. Misc. 42 (1842) ...   .   —   — T. V. N.S.W. —   B.fl.IV,508   M.fr.VI,102;IX,167.
V. gracilis, R. Brown, prodr. 435 (1810)   ...   .   — S.A. T. V. N.S.W. —   B.fl.IV,508   M.fr.VI,103;IX,168.
V. distans, R. Brown, prodr. 435 (1810) ...   .   W.A. S.A. T. V.   —   B.fl.IV,509   M.fr.VI,102.
V. calycina, R. Brown, prodr. 435 (1810) ...   .   W.A. S.A. T. V. N.S.W. Q.   B.fl.IV,509   M.fr.VI,103.
V. plebeja, R. Brown, prodr. 435 (1810)   ...   ...   — T. V. N.S.W. Q.   — B.fl.IV,510   M.fr.VI,102.
V. notabilis, F. v. M. in Benth. Fl. Austr. IV, 511 (1869)   —   — T. V. N.S.W. Q.   — B.fl.IV,511   M.fr.IX,168.
V. serpillifolia, Linné, spec. plant. 12 (1753)   —   —   — V. N.S.W. —   B.fl.IV,511   M.fr.VI,103.
V. peregrina, Linné, spec. plant. 14 (1753)...   ...   ...   W.A. S.A. T. V. N.S.W. Q.   B.fl.IV,511   M.fr.VI,103;IX,168.

RAMPHICARPA, Bentham in Hook. Comp. to the Bot. Mag. I, 368 (1835).
R. longiflora, Bentham in Hook. Comp. Bot. Mag. I, 368 (1835)   —   — — —   — Q. N.A. B.fl.IV,518   M.fr.VI,104;IX,168.

CENTRANTHERA, R. Brown, prodr. fl. Nov. Holland. 438 (1810).
C. hispida, R. Brown, prodr. 438 (1810) ...   ...   —   — — N.S.W. Q. N.A. B.fl.IV,513   M.fr.VI,104;IX,168.

SOPUBIA, Hamilton in D. Don, prodrom. flor. Nepalens. 88 (1825).
S. trifida, Hamilton in D. Don, prodr. fl. Nep. 88 (1825)   —   — — — — Q.   — B.fl.IV,513

BUCHNERA, Linné, hort. Clifort. 501 (1737). (Buchnera, Striga.)
B. tetragona, R. Brown, prodr. 437 (1810)...   —   — — — Q. N.A. B.fl.IV,514   M.fr.IX,168.
B. urticifolia, R. Brown, prodr. 437 (1810)...   ...   —   — — N.S.W. Q. N.A. B.fl.IV,514   M.fr.VI,104.
B. linearis, R. Brown, prodr. 437 (1810) ...   ...   —   — — Q. N.A. D.fl.IV,515
B. tenella, R. Brown, prodr. 437 (1810)   ...   ...   —   — — Q. N.A. B.fl.IV,515
B. gracilis, R. Brown, prodr. 437 (1810) ...   ...   —   — — N.S.W. —   B.fl.IV,515   M.fr.VI,104.
B. ramosissima, R. Brown, prodr. 438 (1810)   ...   —   — — Q. N.A. K.fl.IV,515
B. coccinea, Bentham, Scrophul. Ind. 40 (1835) ...   —   — — — Q.   — B.fl.IV,516
B. parviflora, R. Brown, prodr. 438 (1810)   ...   —   — — — Q.   — B.fl.IV,517   M.fr.VI,104.
B. multiflora, F.v.M.; Striga,Benth.in De Cand.,prodr.X,501 (1836) —   — — — — N.A. B.fl.IV,517   M.fr.VI,104.
B. curvifora, R. Brown, prodr. 438 (1810) ...   ...   —   — — Q. N.A. B.fl.IV,517

EUPHRASIA, Tournefort, instit. 174, t. 78 (1700), from Matthaeus (1478).
E. Brownii, F. v. M., fragm. V, 88 (1865)...   ...   ...   W.A. S.A. T. V. N.S.W. Q.   B.fl.IV,521   M.fr.IX,168.
E. scabra, R. Brown, prodr. 437 (1810)   ...   ...   W.A. S.A. T. V. N.S.W. Q.   B.fl.IV,521   M.fr.IX,168.
E. cuspidata, J. Hooker, Fl. Tasman. I, 298 (1860)   —   — T. —   —   B.fl.IV,522
E. antarctica, Bentham in De Cand., prodr. X, 555 (1846)   —   — — V. N.S.W. —   B.fl.IV,522

OROBANCHEAE.
A. L. de Jussieu in Ann. Mus. V, 25 (1804).

OROBANCHE, Tournefort, inst. 175, t. 81 (1700), from l'Ecluse (1583).
O. Australiana, F. v. M., inedit. ...   .   W.A. S.A. — V. N.S.W. —   B.fl.IV,533

LENTIBULARINAE.
L. C. Richard in Bulliard, Fl. Paris. ed. sec. I, 26 (1796).

UTRICULARIA, Linné, syst. nat. 8 (1735); Linné, gen. plant. 5 (1737).
U. stellaris, Linné, fil. suppl. plant. 86 (1781)   ...   ...   —   — — — Q. N.A. B.fl.IV,525   M.fr.VI,161.
U. flexuosa, Vahl, enum. plant. I, 198 (1804)   ...   ...   W.A. S.A. T. V. N.S.W. Q. N.A. B.fl.IV,525
U. exoleta, R. Brown, prodr. 430 (1810)   ...   ...   —   — — N.S.W. Q. N.A. B.fl.IV,526   M.fr.VI,161.
U. tubulata, F. v. M., fragm. IX, 48 (1875)   ...   ...   —   — — —   —   M.fr.IX,48.
U. albiflora, R. Brown, prodr. 431 (1810)   ...   ...   —   — — —   — B.fl.IV,526
U. pygmaea, R. Brown, prodr. 432 (1810)   ...   ...   —   — — —   — B.fl.IV,526
U. fulva, F. v. M. in Transact. phil. Inst. Vict. III, 63 (1858)...   —   — — — — N.A. B.fl.IV,527   M.fr.VI,160.
U. chrysantha, R. Brown, prodr. 432 (1810)   ...   —   — — — Q. N.A. B.fl.IV,527   M.fr.VI,161.
U. bifida, Linné, spec. plant. 18 (1753)   ...   ...   —   — — —   — B.fl.IV,527
U. cyanea, R. Brown, prodr. 431 (1810)   ...   ...   —   — — N.S.W. Q. N.A. B.fl.IV,527   M.fr.VI,161.
U. lateriflora, R. Brown, prodr. 431 (1810)   ...   ...   —   — T. V. N.S.W. —   B.fl.IV,528   M.fr.VI,162.
U. simplex, R. Brown, prodr. 431 (1810)   ...   ...   W.A. —   — — —   B.fl.IV,528   M.fr.VI,162.
U. monantha, J. Hooker, Fl. Tasman. I, 299 (1860)   —   — T. —   —   B.fl.IV,528
U. dichotoma, Labillard., Nov. Holl. pl. spec. I, 11, t. 8 (1804)   S.A. T. V. N.S.W. —   B.fl.IV,529   M.fr.VI,161.
U. volubilis, R. Brown, prodr. 430 (1810)   ...   ...   W.A. —   — — —   B.fl.IV,529   M.fr.VI,162.
U. Hookeri, Lehmann, nov. stirp. pugill. VIII, 47 (1844)   W.A. —   — — —   B.fl.IV,530   M.fr.VI,162.
U. violacea, R. Brown, prodr. 431 (1810)...   ...   ...   W.A. —   — — —   B.fl.IV,530   M.fr.VI,160.
U. Menziesii, R. Brown, prodr. 431 (1810)   ...   ...   W.A. —   — — —   B.fl.IV,530   M.fr.VI,160.
U. biloba, R. Brown, prodr. 432 (1810)   ...   ...   —   — — N.S.W. —   B.fl.IV,531
U. limosa, R. Brown, prodr. 432 (1810)   ...   ...   —   — — — Q.   B.fl.IV,531
U. Baueri, R. Brown, prodr. 431 (1810)   ...   ...   —   — — — Q.   B.fl.IV,531

POLYPOMPHOLYX, Lehmann in der Bot. Zeitung, Halle, 109 (1844). (Tetralobus.)
P. multifida, F. v. M., fragm. VI, 162 (1868)   ...   ...   W.A. —   — — —   B.fl.IV,532   M.fr.VI,162.
P. tenella, Lehmann, nov. stirp. pugill. VIII, 50 (1844)...   W.A. —   — — —   B.fl.IV,532   M.fr.VI,162.

GESNERIACEAE.
Humboldt, Bonpland & Kunth, nov. gen. et spec. pl. II, 392 (1817).

FIELDIA, Cunningham in Field's New South Wales 363, t. 2 (1825).
F. australis, Cunningham in Field's N.S.W. 364, t. 2 (1825)   ...   — V. N.S.W. —   B.fl.IV,535   M.fr.IX,101.

NEGRIA, F. v. M., fragm. VII, 151 (1871). (Rhabdothamnus partly.)
N. rhabdothamnoides, F. v. M., fragm. VII, 152 (1871)...   —   — — N.S.W. —   — M.fr.VII,152;XI,133.

BAEA, Commerçon in Lamarck, Encycl. méth. I, 401 (1783). (Streptocarpus.)
B. hygroscopica, F. v. M., fragm. IV, 146 (1864)...   ...   —   — — — Q.   — B.fl.IV,535   M.fr.IV,146.

BIGNONIACEAE.
Ventenat, Tabl. du règn. végét. II, 402 (1799).

TECOMA, A. L. de Jussieu, gen. plant. 139 (1789).
T. australis, R. Brown, prodr. 471 (1810) ... ... — S.A. — V. N.S.W. Q. N.A. B.fl.IV,537
T. Hillii, F. v. M., fragm. X, 101 (1877) ... ... — — — — — Q. — — M.fr.X,101;XI,136.
T. jasminoides, Lindley, Bot. Regist. t. 2002 (1838) ... ... — — — — N.S.W. Q. — B.fl.IV,537

DOLICHANDRONE, Fenzl in Ann. nat. hist. third ser. X, 31 (1862). (Spathodea partly, Dolichandra partly.)
D. heterophylla, F. v. M., fragm. IV, 149 (1864).. ... ... — — — — — Q. N.A. B.fl.IV,530 M.fr.IV,149.
D. filiformis, Seemann, Journ. of Bot. I, 226 (1863) ... ... — — — — — N.A. B.fl.IV,539

HAUSSMANNIA, F. v. M., fragm. IV, 148 (1864).
H. jucunda, F. v. M., fragm. IV, 148 (1864) ... ... — — — — — Q. — B.fl.IV,540 M.fr.IV,148.

DIPLANTHERA, R. Brown, prodr. Fl. Nov. Holl. 449 (1810). (Deplanchea, Bulweria.)
D. tetraphylla, R. Brown, prodr. 449 (1810) ... ... — — — — — Q. — B.fl.IV,540

ACANTHACEAE.
A. L. de Jussieu, gen. plant. 102 (1789), from B. de Jussieu (1759).

THUNBERGIA, Retzius, Physiogr. Saellsk. Handl. I, 163 (1776).
T. Arnhemica, F. v. M., fragm. IX, 73 (1875) ... . — — — — — — N.A. — M.fr.IX,73.
T. Powelli, F. v. M. in Wing's South. Sc. Rec. II, 34 (1882) ... — — — — — — N.A. —

NELSONIA, R. Brown, prodr. fl. Nov. Holl. 485 (1810).
N. campestris, R. Brown, prodr. 481 (1810) ... ... — — — — Q. N.A. B.fl.IV,543 M.fr.VI,88;XI,18.

EBERMAYERA, Nees in Wallich, pl. Asiat. rar. III, 75 (1832).
E. glauca, Nees in De Candolle, prodr. XI, 73 (1847) ... ... — — — — Q. N.A. B.fl.IV,544 M.fr.VI,88;XI,18

HYGROPHILA, R. Brown, prodr. Fl. Nov. Holl. 479 (1810).
H. angustifolia, R. Brown, prodr. 479 (1810) ... ... — — — — Q. N.A. B.fl.IV,544 M.fr.VI,88.

STROBILANTHES, Blume, Bijdr. tot de fl. van Nederl. Ind. 796 (1826).
S. Tatei, F. v. M. in transact. Roy. Soc. S. Austr. V, Aug. (1882) — — — — — N.A. —

RUELLIA, Plumier, nov. gen. 12, t. 2 (1703). (Dipteracanthus.)
R. bracteata, R. Brown, prodr. 479 (1810)... ... ... — — — — Q. N.A. B.fl.IV,546 M.fr.VI,91.
R. primulacea, F. v. M. in Benth. Fl. Austr. IV, 546 (1869) ... — S.A. — — — Q. — B.fl.IV,546 M.fr.XI,18.
R. corynotheca, F. v. M. in Benth. Fl. Austr. IV, 546 (1869) ... — — — — — Q. — B.fl.IV,546 M.fr.XI,18.
R. acaulis, R. Brown, prodr. 479 (1810) ... ... — S.A. — N.S.W. Q. — B.fl.IV,547 M.fr.VI,91.
R. spiciflora, F. v. M. in Benth. Fl. Austr. IV, 547 (1869) ... — — — — — Q. — B.fl.IV,547
R. acaulis, R. Brown, prodr. 479 (1810) ... ... — — — — — Q. — B.fl.IV,547

ACANTHUS, Tournefort, inst. 176, t. 81 (1700), from Theophrastos. (Dilivaria.)
A. ilicifolius, Linné, spec. plant. 939 (1753) ... ... — — — — Q. N.A. B.fl.IV,548 M.fr.X,18.

JUSTICIA, Houston in Linné, gen. plant. 4 (1737). (Rostellaria, Rostellularia.)
J. procumbens, Linné, spec. plant. 15 (1753) ... ... — W.A. S.A. — N.S.W. Q. N.A. B.fl.IV,549 M.fr.VI,91;XI,18.
J. Donneyana, F. v. M. in Wing's S. Sc. Record II, 74 (1882) ... — S.A. — N.S.W. — — — M.fr.X,73.
J. hygrophiloides, F. v. M., fragm. VI, 89 (1868)... ... ... — — — — — Q. — B.fl.IV,550 M.fr.XI,101.
J. Kempeana, F. v. M., fragm. XI, 101 (1881) ... ... W.A. S.A. — — — Q. N.A. — M.fr.XI,101.
J. cavernarum, F. v. M., fragm. VI, 91 (1868) ... ... — — — — Q. — B.fl.IV,550 M.fr.VI,91.
J. eranthemoides, F. v. M., fragm. VI, 90 (1868)... ... ... — — — — N.S.W. Q. — B.fl.IV,551 M.fr.VI,90.

GRAPTOPHYLLUM, Nees in Wall. pl. Asiat. rar. III, 102 (1832). (Earlia, Thyrsacanthus.)
G. Earlii, F. v. M., fragm. VI, 87 (1868) ... ... — — — — — Q. — B.fl.IV,551 M.fr.VI,87.
G. ilicifolium, F. v. M. in Benth. Fl. Austr. IV, 552 (1869) ... — — — — — Q. — B.fl.IV,552
G. spinigerum, F. v. M., fragm. XI, 17 (1878) ... ... — — — — — Q. — — M.fr.XI,17.

DICLIPTERA, A. L. de Jussieu in Ann. du Mus. IX, 267 (1807). (Brochosiphon.)
D. glabra, Decaisne in Nouv. Annal. du Mus. III, 388 (1834) ... — — — — — N.A. B.fl.IV,552 M.fr.VI,88;XI,18.
D. apicata, Decaisne in Nouv. Annal. du Mus. III, 389 (1834) ... — — — — — Q. — B.fl.IV,553 M.fr.VI,89.
D. Burmanni, Nees in Wallich, pl. Asiat. rar. III, 112 (1832) ... — — — — — Q. — — M.fr.VII,62.

HYPOESTES, Solander in R. Brown, prodr. 474 (1810).
H. floribunda, R. Brown, prodr. 474 (1810) ... ... — — — — N.S.W. Q. N.A. B.fl.IV,554 M.fr.XI,18.

ERANTHEMUM, Linné, Fl. Zeyl. 6 (1747).
E. variabile, R. Brown, prodr. 477 (1810) ... ... — — — — N.S.W. Q. — B.fl.IV,555 M.fr.VI,91;XI,18.
E. tenellum, Bentham, Fl. Austr. IV, 555 (1869) ... ... — — — — — Q. — B.fl.IV,555

HYDROPHYLLEAE.
R. Brown in Bot. Regist. t. 242 (1817).

HYDROLEA, Linné, spec. plant. ed. sec. 328 (1762).
H. Zeilanica, Vahl, symbol. botan. II, 46 (1791) ... ... — — — — N.A. B.fl.IV,382 M.fr.V,192.
H. spinosa, Linné, spec. plant. ed. sec. 328 (1762) ... ... — — — — — Q. — B.fl.IV,383 M.fr.V,192.

ASPERIFOLIAE.
Haller, enum. stirp. Helv. Praef. 34 (1742).

CORDIA, Plumier, gen. plant. 13, t. 14 (1703).
C. subcordata, Lamarck, Illustr. des genr. t. 91 (1791)... ... — — — — — Q. N.A. B.fl.IV,385 M.fr.VI,114;IX,122,[122.
C. aspera, G. Forster, florul. ins. austr. prodr. 18 (1786)... ... — — — — — Q. — B.fl.IV,386 M.fr.V,103;VI,114;IX,
C. Myxa, Linné, spec. plant. 190 (1753) ... ... — — — — — Q. — B.fl.IV,386 M.fr.I,59;VI,114.

EHRETIA, P. Browne, Civ. and Nat. Hist. of Jamaic. 168, t. 16 (1756).
E. acuminata, R. Brown, prodr. 497 (1810) ... ... — V. N.S.W. Q. — B.fl.IV,387 M.fr.V,21;IX,122;XI,124
E. pilosula, F. v. M., fragm. V, 20 (1865)... ... ... — — — — — Q. — B.fl.IV,388 M.fr.V,20.
E. saligna, R. Brown, prodr. 497 (1810) ... ... — — — — — Q. N.A. B.fl.IV,388 M.fr.V,21;IX,122.

E. membranifolia, R. Brown, prodr. 497 (1810) ... ... — — — — — Q. — B.fl.IV,388 M.fr.V,21;VI,115;IX,122
E. laevis, Roxburgh, Pl. Corom. I, 42, t. 56 (1795) ... ... — — — — — Q. — B.fl.IV,389

TOURNEFORTIA, Linné, syst. nat. 8 (1735); Linné, gen. plant. 55 (1737.)
T. argentea, Linné fil. suppl. plant. 133 (1781) ... ... — — — — — Q. N.A. B.fl.IV,389 M.fr.IV,95;VI,116.
T. mollis, F. v. M., fragm. I, 59 (1858) ... ... — — — — — Q. N.A. B.fl.IV,390 M.fr.VI,116.
T. sarmentosa, Lamarck, Illustr. des genr. I, 416 (1791) ... — — — — — Q. — B.fl.IV,390 M.fr.IV,95;VI,116.

COLDENIA, Linné, amoen. acad. od. prim. 119 (1747). (Lobophyllum.)
C. procumbens, Linné, spec. plant. 125 (1753) ... ... — — — — — Q. N.A. B.fl.IV,391 M.fr.VI,115.

HELIOTROPIUM, Tournefort, inst. 138, t. 57 (1700), from Theophr., Diosc. & Plinins. (Tiaridium, Heliophytum.)
H. Indicum, Linné, spec. plant. 130 (1753) ... ... — N.A. — M.fr.XI,124.
H. plcioptorum, F. v. M., fragm. IX, 121 (1875)... ... ... — S.A. — — — — B.fl.IX,121.
H. curassavicum, Linné, spec. plant. 130 (1753) ... W.A. S.A. — V. N.S.W. Q. N.A. B.fl.IV,393 M.fr.VI,116;IX,121.
H. Europaeum, Linné, spec. plant. 130 (1753) ... S.A. — V. N.S.W. — B.fl.IV,394 M.fr.VI,116. [124.
H. undulatum, Vahl, symbol. bot. I, 13 (1795) ... W.A. S.A. — — N.A. B.fl.IV,394 M.fr.VI,116;IX,121;XI,
H. asperrimum, R. Brown, prodr. 493 (1810) ... W.A. S.A. — V. N.S.W. — B.fl.IV,394 M.fr.VI,116;IX,121.
H. crispatum, F. v. M. in Benth. Fl. Austr. IV, 395 (1869) ... — N.A. B.fl.IV,395
H. fasciculatum, R. Brown, prodr. 494 (1810) ... — — — — N.A. B.fl.IV,395
H. vestitum, Bentham, Fl. Austr. IV, 395 (1869) ... ... — — — N.S.W. — B.fl.IV,395
H. brachygyne, Bentham, Fl. Austr. IV, 396 (1869) ... — — — — Q. — B.fl.IV,396
H. epacrideum, F. v. M. in Benth. Fl. Austr. IV, 396 (1869) ... — N.A. B.fl.IV,396
H. ovalifolium, Forskael, Fl. Aegypt. Arab. 47 (1775) ... S.A. — — Q. N.A. B.fl.IV,396 M.fr.VI,115.
H. strigosum, Willdenow, sp. pl. I, 743 (1797) ... ... — — — Q. N.A. B.fl.IV,397
H. prostratum, R. Brown, prodr. 494 (1810) ... ... — — — Q. N.A. B.fl.IV,397
H. bracteatum, R. Brown, prodr. 493 (1810) ... ... — — — Q. N.A. B.fl.IV,397
H. pauciflorum, R. Brown, prodr. 493 (1810) ... ... — — — Q. N.A. B.fl.IV,398
H. conocarpum, F. v. M. in Benth. Fl. Austr. 398 (1869) ... — — — N.A. B.fl.IV,398
H. filaginoides, Bentham, Fl. Austr. IV, 398 (1869) ... S.A. — — — B.fl.IV,399
H. ventricosum, R. Brown, prodr. 494 (1810) ... ... — — — Q. N.A. B.fl.IV,399
H. tenuifolium, R. Brown, prodr. 494 (1810) ... ... — — — Q. N.A. B.fl.IV,399
H. paniculatum, R. Brown, prodr. 494 (1810) ... ... — — — N.A. B.fl.IV,399
H. Cunninghamii, Bentham, Fl. Austr. IV, 400 (1860) ... — — — N.A. B.fl.IV,400
H. diversifolium, F. v. M. in Benth. Fl. Austr. IV, 400 (1869) ... — — — N.A. B.fl.IV,400

HALGANIA, Gaudichaud in Freyc. in voy. Botan. 448 (1826).
H. solanacea, F. v. M. in Hook. Kew Misc. IX, 21 (1857) ... — — — — N.A. B.fl.IV,401 M.fr.VI,115.
H. littoralis, Gaudichaud in Freyc. voy. Bot. 449, t. 59 (1826) ... W.A. — — — B.fl.IV,401 M.fr.VI,115.
H. corymbosa, Lindley, Bot. Regist. XXV, App. XL (1839) ... W.A. — — — B.fl.IV,402
H. sericiflora, Bentham, Fl. Austr. IV, 402 (1869) ... W.A. — — B.fl.IV,402
H. cyanea, Lindley, Bot. Regist. XXV, App. XL (1839) ... W.A. S.A. — V. N.S.W. Q. N.A. — M.fr.VI,115;IX,122.
H. lavandulacea, Endlicher in Ann. Wien. Mus. II, 205 (1838) ... W.A. S.A. — V. N.S.W. — B.fl.IV,402 M.fr.I,209;VI,115;XI,125
H. integerrima, Endlicher in Ann. Wien. Mus. II, 205 (1838) ... W.A. — — B.fl.IV,402

POLLICHIA, Medikus, Botan. Beobach. 247 (1783). (Trichodesma.)
P. Zeylanica, F. v. M.; Trichodesma, R. Brown, prodr. 496 (1810) W.A. S.A. — N.S.W. Q. N.A. B.fl.IV,404 M.fr.XI,125.
P. latisepalca; Trichodesma, F. v. M., fragm. X, 102 (1877) — — — — — — M.fr.X,121.

MYOSOTIS, Ruppius, Fl. Jenens. 9 (1718), from Dioscorides and Plinius. (Exarrhena.)
M. australis, R. Brown, prodr. 495 (1810) ... ... W.A. S.A. — T. V. N.S.W. — B.fl.IV,405 M.fr.VI,116;IX,122.
M. suaveolens, Poiret, Encycl. méth. suppl. IV, 44 (1816) ... T. V. N.S.W. — B.fl.IV,406 M.fr.VI,115;IX,122.

ERITRICHIUM, Schrader in commentat. Goetting. IV, 186 (1820). (Eritrichium.)
E. Australasicum, A. De Candolle, prodr. X, 134 (1846) ... W.A. S.A. — V. N.S.W. — B.fl.IV,406 M.fr.XI,125.

LAPPULA, Rivinus in Rupp. Fl. Jenens. (1718), from Dalechamp. (Echinospermum.)
L. concava; Echinospermum, F. v. M., fragm. II, 139 (1861) ... W.A. S.A. — N.S.W. — B.fl.IV,407 M.fr.VI,116;IX,121.

ROCHELIA, Reichenbach in Fl. Regensb. bot. Zeit. I, 243 (1824).
R. Maccoya, F. v. M. in Benth. Fl. Austr. IV, 408 (1869) ... — S.A. — V. N.S.W. — B.fl.IV,408 M.fr.I,127;IX,121.

CYNOGLOSSUM, Tournefort, inst. 139, t. 57 (1700), from Dioscorides and Nicandros. [125.
C. latifolium, R. Brown, prodr. 495 (1810)... ... ... — T. V. N.S.W. Q. — B.fl.IV,408 M.fr.VI,115;IX,122;XI,
C. suaveolens, R. Brown, prodr. 495 (1810) ... ... S.A. T. V. N.S.W. — B.fl.IV,409 M.fr.VI,115;IX,122.
C. australe, R. Brown, prodr. 495 (1810) ... ... S.A. T. V. N.S.W. — B.fl.IV,409 M.fr.VI,115;IX,122.
C. Drummondii, Bentham, Fl. Austr. IV, 409 (1869) ... W.A. S.A. — — B.fl.IV,400 M.fr.IX,122.

LABIATAE.
Adanson, Famill. des plant. II, 180 (1763), from B. de Jussieu (1759).
OCIMUM, Tournefort, inst. 203, t. 96 (1700), from Theophrastos and Dioscorides. (Ocymum.)
O. sanctum, Linné, mantiss. plant. 85 (1787) ... ... — — — Q. N.A. B.fl.V,74 M.fr.IV,46.

MOSCHOSMA, Reichenbach, conspect. regn. vegetab. 115 (1828). (Basilicum 1802.)
M. polystachya, Bentham in Wall. pl. As. rar. F, 13 (1831) ... — — N.S.W. Q. N.A. B.fl.V,75 M.fr.VI,108.

ORTHOSIPHON, Bentham, Labiatar. gen. et spec. 25 (1832).
O. stamineus, Bentham in Wall. pl. As. rar. II, 15 (1831) ... ... — — Q. N.A. B.fl.V,76 M.fr.VI,108;IX,161.

PLECTRANTHUS, L'Héritier, stirp. nov. I, 85 (1785).
P. longicornis, F. v. M., fragm. V, 51 (1865) ... ... — — Q. B.fl.V,77 M.fr.V,51.
P. parviflorus, Willdenow, hort. Berol. I, 85 (1800)... ... — S.A. — V. N.S.W. Q. N.A. B.fl.V,78 M.fr.VI,108;IX,161.
P. congestus, R. Brown, prodr. 506 (1810)... ... ... — B.fl.V,78 M.fr.VI,108.
P. foetidus, Bentham, labiat. 35 (1832) ... ... — — — Q. N.A. B.fl.V,78 M.fr.IX,161;XI,135.

COLEUS, Loureiro, Fl. Cochinch. II, 372 (1790).
C. scutellarioides, Bentham in Wall. pl. Asiat. rar. II, 16 (1831) — — — — Q. N.A. B.fl.V,79 M.fr.VI,108.

POGOSTEMON, Desfontaines in Mém. du Mus. II, 154, t. 6 (1815). (Dysophylla.)
P. verticillatus, Hasskarl, catal. hort. Bogor. 131 (1844) ... — — — — Q. N.A. B.fl.V,81 M.fr.V,200;IX,161.

MENTHA, Tournefort, inst. 188, t. 89 (1700), from Hippocrates and Theophrastos.
M. laxiflora, Bentham in De Cand. XII, 174 (1848)	...	...	—	—	—	V.	N.S.W.	—	—	B.fl.V,82	M.fr.VI,100;IX,162.
M. grandiflora, Bentham in Mitch. Trop. Austr. 362 (1848)	...	...	—	—	—	—	N.S.W.	Q.	—	B.fl.V,63	
M. australis, R. Brown, prodr. 505 (1810)	...	...	...	—	S.A.	T.	V.	N.S.W.	Q.	N.A.	B.fl.V,83	M.fr.VI,109;IX,162.
M. gracilis, R. Brown, prodr. 505 (1810)	...	...	...	—	S.A.	T.	V.	N.S.W.	—	—	B.fl.V,83	M.fr.VI,109.
M. saturejoides, R. Brown, prodr. 505 (1810)	...	...	...	W.A.	S.A.	T.	V.	N.S.W.	Q.	—	B.fl.V,84	M.fr.VI,109.

LYCOPUS, Tournefort, inst. 180, t. 83 (1700), from Plinius.
L. australis, R. Brown, prodr. 500 (1810)	...	...	—	S.A.	T.	V.	N.S.W.	Q.	—	B.fl.V,85	M.fr.VI,109.

SALVIA, Linné, gen. plant. 6 (1737), from Plinius.
S. plebeja, R. Brown, prodr. 501 (1810)	...	...	...	—	—	—	V.	N.S.W.	Q.	—	B.fl.V,85	M.fr.VI,100;IX,161.

BRUNELLA, L'Ecluse, rar. stirp. hist. II, 42, 43 (1576), from Brunfels (1536). (Prunella.)
B. vulgaris, A. de Candolle, prodr. XII, 410 (1848)	...	—	—	—	V.	N.S.W.	Q.	—	B.fl.V,87	M.fr.VI,109.

SCUTELLARIA, Hermann, hort. Lugd. bot. cat. (1687), from Cortusi (1591).
S. mollis, R. Brown, prodr. 507 (1810)	...	...	...	—	—	—	V.	N.S.W.	—	—	B.fl.V,88	M.fr.VI,109.
S. humilis, R. Brown, prodr. 507 (1810)	...	...	...	—	S.A.	T.	V.	N.S.W.	Q.	—	B.fl.V,88	M.fr.VI,109.

ANISOMELES, R. Brown, prodr. 503 (1810).
A. salvifolia, R. Brown, prodr. 503 (1810)	...	...	—	—	—	—	—	Q.	N.A.	B.fl.V,89	M.fr.VI,100;IX,162.

LEUCAS, J. Burmann, thesaur. Zeyl. 140 (1737).
L. flaccida, R. Brown, prodr. 505 (1810)	...	...	—	—	—	—	—	Q.	N.A.	B.fl.V,90	M.fr.VI,109.

DEPREMESNILIA, F. v. M., fragm. X, 59 (1876).
D. chrysocalyx, F. v. M., fragm. X, 59 (1876)	...	W.A.	M.fr.X,59.

PROSTANTHERA, Labillardière, Nov. Holl. pl. spec. 18, t. 157 (1806). (Chilodia, Cryphia, Klanderia.)
P. lasiantha, Labillard., Nov. Holl. pl. spec. II, 18, t. 15 (1806)	—	S.A.	T.	V.	N.S.W.	Q.	—	B.fl.V,93	M.fr.VI,105.
P. prunelloides, B. Brown, prodr. 508 (1810)	—	—	—	N.S.W.	—	—	B.fl.V,94	M.fr.VI,106.
P. coerulea, R. Brown, prodr. 508 (1810)	...	—	—	—	N.S.W.	—	—	B.fl.V,94	M.fr.VI,106.
P. ovalifolia, R. Brown, prodr. 509 (1810)	...	—	—	—	N.S.W.	Q.	—	B.fl.V,95	M.fr.VI,106.
P. melissifolia, F. v. M., fragm. I, 19 (1858)	...	—	—	—	V.	—	—	B.fl.V,95	M.fr.I,19,242.
P. incisa, R. Brown, prodr. 509 (1810)	...	—	—	—	N.S.W.	Q.	—	B.fl.V,95	M.fr.VI,107;IX,162.
P. rotundifolia, R. Brown, prodr. 509 (1810)	...	—	S.A.	T.	V.	N.S.W.	—	—	B.fl.V,96	M.fr.VI,107;IX,162.
P. violacea, R. Brown, prodr. 509 (1810)	...	—	—	—	N.S.W.	—	—	B.fl.V,96	M.fr.VI,107;IX,162.
P. incana, Cunningham in Benth. lab. 455 (1836)	...	—	—	—	N.S.W.	—	—	B.fl.V,97	M.fr.IX,162.
P. hirtula, F. v. M., first gen. Report 16 (1853)	...	—	—	—	V.	N.S.W.	—	—	B.fl.V,97	
P. denticulata, R. Brown, prodr. 509 (1810)	...	—	—	—	V.	N.S.W.	—	—	B.fl.V,97	M.fr.VI,105.
P. rugosa, Cunningham in Benth. lab. 456 (1836)	...	—	—	—	N.S.W.	—	—	B.fl.V,98	
P. marifolia, R. Brown, prodr. 500 (1810)	...	—	—	—	N.S.W.	—	—	B.fl.V,98	M.fr.VI,105.
P. rhombea, R. Brown, prodr. 509 (1810)	...	—	—	—	N.S.W.	—	—	B.fl.V,99	
P. spinosa, F. v. M. in Transact. phil. Soc. Vict. I, 48 (1854)	—	S.A.	—	V.	—	—	—	B.fl.V,99	M.fr.VI,108;IX,162.
P. cuneata, Bentham in De Cand., prodr. XII, 560 (1848)	...	—	T.	V.	N.S.W.	—	—	B.fl.V,99	M.fr.VI,108.
P. linearis, R. Brown, prodr. 509 (1810)	...	—	—	—	N.S.W.	—	—	B.fl.V,100	M.fr.VI,106.
P. phylicifolia, F. v. M., fragm. I, 19 (1858)	...	—	—	—	N.S.W.	Q.	—	B.fl.V,100	M.fr.I,19;IX,162.
P. decussata, F. v. M., fragm. I, 126 (1858)	...	—	—	—	V.	—	—	B.fl.V,100	M.fr.I,126.
P. empetrifolia, Sieber in Spreng. syst. cur. post. 226 (1827)	...	—	—	—	N.S.W.	—	—	B.fl.V,101	
P. lithospermoides, F. v. M., fragm. VI, 107 (1868)	...	—	—	—	—	Q.	—	B.fl.V,101	M.fr.VI,107.
P. Behriana, Schlechtendal, Linnaea XX, 610 (1847)	...	—	S.A.	—	V.	N.S.W.	—	—	B.fl.V,102	M.fr.VI,106.
P. Wilkieana, F. v. M., fragm. VIII, 230 (1874)	...	—	S.A.	—	—	—	—	—	—	M.fr.VIII,230.
P. Baxteri, Cunningham in Benth. lab. 452 (1836)	W.A.	S.A.	—	—	—	—	—	B.fl.V,102	M.fr.VI,106.
P. canaliculata, F. v. M., fragm. VI, 105 (1868)	...	W.A.	—	—	—	—	—	B.fl.V,102	M.fr.VI,105.
P. nivea, Cunningham in Benth. lab. 452 (1836)	...	—	—	V.	N.S.W.	Q.	—	B.fl.V,103	M.fr.VI,106.
P. striatiflora, F. v. M. in Schlechtend. Linnaea XXV, 425 (1852)	—	S.A.	—	N.S.W.	Q.	N.A.	B.fl.V,103	M.fr.VI,105;IX,162.
P. saxicola, R. Brown, prodr. 500 (1810)	...	—	—	—	N.S.W.	Q.	—	B.fl.V,104	M.fr.VI,105;IX,162.
P. debilis, F. v. M., fragm. VIII, 147 (1874)	...	—	—	—	V.	—	—	—	M.fr.VIII,147.
P. odoratissima, Bentham in Mitch. Trop. Austr. 291 (1848)	...	—	—	—	—	Q.	—	B.fl.V,104	
P. Eckersleyana, F. v. M., fragm. X, 17 (1876)	...	W.A.	—	—	—	—	—	—	M.fr.X,17.
P. euphrasioides, Bentham in Mitch. Trop. Austr. 360 (1848)	...	—	—	—	N.S.W.	Q.	—	B.fl.V,104	M.fr.VI,105.
P. stauropila, F. v. M., fragm. IX, 73 (1875)	...	—	—	—	N.S.W.	—	—	—	M.fr.IX,73.
P. cryptandroides, Cunningham in Benth. lab. 453 (1836)	...	—	—	—	N.S.W.	—	—	B.fl.V,105	
P. eurybioides, F. v. M. in Transact. phil. Soc. Vict. I, 48 (1854)	—	S.A.	—	—	—	—	—	B.fl.V,105	M.fr.VI,105.
P. ringens, Bentham in Mitch. Trop. Austr. 363 (1848)	...	—	—	—	N.S.W.	Q.	—	B.fl.V,106	
P. Leichhardtii, Bentham, Fl. Austr. V, 106 (1870)	...	—	—	—	—	Q.	—	B.fl.V,106	
P. Walteri, F. v. M., fragm. VII, 108 (1870)	...	—	—	V.	N.S.W.	—	—	—	M.fr.VII,108.
P. coccinea, F. v. M. in Transact. phil. Soc. Vict. I, 48 (1854)	W.A.	S.A.	—	V.	N.S.W.	Q.	—	B.fl.V,106	M.fr.VI,109;IX,162.
P. chlorantha, F. v. M. in Benth. Fl. Austr. V, 108 (1870)	—	S.A.	—	N.S.W.	—	—	B.fl.V,108	M.fr.IX,162;X,18.
P. calycina, F. v. M. in Benth. Fl. Austr. V, 107 (1870)	...	—	S.A.	—	—	—	—	B.fl.V,107	
P. Grylloana, F. v. M., fragm. X, 17 (1876)	...	...	W.A.	—	—	—	—	—	M.fr.X,17.

WRIXONIA, F. v. M., fragm. X, 18 (1876).
W. prostantheroides, F. v. M., fragm. X, 18 (1876)	...	W.A.	M.fr.X,18.

HEMIGENIA, R. Brown, prodr. 502 (1810). (Hemiandra, Colobandra, Atelandra.)
H. macrantha, F. v. M., fragm. I, 210 (1858)	...	W.A.	—	—	—	—	—	B.fl.V,112	M.fr.I,210;XI,19.
H. rigida, Bentham in De Cand. prodr. XII, 563 (1848)	...	W.A.	—	—	—	—	—	B.fl.V,112	
H. ramosissima, Bentham in De Cand. prodr. XII, 563 (1848)	...	W.A.	—	—	—	—	—	B.fl.V,113	
H. microphylla, Bentham in De Cand. prodr. XII, 563 (1848)	...	W.A.	—	—	—	—	—	B.fl.V,113	
H. incana, Bentham in De Cand. prodr. XII, 566 (1848)	...	W.A.	—	—	—	—	—	B.fl.V,113	M.fr.VI,112;IX,162.
H. canescens, Bentham in De Cand. prodr. XII, 566 (1848)	...	W.A.	—	—	—	—	—	B.fl.V,114	
H. podalyrina, F. v. M., fragm. VI, 112 (1868)	...	W.A.	—	—	—	—	—	B.fl.V,114	M.fr.VI,112.
H. platyphylla, Bentham in De Cand. prodr. XII, 566 (1848)	...	W.A.	—	—	—	—	—	B.fl.V,115	
H. obovata, F. v. M., fragm. XI, 10 (1876)	...	W.A.	—	—	—	—	—	—	B.fl.XI,10
H. glabrescens, Bentham in De Cand. prodr. XII, 566 (1848)	...	W.A.	—	—	—	—	—	B.fl.V,115	

H. obtusa, Bentham in De Cand. prodr. XII, 567 (1848)... W.A. — — — — — — B.fl.V,115
H. sericea, Bentham in Hueg. enum. 80 (1837) ... ... W.A. — — — — — — B.fl.V,116 M.fr.VI,112.
H. barbata, Bartling in Lehm. pl. Preiss. I, 360 (1845) ... W.A. — — — — — — B.fl.V,116
H. curvifolia, F. v. M., fragm. I, 210 (1859) ... ... W.A. — — — — — — B.fl.V,117 M.fr.I,210.
H. scabra, Bentham, Fl. Austr. V, 117 (1870) ... W.A. — — — — — — B.fl.V,117
H. humilis, Bentham in De Cand. prodr. V, 117 (1870) ... W.A. — — — — — — B.fl.V,117
H. westringioides, Bentham in De Cand. prodr. XII, 568 (1848) W.A. — — — — — — B.fl.V,117
H. teretiuscula, F. v. M., fragm. VI, 111 (1868) ... ... W.A. — — — — — — B.fl.V,118 M.fr.VI,111.
H. brachyphylla, F. v. M., fragm. X, 19 (1876) ... W.A. — — — — — — — M.fr.X,19.
H. purpurea, R. Brown, prodr. 502 (1810). ... ... — — — N.S.W. — — B.fl.V,118 M.fr.VI,112.
H. cuneifolia, Bentham, Fl. Austr. V, 118 (1870)... ... — — — N.S.W. — — B.fl.V,118
H. Drummondii, Bentham, Fl. Austr. V, 119 (1870) ... W.A. — — — — — — B.fl.V,119
H. pimelifolia, F. v. M., fragm. VI, 112 (1868) ... W.A. — — — — — — B.fl.V,119 M.fr.VI,112.
H. diplanthera, F. v. M., fragm. VI, 111 (1868) ... W.A. — — — — — — B.fl.V,119 M.fr.VI,111.
H. pungens, F. v. M., fragm. XI, 20 (1878) .... W.A. — — — — — — B.fl.V,109
H. leiantha, F. v. M., fragm. XI, 20 (1878) ... W.A. — — — — — — B.fl.V,110
H. loganiacea, F. v. M., fragm. XI, 19 (1878) ... ... W.A. — — — — — — B.fl.V,112 M.fr.VI,113.

MICROCORYS, R. Brown, prodr. 502 (1810). (Anisandra.)
M. longifolia, Bentham in De Cand. prodr. XII, 568 (1848) W.A. — — — — — — B.fl.V,121
M. longiflora, F. v. M., fragm. VI, 113 (1868) ... W.A. — — — — — — B.fl.V,121 M.fr.VI,113.
M. tenuifolia, Bentham, Fl. Austr. V, 122 (1870)... W.A. — — — — — — B.fl.V,122
M. capitata, Bentham in De Cand. prodr. XII, 568 (1848) W.A. — — — — — — B.fl.V,122
M. pimeloides, F. v. M., fragm. I, 156 (1859) ... W.A. — — — — — — B.fl.V,123 M.fr.I,156.
M. subcanescens, Bentham, Fl. Austr. V, 123 (1870) ... W.A. — — — — — — B.fl.V,123
M. ericifolia, Bentham in De Cand. prodr. XII, 560 (1848) W.A. — — — — — — B.fl.V,123
M. glabra, Bentham in De Cand. prodr. XII, 569 (1848)... W.A. — — — — — — B.fl.V,123
M. exserta, Bentham, Fl. Austr. V, 124 (1870) ... W.A. — — — — — — B.fl.V,124
M. virgata, R. Brown, prodr. 502 (1810) ... W.A. — — — — — — B.fl.V,124
M. Macrediana, F. v. M., fragm. VIII, 231 (1874) ... W.A. S.A. — — — — — M.fr.VIII,231.
M. barbata, R. Brown, prodr. 502 (1810) ... ... W.A. — — — — — — B.fl.V,125
M. lenticularis, F. v. M., fragm. VI, 113 (1868) ... W.A. — — — — — — B.fl.V,126 M.fr.VI,113.
M. obovata, Bentham in De Cand. prodr. XII, 569 (1848) W.A. — — — — — — B.fl.V,126
M. purpurea, R. Brown, prodr. 502 (1810)... ... ... W.A. — — — — — — B.fl.V,126

WESTRINGIA, Smith in Kongl. Svenska Vetenskaps Akad. Handlingar 171 (1797).
W. cephalantha, F. v. M., fragm. VI, 110 (1868)... ... W.A. — — — — — — B.fl.V,127 M.fr.VI,110.
W. rosmariniformis, Smith in Vet. Acad. Handl. 171 (1797) ... — T. V. N.S.W. Q. — B.fl.V,128 M.fr.VI,110;IX,162.
W. Dampieri, R. Brown, prodr. 501 (1810) ... W.A. S.A. — — — B.fl.V,129 M.fr.VI,110;IX,163.
W. rigida, R. Brown, prodr. 501 (1810) ... W.A. S.A. — V. N.S.W. — — B.fl.V,129 M.fr.IX,163.
W. angustifolia, R. Brown, prodr. 501 (1810) ... — — T. — — — M.fr.VI,110;IX,163.
W. senifolia, F. v. M. in Transact. phil. Soc. Vict. I, 49 (1854) — — V. — — B.fl.V,130 M.fr.VI,110;IX,163.
W. longifolia, R. Brown, prodr. 501 (1810) ... — — T. V. N.S.W. Q. — B.fl.V,131 M.fr.VI,110;IX,163.
W. glabra, R. Brown, prodr. 501 (1810) ... — — V. N.S.W. Q. — B.fl.V,131 M.fr.VI,110;IX,163.
W. rubiifolia, R. Brown, prodr. 501 (1810)... ... — — T. — — — B.fl.V,131 M.fr.VI,110.

AJUGA, Linné, gen. plant. 167 (1737), from Scribonius (1529).
A. australis, R. Brown, prodr. 503 (1810) ... — S.A. T. V. N.S.W. Q. — B.fl.V,136 M.fr.VI,100.

TEUCRIUM, Tournefort, inst. 207, t. 98 (1700), from Dioscorides.
T. racemosum, R. Brown, prodr. 504 (1810) ... W.A. S.A. — V. N.S.W. Q. N.A. B.fl.V,133 M.fr.VI,109;IX,162.
T. integrifolium, F. v. M. in Benth. Fl. Austr. V, 133 (1870) ... — S.A. — — — Q. N.A. B.fl.V,133 M.fr.IX,162.
T. corymbosum, R. Brown, prodr. 504 (1810) ... — S.A. T. V. N.S.W. Q. — B.fl.V,133 M.fr.VI,110.
T. filifolium, F. v. M. in Benth. Fl. Austr. V, 134 (1870) W.A. — — — — — — B.fl.V,134 M.fr.IX,162.
T. sessiliflorum, Bentham in De Cand. prodr. XI, 580 (1848) W.A. S.A. — V. N.S.W. — — B.fl.V,134 M.fr.VI,110.
T. argutum, R. Brown, prodr. 504 (1810) ... — — — N.S.W. Q. — B.fl.V,135 M.fr.VI,110;IX,162.

VERBENACEAE.

Adanson, Familles des plantes II, 193 (1763), from B. de Jussieu (1759).
SPARTOTHAMNUS, Cunningham in Loudon's bort. Brit. suppl. (1830).
S. junceus, Cunningham in Loud. hort. Brit. 600 (1830)... — — — N.S.W. Q. — B.fl.V,55 M.fr.VI,153;IX,5.
S. teucriiflorus, F. v. M. in Wing's S. Sc. Rec. II, 55 (1882) ... — S.A. — — — N.A. —

NESOGENES, A. de Candolle, prodr. XI, 703 (1847).
N. euphrasioides, A. de Cand., prodr. XI, 703 (1847) ... — — — — Q. — —

LIPPIA, Houston in Linné, syst. nat. 6 (1735); Linné, gen. plant. 347 (1737). (Zapania.)
L. nodiflora, Cl. Richard in Mich. fl. bor. Amer. II, 15 (1803) ... W.A. — — — N.S.W. Q. N.A. B.fl.V,35 M.fr.VI,151;IX,4.
L. geminata, Humboldt, Bonpland et Kunth, nov. gen. et spec. pl. II, 215 (1817) ... ... — — — — Q. — B.fl.V,35 M.fr.VI,151.

VERBENA, Tournefort, inst. 200, t. 94 (1700), from l'Ecluse (1576).
V. officinalis, Linné, spec. plant. 11 (1753)... ... — S.A. T. V. N.S.W. Q. — B.fl.V,36 M.fr.VI,153;IX,5.
V. macrostachya, F. v. M., fragm. I, 60 (1858) ... W.A. — — — — — B.fl.V,36 M.fr.I,60;IX,5.

LACHNOSTACHYS, Hooker, icon. plant. t. 414 (1842). (Walcottia, Pycnolachne.)
L. albicans, Hooker, icon. plant. 414 (1842) ... W.A. — — — — — B.fl.V,38 M.fr.V,159.
L. Cliftoni, F. v. M., fragm. IX, 3 (1875) ... W.A. — — — — — B.fl.V,38 M.fr.IX,3.
L. verbascifolia, F. v. M., fragm. VI, 158 (1868)... W.A. — — — — — B.fl.V,38 M.fr.VI,158;IX,4.
L. ferruginea, Hooker, icon. plant. 415 (1842) ... W.A. — — — — — B.fl.V,39 M.fr.VI,158;IX,4.
L. Walcottii, F. v. M., fragm. II, 140 (1861) ... W.A. — — — — — B.fl.V,39 M.fr.II,140;VI,159.

NEWCASTLIA, F. v. M. in Hook. Kew Misc. IX, 22 (1857).
N. cladotricha, F. v. M. in Hook. Kew Misc. IX, 22 (1857) ... — S.A. — — — N.A. B.fl.V,40 M.fr.I,184;III,21;VI,155.
N. bracteosa, F. v. M., fragm. VIII, 49 (1873) ... — S.A. — — — — M.fr.VIII,230;IX,4.

MYOPORINAE.

R. Brown, prodr. 514 (1810).

MYOPORUM, Banks & Solander in G. Forster, prodr. 44 (1786). (Bertolonia, Andrewsia, Pogonia, Disoon.)

M. tenuifolium, G. Forster, fl. ins. Austr. prodr. 44 (1786)	...	— — V. N.S.W. Q. N.A.	B.fl.V,4	M.fr.VI,149.
M. montanum, R. Brown, prodr. 515 (1810)	...	W.A. S.A. — V. N.S.W. Q. N.A.	B.fl.V,4	M.fr.VI,149.
M. desorti, Cunningham in Hueg. enum. 78 (1837)		W.A. S.A. — V. N.S.W. Q. —	B.fl.V,5	M.fr.VI,149;VII,110.
M. laxiflorum, Bentham, Fl. Austr. V, 6 (1870)	...	— — — — — — Q. —	B.fl.V,6	M.fr.V,23.
M. insulare, R. Brown, prodr. 515 (1810)	...	W.A. S.A. T. V. N.S.W. — —	B.fl.V,6	M.fr.VI,149;VII,110.
M. viscosum, R. Brown, prodr. 516 (1810)... ...	...	— S.A. — V. N.S.W. — —	B.fl.V,5	M.fr.VII,100.
M. serratum, R. Brown, prodr. 516 (1810)... ...	...	W.A. — — — — — — —	B.fl.V,4	M.fr.VI,149.
M. oppositifolium, R. Brown, prodr. 516 (1810) ...	...	W.A. — — — — — — —	B.fl.V,7	M.fr.VI,149.
M. brevipes, Bentham, Fl. Austr. V, 6 (1870)	...	— S.A. — — — — — —	B.fl.V,6	
M. humile, R. Brown, prodr. 516 (1810)	...	W.A. S.A. T. V. N.S.W. — —	B.fl.V,6	M.fr.VI,150.
M. debile, R. Brown, prodr. 516 (1810)	...	— — — — N.S.W. Q. —	B.fl.V,8	M.fr.VI,149.
M. platycarpum, R. Brown, prodr. 516 (1810) ...	...	W.A. S.A. — V. N.S.W. — —	B.fl.V,7	M.fr.VI,150.
M. Batese, F. v. M. in Proceed. Linn. Soc. N.S.W. VI, 792 (1881)	— — — — N.S.W. — —		M.fr.XI,139.
M. floribundum, Cunningham in Hueg. enum. 78 (1837)	— — — V. N.S.W. — —	B.fl.V,8	M.fr.I,126;VI,150.
M. Beckeri, F. v. M. in Benth. Fl. Austr. V, 7 (1870) ...	...	W.A. — — — — — — —	B.fl.V,7	M.fr.I,126,244;VI,150.
M. salsoloides, Turcz. in Bull. de Mosc. XXXVI, 226 (1863) ...	W.A. — — — — — — —	B.fl.V,8	M.fr.I,126;VI,150.

EREMOPHILA, R. Brown, prodr. 518 (1810). (Pholidia, Stenochilus, Eremodendron, Pseudopholidia, Sentis, Duttonia, Pholidiopsis.)

E. Mackinlayi, F. v. M., fragm. IV, 80 (1864) ...	W.A. — — — — — — —	B.fl.V,17	M.fr.IV,80.
E. Bowmani, F. v. M., fragm. II, 139 (1863)	— S.A. — — N.S.W. Q. —	B.fl.V,18	M.fr.II,139.
E. strongylophylla, F. v. M., fragm. X, 87 (1876) ...	W.A. S.A. — — — — — —		M.fr.X,87.
E. leucophylla, Bentham, Fl. Austr. V, 18 (1870) ...	W.A. — — — — — — —	B.fl.V,18	
E. Turtoni, F. v. M., fragm. X, 87 (1876)	W.A. S.A. — — — — — —		M.fr.X,87.
E. Forresti, F. v. M., fragm. VII, 49 (1869)	W.A. — — — — — — —	B.fl.V,18	M.fr.VII,49.
E. criocalyx, F. v. M., fragm. I, 236 (1859)	W.A. — — — — — — —	B.fl.V,19	M.fr.I,236.
E. Maitlandi, F. v. M. in Benth. Fl. Austr. V, 19 (1870) ...	W.A. — — — — — — —	B.fl.V,19	
E. rotundifolia, F. v. M., fragm. I, 207 (1859)	— S.A. — — — — — —	B.fl.V,19	M.fr.I,207.
E. oppositifolia, R. Brown, prodr. 518 (1810)	— S.A. — V. N.S.W. — —	B.fl.V,20	
E. Mitchelli, Bentham in Mitch. Trop. Austr. 31 (1848)	— S.A. — — N.S.W. Q. —	B.fl.V,21	M.fr.VI,149.
E. Paisleyi, F. v. M., Rep. Babb. Exped. 17 (1858) ...	— S.A. — — — — — —	B.fl.V,20	
E. Sturtii, R. Brown, App. to Sturt's Exped. 22 (1849)...	— S.A. — — N.S.W. — —	B.fl.V,21	M.fr.VI,149.
E. oxilifolia, F. v. M., fragm. X, 88 (1876)	— S.A. — — — — — —		M.fr.X,88.
E. Dempsteri, F. v. M., fragm. X, 60 (1876)	W.A. — — — — — — —		M.fr.X,60.
E. Gibsoni, F. v. M., fragm. VIII, 227 (1874)	W.A. — — — — — — —		M.fr.VIII,227.
E. Berryi, F. v. M., fragm. VIII, 228 (1874)	W.A. S.A. — — — — — —		M.fr.VIII,228;X,88.
E. Clarkei, F. v. M., fragm. I, 208 (1859)...	W.A. — — — — — — —	B.fl.V,21	M.fr.I,208.
E. Gilesii, F. v. M., fragm. VIII, 49 (1873)	— S.A. — — — — — N.A.		M.fr.VIII,49.
E. Latrobei, F. v. M. in Proceed. Roy. Soc. Tasm. III, 294 (1858)	W.A. S.A. — — N.S.W. Q. N.A.	B.fl.V,22	M.fr.I,125.
E. gracillifera, F. v. M., fragm. I, 208 (1858)	W.A. — — — — — — —	B.fl.V,23	M.fr.I,208.
E. longifolia, F. v. M. in Proceed. Roy. Soc. Tasm. III, 295 (1858)	W.A. S.A. — V. N.S.W. Q. N.A.	B.fl.V,23	M.fr.VI,148.
E. Drummondii, F. v. M., fragm. VI, 147 (1865)... ...	W.A. — — — — — — —	B.fl.V,24	M.fr.VI,147.
E. Hughesii, F. v. M., fragm. VIII, 228 (1874)	W.A. — — — — — — —		M.fr.VIII,228.
E. polyclada, F. v. M. in Proceed. Roy. Soc. Tasm. III, 294 (1858)	— S.A. — V. N.S.W. Q. —	B.fl.V,24	
E. bignoniflora, F. v. M. in Proceed. Roy. Soc. Tasm. III, 294 (1858)	— S.A. — V. N.S.W. Q. N.A.	B.fl.V,25	M.fr.VI,148.
E. Freelingii, F. v. M. in Proceed. Roy. Soc. Tasm. III, 295 (1858)	— S.A. — V. N.S.W. — —	B.fl.V,25	M.fr.XI,51.
E. Fraseri, F. v. M., fragm. XI, 51 (1878)... ...	W.A. — — — — — — N.A.		M.fr.XI,51.
E. Macdonnelli, F. v. M., Rep. Babb. Exped. 18 (1858)...	— S.A. — — — — — —	B.fl.V,22	M.fr.VI,150.
E. Goodwinii, F. v. M., Rep. Babb. Exped. 17 (1859) ...	— S.A. — — N.S.W. — —	B.fl.V,27	M.fr.VI,148.
E. Elderi, F. v. M., fragm. VIII, 228 (1874)	W.A. S.A. — — — — — —		M.fr.VIII,228.
E. Willsii, F. v. M., fragm. III, 21, t. 20 (1862) ...	W.A. S.A. — — — — — —	B.fl.V,26	M.fr.III,21.
E. platycalyx, F. v. M., fragm. V, 109 (1866)	W.A. — — — — — — —	B.fl.V,26	M.fr.V,109.
E. viscida, Endlicher, nov. stirp. dec. 51 (1839)	W.A. — — — — — — —	B.fl.V,26	
E. Brownii, F. v. M. in Proceed. Roy. Soc. Tasm. III, 297 (1858)	W.A. S.A. — V. N.S.W. Q. —	B.fl.V,27	M.fr.VI,148.
E. subfloccosa, Bentham, Fl. Austr. V, 28 (1870)	W.A. — — — — — — —	B.fl.V,28	M.fr.I,208.
E. Oldfieldii, F. v. M., fragm. I, 208 (1859)	W.A. — — — — — — —	B.fl.V,28	M.fr.VI,148.
E. maculata, F. v. M., Rep. Babb. Exped. 16 (1858) ...	— S.A. — — N.S.W. — —	B.fl.V,29	
E. denticulata, F. v. M., fragm. I, 125 (1859)	W.A. — — — — — — —	B.fl.V,29	M.fr.IV,170.
E. latifolia, F. v. M. in Schlechtend. Linnaea XXV, 428 (1852)...	W.A. — — — — N.S.W. — —	B.fl.V,30	M.fr.VI,148.
E. alternifolia, R. Brown, prodr. 518 (1810)	— S.A. — — N.S.W. — —	B.fl.V,30	M.fr.VI,148.
E. Dalyana, F. v. M., fragm. V, 22 (1865)...	— S.A. — — N.S.W. — —	B.fl.V,10	M.fr.V,22.
E. scoparia, F. v. M. in Proceed. Roy. Soc. Tasm. III, 296 (1858)	— S.A. — — N.S.W. — —	B.fl.V,10	M.fr.V,22;VI,148.
E. Dellaserti, F. v. M., fragm. V, 103, t. 42 (1865) ...	W.A. — — — — N.S.W. — —	B.fl.V,11	M.fr.V,189.
E. crassifolia, F. v. M. in Proceed. Roy. Soc. Tasm. III, 297 (1858)	— S.A. — — — — — —	B.fl.V,11	M.fr.VI,148.
E. rostrosa, F. v. M. in Proceed. Roy. Soc. Tasm. III, 296 (1858)	W.A. — — — — — — —	B.fl.V,12	
E. Behrii, F. v. M. in Proceed. Roy. Soc. Tasm. III, 296 (1858)	— S.A. — — — — — —	B.fl.V,12	
E. Woollsiana, F. v. M., fragm. I, 125, t. 7 (1859) ...	W.A. — — — — — — —	B.fl.V,12	M.fr.VI,151.
E. Christophori, F. v. M., fragm. IX, 120 (1875)... ...	— S.A. — — — — — —	B.fl.V,12	M.fr.IX,198.
E. Weldii, F. v. M., fragm. VII, 109 (1870)	— S.A. — — — — — —	B.fl.V,12	M.fr.IX,120.
E. brevifolia, F. v. M.; Myoporum, Bartling in pl. Preiss. I, 350 (1845)	W.A. — — — — — — —	B.fl.V,12	M.fr.VI,150.
E. imbricata, F. v. M.; Pholidia, Bentham, Fl. Austr. V, 13 (1870)	W.A. — — — — — — —	B.fl.V,13	
E. densifolia, F. v. M., fragm. I, 160 (1861)	W.A. — — — — — — —	B.fl.V,13	M.fr.II,160.
E. gibbosifolia, F. v. M., Rep. Babb. Exped. 18 (1858) ...	— S.A. — V. — — — —	B.fl.V,13	M.fr.VI,150.
E. divaricata, F. v. M. in Proceed. Roy. Soc. Tasm. III, 293 (1858)	— S.A. — V. N.S.W. — —	B.fl.V,14	M.fr.I,126;VI,150.
E. microtheca, F. v. M. in Benth. Fl. Austr. V, 14 (1870)	W.A. — — — — — — —	B.fl.V,14	M.fr.II,160.
E. adenotricha, F. v. M. in Benth. Fl. Austr. V, 15 (1870)	W.A. — — — — — — —	B.fl.V,14	
E. santalina, F. v. M. in Proceed. Roy. Soc. Tasm. III, 295 (1858)	— S.A. — — — — — —	B.fl.V,15	
E. Youngii, F. v. M., fragm. X, 16 (1876)...	W.A. — — — — — — —		M.fr.X,16.

PEDALINAE.

R. Brown, prodr. Fl. Nov. Holl. 510 (1810).

JOSEPHINIA, Ventenat, Jardin de la Malmaison, 103 t. 67 (1804).

J. grandiflora, R. Brown, prodr. 520 (1810) ... ... — — — — — Q. — B.fl.IV,556
J. imperatricis, Ventenat, Jard. de la Malm. t. 67 (1804) ... — — — — — — N.A. B.fl.IV,557 M.fr.VI,103.
J. Eugeniae, F. v. M. in Hook. Kew Misc. IX, 370, t. 11 (1857) W.A. S.A. — — Q. N.A. B.fl.IV,557 M.fr.VI,163.

ERICACEAE.

A. L. de Jussieu, gen. plant. 159 (1789).

PERNETTYA, Gaudichaud in Ann. des sc. nat. V, 102 (1825).

P. Tasmanica, J. Hooker in Lond. Journ. of Bot. VI, 268 (1847) — — T. — — — — B.fl.IV,140

GAULTIERA, Kalm in Linné, amoen. acad. III, 14, fig. 6 (1751). (Gaultheria.)
G. hispida, R. Brown, prodr. 559 (1810) ... ... — T. V. N.S.W. — — B.fl.IV,141
G. lanceolata, J. Hooker in Lond. Journ. of Bot. VI, 267 (1847) — — T. — — — — B.fl.IV,141
G. antipoda, G. Forster, florul. ins. Austr. pr. 34 (1786)... ... — — T. — — — — B.fl.IV,142

WITTSTEINIA, F. v. M., fragm. II, 136 (1861).
W. vacciniacea, F. v. M., fragm. II, 136 (1861) ... ... — V. — — -- — B.fl.IV,139 M.fr.III,166.

EPACRIDEAE.

R. Brown, prodr. 535 (1810).

STYPHELIA, Solander in G. Forster, fl. ins. Austr. prodr. 13 (1786). (Epacris, Forst. (1770) partly, Stiphelia, Ardisia, Peros, Perojoa, Veutenatia, Cyathodes, Stenanthera, Astroloma, Leucopogon, Melichrus, Acrotriche, Monotoca, Soleniscia, Stomarrhena, Pentapthrus, Mesotriche, Phanerandra, Froebelia, Pentaptelion, Androstoma, Cyathopsis, Lissanthe partly.)

S. adscendens, R. Brown, prodr. 537 (1810) ... ... — S.A. T. V. — — M.fr.VI,36;XI,122.
S. longifolia, R. Brown, prodr. 537 (1810) ... ... — — N.S.W. — — B.fl.IV,147
S. laeta, R. Brown, prodr. 537 (1810) ... ... — — N.S.W. — — B.fl.IV,147
S. triflora, Andrews, Bot. Reposit. t. 72 (1799) ... ... — — N.S.W. Q. — B.fl.IV,147 M.fr.XI,122.
S. viridis, Andrews, Bot. Reposit. t. 312 (1806) ... ... — — N.S.W. Q. — B.fl.IV,148
S. Hainesii, F. v. M., fragm. IV, 96, t. 28 (1864)... ... ... W.A. — — — — B.fl.IV,148 M.fr.VIII,55.
S. tubiflora, Smith, Specim. of the Bot. of New Holl. 45, t. 14 (1793) — — N.S.W. — — B.fl.IV,148 M.fr.VI,46;VIII,54.
S. tenuiflora, Lindley, Bot. Regist. XXV, App. XXV (1839) W.A. — — — — B.fl.IV,148 M.fr.VI,46;VIII,54.
S. melaleucoides, F. v. M., fragm. IV, 97 (1864) ... ... W.A. — — — — B.fl.IV,149 M.fr.VI,30.
S. pusilliflora, F. v. M., fragm. IV, 105 (1864) ... ... — S.A. — — — — B.fl.IV,149 M.fr.I,178;VI,31.
S. Leucopogon, F. v. M., fragm. IV, 97 (1864) ... ... W.A. — — — — B.fl.IV,149 M.fr.III,143;VI,31.
S. lasionema, F. v. M., fragm. VI, 40 (1867) ... ... W.A. — — — — B.fl.IV,153 M.fr.VIII,54.
S. macrocalyx, F. v. M., fragm. VI, 37 (1867) ... ... W.A. — — — — B.fl.IV,153 M.fr.VI,37.
S. xerophylla, F. v. M., fragm. VI, 38 (1867) ... ... W.A. — — — — B.fl.IV,153 M.fr.VI,38.
S. pentapogonea, F. v. M., fragm. VI, 36 (1867) ... ... W.A. — — — — B.fl.IV,153 M.fr.VI,30.
S. prostrata, F. v. M.; Astroloma, R. Brown, prodr. 538 (1810) W.A. — — — — B.fl.IV,154
S. tecta, Sprengel, syst. I, 657 (1825) ... ... W.A. — — — — B.fl.IV,154 M.fr.VI,37.
S. Candolleana, F. v. M., fragm. VI, 38 (1867) ... ... W.A. — — — — B.fl.IV,154 M.fr.VIII,54.
S. microdonta; Astroloma,F.v.M.inBenth.Fl.Austr.IV,155(1869) W.A. — — — — B.fl.IV,155
S. pallida, Sprengel, syst. I, 658 (1825) ... ... W.A. — — — — B.fl.IV,155 M.fr.VI,37.
S. compacta, Sprengel, syst. I, 657 (1825) ... ... W.A. — — — — B.fl.IV,155 M.fr.VI,38.
S. humifusa, Persoon, synops. plant. I, 174 (1805) ... W.A. S.A. T. V. N.S.W. — — B.fl.IV,156 M.fr.VI,37;VIII,54.
S. Epacridis, F. v. M., fragm. VI, 38 (1867) ... ... W.A. — — — — B.fl.IV,156 M.fr.VI,38.
S. Drummondii, F. v. M., fragm. VI, 37 (1867) ... ... W.A. — — — — B.fl.IV,157 M.fr.VI,37.
S. microcalyx, F. v. M., fragm. VI, 37 (1867) ... ... W.A. — — — — B.fl.IV,157 M.fr.VI,37.
S. Baxteri, F. v. M., fragm. VI, 37 (1867) .. ... W.A. — — — — H.fl.IV,158 M.fr.VI,37. [122.
S. Sonderi, F. v. M., fragm. VI, 36 (1867)... ... — S.A. — V. N.S.W. — — B.fl.IV,158 M.fr.VIII,54;IX,48;XI,
S. longiflora, F. v. M., fragm. VIII, 54 (1873) ... ... — — — V. — — B.fl.IV,158 M.fr.VIII,54.
S. pinifolia, F. v. M., fragm. VI, 36 (1867) ... ... — — T. V. N.S.W. — — B.fl.IV,159 M.fr.VI,30.
S. procumbens, Persoon, synops. I, 174 (1805) ... ... — — — N.S.W. Q. — B.fl.IV,162 M.fr.VI,38;VIII,55.
S. urceolata, F. v. M., fragm. VI, 38 (1868) ... ... — — V. N.S.W. Q. — B.fl.IV,162 M.fr.VIII,54.
S. Billardieri, F. v. M., fragm. VI, 43 (1867) ... ... — — T. — — — B.fl.IV,169 M.fr.VIII,54.
S. stramiuea, Sprengel, syst. I, 656 (1825)... ... ... — — T. — — — B.fl.IV,169 M.fr.VI,43.
S. Hookeri, F. v. M., fragm. VI, 44 (1867) ... ... — — T. — — — B.fl.IV,169 M.fr.VI,44;VIII,54.
S. dealbata, Sprengel, syst. I, 650 (1825) ... ... — — T. — — — B.fl.IV,170 M.fr.VI,44;VIII,54.
S. abietina, Labillardière, Nov. Holl. pl. spec. I, 48, t. 68 (1804) — — T. — — — B.fl.IV,170 M.fr.VI,43.
S. Oxycedrus, Labillardière, Nov. Holl. pl. spec. I, 49, t. 69 (1804) — — T. V. — — B.fl.IV,170 M.fr.VI,43.
S. parvifolia, F. v. M. in papers of R. S. Tasm. 86 (1874) — — T. — — — B.fl.IV,171
S. sapida, F. v. M., fragm. VI, 42 (1867) ... ... — — — N.S.W. — — B.fl.IV,175 M.fr.VI,42;VIII,54.
S. strigosa, Smith, Specim. of the Bot. of New Holl. 49 (1793)... — S.A. T. V. N.S.W. Q. — B.fl.IV,176 M.fr.VI,42;IX,48;XI,122.
S. verticillata, Sprengel, syst. I, 656 (1825) ... ... W.A. — — — — B.fl.IV,184 M.fr.VI,43;VIII,55.
S. interrupta, Sprengel, syst. I, 656 (1825)... ... ... W.A. — — — — B.fl.IV,184
S. amplexicaulis, Rudge in Transact. Linn. Soc. VIII, 292, t. 8... — — N.S.W. — — B.fl.IV,185 M.fr.VI,44.
S. alternifolia, Sprengel, syst. I, 635 (1825) ... ... W.A. — — — — B.fl.IV,185
S. lanceolata, Smith, Specim. of the Bot. of New Holl. (1804)... — T. V. N.S.W. Q. — B.fl.IV,185 M.fr.VI,43.
S. Richei, Labillardière, Nov. Holl. pl. spec. I, 44, t. 60 (1804)... W.A. S.A. T. V. N.S.W. — — B.fl.IV,186 M.fr.VI,42;VIII,54.
S. australis, F. v. M., fragm. VI, 43 (1867) ... ... W.A. — — — — B.fl.IV,187 M.fr.VI,31.
S. capitellata, F. v. M., fragm. VI, 31 (1867) ... ... W.A. — — — — B.fl.IV,187 M.fr.VI,31.
S. revoluta, Sprengel, syst. I, 657 (1825) ... ... W.A. — — — — B.fl.IV,188 M.fr.VI,47.
S. grandiuscula, F. v. M., fragm. VI, 47 (1867) ... ... W.A. — — — — B.fl.IV,188 M.fr.VI,47.
S. reflexa, Sprengel, syst. I, 635 (1825) ... ... W.A. — — — — B.fl.IV,188 M.fr.VI,32.
S. corifolia, F. v M.; Leucopogon, Endl., nov. stirp. dec. 15 (1839) W.A. — — — — B.fl.IV,188
S. distans, Sprengel, syst. I, 655 (1825) ... ... W.A. — — — — B.fl.IV,189 M.fr.VI,32.

106

Species	W.A.	S.A.	T.	V.	N.S.W.	Q.		Ref.	M.fr.
S. gibbosa, F. v. M.; Leucopogon, Stachegleew in Bull. de Mosc. XXXII, 12 (1859)	W.A.	—	—	—	—	—	—	B.fl.IV,189	
S. thymifolia, F. v. M.; Leucopogon, Lindley in Benth. Fl. Austr. IV, 189 (1869)	—	—	—	V.	—	—	—	B.fl.IV,189	
S. cordata, F. v. M.; Leucop., Sonder in pl. Preiss. I, 313 (1845)	W.A.	—	—	—	—	—	—	B.fl.IV,190	
S. Boeslaea, F. v. M., fragm. VI, 47 (1867)	W.A.	—	—	—	—	—	—	B.fl.IV,190	M.fr.VI,47.
S. hirsuta, F. v. M., fragm. VI, 31 (1867)	W.A.	—	—	—	—	—	—	B.fl.IV,191	M.fr.VI,31.
S. collina, Labillardière, Nov. Holl. pl. spec. I, 47, t. 65 (1804)	—	S.A.	T.	V.	N.S.W.	—	—	B.fl.IV,101	M.fr.VI,45;VIII,54.
S. glacialis, F. v. M.; Leucopogon, Lindl. in Mitch. three Exped. II, 127 (1838)	—	—	—	V.	—	—	—	B.fl.IV,191	
S. compacta, F. v. M.; Leucopogon, Stachegl. in Bull. de Mosc. XXXII, 13 (1859)	W.A.	—	—	—	—	—	—	B.fl.IV,192	
S. squarrosa, F. v. M., fragm. VI, 31 (1867)	W.A.	—	—	—	—	—	—	B.fl.IV,192	M.fr.VI,31.
S. microphylla, Sprengel, syst. I, 656 (1825)	—	—	—	—	N.S.W.	—	—	B.fl.IV,192	M.fr.VI,31.
S. tetragona, F. v. M., fragm. VI, 31 (1867)	W.A.	—	—	—	—	—	—	B.fl.IV,193	M.fr.VI,31.
S. phyllostachys, F. v. M.; Leuc., Benth., fl. Austr. IV, 193 (1869)	W.A.	—	—	—	—	—	—	B.fl.IV,193	M.fr.VI,32.
S. glabella, Sprengel, syst. I, 655 (1825)	W.A.	—	—	—	—	—	—	B.fl.IV,194	M.fr.VI,32.
S. semiopposita, F. v. M., fragm. VI, 32 (1867)	W.A.	—	—	—	—	—	—	D.fl.IV,194	M.fr.VI,49.
S. florulenta, F. v. M.; Leuc., Bentham, Fl. Austr. IV, 194 (1869)	W.A.	—	—	—	—	—	—	B.fl.IV,194	
S. striata, Sprengel, syst. I, 656 (1825)	W.A.	S.A.	—	—	—	—	—	B.fl.IV,194	
S. lasiostachya, F. v. M.; Leucopogon, Stachegl. in Bull. de Mosc. XXXII, 11 (1859)	W.A.	—	—	—	—	—	—	B.fl.IV,195	
S. carinata, Sprengel, syst. I, 658 (1825)	W.A.	—	—	—	—	—	—	B.fl.IV,195	
S. opponens, F. v. M., fragm. VI, 48 (1867)	W.A.	—	—	—	—	—	—	B.fl.IV,196	M.fr.VI,48.
S. oppositifolia, F. v. M. fragm. VI, 32 (1867)	W.A.	—	—	—	—	—	—	B.fl.IV,196	M.fr.VI,32.
S. tamariscina, Sprengel, syst. I, 656 (1825)	W.A.	—	—	—	—	—	—	B.fl.IV,197	
S. bracteolaris, F. v. M.; Leuc., Benth., Fl. Austr. IV, 197 (1869)	W.A.	—	—	—	—	—	—	B.fl.IV,197	
S. blepharophylla, F. v. M. fragm. VI, 34 (1867)	W.A.	—	—	—	—	—	—	B.fl.IV,197	M.fr.VI,34.
S. tenuis, F. v. M.; Leucop., De Candolle prodr. VII, 744 (1833)	W.A.	—	—	—	—	—	—	B.fl.IV,197	
S. gnaphalioides, F. v. M.; Leucopogon, Stachegl. in Bull. de Mosc. XXXII, 14 (1859)	W.A.	—	—	—	—	—	—	B.fl.IV,198	
S. concurva, F. v. M., fragm. VI, 36 (1867)	—	S.A.	—	—	—	—	—	B.fl.IV,198	M.fr.III,144.
S. Gilbertii, F. v. M.; Leucopogon, Stachegl. in Bull. de Mosc. XXII, 15 (1859)	W.A.	—	—	—	—	—	—	D.fl.IV,198	
S. gracilis, Sprengel, syst. I, 658 (1825)	W.A.	—	—	—	—	—	—	B.fl.IV,199	
S. acicularis, F. v. M.; Leucopogon, Benth., Fl. Austr. IV, 199 (1869)	W.A.	—	—	—	—	—	—	B.fl.IV,199	
S. cryptantha, F. v. M.; Leuc., Benth., Fl. Austr. IV, 199 (1869)	W.A.	—	—	—	—	—	—	B.fl.IV,199	
S. gracillima, F. v. M., fragm. VI, 34 (1867)	W.A.	—	—	—	—	—	—	B.fl.IV,200	M.fr.VI,34.
S. cymbiformis, F. v. M., fragm. VI, 34 (1867)	W.A.	—	—	—	—	—	—	D.fl.IV,200	M.fr.VI,34.
S. apiculata, Sprengel, syst. I, 656 (1825)	W.A.	—	—	—	—	—	—	B.fl.IV,201	M.fr.IV,103;VI,31.
S. pulchella, Sprengel, syst. I, 659 (1825)	W.A.	—	—	—	—	—	—	B.fl.IV,201	M.fr.VI,31.
S. virgata, Labillardière, Nov. Holl. pl. spec. I, 46, t. 64 (1804)	—	S.A.	T.	V.	N.S.W.	—	—	B.fl.IV,201	M.fr.VI,42.
S. pulchella, F. v. M., fragm. VI, 34 (1867)	W.A.	—	—	—	—	—	—	B.fl.IV,202	M.fr.VI,34.
S. polymorpha, F. v. M., fragm. VI, 31 (1867)	W.A.	—	—	—	—	—	—	B.fl.IV,202	M.fr.VI,31.
S. assimilis, F. v. M.; Leucopogon, R. Brown, prodr. 545 (1810)	W.A.	—	—	—	—	—	—	B.fl.IV,202	M.fr.IV,100;VI,32.
S. Oldfieldii, F.v.M.; Leucopogon,Bentham,Fl.Austr.IV,203(1869)	W.A.	—	—	—	—	—	—	B.fl.IV,203	
S. cucullata, Sprengel, syst. I, 656 (1825)	W.A.	—	—	—	—	—	—	B.fl.IV,204	M.fr.VI,34.
S. sprengelioides, F. v. M.; Leuc., Sonder in pl. Preiss. I, 319 (1845)	W.A.	—	—	—	—	—	—	B.fl.IV,204	M.fr.VI,34.
S. obtusata, F. v. M.; Leuc., Sonder in pl. Preiss. I, 313 (1845)	W.A.	—	—	—	—	—	—	B.fl.IV,204	
S. fimbriata, F. v. M.; Leucopogon, Stachegl. in Bull. de Mosc. XXXII, 17 (1859)	W.A.	—	—	—	—	—	—	B.fl.IV,204	
S. ozothamnoides, Leuc., F. v. M. in Benth. fl. Austr. IV, 205 (1869)	W.A.	—	—	—	—	—	—	B.fl.IV,204	M.fr.VI,155.
S. plumuliflora, F. v. M., fragm. VI, 29 (1867)	W.A.	—	—	—	—	—	—	B.fl.IV,205	M.fr.XI,122.
S. unilateralis, F. v. M.; Leucopogon, Stachegl. in Bull. de Mosc. XXXII, 19 (1859)	W.A.	—	—	—	—	—	—	B.fl.IV,205	
S. montana, F. v. M., fragm. VI, 45 (1867)	—	—	T.	V.	N.S.W.	—	—	B.fl.IV,206	M.fr.VI,45.
S. Macraei, F. v. M., fragm. VI, 46 (1867)	—	—	—	V.	N.S.W.	—	—	B.fl.IV,206	M.fr.VI,46;VIII,54.
S. pleurandroides, F. v. M., fragm. VI, 32 (1867)	W.A.	—	—	—	—	—	—	B.fl.IV,206	M.fr.III,143.
S. linifolia, F. v. M., fragm. VI, 36 (1867)	—	—	—	—	N.S.W.	Q.	—	B.fl.IV,207	M.fr.VI,36.
S. pluriloculata, F. v. M., fragm. VI, 33 (1867)	—	—	—	—	N.S.W.	Q.	—	B.fl.IV,207	M.fr.I,37.
S. plciosperma, F. v. M., fragm. VI, 41 (1867)	—	—	—	—	N.S.W.	Q.	—	B.fl.IV,207	M.fr.VI,41.
S. rubicunda, F. v. M., fragm. VI, 33 (1867)	W.A.	—	—	—	—	—	—	B.fl.IV,208	M.fr.IV,99.
S. attenuata, F. v. M.; Leucopogon, Cunningham in Field's N.S.W. 341 (1825)	—	—	—	—	N.S.W.	—	—	B.fl.IV,208	
S. conferta, F. v. M.; Leuc., Bentham, Fl. Austr. IV, 208 (1869)	—	—	—	—	N.S.W.	—	—	B.fl.IV,208	
S. mutica, F. v. M., fragm. VI, 45 (1867)	—	—	—	—	N.S.W.	—	—	B.fl.IV,209	M.fr.VI,45.
S. ericoides, Smith, Specim. of the Bot. of New Holl. 48 (1793)	—	S.A.	T.	V.	N.S.W.	Q.	—	B.fl.IV,209	M.fr.VI,45.
S. brevicuspis, F. v. M.; Leuc., Benth., Fl. Austr. IV, 210 (1869)	W.A.	—	—	—	—	—	—	B.fl.IV,210	
S. propinqua, Sprengel, syst. I, 658 (1825)	W.A.	—	—	—	—	—	—	B.fl.IV,210	M.fr.VI,34.
S. subulifolia, F. v. M., fragm. VI, 33 (1825)	W.A.	—	—	—	—	—	—	B.fl.IV,211	M.fr.IV,103.
S. Allittii, F. v. M., fragm. VI, 34 (1867)	W.A.	—	—	—	—	—	—	B.fl.IV,211	M.fr.IV,103.
S. racemulosa, F. v. M., fragm. VI, 33 (1867)	W.A.	—	—	—	—	—	—	B.fl.IV,211	M.fr.VI,33.
S. pendula, Sprengel, syst. I, 657 (1825)	W.A.	—	—	—	—	—	—	B.fl.IV,212	M.fr.VI,33.
S. concinna, F. v. M.; Leuc., Bentham, Fl. Austr. IV, 212 (1869)	W.A.	—	—	—	—	—	—	B.fl.IV,212	
S. margarodes, Sprengel, syst. I, 657 (1825)	—	—	—	—	N.S.W.	Q.	—	B.fl.IV,213	M.fr.VI,33.
S. flavescens, F. v. M., fragm. VI, 33 (1867)	—	—	—	—	N.S.W.	—	—	B.fl.IV,213	M.fr.VI,100.
S. blepharolepis, F. v. M., 48 (1867)	—	—	—	—	N.S.W.	—	—	B.fl.IV,214	M.fr.VI,48.
S. esquamata, Sprengel, syst. I, 655 (1825)	—	—	—	—	N.S.W.	—	—	B.fl.IV,214	
S. rotundifolia, Sprengel, syst. I, 655 (1825)	W.A.	—	—	—	—	—	—	B.fl.IV,214	
S. cordifolia, F. v. M., fragm. VIII, 54 (1873)	W.A.	S.A.	—	V.	N.S.W.	—	—	B.fl.IV,214	M.fr.VI,45.
S. megacarpa, F. v. M., fragm. VI, 32 (1867)	W.A.	—	—	—	—	—	—	B.fl.IV,215	M.fr.IV,102.

S. ruscifolia, Sprengel, syst. I, 656 (1825) ... ... — — — — — Q. — B.fl.IV,215
S. imbricata, Sprengel, syst. I, 656 (1825)... ... ... — — — — — Q. — B.fl.IV,215
S. cuspidata, Sprengel, syst. I, 657 (1825)... ... ... — — — — — Q. — B.fl.IV,216
S. leptospermoides, Sprengel, syst. I, 659 (1825) .. .. — — — .. — Q. — B.fl.IV,216
S. acuminata, Sprengel, syst. I, 659 (1825)... ... ... — — — — — N.A. B.fl.IV,216
S. flexifolia, Sprengel, syst. I, 659 (1825) ... ... — — — — — Q. — B.fl.IV,217
S. biflora, Sprengel, syst. I, 659 (1825) ... ... — — — V. N.S.W. Q. — B.fl.IV,217 M.fr.IV,104.
S. setigera, Sprengel, syst. I, 650 (1825) ... ... — — — — N.S.W. — B.fl.IV,217 M.fr.VI,45.
S. exolaea, F. v. M., fragm. VI, 34 (1807)... ... ... — — — — — N.S.W. — B.fl.IV,217 M.fr.VI,34.
S. Fraseri, F. v. M.; Lencop., Cunn. in Ann. nat. hist. II. 47 (1840) — T. V. N.S.W. — B.fl.IV,218 M.fr.IV,105;VI,46.
S. hirtella, F. v. M.; Leucopogon in Benth. Fl. Austr. IV, 218 (1869) — S.A. — — — — B.fl.IV,218
S. Oldfieldii, F.v.M.; Leuc. ovalifolius, Sond. in pl. Preiss.I,324(1845)W.A. — — — — — B.fl.IV,218
S. erubescens, F. v. M., fragm. VI, 33 (1867) ... W.A. — — — — — B.fl.IV,219 M.fr.IV,102.
S. lissanthoides, F. v. M., fragm. VI, 33 (1867) ... W.A. — — — — — B.fl.IV,219 M.fr.IV,101.
S. stricta, F. v. M.; Leucopogon, Bentham, Fl. Austr. IV, 219(1869)W.A. — — — — — B.fl.IV,219
S. Mitchellii, F. v. M.; Leucopogon, Benth. Fl. Austr. IV, 220(1869) — — — — — Q. — B.fl.IV,220
S. juniperina, Sprengel, syst. I, 658 (1825)... ... ... — — — V. N.S.W. Q. — B.fl.IV,220 M.fr.IV,104;VI,46.
S. rufa, F. v. M., fragm. VI, 46 (1867) ... ... S.A. — V. N.S.W. — B.fl.IV,221 M.fr.VIII,54;IX,48.
S. concetephioides, F. v. M., fragm. VI, 34 (1867) ... W.A. — — — — — B.fl.IV,221 M.fr.VI,34.
S. deformis, Sprengel, syst. I, 658 (1825) ... ... — — — — N.S.W. — B.fl.IV,221
S. pogonocalyx;Leucop., F.v.M. in Benth. Fl. Austr. IV, 222 (1869)W.A. — — — — — B.fl.IV,222
S. breviflora; Leucopogon, F. v. M., fragm. IV, 102 (1864) ... W.A. — — — — — B.fl.IV,222 M.fr.IV,102.
S. dura, F. v. M.; Leucopogon, Bentham, Fl. Austr. IV, 222 (1869)W.A. — — — — — B.fl.IV,222
S. multiflora, Sprengel, syst. I, 658 (1825)... ... W.A. — — — — — B.fl.IV,222 M.fr.VI,32.
S. appressa, Sprengel, syst. I, 658 (1825) ... ... — — — N.S.W. — B.fl.IV,223
S. Neo-Anglica;Leucop., F. v. M. in Benth. Fl. Austr. IV, 222(1869) — — — N.S.W. Q. — B.fl.IV,223
S. obtecta, F. v. M.; Leucopogon, Benth. Fl. Austr. IV, 223 (1869) W.A. — — — — — B.fl.IV,223
S. crassiflora, F. v. M., fragm. VI, 40 (1867) ... ... W.A. — — — — — B.fl.IV,224 M.fr.VI,40.
S. strongylophylla, F. v. M., fragm. VI, 33 (1867) ... W.A. — — — — — B.fl.IV,224 M.fr.IV,101.
S. crassifolia, F. v. M., fragm. VI, 33 (1867) ... W.A. — — — — — B.fl.IV,224 M.fr.VI,34.
S. corynocarpa, F. v. M.; Leucop., Sonder in pl. Preiss. I, 322 (1845)W.A. — — — — — B.fl.IV,225 M.fr.VI,33.
S. Woodsii; Leucopogon, F. v. M., fragm. I, 178 (1859)... W.A. S.A. — V. — B.fl.IV,225
S. leptantha, F. v. M.; Leucop., Bentham, Fl. Austr. IV, 225 (1869)W.A. — — — — — B.fl.IV,225
S. divaricata, Sprengel, syst. I, 658 (1825) ... ... — — — N.S.W. Q. — B.fl.IV,226 M.fr.VI,44.
S. aggregata, Sprengel, syst. I, 657 (1825)... ... ... — — — N.S.W. Q. — B.fl.IV,227 M.fr.VI,44;VIII,55.
S. serrulata, Labillardiere, Nov. Holl. pl. spec. I, 45, t. 62 (1804) — S.A. T. V. N.S.W. — B.fl.IV,227 M.fr.VI,44.
S. patula, Sprengel, syst. I, 657 (1825) ... ... — S.A. — — — B.fl.IV,227 M.fr.VI,44;XI,122.
S. ovalifolia, Sprengel, syst. I, 656 (1825) ... ... W.A. S.A. — V. — B.fl.IV,228 M.fr.VI,44.
S. ramiflora, Sprengel, syst. I, 659 (1825) ... ... W.A. S.A. — — — B.fl.IV,228 M.fr.
S. depressa, Sprengel, syst. I, 655 (1825) ... ... W.A. S.A. — — — B.fl.IV,228 M.fr.VI,44.
S. fasciculiflora, F. v. M., fragm. VIII, 55 (1873)... — S.A. — — — B.fl.IV,229 M.fr.VI,44.
S. elliptica, Smith, Specim. of the Bot. of New Holl. 49 (1793)... — T. V. N.S.W. Q. — B.fl.IV,230 M.fr.VI,58.
S. glauca, Labillardiere, Nov. Holl. pl. spec. I, 45, t. 61 (1804) — T. — — — B.fl.IV,230
S. scoparia, Smith, Specim. of the Bot. of New Holl. 49 (1793)... — — T. V. N.S.W. Q. — B.fl.IV,231 M.fr.VI,55;XI,122.
S. ledifolia, F. v. M.; Monotoca, Cunningham in De Cand., prodr.
    VII, 756 (1839) ... ... — — — N.S.W. — B.fl.IV,231
S. empetrifolia, F. v. M. in Papers of R. S. Tasm. 86 (1874) ... — T. — — B.fl.IV,231
S. oligarrhenoides, F. v. M., fragm. IX, 47 (1875) ... W.A. — — — — — M.fr.IX,47.
S. minutiflora, F. v. M., fragm. XI, 122 (1881) ... ... W.A. — — — — — B.fl.IV,231 M.fr.XI,122.

OLIGARRHENA, R. Brown, prodr. Fl. Nov. Holl. 549 (1810).
O. micrantha, R. Brown, prodr. 450 (1810)... ... W.A. — — — — — B.fl.IV,232

NEEDHAMIA, R. Brown, prodr. Fl. Nov. Holl. 549 (1810).
N. pumilio, R. Brown, prodr. 549 (1810) ... ... W.A. — — — — — B.fl.IV,174

BRACHYLOMA, Sonder in Lehm. pl. Preiss. I, 304 (1845). (Lobopogon, Lissanthe partly.)
B. daphnoides, Bentham, Fl. Austr. IV, 173 (1869) — S.A. T. V. N.S.W. Q. — B.fl.IV,173 M.fr.VI,42;VIII,55.
B. ciliatum, Bentham, Fl. Austr. IV, 173 (1869) ... — S.A. T. V. — B.fl.IV,173 M.fr.VIII,55.
B. depressum, Bentham, Fl. Austr. IV, 173 (1869) ... — T. V. — B.fl.IV,173 M.fr.I,36;VI,42.
B. Scortechinii, F. v. M., fragm. XI, 121 (1881) ... — — — — Q. — M.fr.XI,121.
B. Preissii, Sonder in Lehm. pl. Preiss. I, 305 (1845) ... W.A. — — — — — B.fl.IV,172 M.fr.VI,39;VIII,55.
B. concolor, F. v. M., fragm. VI, 39 (1867) ... W.A. — — — — — B.fl.IV,172 M.fr.VI,98.
B. ericoides, Sonder in Schlecht. Linnaea XXVI, 247 (1853) ... — S.A. — V. N.S.W. — B.fl.IV,172 M.fr.I,98;VI,39;XI,46.

CONOSTEPHIUM, Bentham in Hueg. enum. 76 (1837). (Conostephiopsis.)
C. pendulum, Bentham in Hueg. enum. 76 (1837)... W.A. — — — — — B.fl.IV,100 M.fr.VI,40.
C. minus, Lindley, Bot. Regist. XXV, App. XXV (1839) ... W.A. — — — — — B.fl.IV,160 M.fr.VI,40.
C. Roei, Bentham, Fl. Austr. IV, 100 (1869) ... W.A. — — — — — B.fl.IV,160 M.fr.VIII,53.
C. Preissii, Sonder in Lehm. pl. Preiss. I, 304 (1845) ... W.A. — — — — — B.fl.IV,161 M.fr.VI,40.
C. planifolium, F. v. M., fragm. VI, 30 (1867) ... W.A. — — — — — B.fl.IV,161 M.fr.VI,30.

COLEANTHERA, Stschegleew in Bull. Soc. Mosc. XXXII, 4 (1859). (Michiea.)
C. coelophylla, Bentham, Fl. Austr. IV, 150 (1869) ... W.A. — — — — — B.fl.IV,150
C. myrtoides, Stschegleew in Bull. de Mosc. XXXII, 4 (1859) ... W.A. — — — — — B.fl.IV,150 M.fr.VI,80.
C. virgata, Stschegleew in Bull. de Mosc. XXXII, 5 (1859) ... W.A. — — — — — B.fl.IV,151

TROCHOCARPA, R. Brown, prodr. 548 (1810). (Decaspora, Pentachondra.)
T. laurina, R. Brown, prodr. 548 (1810) ... ... — — — N.S.W. Q. — B.fl.IV,166 M.fr.VI,57;IX,48.
T. disticha, Sprengel, syst. I, 660 (1825) ... ... — — T. — — B.fl.IV,166 M.fr.VI,57;VIII,55.
T. thymifolia, Sprengel, syst. I, 660 (1825) ... ... — T. — — B.fl.IV,166 M.fr.VI,57;VIII,55.
T. Clarkei, F. v. M., fragm. VI, 57 (1867)... ... ... — V. — B.fl.IV,167 M.fr.VI,57.
T. Gunnii, Bentham, fl. Austr. IV, 167 (1869) ... ... — — T. — — B.fl.IV,167 M.fr.VIII,55.
T. parviflora, Bentham, Fl. Austr. IV, 167 (1869) ... W.A. — — — — — B.fl.IV,167

T. involucrata, F. v. M., fragm. VI, 57 (1867)	—	—	T.	—	—	—	— B.fl.IV,164	M.fr.VIII,56.
T. pumila, F. v. M., fragm. VI, 57 (1867)	—	—	T.	V.	N.S.W.	—	— B.fl.IV,164	M.fr.VI,57.
T. ericifolia, F. v. M. in Papers R. S. Tasm. 86 (1874)	—	—	T.	—	—	—	— B.fl.IV,164	
T. verticillata, F. v. M. in Papers R. S. Tasm. 86 (1874)	—	—	T.	—	—	—	— B.fl.IV,164	

EPACRIS, Cavanilles, icon. et descr. pl. IV, 25, t. 344 et 345 (1797). (Archeria.)

E. micranthera, F. v. M., fragm. VI, 72 (1867)	—	—	T.	—	—	—	— B.fl.VI,246	M.fr.VIII,56.
E. hirtella, J. Hooker in Lond. Journ. of Bot. VI, 271 (1847)	—	—	T.	—	—	—	— B.fl.VI,246	M.fr.VI,71.
E. longiflora, Cavanilles, icon. et descr. pl. IV, 25, t. 344 (1797)	—	—	—	—	N.S.W.	—	— B.fl.IV,234	M.fr.VI,70;VIII,56.
E. reclinata, Cunningham in Benth. Fl. Austr. IV, 234 (1869)	—	—	—	—	N.S.W.	—	— B.fl.IV,234	M.fr.VIII,56. [122.
E. impressa, Labillardière, Nov. Holl. pl. spec. I, 43, t. 58 (1804)	—	S.A.	T.	V.	N.S.W.	—	— B.fl.IV,235	M.fr.VI,70;VIII,56;XI,
E. sparsa, R. Brown, prodr. 551 (1810)	—	—	—	—	N.S.W.	—	— B.fl.IV,235	M.fr.VIII,53.
E. petrophila, J. Hooker, Fl. Tasman. I, 261 (1860)	—	—	T.	V.	N.S.W.	—	— B.fl.IV,236	
E. rigida, Sieber in Spreng. cur. poster. 64 (1827)	—	—	—	—	N.S.W.	—	— B.fl.IV,236	
E. coriacea, Cunningham in Cand. prodr. VII, 763 (1839)	—	—	—	—	N.S.W.	—	— B.fl.IV,236	
E. crassifolia, R. Brown, prodr. 551 (1810)	—	—	—	—	N.S.W.	—	— B.fl.IV,237	
E. robusta, Bentham, Fl. Austr. IV, 237 (1869)	—	—	—	V.	N.S.W.	—	— B.fl.IV,237	
E. obtusifolia, Smith, Exot. Bot. I, 77, t. 40 (1804)	—	S.A.	T.	V.	N.S.W.	Q.	— B.fl.IV,237	M.fr.VI,71;VIII,56.
E. myrtifolia, Labillardière, Nov. Holl. pl. spec. I, 41, t. 55 (1804)	—	—	T.	—	—	—	— B.fl.IV,238	
E. exserta, R. Brown, prodr. 551 (1810)	—	—	T.	—	—	—	— B.fl.IV,238	
E. mucronulata, R. Brown, prodr. 552 (1810)	—	—	T.	—	—	—	— B.fl.IV,238	
E. lanuginosa, Labillardière, Nov. Holl. pl. spec. I, 42, t. 57 (1804)	—	S.A.	T.	V.	—	—	— B.fl.IV,238	M.fr.VIII,56;XI,122.
E. paludosa, R. Brown, prodr. 551 (1810)	—	—	—	V.	N.S.W.	—	— B.fl.IV,239	M.fr.VIII,56.
E. Calvertiana, F. v. M., fragm. VIII, 53 (1874)	—	—	—	—	N.S.W.	—	—	M.fr.VIII,53.
E. heteronema, Labillardière, Nov. Holl. pl. spec. I, 42, t. 56 (1804)	—	—	T.	V.	N.S.W.	—	— B.fl.IV,239	M.fr.VI,71;VIII,56.
E. serpillifolia, R. Brown, prodr. 240 (1810)	—	—	T.	V.	N.S.W.	—	— B.fl.IV,240	
E. microphylla, R. Brown, prodr. 550 (1810)	—	—	—	S.A.	T.	V.	N.S.W. Q.	— B.fl.IV,240	
E. acuminata, Bentham, Fl. Austr. IV,'240 (1869)	—	—	—	T.	—	—	— B.fl.IV,240	
E. apiculata, Cunningham in Field, N.S. Wales, 340 (1825)	—	—	—	—	N.S.W.	—	— B.fl.IV,241	M.fr.VI,71;XI,122.
E. pulchella, Cavanilles, icon. IV, 26, t. 345 (1797)	—	—	—	—	N.S.W.	Q.	— B.fl.IV,241	M.fr.IV,127;VIII,56.
E. purpurascens, R. Brown, prodr. 550 (1810)	—	—	—	—	N.S.W.	—	— B.fl.IV,241	

WOOLLSIA, F. v. M., fragm. VIII, 52 (1872). (Lysincma partly.)

W. pungens, F. v. M., fragm. VIII, 52 (1872)	—	—	—	—	N.S.W.	Q.	— B.fl.IV,243	M.fr.IV,126;VI,70,86; VIII,55;IX,48;XI,122.

LYSINEMA, R. Brown, prodr. 552 (1810).

L. lasianthum, R. Brown, prodr. 552 (1810)	W.A.	—	—	—	—	—	— B.fl.IV,243	M.fr.III,142,VI,70;VIII, 55. [122.
L. conspicuum, R. Brown, prodr. 552 (1810)	W.A.	—	—	—	—	—	— B.fl.IV,243	M.fr.VI,70;VIII,55;XI,
L. ciliatum, R. Brown, prodr. 552 (1810)	W.A.	—	—	—	—	—	— B.fl.IV,243	M.fr.VI,69,70;VIII,55.
L. fimbriatum, F. v. M., fragm. IV, 125 (1864)	W.A.	—	—	—	—	—	— B.fl.IV,244	M.fr.VI,70.
L. elegans, Sonder in Lehm. pl. Preiss. I, 327 (1845)	W.A.	—	—	—	—	—	— B.fl.IV,244	M.fr.VI,70.

PRIONOTES, R. Brown, prodr. fl. Nov. Holl. 553 (1810).

P. cerinthoides, R. Brown, prodr. 553 (1810)	—	—	T.	—	—	—	— B.fl.IV,246	M.fr.VI,69;XI,122.

COSMELIA, R. Brown, prodr. fl. Nov. Holl. 553 (1810).

C. rubra, R. Brown, prodr. 553 (1810)	W.A.	—	—	—	—	—	— B.fl.IV,247	M.fr.VI,64.

PONCELETIA, R. Brown, prodr. fl. Nov. Holl. 554 (1810).

P. sprengelioides, R. Brown, prodr. 554 (1810)	—	—	—	—	N.S.W.	Q.	— B.fl.IV,248	M.fr.I,39;VI,60.
P. monticola, Cunningham in De Cand. prodr. VII, 768 (1843)	—	—	—	—	N.S.W.	—	— B.fl.IV,249	

SPRENGELIA, Smith in Svensk. Vetens. Acad. Handling. 260 (1794). (Poiretia.)

S. incarnata, Smith in Vetens. Acad. Handl. 260, t. 8 (1794)	—	—	T.	V.	N.S.W.	—	— B.fl.IV,249	M.fr.L,39;VI,54;IX,48; [VIII,56;XI,122.

ANDERSONIA, R. Brown, prodr. 553 (1810). (Atherocephala, Homalostoma, Sphincterostoma.)

A. coloesca, F. v. M., fragm. VI, 63 (1867)	W.A.	—	—	—	—	—	— B.fl.IV,251	M.fr.VI,63.
A. patricia, F. v. M., fragm. VI, 79 (1868)	W.A.	—	—	—	—	—	— B.fl.IV,251	M.fr.VI,70.
A. grandiflora, Stachegleew in Bull. de Mosc. XXXII, 21 (1859)	W.A.	—	—	—	—	—	— B.fl.IV,252	M.fr.VI,62.
A. setifolia, Bentham, Fl. Austr. IV, 252 (1869)	W.A.	—	—	—	—	—	— B.fl.IV,252	M.fr.VIII,51.
A. involucrata, Sonder in Lehm. pl. Preiss. I, 331 (1845)	W.A.	—	—	—	—	—	— B.fl.IV,252	M.fr.VI,62.
A. homalostoma, Bentham, Fl. Austr. IV, 253 (1869)	W.A.	—	—	—	—	—	— B.fl.IV,253	M.fr.VIII,56.
A. sprengelioides, R. Brown, prodr. 554 (1810)	W.A.	—	—	—	—	—	— B.fl.IV,253	M.fr.VI,64.
A. latifora, F. v. M., fragm. VI, 61 (1867)	W.A.	—	—	—	—	—	— B.fl.IV,253	M.fr.VI,61.
A. gracilis, De Candolle, prodr. VII, 767 (1843)	W.A.	—	—	—	—	—	— B.fl.IV,254	
A. aristata, Lindley, Bot. Regist. XXV, App. XXV (1839)	W.A.	—	—	—	—	—	— B.fl.IV,254	M.fr.VI,63.
A. macronema, F. v. M., fragm. VIII, 51 (1873)	W.A.	—	—	—	—	—	—	M.fr.VIII,51.
A. parvifolia, R. Brown, prodr. 554 (1810)	W.A.	—	—	—	—	—	— B.fl.IV,254	M.fr.IV,126;VI,62.
A. depressa, R. Brown, prodr. 554 (1810)	W.A.	—	—	—	—	—	— B.fl.IV,255	
A. coerulea, R. Brown, prodr. 554 (1810)	W.A.	—	—	—	—	—	— B.fl.IV,255	M.fr.VI,64.
A. subulata, Bentham, Fl. Austr. IV, 256 (1869)	W.A.	—	—	—	—	—	— B.fl.IV,256	
A. heterophylla, Sonder in Lehm. pl. Preiss. I, 333 (1845)	W.A.	—	—	—	—	—	— B.fl.IV,256	M.fr.IV,124.
A. brachyanthera, F. v. M., fragm. VI, 61 (1867)	W.A.	—	—	—	—	—	— B.fl.IV,256	M.fr.VI,61.
A. brevifolia, Sonder in Lehm. pl. Preiss. I, 332 (1845)	W.A.	—	—	—	—	—	— B.fl.IV,256	M.fr.VI,61.
A. variegata, Sonder in Lehm. pl. Preiss. I, 334 (1845)	W.A.	—	—	—	—	—	— B.fl.IV,257	
A. micrantha, R. Brown, prodr. 554 (1810)	W.A.	—	—	—	—	—	— B.fl.IV,257	M.fr.VI,61.

RICHEA, R. Brown, prodr. 555 (1810). (Cystenthe, Pilitis.)

R. sprengelioides, F. v. M., fragm. VI, 68 (1867)	—	—	T.	—	—	—	— B.fl.IV,258	M.fr.I,38;VI,68.
R. procera, F. v. M., fragm. VI, 68 (1867)	—	—	T.	—	—	—	— B.fl.IV,259	M.fr.I,38;VI,68.
R. acerosa, F. v. M., fragm. VI, 69 (1867)	—	—	T.	—	—	—	— B.fl.IV,259	M.fr.I,36;VI,69.
R. Milligani, F. v. M., fragm. VI, 69 (1867)	—	—	T.	—	—	—	— B.fl.IV,259	M.fr.I,38;VIII,56.
R. Gunnii, J. Hooker in Lond. Journ. VI, 273 (1847)	—	—	T.	V.	—	—	— B.fl.IV,260	M.fr.VI,67;XI,122.
R. scoparia, J. Hooker in Lond. Journ. VI, 273 (1847)	—	—	T.	—	—	—	— B.fl.IV,260	M.fr.VI,68.

R. dracophylla, R. Brown, prodr. 555 (1810) ... ... — — T. — — — — B.fl.IV,260 M.fr.VI,67.
R. pandanifolia, J. Hooker, Fl. Antarct. I, 50 (1844) .. ... — — T. — — — — B.fl.IV,261 M.fr.VI,67;IX,48,
  DRACOPHYLLUM, Labillardière, Voy. II, 211, t. 40 (1798). (Epacris, Forst. 1776 partly.)
D. Milligani, Hooker, icon. t. 845 (1852) ... ... — — T. — — — — B.fl.IV,262 M.fr.VI,65.
D. Fitzgeraldi, F. v. M., fragm. VII, 27 (1869) ... ... — — — N.S.W. — — — M.fr.X,110.
D. secundum, R. Brown, prodr. 556 (1810) ... ... — — — N.S.W. — B.fl.IV,262 M.fr.VI,65;VIII,56.
D. minimum, F. v. M., fragm. I, 39 (1855) ... ... — — T. — — — — B.fl.IV,265 M.fr.VI,65;VII,56.
  SPHENOTOMA, Sweet, Flora Australasica, t. 44 (1823).
S. squarrosum, G. Don, gen. syst. III, 785 (1834) ... ... W.A. — — — — — B.fl.IV,263 M.fr.VI,66.
S. Drummondii, F. v. M.; Dracophyllum, Benth. Fl. Austr. IV,
  263 (1869) ... ... W.A. — — — — — B.fl.IV,263
S. dracophylloides, Sonder in Lehm. pl. Preiss. I, 33 (1844) ... W.A. — — — — — B.fl.IV,264 M.fr.VI,65.
S. capitatum, Lindley, Bot. Regist. t. 1515 (1832) ... W.A. — — — — — B.fl.IV,264
S. gracile, Sweet, Fl. Austral. t. 44 (1828).. ... ... W.A. — — — — — B.fl.IV,264 M.fr.VI,66.
S. parviflorum ; Dracophyllum, F. v. M. in Benth. Fl. Austr.
  IV, 265 (1869) ... ... W.A. — — — — — B.fl.IV,265 M.fr.VI,66.

APETALEAE GYMNOSPERMEAE.

F. v. M. in Woolls's plants of the neighb. of Sydney 40 (1880).

CONIFERAE.

Haller, enum. stirp. Helv. I, 145 (1742).
  DAMMARA, Rumphius, herb. Amboin. II, 174, t. 54 (1741). (Agathis.)
D. robusta, C. Moore in Transact. pharm. Soc. Vict. II, 174 (1860) — — — — — Q. — B.fr.VI,244 M.fr.XI,104.
  ARAUCARIA, A. L. de Jussieu, gen. plant. 413 (1789). (Colymbea, Columbea, Eutassa, Eutacta, Altingia.)
A. Cunninghamii, Aiton in Sweet. hort. Brit. 475 (1827) ... — — — — N.S.W. Q. — B.fl.VI,243
A. Bidwilli, Hooker, Lond. Journ. of Bot. II, 503, t. 18 (1850)... — — — — Q. — B.fl.VI,243
A. excelsa, R. Brown in Ait. hort. Kew. sec. ed. V, 412 (1813)... — — — — N.S.W. — — M.fr.IX,169.
  ATHROTAXIS, D. Don in Transact. Linn. Soc. XVIII, 171 (1839). (Arthrotaxis, Sequoia.)
A. cupressoides, D. Don in Trans. Linn. Soc. XVIII, 173, t. 13(1839) — — T. — — — — B.fl.VI,242
A. laxifolia, Hooker, icon. plant. t. 573 (1843) ... — — T. — — — — B.fl.VI,242
A. sclaginoides, D. Don in Trans. Linn. Soc. XVIII, 172, t. 14 (1839) — — T. — — — — B.fl.VI,242 M.fr.XI,104.
  CALLITRIS, Ventenat, decas generum novorum 10 (1808). (Frenela, Actinostrobus, Leichhardtia, Octoclinis.)
C. Macleayana, F. v. M. in Rep. Burdck. Exped. 17 (1860) ... — — — N.S.W. Q. — B.fl.VI,235 M.fr.XI,104.
C. Parlatorei, F. v. M. in Seemann Journ. of Bot. 267 (1866) ... — — — N.S.W. Q. — B.fl.VI,235 M.fr.V,186;XI,104.
C. Roei, Bentham & J. Hooker, gen. plant. III, 424 (1880) ... W.A. — — — — — B.fl.VI,236
C. Drummondi, Bentham & J. Hooker, gen. plant. III, 424 (1880) W.A. — — — — — B.fl.VI,236
C. verrucosa, R. Brown in Mém. du Mus. Par. XIII, 74 (1826) W.A. S.A. — V. N.S.W. Q. N.A. B.fl.VI,236 M.fr.V,237.
C. columellaris, F. v. M., fragm. V, 198 (1866) ... — — — N.S.W. Q. — M.fr.V,198.
C. Muelleri, Bentham & J. Hooker, gen. plant. III, 424 (1880)... — — — N.S.W. — — B.fl.VI,237
C. cupressiformis, Ventenat, decas gen. nov. 10 (1808) ... — S.A. T. V. N.S.W. Q. — B.fl.VI,237
C. calcarata, R. Brown in Mém. du Mus. Par. XIII, 74 (1826)... — — — V. N.S.W. Q. — B.fl.VI,238
C. oblonga, L. Cl. Richard, comment. de conif. 49, t. 18 (1826) — — T. — — — — B.fl.VI,238
C. Actinostrobus, F. v. M., Rep. Burdck. Exped. 19 (1860) ... W.A. — — — — — B.fl.VI,240
C. acuminata, F. v. M.; Actinostrobus, Parlatore, enum. sem. hort.
  Florent. 25 (1862) ... ... W.A. — — — — — B.fl.VI,240 M.fr.XI,104.
  FITZROYA, J. Hooker in Journ. of the hort. Soc. VI, 264 (1851). (Diselma.)
F. Archeri, Bentham & J. Hooker, gen. pl. III, 426 (1880) ... — T. — — — — B.fl.VI,240
  PHEROSPHAERA. Archer in Hooker's Kew Misc. II, 52 (1850). (Dacrydium partly.)
P. Hookeriana, Archer in Hooker, Kew Misc. II, 52 (1850) ... — — T. — — — — B.fl.VI,245 M.fr.XI,103.
P. Fitzgeraldi, F. v. M. in J. Hooker, icon. pl. XIV, t. 1383(1882) — — — N.S.W. — — M.fr.XI,102,138.
  MICROCACHRYS, J. Hooker in Lond. Journ. of Bot. IV, 149 (1845). (Dacrydium partly.)
M. tetragona, J. Hooker, Fl. Tasman. I, 358, t. 100 (1860) — — T. — — — — B.fl.VI,241
  DACRYDIUM, Solander in ? Forster, pl. escul. insul. ocean. austr. 80 (1786).
D. Franklinii, J. Hooker in Lond. Journ. of Bot. IV, 152, t. 6 (1845) — — T. — — — — B.fl.VI,245
  NAGEIA, Gaertner, de fructib. t. 39 (1788). (Podocarpus.)
N. elata, F. v. M.; Podoc., R. Brown in Mém. de Mus. XIII, 75(1826) — — — N.S.W. Q. — B.fl.VI,247
N. spinulosa, F. v. M.; Podocarpus, R. Brown in Mém. du Mus.
  XIII, 75 (1826) ... ... — — — N.S.W. — — B.fl.VI,247
N. Drouyniana; Podocarpus, F. v. M., fragm. IV, 86 (1864) ... W.A. — — — — — B.fl.VI,247 M.fr.VI,251;XI,104.
N. alpina, F. v. M. in papers Roy. Soc. of Tasm. 23 (1879) ... — — T. V. N.S.W. — — B.fl.VI,248 M.fr.XI,104.
  THALAMIA, Sprengel, Anleit. zur Kenntn. der Gew. zweite Ausg. II, 218 (1817). (Phyllocladus.)
T. asplenifolia, Sprengel, Anl. zur Kenntn. der Gew. II, 218 (1817) — — T. — — — — B.fl.VI,246 M.fr.XI,104.

CYCADEAE.

L. C. Richard in Persoon, synops. 630 (1807).
  CYCAS, Linné, hort. Cliffort. 482 (1737).
C. media, R. Brown, prodr. 348 (1810) ... ... — — — — Q. N.A. B.fl.VI,249 M.fr.VIII,171.
C. Normanbyana, F. v. M., fragm. VIII, 169 (1874) ... — — — — Q. — M.fr.VIII,169.
C. Kennedyana, F. v. M. in Melb. Chemist, March (1882) — — — — Q. —
C. Cairnsiana, F. v. M., fragm. X, 63 (1876) ... ... — — — — — M.fr.X,63,121.

ENCEPHALARTOS, Lehmann, pugill. VI, 9, t. 1, et 3 (1834). (Macrozamia, Lepidozamia, Catakidozamia, Zamia partly.)
E. Pauli Guilielmi, F.v. M. in Trans. Pharm. Soc. Vict. II, 91 (1859) — — — — N.S.W. Q. — B.fl.VI,251 M.fr.I,86,243;IX,124.
E. tridentatus, Lehmann, pugill. VI, 13 (1834) ... — — — — N.S.W. — — B.fl.VI,253 M.fr.III,38;XI,125.
E. spiralis, Lehmann, pugill. VI, 13 (1834)... ... ... — — — — N.S.W. Q. — B.fl.VI,251 M.fr.1,41;II.179;VIII,173
E. Douglasii, F. v. M.; Macrozamia, Hill (inedit.) ... ... — — — — — Q. — — [167;VIII,173;XI,126.
E. Fraseri, Miquel in Versl. K. Akad. Wet. Amst. XV, 368 (1863) W.A. — — — — — — B.fl.VI,252 M.fr.I,41,243;II,179;111,
E. Moorei, F. v. M., fragm. XI, 125 (1881)... ... — — — — — Q. — — M.fr.I,243;XI,125.
E. Macdonnelli, F.v.M.inVeral. K.Akad.Wet.Amst.XV, 376 (1863) — S.A. — — — — — — M.fr.IX,124;XI,126.[124.
E. Denisonii, F. v. M. in Journ. Pharm. Soc. Vict. II, 90 (1859) — — — — N.S.W. Q. — B.fl.VI,253 M.fr.1,41;VIII,173;IX,

BOWENIA, Hooker, Bot. Magaz. t. 5398 (1863).
B. spectabilis, Hooker, Bot. Mag. t. 5398 (1863) ... ... — — — — — Q. — B.fl.VI,254 M.fr.V,171.

MONOCOTYLEDONEAE.
Ray, meth. plant. nov. (1682).

CALYCEAE PERIGYNAE.
F. v. M. in Woolls's plants of the neighb. of Sydney 41 (1880).

ORCHIDEAE.
Haller, enum. stirp. Helv. praefat. 33 (1742).

PHOLIDOTA, Lindley in Hooker, Exotic Flora II, t. 138 (1825).
P. imbricata, Lindley in Hook. Exot. Fl. II, t. 138 (1825) ... — — — — — Q. — B.fl.VI,290 M.fr.IV,163;V,105.
STURMIA, Reichenbach in Moessler's Handb. II, 1552 (1828). (Liparis, Richard 1818 not of Zoologists 1738.)
S. reflexa, F. v. M., fragm. II, 72 (1861) ... ... — — — — N.S.W. Q. — B.fl.VI,273 M.fr.III,165;V,102.
S. angustilabris, F. v. M., fragm. IV, 164 (1864)... ... ... — — — — — Q. — B.fl.VI,273 M.fr.IX,50.
S. coelogynoides, F. v. M., fragm. II, 71 (1861) ... ... — — — — N.S.W. Q. — B.fl.VI,273 M.fr.IX,50.
S. habenarina, F. v. M., fragm. IV, 131 (1864) ... ... — — — — — Q. — B.fl.VI,273 M.fr.V,102;X1,53.

OBERONIA, Lindley, gen. and spec. of orchid. pl. 15 (1830). (Titania.)
O. iridifolia, Lindley, gen. and spec. of orchid. pl. 15 (1830) ... — — — — N.S.W. Q. — B.fl.VI,274 [88.
O. palmicola, F. v. M., fragm. II, 24 (1860) ... ... — — — — N.S.W. Q. — B.fl.VI,274 M.fr.II,24;VII,30; XI,53,

PHREATIA, Lindley, gen. and spec. of orchid. pl. 63 (1830). (Plexaure.)
P. Limenophylax, G. Reichenbach in Seem. Bonpl. 54 (1857) .. — — — — — Q. — B.fl.VI,290 M.fr.X,64.

DENDROBIUM, Swartz in nov. act. Upsal. VI, 82 (1799). (Thelychiton partly, Pedilonum, Coelandria.)
D. bigibbum, Lindley in Paxt. Flower Gard. III, 25 (1853) ... — — — — — Q. N.A. B.fl.VI,277 M.fr.VI,110;XI,87.
D. dicuphum, F. v. M., fragm. VIII, 28 (1871) ... ... — — — — — N.A. B.fl.VI,277 M.fr.VI,119.
D. Sumneri, F. v. M., fragm. VI, 94 (1868) ... ... — — — — — Q. — B.fl.VI,278 M.fr.VI,94.
D. superbiens, G. Reichenbach in Garden. Chron. 516 (1876) ... — — — — — Q. — — M.fr.XI,22.
D. Phalaenopsis, Fitzgerald in Gardeners' Chronicle, July (1880) — — — — — — — B.fl.VI,278 M.fr.I,87.
D. undulatum, R. Brown, prodr. 332 (1810) ... ... — — — — — — — B.fl.VI,279
D. Johannis, G. Reichenbach in Garden. Chron. 800 (1865) ... — — — — — — — B.fl.VI,279
D. speciosum, Smith, Exot. Botany I, 17, t. 10 (1804) ... — — — V. N.S.W. Q. — B.fl.VI,279 M.fr.VIII,240.
D. falconirostre, Fitzgerald in Sydney Morn. Herald, Nov. 18 (1876) — — — — N.S.W. — — — M.fr.X,87.
D. tetragonum, Cunningham in Bot. Regist. XXV, Misc. 33 (1839) — — — — N.S.W. Q. — B.fl.VI,279 M.fr.I,87;VI,119.
D. aemulum, R. Brown, prodr. 333 (1810)... ... — — — — N.S.W. Q. — B.fl.VI,280 M.fr.I,213;XI,87.
D. brachypus, G. Reichenbach in Garcke, Linnaca XLI, 42 (1876) — — — — N.S.W. — — —
D. gracilicaule, F. v. M., fragm. I, 179 (1859) ... ... — — — — N.S.W. Q. — B.fl.VI,281 M.fr.IV,171.
D. Moorei, F. v. M., fragm. VII, 29 (1870) ... ... — — — — N.S.W. — — B.fl.VI,281 M.fr.XI,54.
D. agrostophyllum, F. v. M., fragm. VIII, 28 (1871) ... — — — — N.S.W. — — B.fl.VI,282 M.fr.VIII,173.
D. Baileyi, F. v. M., fragm. VIII, 173 (1874) ... ... — — — — — Q. — — M.fr.VIII,173.
D. Smilliae, F. v. M., fragm. VI, 94 (1868) ... ... — — — — — Q. — B.fl.VI,282 M.fr.VII,93;XI,87.
D. ophioglossum, G. Reichenbach in Journ. L. Soc. XV, 113 (1876) — — — — — Q. — — M.fr.XI,54.
D. canaliculatum, R. Brown, prodr. 333 (1810) ... ... — — — — — N.A. B.fl.VI,282 M.fr.III,126.
D. Foelschcel, F. v. M. in Wing's S. Sc. Record, Oct. (1882) ... — — — — — N.A.
D. monophyllum, F. v. M., fragm. I, 180 (1859) ... ... — — — N.S.W. Q. — B.fl.VI,283 M.fr.VII,64.
D. hispidum, A. Richard, sert. Astrol. 13, t. 5 (1833) ... — — — — N.S.W. Q. — B.fl.VI,283 M.fr.VII,30.
D. cucumerinum, MacLeay in Bot. Reg., new ser. V, Misc. 58 (1842)— — — — N.S.W. Q. — B.fl.VI,283 M.fr.III,59;VIII,151,
D. pugioniforme, Cunningham in Bot. Reg. XXV, Misc. 33 (1839) — — — — N.S.W. Q. — B.fl.VI,284 M.fr.I,180.
D. rigidum, R. Brown, prodr. 333 (1810) ... ... — — — — — Q. — B.fl.VI,284 M.fr.VIII,248;XI,87.
D. linguiforme, Swartz in Svensk. Vetensk. Ac. Handl. 247 (1800) — — — — N.S.W. Q. — B.fl.VI,284 M.fr.I,80.
D. teretifolium, R. Brown, prodr. 333 (1810) ... — — — — N.S.W. Q. — B.fl.VI,285 M.fr.X,122.
D. Fairfaxii, Fitzgerald & F. v. M. in Sydney Mail 300, Nov. (1872) — — — N.S.W. — — B.fl.VI,285 M.fr.X,122.
D. striolatum, G. Reichenbach in Hamburg Gartenz. 313 (1857) — T. V. N.S.W. — — B.fl.VI,285 M.fr.I,88;VII,30,
D. Mortii, F. v. M., fragm. II, 93 (1860) ... ... — — — — N.S.W. Q. — B.fl.VI,286 M.fr.II,93.
D. Beckleri, F. v. M., fragm. V, 95 (1865) ... ... — — — — N.S.W. Q. — B.fl.VI,286

BULBOPHYLLUM, Petit-Thouars, Orchid. Afric. t. 95 ct 108 (1822). (Bolbophyllum, Thelychiton partly.)
B. Lichenastrum, F. v. M., fragm. VII, 60 (1869) ... ... — — — — — Q. — B.fl.VI,287 M.fr.X,119.
B. Baileyi, F. v. M., fragm. IX, 5 (1875) ... ... — — — — — Q. — — M.fr.IX,5.
B. nematopodum, F. v. M., fragm. VIII, 30 (1873) ... ... — — — — — Q. — B.fl.VI,287 M.fr.VIII,130.

B. Shepherdi, F. v. M., fragm. III, 40 (1862) ... ... — — — N.S.W. Q. — B.fl.VI,288 M.fr.I,190.
B. Taylori, F. v. M., fragm. VIII, 150 (1874) ... ... — — — — Q. — — M.fr.VIII,150.
B. aurantiacum, F. v. M., fragm. III, 39 (1862) ... ... — — — N.S.W. Q. — B.fl.VI,288 M.fr.VII,98.
B. argyropus, G. Reichenbach in Garcke, Linnaea XLI, 42 (1876) — — — N.S.W. — — — —
B. exiguum, F. v. M., fragm. II, 72 (1861)... ... ... — — — N.S.W. Q. — B.fl.VI,288 M.fr.XI,22.
B. minutissimum, F. v. M., fragm. XI, 53 (1878)... ... ... — — — N.S.W. Q. — — M.fr.V,85;XI,88.
B. Prenticei, F. v. M. in Wing's South. Sc. Record I, 173 (1881) — — — — — Q. — — B.fl.X,136.
B. Eliseo, F. v. M., fragm. VI, 120 (1868)... ... ... — — — N.S.W. Q. — B.fl.VI,289 M.fr.VIII,28.

SARCOCHILUS, R. Brown, prodr. 332 (1810). (Thrixpermum, Gunnia, Cleisostoma.)
S. erectus; Cleisostoma, Fitzgerald, Austr. Orchids part 4 (1877) — — — N.S.W. — — — M.fr.XI,52,139.
S. tridentatus, G. Reichenbach in Walp. Annal. VI, 500 (1861) — — — N.S.W. Q. — B.fl.VI,296 M.fr.I,102;II,181,VII,98.
S. Armitii, F. v. M., fragm. IX, 49 (1875).. ... ... — — — — Q. — — M.fr.IX,49.
S. Beckleri; Cleisostoma, F.v.M. in Benth. Fl. Austr. VI, 296(1870) — — — N.S.W. — — B.fl.VI,296 M.fr.VIII,248.
S. brevilabris; Cleisostoma, F. v. M., fragm. XI, 87 (1880) ... — — — — Q. — — M.fr.VIII,248;XI,139.
S. Macphersoni, F. v. M., fragm. VIII, 248 (1874) ... ... — — — — Q. — B.fl.VI,297 M.fr.VII,96.
S. Hartmanni, F. v. M., fragm. VIII, 248 (1874) .. .. — — — — Q. — — M.fr.VIII,248.
S. divitiflorus, F. v. M. in Benth. Fl. Austr. VI, 292 (1870) ... — — — N.S.W. Q. — B.fl.VI,292 M.fr.IX,50;XI,22.
S. falcatus, R. Brown, prodr. 332 (1810) ... ... — — — N.S.W. Q. — B.fl.VI,293 M.fr.I,89;VII,97;X,50.
S. Fitzgeraldi, F. v. M., fragm. VII, 115 (1870) ... ... — — — N.S.W. Q. — B.fl.VI,293 M.fr.I,203 M.fr.VII,115.
S. olivaceus, Lindley, Bot. Regist. Misc. XXV, 32 (1839) ... — — — N.S.W. Q. — B.fl.VI,293 M.fr.I,90,191;VII,97.
S. parviflorus, Lindley, Bot. Regist. Misc. XXIV, 34 (1838) ... — T. V. N.S.W. — — B.fl.VI,294 M.fr.I,89;VII,98.
S. Cecillae, F. v. M., fragm. V, 42, t. 42 (1865) ... ... — — — — Q. — B.fl.VI,294 M.fr.VII,98;XI,22.
S. Hillii, F. v. M., fragm. II, 94 (1861) ... ... — — — N.S.W. Q. — B.fl.VI,295 M.fr.I,88;III,39;VII,98.
S. phyllorrhizus, F. v. M., fragm. V, 201 (1866) ... ... — — — — Q. — B.fl.VI,295 M.fr.VII,30.

TAENIOPHYLLUM, Blume, Bijdr. 355 (1825).
T. Muelleri, Lindley in Benth. Fl. Austr. VI, 291 (1870) ... — — — — — Q. — B.fl.VI,291

ORNITHOCHILUS, Wallich in Lindley, gen. and spec. of orchid. plants 242 (1833). (Saccolabium partly.)
O. Hillii,.Bentham in Journ. Linn. Soc. XVIII, 334 (1880) ... — — — N.S.W. Q. — B.fl.VI,298 M.fr.VII,97.

GEODORUM, Jackson in Andrews, Botan. Reposit. t. 626 (1810). [31.
G. pictum, Lindley, gen. and spec. of orchid. pl. 175 (1833) ... — — — — Q. N.A. B.fl.VI,299 M.fr.III,24;V,102;VIII,

EULOPHIA, R. Brown in Edwards, Bot. Regist. 686 (1822). (Eulophus 1822.)
E. venosa, G. Reichenbach in Benth. Fl. Austr. VI, 300 (1870) — — — — Q. N.A. B.fl.VI,300 M.fr.I,61.
E. Fitzalani, F. v. M., fragm. VIII, 30 (1872) ... ... — — — — Q. — B.fl.VI,300 M.fr.VII,30.

DIPODIUM, R. Brown, prodr. 330 (1810).
D. punctatum, R. Brown, prodr. 331 (1810) ... ... — S.A. T. V. N.S.W. Q. N.A. B.fl.VI,301 M.fr.X,64.
D. ensifolium, F. v. M., fragm. V, 42 (1865) ... ... — — — — Q. — B.fl.VI,301 M.fr.X,64.

CYMBIDIUM, Swartz in nov. act. Upsal. VI, 70 (1799).
C. Hillii, F. v. M. in Regel's Gartenflora, 138 (1879) ... ... — — — — Q. — — M.fr.XI,88.
C. canaliculatum, R. Brown, prodr. 331 (1810) ... ... — S.A. — N.S.W. Q. N.A. B.fl.VI,302 M.fr.XI,89.
C. albuciflorum, F. v. M., fragm. I, 188 (1859) ... ... — — — N.S.W. Q. — B.fl.VI,303 M.fr.I,188.
C. suave, R. Brown, prodr. 331 (1810) ... ... — — — N.S.W. Q. — B.fl.VI,303 M.fr.I,187.

SPATHOGLOTTIS, Blume, Bijdr. 400 (1825).
S. Paulinae, F. v. M., fragm. VII, 95 (1868) ... ... — — — — Q. — B.fl.VI,304 M.fr.X,64.

PHAJUS, Loureiro, Fl. Cochinch. II, 529 (1790).
P. grandifolius, Loureiro, Fl. Cochinch. II, 529 (1790) ... ... — — — N.S.W. Q. — B.fl.VI,304 M.fr.I,42;IV,163;X,6

CALANTHE, R. Brown in Edwards, Bot. Regist. 573 (1821).
C. veratrifolia, R. Brown in Bot. Regist. 573 (1821) ... ... — — — N.S.W. Q. — B.fl.VI,305 M.fr.VIII,30;IX,50.

GALEOLA, Loureiro, Fl. Cochinch. II, 520 (1790). (Erythrorchis, Ledgeria.) [XI,105.
G. cassythoides, G. Reichenbach, Xen. Orchid. II, 77 (1869) ... — — — N.S.W. Q. — B.fl.VI,307 M.fr.II,167;V,102;XI,127.
G. foliata, F. v. M., fragm. VIII, 31 (1872) ... ... — — — N.S.W. Q. — B.fl.VI,307 M.fr.II,167;V,102;XI,127.

EPIPOGUM, Gmelin, Fl. Sibir. I, 11, t. 2 (1747).
E. nutans, Lindley in Journ. Linn. Soc. I, 177 (1856) ... ... — — — N.S.W. Q. — B.fl.VI,308 M.fr.IX,50;XI,105.

GASTRODIA, R. Brown, prodr. 330 (1810).
G. sesamoides, R. Brown, prodr. 330 (1810) ... ... — — T. V. N.S.W. Q. — B.fl.VI,309 M.fr.VIII,151;X,64.

POGONIA. A. L. de Jussieu, gen. plant. 65 (1789).
P. uniflora, F. v. M., fragm. V, 201 (1866)... ... ... — — — — Q. — B.fl.VI,310 M.fr.V,201.
P. holochila, F. v. M., fragm. V, 200 (1866) ... ... — — — — Q. — B.fl.VI,310 M.fr.V,200.
P. Dallachyana, F. v. M. in Benth. Fl. Austr. VI, 310 (1870) ... — — — — Q. — B.fl.VI,310
P. pachystomoides, F. v. M., fragm. VIII, 174 (1874) ... ... — — — — Q. — — M.fr.VIII,174.

CORYMBORCHIS, Petit-Thouars, hist. des plant. Orchid. recueill. sur les isles d'Afriq. t. 37-38 (1822). (Corymbis.)
C. veratrifolia, Blume, Collect. des Orchidées 125, t. 42 & 43 (1838) — — — — Q. — B.fl.VI,311 M.fr.VIII,30;X,64.

ETAERIA, Blume, Bijdr. 409 (1825). (Hetaeria, Ramphidia.)
E. tenuis, Bentham in Journ. Linn. Soc. XVIII, 345 (1880) ... — — — — Q. — B.fl.VI,312

MICROSTYLIS, Nuttall, Genera of North American plants II, 196 (1818).
M. Bernaysii, F. v. M., fragm. XI, 21 (1878) ... ... — — — — Q. — — M.fr.XI,21.

GOODYERA, R. Brown in Aiton, hort. Kew. sec. ed. V, 197 (1813).
G. viridiflora, Blume, Coll. des Orchid. 41, t. 9 (1858) ... ... — — — — Q. — B.fl.VI,313 M.fr.VIII,29. [04.
G. polygonoides, F. v. M., fragm. VIII, 20 (1872) ... ... — — — — Q. — B.fl.VI,313 M.fr.VII,30;VIII,29;X,

SPIRANTHES, L. C. Richard in Mém. du Mus. IV, 40 (1818). (Gyrostachis, 1807.)
S. australis, Lindley, Bot. Regist. 823 (1824) ... ... — S.A. T. V. N.S.W. Q. — B.fl.VI,314 M.fr.X,65.

THELYMITRA, R. & G. Forster, char. gen. 97, t. 49 (1776). (Macdonaldia.)
T. ixioides, Swartz in Kong. Svensk. Acad. Handl. 228, t. 3 (1800) W.A. S.A. T. V. N.S.W. Q. — B.fl.VI,317
T. circumsepta, Fitzgerald, Austral. Orchid., part 4 (1877) ... — — — N.S.W. — — — —

Species	W.A.	S.A.	T.	V.	N.S.W.	Q.	B.fl.	M.fr.
T. criuita, Lindley, Bot. Regist. XXV, App. XLIX (1839)	W.A.	—	—	—	—	—	B.fl.VI,319	M.fr.VI,84.
T. aristata, Lindley, gen. and spec. of orchid. pl. 521 (1840)	W.A.	S.A.	T.	V.	N.S.W.	—	B.fl.VI,319	
T. longifolia, R. & G. Forster, char. gen. 98, t. 49 (1776)	W.A.	S.A.	T.	V.	N.S.W.	Q.	B.fl.VI,319	
T. villosa, Lindley, Bot. Regist. XXV, App. XLIX (1839)	W.A.	—	—	—	—	—	B.fl.VI,320	M.fr.V,04,214.
T. tigrina, R. Brown, prodr. 315 (1810)	W.A.	—	—	—	—	—	B.fl.VI,320	
T. fusco-lutea, R. Brown, prodr. 315 (1810)	W.A.	S.A.	—	—	—	—	B.fl.VI,321	M.fr.V,94.
T. stellata, Lindley, Bot. Regist. XXV, App. XLIX (1839)	W.A.	—	—	—	—	—	B.fl.VI,321	
T. carnea, R. Brown, prodr. 314 (1810)	—	S.A.	T.	V.	N.S.W.	—	B.fl.VI,322	M.fr.V,98.
T. flexusa, Endlicher, nov. stirp. decad. 23 (1839)	W.A.	S.A.	T.	V.	—	—	B.fl.VI,322	
T. antennifera, J. Hooker, Fl. Tasman. II, 4, t. 101 (1860)	W.A.	S.A.	T.	V.	—	—	B.fl.VI,322	
T. Macmillani, F. v. M., fragm. V, 93 (1865)	—	—	—	V.	—	—	B.fl.VI,323	M.fr.VI,93.
T. Mackibbinii, F. v. M. in Melb, Chemist 44 (1881)	W.A.	S.A.	—	V.	—	—	—	M.fr.XI,139.
T. variegata, Lindley in Benth. Fl. Austr. VI, 323 (1870)	W.A.	—	—	—	—	—	B.fl.VI,323	M.fr.V,97,98.
T. venosa, R. Brown, prodr. 314 (1870)	—	—	—	—	N.S.W.	—	B.fl.VI,323	
T. cyanea, Lindley in Benth. Fl. Austr. VI, 323 (1870)...	—	—	T.	—	—	—	B.fl.VI,323	
EPIBLEMA, R. Brown, prodr. fl. Nov. Holl. 315 (1810).								
E. grandiflorum, R. Brown, prodr. 315 (1810)	W.A.	—	—	—	—	—	B.fl.VI,324	M.fr.XI,21.
DIURIS, Smith in Transact. Linn. Soc. IV, 222 (1798).								
D. alba, R. Brown, prodr. 316 (1810)	—	—	—	—	N.S.W.	Q.	B.fl.VI,325	M.fr.V,102.
D. punctata, Smith, Exot. Bot. I, 13, t. 8 (1804)...	—	S.A.	—	V.	N.S.W.	—	B.fl.VI,326	M.fr.V,102.
D. secundiflora, Fitzgerald, Austral. Orchids, part 4 (1877)	—	—	—	—	N.S.W.	—	B.fl.VI,327	M.fr.V,172.
D. aurea, Smith, Exot. Bot. 15, t. 9 (1804)	—	—	—	—	N.S.W.	Q.	B.fl.VI,327	M.fr.V,102.
D. palustris, Lindley, gen. and spec. of orchid. pl. 507 (1840)	—	S.A.	T.	V.	—	—	B.fl.VI,327	M.fr.V,102.
D. maculata, Smith, Exot. Bot. I, 57, t. 30 (1804)	—	S.A.	T.	V.	N.S.W.	Q.	B.fl.VI,327	M.fr.V,102,173.
D. pedunculata, R. Brown, prodr. 316 (1810)	—	S.A.	T.	V.	N.S.W.	—	B.fl.VI,328	M.fr.V,102.
D. pallens, Bentham, Fl. Austr. VI, 329 (1870)	—	—	—	—	N.S.W.	—	B.fl.VI,329	
D. abbreviata, F. v. M. in Benth. Fl. Austr. VI, 329 (1870)	—	—	—	—	N.S.W.	Q.	B.fl.VI,329	
D. setacea, R. Brown, prodr. 316 (1810)	W.A.	—	—	—	—	—	B.fl.VI,329	M.fr.V,101.
D. emarginata, R. Brown, prodr. 316 (1810)	W.A.	—	—	—	—	—	B.fl.VI,330	
D. sulphurea, R. Brown, prodr. 316 (1810)	—	S.A.	T.	V.	N.S.W.	—	B.fl.VI,330	M.fr.V,102,173.
D. aequalis, F. v. M. in Fitzgerald, Austral. orch. part 2 (1876)	—	—	—	—	N.S.W.	—		M.fr.X,65.
D. dendrobioides, Fitzgerald, Austral. Orchids, part 7 (1882)	—	—	—	—	N.S.W.	—		
D. longifolia, R. Brown, prodr. 316 (1810)...	W.A.	S.A.	T.	V.	—	—	B.fl.VI,331	M.fr.V,101,172.
D. pauciflora, R. Brown, prodr. 316 (1810)	W.A.	—	—	—	—	—	B.fl.VI,331	
ORTHOCERAS, R. Brown, prodr. fl. Nov. Holl. 317 (1810).								
O. strictum, R. Brown, prodr. 317 (1810)	—	S.A.	—	V.	N.S.W.	—	B.fl.VI,332	
CALOCHILUS, R. Brown, prodr. fl. Nov. Holl. 320 (1810).								
C. campestris, R. Brown, prodr. 320 (1810)	—	—	T.	V.	N.S.W.	Q.	B.fl.VI,315	M.fr.VIII,151.
C. Robertsoni, Bentham, Fl. Austr. VI, 315 (1870)	—	—	T.	V.	—	—	B.fl.VI,315	
C. paludosus, R. Brown, prodr. 320 (1810)...	—	S.A.	—	V.	N.S.W.	Q.	B.fl.VI,316	
CRYPTOSTYLIS, R. Brown, prodr. 317 (1810). (Zosterostylis.)								
C. longifolia, R. Brown, prodr. 317 (1810)	—	—	T.	V.	N.S.W.	Q.	B.fl.VI,333	M.fr.XI,21.
C. ovata, R. Brown, prodr. 317 (1810)	W.A.	—	—	—	—	—	B.fl.VI,334	
C. erecta, R. Brown, prodr. 317 (1810)	—	—	—	—	N.S.W.	—	B.fl.VI,334	
C. leptochila, F. v. M. in Benth. Fl. Austr. VI, 324 (1870)	—	—	—	—	N.S.W.	—	B.fl.VI,334	
PRASOPHYLLUM, R. Brown, prodr. 317 (1810). (Genoplesium.)								
P. australe, R. Brown, prodr. 318 (1810)	—	S.A.	T.	V.	—	—	B.fl.VI,337	
P. flavum, R. Brown, prodr. 318 (1810)	—	—	—	—	N.S.W.	—	B.fl.VI,337	
P. elatum, R. Brown, prodr. 318 (1810)	W.A.	S.A.	T.	V.	N.S.W.	—	B.fl.VI,337	M.fr.V,100.
P. brevilabre, J. Hooker, Fl. Tasman II, 11, t. 110 (1860)	—	—	T.	V.	N.S.W.	Q.	B.fl.VI,338	M.fr.V,101.
P. hians, G. Reichenbach, Beitr. zur syst. Pflanzenk. 61 (1871)...	W.A.	—	—	—	—	—	B.fl.VI,338	
P. patens, R. Brown, prodr. 318 (1810)	—	S.A.	T.	V.	N.S.W.	Q.	B.fl.VI,339	M.fr.V,101.
P. fuscum, R. Brown, prodr. 318 (1810)	—	S.A.	T.	V.	N.S.W.	Q.	B.fl.VI,339	M.fr.V,101.
P. cyphochilum, Bentham, Fl. Austr. VI, 340 (1870)	W.A.	—	—	—	—	—	B.fl.VI,340	
P. ovale, Lindley, Bot. Regist. XXV, App. LIV (1839)	W.A.	—	—	—	—	—	B.fl.VI,341	
P. macrostachyum, R. Brown, prodr. 318 (1810)	W.A.	—	—	—	—	—	B.fl.VI,341	
P. Fimbria, G. Reichenbach, Beitr. zur syst. Pflanzenk. 60 (1871)	W.A.	—	—	—	—	—	B.fl.VI,341	
P. striatum, R. Brown, prodr. 318 (1810)	—	—	—	—	N.S.W.	—	B.fl.VI,342	M.fr.V,100.
P. parviflorum, Lindley, Bot. Regist. XXV, App. LIV (1839)...	W.A.	—	—	—	—	—	B.fl.VI,342	M.fr.V,101.
P. gibbosum, R. Brown, prodr. 331 (1810)	W.A.	—	—	—	—	—	B.fl.VI,342	M fr.V,100.
P. cucullatum, G. Reichenbach, Beitr. zur syst. Pflanzenk. 59 (1871)	W.A.	—	—	—	—	—	B.fl.VI,343	
P. nigricans, R. Brown, prodr. 319 (1810)	—	S.A.	T.	V.	N.S.W.	—	B.fl.VI,343	M.fr.IX,50.
P. rufum, R. Brown, prodr. 319 (1810)	—	—	T.	V.	N.S.W.	Q.	B.fl.VI,344	M.fr.V,160.
P. brachystachyum, Lindley, gen. and spec. of orchid. pl. 513 (1840)	--	—	T.	—	—	—	B.fl.VI,344	M.fr.V,100.
P. despectans, J. Hooker, Fl. Tasman. II, 13, t. 113 (1860)	—	S.A.	T.	V.	—	—	B.fl.VI,345	M.fr.V,100.
P. fimbriatum, R. Brown, prodr. 319 (1810)	—	—	—	—	N.S.W.	—	B.fl.VI,345	M.fr.V,100.
P. Archeri, J. Hooker, Fl. Tasman. II, 14, t. 113 (1860)...	—	—	T.	V.	—	—	B.fl.VI,345	M.fr.V,100.
P. intricatum, C. Stuart in Benth. Fl. Austr. VI, 346 (1870)	—	—	T.	—	—	—	B.fl.VI,346	
P. Woollsii, F. v. M., fragm. V, 100 (1866)	—	—	—	—	N.S.W.	—	B.fl.VI,346	M.fr.V,100.
MICROTIS, R. Brown, prodr. fl. Nov. Holl. 320 (1810).								
M. porrifolia, R. Brown, prodr. 320 (1810)	W.A.	S.A.	T.	V.	N.S.W.	Q.	B.fl.VI,347	M.fr.X,65;XI,21.
M. media, R. Brown, prodr. 321 (1810)	W.A.	—	—	—	—	—	B.fl.VI,348	
M. alba, R. Brown, prodr. 321 (1810)	W.A.	—	—	—	—	—	B.fl.VI,348	
M. pulchella, R. Brown, prodr. 321 (1810)	W.A.	—	—	—	—	—	B.fl.VI,349	
M. atrata, Lindley, Bot. Regist. XXV, App. LIV (1839)	W.A.	S.A.	—	V.	—	—	B.fl.VI,349	M.fr.I,90,244;X,65.
CORYSANTHES, R. Brown, prodr. 328 (1810). (Coryhas, Nematoceras.)								
C. unguiculata, R. Brown, prodr. 328 (1810)	—	—	—	—	N.S.W.	—	B.fl.VI,350	
C. pruinosa, R. Cunningham in N.S. Wales, Magaz. n, 1 (1833)	W.A.	S.A.	T.	V.	N.S.W.	—	—	M.fr.IX,48.

113

```
C. fimbriata, R. Brown, prodr. 328 (1810) ...    ...    ...    ...    — — — — N.S.W. —    R.fl.VI,351    M.fr.X,65.
C. bicalcarata, R. Brown, prodr. 328 (1810)      ...    ...    ...    — — — — N.S.W. Q.    B.fl.VI,351

      PTEROSTYLIS, R. Brown, prodr. fl. Nov. Holl. 326 (1810).
P. ophioglossa, R. Brown, prodr. 326 (1810)       ...    ...    — — — — N.S.W. Q.    B.fl.VI,354
P. concinna, R. Brown, prodr. 326 (1810) ...    ...    ...    — S.A. — V. N.S.W. Q.    B.fl.VI,355    M.fr.V,99;XI,105.
P. curta, R. Brown, prodr. 326 (1810)     ...    ...    ...    — S.A. T. V. N.S.W. Q.    B.fl.VI,355    M.fr.V,99.
P. acuminata, R. Brown, prodr. 326 (1810)...    ...    ...    — — — V. N.S.W. —    B.fl.VI,355
P. nutans, R. Brown, prodr. 327 (1810)     ...    ...    ...    — S.A. T. V. N.S.W. Q.    B.fl.VI,356    M.fr.V,99.
P. podaglossa, Fitzgerald, Austral. Orchids, part 3 (1877)    ...    — — N.S.W. —    —
P. pedunculata, R. Brown, prodr. 327 (1810)    ...    ...    — S.A. T. V. N.S.W. —    B.fl.VI,350    M.fr.VIII,240;XI,105.
P. nana, R. Brown, prodr. 327 (1810)     ...    ...    ...    W.A. S.A. T. V. —    B.fl.VI,357    M.fr.V,99.
P. cucullata, R. Brown, prodr. 327 (1810) ...    ...    ...    — S.A. T. V. —    B.fl.VI,357    M.fr.V,99.
P. grandiflora, R. Brown, prodr. 327 (1810)    ...    ...    — — — — N.S.W. —    R.fl.VI,358    M.fr.V,99;XI,105.
P. truncata, Fitzgerald, Austral. Orchids, part 4 (1878) ..    ...    — — — N.S.W. —    —
P. reflexa, R. Brown, prodr. 327 (1810)     ...    ...    W.A. S.A. — V. N.S.W. —    B.fl.VI,359    M.fr.XI,105.
P. praecox, Lindley, gen. and spec. of Orchid. pl. 388 (1840)    ...    — S.A. T. V. N.S.W. —    B.fl.VI,359
P. obtusa, R. Brown, prodr. 327 (1810)    ...    ...    — T. V. N.S.W. —    B.fl.VI,360    M.fr.IX,50;XI,53.
P. recurva, Bentham, Fl. Austr. VI, 360 (1870) ...    ...    W.A.    — — — —    B.fl.VI,360
P. parviflora, R. Brown, prodr. 327 (1810)    ...    ...    — T. V. N.S.W. Q.    B.fl.VI,361    M.fr.V,100.
P. barbata, Lindley, Bot. Regist. XXV, App. LIII (1839)    ...    W.A. S.A. T. V. N.S.W. —    B.fl.VI,362    M.fr.VIII,151;XI,22,105.
P. turfosa, Endlicher in Lehm. pl. Preiss. II, 5 (1846)    ...    W.A.    — — — —    B.fl.VI,362
P. mutica, R. Brown, prodr. 328 (1810)     ...    ...    — S.A. T. V. N.S.W. Q.    B.fl.VI,362    M.fr.V,100.
P. rufa, R. Brown, prodr. 327 (1810)     ...    ...    W.A. S.A. T. V. N.S.W. Q.    B.fl.VI,363    M.fr.V,100.
P. Daintreyana, F. v. M. in Benth. Fl. Austr. VI, 360 (1870) ...    — — N.S.W. —    B.fl.VI,360
P. longifolia, R. Brown, prodr. 327 (1810)    ...    ...    — S.A. T. V. N.S.W. —    B.fl.VI,364    M.fr.V,100.
P. vittata, Lindley, Bot. Regist. XXV, App. LIII (1839)    ...    W.A. S.A. — V.    — — —    B.fl.VI,364    M.fr.XI,126.

      CALEYA, R. Brown in Aiton, Hort. Kew. sec. ed. V, 204 (1813).    (Caleana, 1810.)
C. major, R. Brown, prodr. 329 (1810)     ...    ...    ...    — T. V. N.S.W. Q.    B.fl.VI,366    M.fr.XI,105.
C. minor, R. Brown, prodr. 329 (1810)     ...    ...    ...    — T. — N.S.W. —    B.fl.VI,366
C. nigrita, Lindley, Bot. Reg. XXV, App. LIV (1839) ...    ...    W.A.    — — — —    B.fl.VI,366
C. Sullivani, F. v. M. in Melb. Chemist 44 (1882)    ...    ...    — V.    — — —

      DRAKAEA, Lindley, Bot. Regist. XXV, App. LV (1839).    (Spiculaea, Arthrochilus.)
D. elastica, G. Reichenbach, Beitr. zur syst. Pflanzenk. 68 (1871)    W.A.    — — — —    B.fl.VI,367
D. irritabilis, G. Reichenbach, Beitr. zur syst. Pflanzenk. 68 (1871)    — — — — N.S.W. Q.    B.fl.VI,368    M.fr.VIII,151;XI,53.
D. elastica, Lindley, Bot. Regist. XXV, App. LV (1839)    ...    W.A.    — — — —    B.fl.VI,368

      ACIANTHUS, R. Brown, prodr. fl. Nov. Holl. 321 (1810).
A. caudatus, R. Brown, prodr. 321 (1810)...    ...    ...    — T. V. N.S.W. —    B.fl.VI,369
A. fornicatus, R. Brown, prodr. 321 (1810)    ...    ...    — — N.S.W. Q.    B.fl.VI,370
A. exsertus, R. Brown, prodr. 321 (1810)    ...    ...    — S.A. T. V. N.S.W. —    B.fl.VI,370    M.fr.XI,53,105.
A. viridis, J. Hooker, Fl. Tasman. II, 372 (1860)...    ...    — T. — — —    R.fl.VI,371    M.fr.IX,50.

      CYRTOSTYLIS, R. Brown, prodr. flor. Nov. Holl. 322 (1810).
C. reniformis, R. Brown, prodr. 322 (1810)    ...    ...    W.A. S.A. T. V. N.S.W. Q.    B.fl.VI,376

      LYPERANTHUS, R. Brown, prodr. 325 (1810).    (Burnettia).
L. nigricans, R. Brown, prodr. 325 (1810) ...    ...    ...    W.A. S.A. T. V. N.S.W. —    B.fl.VI,374    M.fr.VII,134;X,05.
L. Burnettii, F. v. M., fragm. V, 96 (1865)    ...    ...    — T. V.    — — —    B.fl.VI,375    M.fr.V,96;VII,134.
L. Forrestii, F. v. M. in Wing's S. Sc. Rec. II, 55 (1882)    ...    W.A.    — — — —
L. ellipticus, R. Brown, prodr. 325 (1810)...    ...    ...    — — — — N.S.W. —    B.fl.VI,374    M.fr.VII,133.

      ERIOCHILUS, R. Brown, prodr. fl. Nov. Holl. 323 (1810).
E. autumnalis, R. Brown, prodr. 323 (1810)    ...    ...    — S.A. T. V. N.S.W. Q.    B.fl.VI,372    M.fr.XI,22,105.
E. scaber, Lindley, Bot. Regist. XXV, App. LIII (1839)    ...    W.A.    — — — —    B.fl.VI,372
E. tenuis, Lindley, Bot. Regist. XXV, App. LIII (1839)    ...    W.A.    — — — —    B.fl.VI,372
E. dilatatus, Lindley, Bot. Regist. XXV, App. LIII (1839)    ...    W.A.    — — — —    B.fl.VI,372
E. multiflorus, Lindley, Bot. Regist. XXV, App. LIII (1839) ...    W.A.    — — — —    B.fl.VI,373    M.fr.XI,105.
E. fimbriatus, F. v. M. in Wing's S. Sc. Record II, 152 (1882)    ...    W.A.    — — — —    B.fl.VI,377    M.fr.XI,104.

      CALADENIA, R. Brown, prodr. 325 (1810).    (Leptoceras, Adenochilus.)
C. Menziesii, R. Brown, prodr. 325 (1810) ...    ...    W.A. S.A. T. V.    — — —    B.fl.VI,377    M.fr.V,101;XI,105.
C. Nortoni; Adenochilus, Fitzgerald, Austral. Orch. part 2 (1876)    — — — N.S.W. —    —    M.fr.X,64.
C. Cairnsiana, F. v. M., fragm. VII, 31 (1869)    ...    W.A.    — — — —    B.fl.VI,378    M.fr.VII,31.
C. multiclavia, G. Reichenbach, Beitr. zur syst. Pflanzenk. 64 (1871)    W.A.    — — — —    B.fl.VI,378
C. discoidea, Lindley, Bot. Regist. XXV, App. LII (1839)    ...    W.A.    — — — —    B.fl.VI,382
C. Patersoni, R. Brown, prodr. 324 (1810)    ...    W.A. S.A. T. V. N.S.W. —    B.fl.VI,382    M.fr.V,101.
C. Drummondii, Bentham, Fl. Austr. VI, 383 (1873)    ...    W.A.    — — — —    B.fl.VI,383
C. hirta, Lindley, Bot. Regist. XXV, App. LII (1839)    ...    W.A.    — — — —    B.fl.VI,383
C. Roei, Bentham, Fl. Austr. VI, 383 (1873)    ...    ...    W.A.    — — — —    B.fl.VI,383
C. Barbarossae, G. Reichenbach, Beitr. z. syst. Pflanzenk. 64 (1871) W.A.    — — — —    B.fl.VI,383
C. flava, R. Brown, prodr. 324 (1810)     ...    ...    ...    W.A.    — — — —    B.fl.VI,384    M.fr.V,101;VI,83;XI,105.
C. latifolia, R. Brown, prodr. 324 (1810)    ...    ...    W.A. S.A. T. V. N.S.W. —    B.fl.VI,384    M.fr.V,101;XI,105.
C. reptans, Lindley, Bot. Regist. XXV, App. LII (1839)    ...    W.A.    — — — —    B.fl.VI,385
C. suaveolens, G. Reichenbach, Beitr. zur syst. Pflanzenk. 67 (1871)    — T. V. N.S.W. —    B.fl.VI,385    M.fr.V,98.
C. serrata, G. Reichenbach, Beitr. zur syst. Pflanzenk. 67 (1871) W.A.    — — — —    B.fl.VI,386    M.fr.V,99.
C. carnea, R. Brown, prodr. 324 (1810)    ...    ...    — S.A. T. V. N.S.W. —    B.fl.VI,386    M.fr.V,101.
C. congesta, R. Brown, prodr. 324 (1810)    ...    ...    — T. V. N.S.W. —    B.fl.VI,387
C. aphylla, Bentham, Fl. Austr. VI, 387 (1873) ...    ...    W.A.    — — — —    B.fl.VI,387
C. coerulea, R. Brown, prodr. 324 (1810)    ...    ...    — S.A. T. V. N.S.W. —    B.fl.VI,389    M.fr.V,101;XI,22.
C. deformis, R. Brown, prodr. 324 (1810) ...    ...    ...    — S.A. T. V. N.S.W. —    B.fl.VI,389    M.fr.V,101;XI,105.
C. sericea, Lindley, Bot. Regist. XXV, App. LII (1839)    ...    W.A.    — — — —    B.fl.VI,389
C. gemmata, Lindley, Bot. Regist. XXV, App. LII (1839)    ...    W.A.    — — — —    B.fl.VI,389    M.fr.V,101.
```

P

114

C. ixioides, Lindley, Bot. Regist. XXV, App. LII (1839) .. W.A. — — — — — — B.fl.VI,389
C. unita, Fitzgerald in Gardeners' Chronicle 461 (1882) ... ... W.A. — — — — — — —

CHILOGLOTTIS, R. Brown, prodr. fl. 322 (1810).
C. diphylla, R. Brown, prodr. 323 (1810) ... — ... — T. V. N.S.W. Q. — B.fl.VI,390 M.fr.X,116.
C. trapeziformis, Fitzgerald, Austral. Orchids, part 3 (1877) ... — — — N.S.W. — — — M.fr.X,116.
C. formicifera, Fitzgerald, Austral. Orchids, part 3 (1877) ... — — — N.S.W. — —. — M.fr.X,116. [116;XI,22.
C. Gunnii, Lindley, gen. and spec. Orchid. 387 (1840) ... ... — T. V. — — B.fl.VI,391 M.fr.VIII,151; IX,50; X,

GLOSSODIA, R. Brown, prodr. 325 (1810).
G. major, R. Brown, prodr. 326 (1810) ... ... — S.A. T. V. N.S.W. Q. — B.fl.VI,392
G. minor, R. Brown, prodr. 326 (1810) ... ... — — V. N.S.W. Q. — B.fl.VI,392
G. Brunonis, Endlicher, nov. stirp. dec. 16 (1839) ... W.A. — — — — — B.fl.VI,393 M.fr.VI,83.
G. intermedia, Fitzgerald in Gardeners' Chron. 402 (1882) ... W.A. — — — — — — —
G. emarginata, Lindley, gen. and spec. of Orchid. pl. 424 (1840) W.A. — — — — — B.fl.VI,393 M.fr.XI,22.

HABENARIA, Willdenow, spec. plant. IV, 44 (1805).
H. trinervis, Wight, icon. plant. Ind. or. t. 1701 (1852)... ... — — — — — N.A. B.fl.VI,394
H. elongata, R. Brown, prodr. 313 (1810) ... ... — — — — Q. N.A. B.fl.VI,394 M.fr.VII,15.
H. graminea, Lindley, gen. and spec. of orchid. pl. 318 (1840) ... — — — — Q. N.A. B.fl.VI,394 M.fr.VII,16.
H. ochroleuca, R. Brown, prodr. 313 (1810) ... ... — — — — — N.A. B.fl.VI,395 M.fr.V,102.
H. xanthantha, F. v. M., fragm. VII, 16 (1869) ... ... — — — — Q. — B.fl.VI,395 M.fr.VII,16.

APOSTASIA, Blume, Bijdr. 423 (1825). (Niemeyera.)
A. stylidioides, G. Reichenbach in Benth. Fl. Austr. VI, 396 (1873) — — — — Q. — B.fl.VI,396

SCITAMINEAE.
Linné, philos. botan. 27 (1751).

MUSA, l'Ecluse, éxoticorum libri decem 229 (1605).
M. Banksii, F. v. M., fragm. IV, 132 (1864) ... ... — — — — Q. — B.fl.VI,261 M.fr.IV,132.
M. Fitzalani, F. v. M., fragm. IX, 188 (1875) ... ... — — — — — Q. — M.fr.IX,188.
M. Hillii, F. v. M., fragm. IX, 189 (1875)... ... — — — — — Q. — M.fr.IX,109.

CURCUMA, Linné, gen. plant. 332 (1737), from Hermann (1687).
C. Australasica, J. Hooker, Bot. Mag. t. 5620 (1867) ... — — — — — Q. — B.fl.VI,263 M.fr.VIII,26.

AMOMUM, Linné, gen. plant. 330 (1737).
A. Dallachyi, F. v. M., fragm. VIII, 25 (1872) ... ... — — — — — Q. — B.fl.VI,263 M.fr.VIII,25.

ELETTARIA, White and Matton in Transact. Linn. Soc. X, 249 (1809), from Rheede (1692).
E. Scottiana, F. v. M., fragm. VIII, 24 (1872) ... ... — — — — — Q. — B.fl.VI,264 M.fr.VIII,24.

ALPINIA, Linné, gen. plant. 332 (1737). (Hellenia.)
A. racemigera, F. v. M., fragm. VIII, 27 (1872) ... ... — — — — — Q. — B.fl.VI,265 M.fr.VIII,27.
A. caerulea, Bentham, Fl. Austr. VI, 265 (1873) ... ... — — — — N.S.W. Q. — B.fl.VI,265 M.fr.VIII,26.
A. arctiflora, F. v. M., fragm. VIII, 23 (1872) ... ... — — — — — Q. — B.fl.VI,266 M.fr.VIII,25.

COSTUS, Linné, Bauhin, pinax, 36 (1623).
C. Potierae, F. v. M., fragm. IV, 164 (1864) ... ... — — — — — Q. — B.fl.VI,296 M.fr.IV,164;VIII,27.

TAPEINOCHEILOS, Miquel in annal. mus. bot. Lugdun. IV, 101, t. 4 (1869).
T. pungens, Miquel in ann. mus. Lugd. IV, 101, t. 4 (1869) ... — — — — — Q. — B.fl.VI,297 M.fr.VIII,26.

IRIDEAE.
Ventenat, Tableau du règne végétal II, 188 (1799).

DIPLARRHENA, Labillardière, Relation du voy. à la rech. de La Per. I, 275, t. 15 (1799).
D. Moraea, Labillardière, Voy. I, 157, t. 15 (1799) ... — — — T. V. N.S.W. — B.fl.VI,400 M.fr.VII,94.

PATERSONIA, R. Brown, in Bot. Magaz. t. 1041 (1807). (Genosiris, 1804.)
P. glauca, R. Brown, prodr. 304 (1810) ... — — S.A. T. V. N.S.W. — B.fl.VI,402 M.fr.VII,36.
P. longiscapa, Sweet, Fl. Austral. t. 39 (1828) ... ... — S.A. T. V. — — B.fl.VI,402 M.fr.VII,36.
P. occidentalis, R. Brown, prodr. 304 (1810) ... ... W.A. — — — — — B.fl.VI,403 M.fr.VII,31.
P. umbrosa, Endlicher in Lehm. pl. Preiss. II, 31 (1846) ... W.A. — — — — — B.fl.VI,404 M.fr.VII,32.
P. xanthina, F. v. M., fragm. I, 214 (1859) ... ... W.A. — — — — — B.fl.VI,404 M.fr.I,214.
P. limbata, Endlicher in Lehm. pl. Preiss. II, 29 (1846)... ... W.A. — — — — — B.fl.VI,404
P. juncea, Lindley, Bot. Regist. XXV, App. LVIII (1839) ... W.A. — — — — — B.fl.VI,404 M.fr.VII,33.
P. Maxwelli, F. v. M. in Benth. fl. Austr. VI, 405 (1873) ... W.A. — — — — — B.fl.VI,405 M.fr.VII,33.
P. pygmaea, Lindley, Bot. Regist. XXV, App. LVIII (1839) ... W.A. — — V. N.S.W. Q. — B.fl.VI,405 M.fr.VII,33.
P. sericea, R. Brown in Bot. Mag. t. 1041 (1807)... ... ... W.A. — — V. N.S.W. Q. — B.fl.VI,406 M.fr.VII,35.
P. lanata, R. Brown, prodr. 303 (1810) ... ... W.A. — — — — — B.fl.VI,406 M.fr.VII,35.
P. rudis, Endlicher in Lehm. pl. Preiss. II, 29 (1846) ... W.A. — — — — — B.fl.VI,407
P. macrantha, Bentham, Fl. Austr. VI, 407 (1873) ... ... W.A. — — — — — B.fl.VI,407
P. glabrata, R. Brown, prodr. 304 (1810) ... ... W.A. — — V. N.S.W. — B.fl.VI,407 M.fr.VII,35.
P. Drummondii, F. v. M. in Benth. Fl. Austr. VI, 407 (1873) ... W.A. — — — — — B.fl.VI,407 M.fr.XI,23.
P. inaequalis, Bentham, Fl. Austr. VI, 408 (1873) ... ... W.A. — — — — — B.fl.VI,408
P. graminea, Bentham, Fl. Austr. VI, 408 (1873) ... ... W.A. — — — — — B.fl.VI,408
P. babianoides, Bentham, Fl. Austr. VI, 408 (1873) ... ... W.A. — — — — — B.fl.VI,408

IRIS, Tournefort, inst. 358, t. 186, 188 (1700), from Hippocrates, Dioscorides and Plinius. (Moraea partly.)
I. Robinsoniana, F. v. M., fragm. VII, 153 (1870) ... ... — — — — N.S.W. — B.fl.VI,153 M.fr.VII,153.

SISYRINCHIUM, Linné, gen. plant. 273 (1737). (Libertia, Orthrosanthus, Renealmia, Nematostigma.)
S. paniculatum, R. Brown, prodr. 305 (1810) ... ... — T. V. N.S.W. — B.fl.VI,413 M.fr.VII,91.
S. pulchellum, R. Brown, prodr. 305 (1810) ... ... — T. V. N.S.W. — B.fl.VI,413 M.fr.VII,92.
S. cyaneum, Lindley, Bot. Regist. t. 1090 (1827)... ... W.A. S.A. — — — — B.fl.VI,410 M.fr.VII,92.
S. polystachyum, F. v. M.; Orthrosanthus, Benth. Fl. Austr. VI, 411 (1873) ... ... W.A. — — — — — B.fl.VI,411

CAMPYNEMA, Labillardière, Nov. Holl. pl. spec. I, 93, t. 121 (1804).
C. linearc, Labillardière, Nov. Holl. pl. spec. I, 93, t. 121 (1804) — — T. — — — — B.fl.VI,415 M.fr.VI,96.

BURMANNIACEAE.

Sprengel, syst. veg. I, 125 (1825).

BURMANNIA, Linné, syst. nat. 8 (1735); Linné, gen. plant. 100 (1737).
B. disticha, Linné, spec. plant. 287 (1753)... ... ... — — — — N.S.W. Q. N.A. B.fl.VI,397
B. juncea, Solander in R. Brown, prodr. 265 (1810) ... ... — — — — — Q. N.A. B.fl.VI,397

TACCACEAE.

Presl, reliq. Haenk. I, 149 (1828).

TACCA, R. et G. Forster, char. gen. 65 (1776).
T. pinnatifida, R. & G. Forster, char. gen. 65 (1776) ... — — — — — Q. N.A. B.fl.VI,458

DIOSCORIDEAE.

Du Mortier, Analyse des familles des plantes 57 (1829), from R. Brown (1810).

DIOSCOREA, Plumier, nov. plant. Americ. gen. 9, t. 26 (1703). (Dioscoridea.)
D. hastifolia, Nees in Lehm. pl. Preiss. II, 33 (1846) ... ... W.A. — — — — B.fl.VI,461 M.fr.VII,80;XI,20.
D. transversa, R. Brown, prodr. 295 (1810) ... ... — — — N.S.W. Q. N.A. B.fl.VI,460 M.fr.VII,80.
D. sativa, Linné, spec. plant. 1033 (1753) ... ... — — Q. N.A. B.fl.VI,461 M.fr.VII,81.

PETERMANNIA, F. v. M., fragm. II, 92 (1860).
P. cirrosa, F. v. M., fragm. II, 93 (1860) ... ... — — — — N.S.W. — — B.fl.VI,462 M.fr.II,176.

HYDROCHARIDEAE.

Lamarck and De Candolle, Flore Française, III, 265 (1805).

HYDROCHARIS, Linné, syst. nat. 8 (1735); Linné, gen. plant. 308 (1737).
H. morsus ranae, Linné, spec. plant. 1036 (1753)... — — — — — Q. — B.fl.VI,256 M.fr.VI,109.
E. ENHALUS, L. C. Richard in Mém. de l'Inst. II, 64 (1811).
E. Koenigii, Richard in Mém. de l'Inst. II, 64 (1811) ... ... — — — Q. — M.fr.VIII,210
HALOPHILA, Du Petit-Thouars, gen. nov. Madag. n. 6 (1806).
H. ovata, Gaudichaud, Bot. Voy. Freyc. 470, t. 40 (1826) ... — S.A. T. V. N.S.W. Q. — B.fl.VII,182 M.fr.VIII,219.
H. spinulosa, Ascherson in Neumayer, Anleit. zu wiss. Beob. 368 (1875)— — — Q. — B.fl.VII,183 M.fr.VIII,219.
OTTELIA, Persoon, synops. plant. I, 400 (1805). (Damasonium partly.)
O. alismoides, Persoon, synops. plant. I, 400 (1805) ... — — Q. N.A. B.fl.VI,257 M.fr.VI,109.
O. ovalifolia, L. C. Richard in Mém. de l'Inst. II, 78 (1811) ... W.A. S.A. — V. N.S.W. Q. N.A. B.fl.VI,257
BLYXA, Noronha in Du Petit-Thouars, gen. nov. Madag. n. 14 (1806).
B. Roxburghii, L. C. Richard, Mém. de l'Inst. II, 77, t. 5 (1811) — S.A. — — Q. N.A. B.fl.VI,258
VALLISNERIA, Micheli, nov. plant. gen. 12, t. 10 (1729).
V. spiralis, Linné, spec. plant. 1015 (1753)... ... ... — S.A. T. V. N.S.W. Q. N.A. B.fl.VI,259
HYDRILLA, L. C. Richard in Mém. de l'Inst. II, 61 (1811).
H. verticillata, Caspary in Bot. Zeitung, Dec. (1856) ... ... — S.A. — V. N.S.W. Q. N.A. B.fl.VI,259

AMARYLLIDEAE.

J. St. Hilaire, Expos. des famill. natur. des pl. I, 134, t. 21 (1805).

HAEMODORUM, Smith in Transact. Linn. Soc. 213 (1798).
H. distichophyllum, Hooker, icon. plant. t. 866 (1852) ... ... — — T. — — B.fl.VI,419 M.fr.VII,118;IX,51.
H. brevicaule, F. v. M., fragm. I, 64 (1858) ... ... — — — — N.A. B.fl.VI,419 M.fr.VII,118.
H. sparsiflorum, F. v. M., fragm. VII, 117 (1870) ... ... W.A. — — — — B.fl.VI,420 M.fr.VII,117.
H. spicatum, R. Brown, prodr. 300 (1810) ... ... W.A. — — — — B.fl.VI,420 M.fr.VII,118.
H. brevisepaleum, Bentham, Fl. Austr. VI, 420 (1873) ... W.A. — — — — B.fl.VI,420
H. paniculatum, Lindley, Bot. Regist. XXV, App. XLIV (1839) W.A. — — — — B.fl.VI,420
H. laxum, R. Brown, prodr. 300 (1810) ... ... — — — — B.fl.VI,421 M.fr.VII,118.
H. simplex, Lindley, Bot. Regist. XXV, App. XLIV (1839) ... W.A. — — — — B.fl.VI,421
H. simulans, F. v. M., fragm. VII, 117 (1870) ... ... — — — — B.fl.VI,421 M.fr.VII,117.
H. planifolium, R. Brown, prodr. 300 (1810) ... ... — — N.S.W. Q. — B.fl.VI,422 M.fr.VII,118.
H. teretifolium, R. Brown, prodr. 300 (1810) ... ... — — N.S.W. — — B.fl.VI,422 M.fr.VII,118.
H. coccineum, R. Brown, prodr. 300 (1810) ... ... — — — Q. N.A. B.fl.VI,422
H. subvirens, F. v. M., fragm. I, 63 (1858) ... ... — — — — N.A. B.fl.VI,423 M.fr.I,63.
H. ensifolium, F. v. M., fragm. I, 64 (1858) ... ... — — — Q. N.A. B.fl.VI,423 M.fr.VII,118.
H. parviflorum, Bentham, Fl. Austr. VI, 423 (1873) ... ... — — — — N.A. B.fl.VI,423
H. leptostachyum, Bentham, Fl. Austr. VI, 423 (1873) ... — — — — N.A. B.fl.VI,423
H. tenuifolium, Cunningham in Benth. Fl. Austr. VI, 425 (1873) — — — — Q. — B.fl.VI,423

PHLEBOCARYA, R. Brown, prodr. 301 (1810).
P. ciliata, R. Brown, prodr. 301 (1810) ... ... W.A. — — — — B.fl.VI,424 M.fr.VIII,23.
P. pilosissima, F. v. M. in Benth. fl. Austr. VI, 425 (1873) W.A. — — — — B.fl.VI,425 M.fr.VIII,23.
P. filifolia, F. v. M. in Benth. fl. Austr. VI, 425 (1873)... W.A. — — — — B.fl.VI,425 M.fr.VIII,23.

TRIBONANTHES, Endlicher, nov. stirp. dec. I, 27 (1839).
T. brachypetala, Lindley Bot. Regist. XXV, App. XLIV (1839) W.A. _ _ _ — — — — B.fl.VI,426 M.fr.VIII,23.
T. uniflora, Lindley, Bot. Regist. XXV, App. XLIV (1839) ... W.A. _ _ _ — — — — B.fl.VI,427 M.fr.VIII,23.
T. australis, Endlicher, nov. stirp. dec. I, 27 (1839) ... ... W.A. _ _ _ — — — — B.fl.VI,427 M.fr.VIII,23.
T. variabilis, Lindley, Bot. Regist. XXV, App. XLIV (1839) ... W.A. _ _ _ — — — — B.fl.VI,427
T. longipetala, Lindley, Bot. Regist. XXV, App. XLIV (1839) W.A. _ _ _ — — — — B.fl.VI,427 M.fr.VIII,23.

CONOSTYLIS, R. Brown, prodr. 300 (1810). (Blancoa, Androstemma.)
C. breviscapa, R. Brown, prodr. 301 (1810) ... ... W.A. _ _ — — — — — B.fl.VI,430
C. vaginata, Endlicher in Lehm. pl. Preiss. II, 23 (1846) ... W.A. _ _ _ — — — — B.fl.VI,431 M.fr.VIII,19.
C. petrophiloides, F. v. M. in Benth. Fl. Austr. VI, 431 (1873) W.A. _ _ _ — — — — B.fl.VI,431
C. setosa, Lindley, Bot. Regist. XXV, App. XLIV, t. 6 (1839) W.A. _ _ _ — — — — B.fl.VI,431 M.fr.VIII,19.
C. aurea, Lindley. Bot. Regist. XXV, App. XLIV (1839) ... W.A. _ _ _ — — — — B.fl.VI,432 M.fr.VIII,19.
C. melanopogon, Endlicher in Lehm. pl. Preiss. II, 18 (1846) ... W.A. _ _ _ — — — — B.fl.VI,432 M.fr.VIII,20.
C. setigera, R. Brown, prodr. 300 (1810) ... ... W.A. _ _ _ — — — — B.fl.VI,433 M.fr.VIII,20.
C. Psyllium, Endlicher in Lehm. pl. Preiss. II, 21 (1846) .. W.A. _ _ _ — — — — B.fl.VI,433
C. villosa, Bentham, Fl. Austr. VI, 433 (1873) ... ... W.A. _ _ _ — — — — B.fl.VI,433
C. Drummondii, Bentham, Fl. Austr. VI, 433 (1873) ... ... W.A. _ _ _ — — — — B.fl.VI,433
C. involucrata, Endlicher in Lehm. pl. Preiss. II, 23 (1846) W.A. _ _ _ — — — -- B.fl.VI,434
C. gladiata, Bentham, Fl. Austr. VI, 434 (1873) ... ... W.A. _ _ - - — — -- B.fl.VI,434
C. seorsiflora, F. v. M., fragm. I, 158 (1859) ... . W.A. _ _ _ — — — — B.fl.VI,435 M.fr.VIII,19.
C. teretiuscula, F. v. M., fragm. VIII, 18 (1872) ... ... W.A. _ _ _ — — — — B.fl.VI,435 M.fr.VIII,18.
C. stylidioides, F. v. M., fragm. VIII, 17 (1872) ... ... W.A. _ _ _ — — — — B.fl.VI,435 M.fr.VIII,17.
C. prolifera, Bentham, Fl. Austr. VI, 436 (1839) ... ... W.A. _ _ _ — — — — B.fl.VI,436
C. racemosa, Bentham, Fl. Austr. VI, 436 (1839) .. ... W.A. _ _ _ — — — -- B.fl.VI,436
C. candicans, Endlicher, nov. stirp. dec. 20 (1839) ... W.A. _ _ _ — — — — B.fl.VI,436 M.fr.VI,255.
C. dealbata, Lindley, Bot. Regist. XXV, App. XLV (1839) ... W.A. _ _ _ — — — — B.fl.VI,436
C. Preissii, Endlicher in Lehm. pl. Preiss. II, 18 (1846) ... W.A. _ _ _ — — — — B.fl.VI,437 M.fr.VIII,20.
C. bracteata, Endlicher in Lehm. pl. Preiss. II, 16 (1846) W.A. _ _ _ — — — — B.fl.VI,437 M.fr.VIII,20.
C. filifolia, F. v. M., fragm. VIII, 16 (1872) ... ... W.A. _ _ _ — — — — B.fl.VI,438 M.fr.VIII,18.
C. apiculigera, F. v. M. in Benth. Fl. Austr. VI, 438 (1873) ... W.A. _ _ _ — — — — B.fl.VI,438
C. bromelioides, Endlicher in Lehm. pl. Preiss. II, 18 (1846) ... W.A. _ _ _ — — — — B.fl.VI,438 M.fr.VIII,19.
C. aculeata, R. Brown, prodr. 300 (1810) ... ... W.A. _ _ _ — — — — B.fl.VI,439
C. laxiflora, Bentham, Fl. Austr. VI, 439 (1873) ... ... W.A. _ _ _ — — — — B.fl.VI,439
C. cymosa, F. v. M. in Benth. Fl. Austr. VI, 439 (1873)... ... W.A. _ _ _ — — — — B.fl.VI,439 M.fr.VIII,19.
C. serrulata, R. Brown, prodr. 300 (1810) ... ... W.A. _ _ _ — — — — B.fl.VI,440 M.fr.VIII,20.
C. caricina, Lindley, Bot. Regist. XXV, App. XLV (1839) ... W.A. _ _ _ — — — — B.fl.VI,440 M.fr.VIII,19.
C. Androstemma, F. v. M., fragm. VIII, 19 (1873) ... ... W.A. _ _ _ — — — — M.fr.VIII,19.
C. Bealiana, F. v. M., fragm. IX, 50 (1875) ... ... W.A. _ _ _ — — — — M.fr.IX,50.
C. canescens, F. v. M., fragm. VIII, 19 (1873) ... ... W.A. _ _ _ — — — B.fl.VI,441 M.fr.VIII,19.

AGONIZANTHOS, Labillardière, Relat. du voy. à la rech. de La Perouse I, 411 (1798). (Anigozanthus, Macropidia, Macropodia, Schwaegrichenia.)
A. rufa, Labillardière, Voy. I, 411, t. 22 (1798) ... W.A. _ _ _ — — — — B.fl.VI,442 M.fr.VIII,21.
A. pulcherrima, Hooker, Bot. Mag. t. 4180 (1845) ... W.A. _ _ _ — — — — B.fl.VI,443 M.fr.VIII,21.
A. flavida, Redouté et De Candolle, les Liliacées t. 176 (1807) ... W.A. _ _ _ — — — — B.fl.VI,444 M.fr.VIII,21.
A. Preissii, Endlicher in Lehm. pl. Preiss. II, 26 (1846)... ... W.A. _ _ _ — — — — B.fl.VI,444 M.fr.VIII,23.
A. humilis, Lindley, Bot. Regist. XXV, App. XLVI (1839) ... W.A. _ _ _ — — -- B.fl.VI,444 M.fr.VIII,21.
A. viridis, Endlicher in Lehm. pl. Preiss. II, 25 (1846) ... W.A. _ _ _ — — -- -- B.fl.VI,445 M.fr.VIII,22.
A. Manglesii, D. Don in Sweet, Brit. Fl. Gard. sec. ser. t. 265 (1836) W.A. _ _ _ — — — — B.fl.VI,445 M.fr.VIII,22.
A. bicolor, Endlicher in Lehm. pl. Preiss. II, 26 (1846) ... ... W.A. _ _ _ — — — — B.fl.VI,447 M.fr.VIII,22.
A. fuliginosus, Hooker, Bot. Mag. t. 4291 (1847) ... ... W.A. _ _ _ — — — — B.fl.VI,447 M.fr.VIII,20.

CURCULIGO, Gaertner, de fruct. I, 63, t. 16 (1788).
C. recurvata, Aiton, hort. Kew. sec. ed. II, 253 (1811) ... ... _ _ _ — Q. — B.fl.VI,448 M.fr.VIII,27.
C. ensifolia, R. Brown, prodr. 290 (1810) ... ... — _ _ N.S.W. Q. N.A. B.fl.VI,448 M.fr.VI,96.

HYPOXIS, Linné, syst. ed decim. 986 (1759).
H. hygrometrica, Labillardière, Nov. Holl. pl. spec. I, 82, t. 108 (1804) — S.A. T. V. N.S.W. Q. — B.fl.VI,449 M.fr.VI,96.
H. glabella, R. Brown, prodr. 280 (1810) ... ... W.A. S.A. T. V. N.S.W. — — B.fl.VI,450 M.fr.VI,96.
H. occidentalis, Bentham, Fl. Austr. VI, 451 (1873) ... ... W.A. _ _ _ — — — B.fl.VI,451
H. marginata, R. Brown, prodr. 289 (1810) ... ... _ _ _ — N.A. B.fl.VI,451

DORYANTHES, Correa in Transact. Linn. Soc. VI, 211 (1802).
D. excelsa, Correa in Transact. Linn. Soc. VI, 211, t. 23 (1802) ... _ _ _ — N.S.W. Q. — B.fl.VI,452 M.fr.IX,51.
D. Palmeri, W. Hill in Benth. Fl. Austr. VI, 453 (1873) ... ... _ _ _ — N.S.W. Q. — B.fl.VI,452 M.fr.IX,51.

CRINUM, Linné, gen. plant. 97 (1737).
C. venosum, R. Brown, prodr. 297 (1810) ... ... _ _ _ — Q. N.A. B.fl.VI,453 M.fr.III,22.
C. Asiaticum, Linné, spec. plant. 292 (1753) ... ... _ _ _ — Q. N.A. B.fl.VI,454 M.fr.III,22.
C. brachyandrum, Herbert in Bot. Mag. t. 2121 (1820) ... ... _ _ _ — Q. — — M.fr.XI,139.
C. angustifolium, R. Brown, prodr. 297 (1810) ... ... _ _ _ — Q. — — M.fr.XI,139.
C. flaccidum, Herbert in Bot. Mag. t. 2133 (1820) ... ... S.A. _ — V. N.S.W. Q. — B.fl.VI,454 M.fr.III,23.
C. uniflorum, F. v. M., fragm. III, 23 (1862) ... ... _ _ _ — Q. N.A. B.fl.VI,454 M.fr.VI,251.
C. pedunculatum, R. Brown, prodr. 297 (1810) ... ... _ _ _ — N.S.W. Q. — B.fl.VI,455 M.fr.III,23.

EURYCLES, Salisbury in Transact. of the Hort. Soc. I, 337 (1812).
E. silvestris, Salisbury in Transact. Hort. Soc. I, 337 (1812) ... _ _ _ — Q. — B.fl.VI,456
E. Cunninghamii, Aiton in Loud. Hort. Brit. Suppl. 588 (1832) ... _ _ _ — N.S.W. Q. — B.fl.VI,456

CALOSTEMMA, R. Brown, prodr. fl. Nov. Holl. 297 (1810).
C. album, R. Brown, prodr. 296 (1810) ... ... _ _ _ — N.A. B.fl.VI,458
C. purpureum, R. Brown, prodr. 298 (1810) ... ... S.A. _ — V. N.S.W. — -- B.fl.VI,457
C. luteum, Sims, Bot. Mag. t. 2101 (1819)... ... ... S.A. - - V. N.S.W. Q. — B.fl.VI,457

117

CALYCEAE HYPOGYNAE.
F. v. M. in Woolls's plants of the neighb. of Sydney 44 (1860).

ROXBURGHIACEAE.
Lindley in Wallich, pl. Asiat. rar. III, 50 (1832).
ROXBURGHIA, Jones in Roxburgh, pl. Corom. I, 29, t. 32 (1795). (Stemona 1790.)
R. javanica, Kunth, enum. V, 288 (1850)... ... ... — — — — — Q. N.A. B.fl.VII,1 M.fr.VII,81

LILIACEAE.
Haller, enum. stirp. Helv. I, 270 (1742).
PHORMIUM, R. & G. Forster, charact. gener. 47, t. 24 (1776). (Chlamydia.)
P. tenax, R. & G. Forster, char. gen. 47, t. 24 (1776) ... ... — — N.S.W. — — —
SMILAX, Tournefort, inst. 654, t. 421 (1700), from Theophrastos and Dioscorides.
S. glycyphylla, Smith in White, Voy. to N.S.Wales 230 (1790) — — — — N.S.W. Q. — B.fl.VII,7 M.fr.VII,77.
S. australis, R. Brown, prodr. 293 (1810) ... ... — — — V. N.S.W. Q. N.A. B.fl.VII,7 M.fr.VII,78.
RHIPOGONUM, R. & G. Forster, char. gen. 49, t. 25 (1776). (Ripogonum.)
R. album, R. Brown, prodr. 293 (1810) ... ... — — — V. N.S.W. Q. — B.fl.VII,9 M.fr.VII,79.
R. discolor, F. v. M., fragm. VII, 78 (1870) ... ... — — — — N.S.W. Q. — B.fl.VII,9 M.fr.VII,79.
R. Fawcettianum, F. v. M. in Benth. Fl. Austr. VII, 9 (1878)... — — — — N.S.W. Q. — B.fl.VII,9 M.fr.VII,80.
R. Elseyanum, F. v. M., fragm. I, 44 (1858) ... ... — — — — N.S.W. Q. — B.fl.VII,10 M.fr.VII,80.
FLAGELLARIA, Linné, amoen. acad. 306 (1747).
F. indica, Linné, spec. plant. 333 (1753) ... ... — — — — — N.S.W. Q. N.A. B.fl.VII,10
DRYMOPHILA, R. Brown, prodr. fl. Nov. Holl. 292 (1810).
D. cyanocarpa, R. Brown, prodr. 292 (1810) ... ... — T. V. N.S.W. — — R.fl.VII,12 M.fr.VII,73;IX,101.
D. Moorei, Baker in Journ. Linn. Soc. XIV, 571 (1875)... ... — — N.S.W. — — B.fl.VII,13 M.fr.IX,100.
DIANELLA, Lamarck, Encyclop. méthod. II, t. 250 (1786).
D. tasmanica, J. Hooker, fl. Tasman. II, 57, t. 133 (1860) ... — — T. V. N.S.W. — — B.fl.VII,14 M.fr.VI,121.
D. longifolia, R. Brown, prodr. 280 (1810)... ... ... — S.A. T. V. N.S.W. Q. — B.fl.VII,14 M.fr.VI,122.
D. revoluta, R. Brown, prodr. 280 (1810) ... ... W.A. S.A. T. V. N.S.W. Q. — B.fl.VII,15 M.fr.VI,121;XI,20.
D. caerulea, Sims, Bot. Magaz. t. 505 (1790) ... ... — — — N.S.W. Q. — B.fl.VII,16 M.fr.VI,123.
D. ensifolia, De Candolle & Redouté, Liliac. t. 1 (1802)... — — — — — Q. N.A. B.fl.VII,16 M.fr.VI,123.
ASPARAGUS, Tournefort, inst. 300, t. 154 (1700), from Theophrastos, Dioscorides and Plinius. (Asparagopsis.)
A. racemosus, Willdenow, spec. plant. II, 152 (1799) ... — — — — — Q. N.A. B.fl.VII,17 M.fr.VII,73.
EUSTREPHUS, R. Brown, prodr. 281 (1810).
E. Brownii, F. v. M., fragm. VII, 73 (1870) ... ... — — — — V. N.S.W. Q. — B.fl.VII,18 M.fr.VII,73.
GEITONOPLESIUM, Cunningham in Bot. Magaz. t. 3131 (1832). (Luzuriago partly.)
G. cymosum, Cunningham in Bot. Mag. t. 3131 (1832) ... — — — V. N.S.W. Q. — B.fl.VII,19 M.fr.VII,74.
DRACAENA, Vandelli, dissertatio de Dracaena t. 1 (1762). [190.
D. angustifolia, Roxburgh, fl. Indic. II, 155 (1832) ... — — — — — Q. N.A. B.fl.VII,20 M.fr.V,104;VI,120;IX,
CORDYLINE, Commerçon in R. Hedwig, gen. plant. 243 (1806). (Charlwoodia.)
C. terminalis, Kunth in act. acad. Berol. 30 (1820) ... — N.S.W. Q. — B.fl.VII,21 M.fr.V,196.
C. Murchisoniae, F. v. M., fragm. V, 195 (1866) ... ... — — — Q. — B.fl.VII,22 M.fr.V,195.
C. stricta, Endlicher in Ann. Wien. Mus. I, 162 (1838) ... — — N.S.W. Q. — B.fl.VII,22 M.fr.V,195;VI,121.
C. Baueri, J. Hooker in Gardeners' Chronicle 792 (1860) — — N.S.W. — — —
BLANDFORDIA, Smith, Exot. Bot. I, 5, t. 4 (1804).
B. marginata, Herbert in Bot. Regist. Misc. 84 (1842) ... — T. — — B.fl.VII,23 M.fr.VII,71.
B. grandiflora, R. Brown, prodr. 296 (1810) ... ... — N.S.W. — — B.fl.VII,23
B. nobilis, Smith, Exot. Bot. I, 5, t. 4 (1804) ... ... — N.S.W. — — B.fl.VII,24 M.fr.VII,70.
B. flammea, Hooker, Bot. Mag. t. 4810 (1854) ... ... — N.S.W. — — B.fl.VII,24
HEWARDIA, Hooker, Icon. plant. 858 (1852).
H. tasmanica, Hooker, icon. plant. t. 858 (1852)... ... — T. — — — — B.fl.VII,25 M.fr.VII,72.
ASTELIA, Banks & Solander in R. Brown, prodr. 291 (1810). (Hamelinia.)
A. alpina, R. Brown, prodr. 291 (1810) ... ... — T. V. N.S.W. — — B.fl.VII,11
MILLIGANIA, J. Hooker in Kew misc. V, 296, t. IX (1853).
M. densiflora, J. Hooker, Fl. Tasman. II, 62 (1860) ... — — T. — — B.fl.VII,26
M. longifolia, J. Hooker in Kew misc. V, 296, t. IX (1853) ... — — T. — — B.fl.VII,26
M. Johnstoni, F. v. M. in Benth. Fl. Austr. VII, 26 (1878) ... — — T. — — B.fl.VII,26
M. stylosa, F. v. M. in papers of the R. S. of Tasm. 11 (1876) ... — — T. — — B.fl.VII,27
WURMBEA, Thunberg, nov. gen. I, 18 (1781). (Anguillaria, Melanthium partly, Plees partly.)
W. tubulosa, Bentham, Fl. Austr. VII, 28 (1878)... ... — W.A. — — B.fl.VII,28 M.fr.VII,75.
W. dioica, F. v. M., fragm. X, 119 (1877) ... ... — — S.A. T. V. N.S.W. Q. — B.fl.VII,30 M.fr.VII,74.
IPHIGENIA, Kunth, enum. plant. omn. hucusq. cognit. IV, 212 (1843).
I. indica, Kunth, enum. plant. IV, 213 (1843) ... ... — — — — — Q. N.A. B.fl.VII,31 M.fr.VII,74;IX,193.
SCHELHAMMERA, R. Brown, prodr. 273 (1810).
S. undulata, R. Brown, prodr. 274 (1810) ... ... — — — V. N.S.W. Q. — B.fl.VII,31 M.fr.VII,71.
S. multiflora, R. Brown, prodr. 274 (1810)... ... ... — — — — N.S.W. Q. — B.fl.VII,32 M.fr.VII,72.
KREYSSIGIA, Reichenbach, iconogr. bot. exotic. III, 11, t. 229 (1829). (Kreysigia, Tripladenia.)
K. Cunninghami, F. v. M., fragm. VII, 71 (1870) ... ... — — — N.S.W. Q. — B.fl.VII,32 M.fr.VII,71.
BURCHARDIA, R. Brown, prodr. 272 (1810).
B. umbellata, R. Brown, prodr. 272 (1810)... ... ... W.A. S.A. T. V. N.S.W. Q. — B.fl.VII,33 M.fr.VII,77.

118

BULBINE, Linné, hort. Cliffort. 122 (1737).   (Anthericum partly, Chrysobactron.)
B. bulbosa, Haworth, revis. pl. succul. 33 (1821)... ... ... — S.A. T. V. N.S.W. Q. — B.fl.VII,34 M.fr.VII,70;XI,27.
B. semibarbata, Haworth, rev. pl. succ. 33 (1821)... ... ... W.A. S.A. T. V. N.S.W. Q. — B.fl.VII,35 M.fr.VII,70.
AGROSTOCRINUM, F. v. M., fragm. phytogr. Austr. II, 95 (1860).
A. stypandroides, F. v. M., fragm. II, 95 (1860) ... ... W.A. — — — — — — — B.fl.VII,36 M.fr.VII,65.
THYSANOTUS, R. Brown, prodr. 282 (1810).   (Chlamysspermum.)
T. multiflorus, R. Brown, prodr. 285 (1810) ... ... W.A. — — — — — — B.fl.VII,38 M.fr.VII,69.
T. triandrus, R. Brown, prodr. 284 (1810)... ... ... W.A. — — — — — — B.fl.VII,39 M.fr.VII,69.
T. glaucus, Endlicher in Lehm. pl. Preiss. II, 33 (1846)... ... W.A. — — — — — — B.fl.VII,39
T. Drummondii, Baker in Journ. Linn. Soc. XV, 341 (1876) ... W.A. — — — — — — B.fl.VII,39
T. pauciflorus, R. Brown, prodr. 285 (1810) ... ... W.A. — — — — — — B.fl.VII,39
T. asper, Lindley, Bot. Regist. XXV, App. LVIII (1839) ... W.A. — — — — — — B.fl.VII,39
T. chrysantherus, F. v. M. in Bentham, flor. Ilongk. 372 (1861) ... — — — — — N.A. B.fl.VII,40 M.fr.V,202.
T. exiliflorus, F. v. M. in transact. R. S. of S. Austr. IV, 112 (1881) — S.A. — — — — — — M.fr.XI,139.
T. isantherus, R. Brown, prodr. 283 (1810) ... ... — — — — — — — B.fl.VII,40
T. tenellus, Endlicher in Lehm. pl. Preiss. II, 37 (1846)... ... W.A. — — — — — — B.fl.VII,40
T. scaber, Endlicher in Lehm. pl. Preiss. II, 37 (1846) ... ... W.A. — — — — — — B.fl.VII,41
T. tuberosus, R. Brown, prodr. 282 (1810) ... ... — S.A. — V. N.S.W. Q. N.A. B.fl.VII,41 M.fr.VII,69.
T. exasperatus, F. v. M., fragm. I, 21 (1858) ... ... — S.A. — — — — — B.fl.VII,42 M.fr.I,21.
T. thyrsoideus, Baker in Journ. Linn. Soc. XV, 336 (1876) ... W.A. — — — — — — B.fl.VII,42
T. Baueri, R. Brown, prodr. 283 (1810) ... ... — S.A. — V. N.S.W. — — B.fl.VII,42 M.fr.I,22;VII,69.
T. Patersoni, R. Brown, prodr. 284 (1810)... ... ... W.A. S.A. T. V. N.S.W. — — B.fl.VII,43 M.fr.VII,69;XI,20.
T. junceus, R. Brown, prodr. 283 (1810) ... ... — — — N.S.W. — — B.fl.VII,43 M.fr.VII,69.
T. dichotomus, R. Brown, prodr. 284 (1810) ... ... W.A. S.A. — — — — — B.fl.VII,44 M.fr.VII,69.
T. arbuscula, Baker in Journ. Linn. Soc. XV, 339 (1876) ... W.A. — — — — — — B.fl.VII,45
T. anceps, Lindley, Bot. Regist. XXV, App. LVIII (1839) ... W.A. — — — — — — B.fl.VII,45
HODGSONIOLA, F. v. M., fragm. II, 176 (1861).   (Hodgsonia.)
H. juncifolmis, F. v. M., fragm. II, 176 (1861) ... ... W.A. — — — — — B.fl.VII,46 M.fr.II,95,176.
CAESIA, R. Brown, prodr. fl. Nov. Holl. 277 (1810).
C. vittata, R. Brown, prodr. 277 (1810) ... ... — S.A. T. V. N.S.W. Q. — B.fl.VII,47 M.fr.I,63;VII,68.
C. chlorantha, F. v. M., fragm. I, 63 (1858) ... ... — — — — Q. N.A. — M.fr.I,63.
C. parviflora, R. Brown, prodr. 277 (1810)... ... ... — S.A. T. V. N.S.W. Q. — B.fl.VII,47 M.fr.VII,67.
C. rigidifolia, F. v. M., fragm. X, 48 (1876) ... ... W.A. — — — — — — B.fl.VII,47 M.fr.X,48.
C. setifera, Baker in Journ. Linn. Soc. XV, 359 (1876) ... ... — — — — — — N.A. B.fl.VII,47
CHAMAESCILLA, F. v. M., fragm. VII, 68 (1870).   (Caesia partly.)
C. corymbosa, F. v. M., fragm. VII, 68 (1870) ... W.A. S.A. T. V. N.S.W. — — B.fl.VII,48 M.fr.VII,68.
C. spiralis, F. v. M., fragm. VII, 66 (1870) ... ... W.A. — — — — — — B.fl.VII,48 M.fr.VII,68.
CORYNOTHECA, F. v. M., fragm. VII, 68 (1870).   (Caesia partly.)
C. lateriflora, F. v. M., fragm. VII, 68 (1870) ... ... W.A. S.A. — V. N.S.W. Q. N.A. B.fl.VII,49 M.fr.VII,68.
C. dichotoma, F. v. M., fragm. VII, 68 (1870) ... ... W.A. — — — — — — B.fl.VII,50 M.fr.I,215;XI,20.
C. acanthoclada, F. v. M., fragm. VII, 68 (1870)... ... ... W.A. — — — — — — B.fl.VII,50 M.fr.I,215;XI,20.
TRICHORYNE, R. Brown, prodr. fl. Nov. Holl. 278 (1810).
T. platyptera, G. Reichenbach, Beitr. zur syst. Pflanzenk. 72 (1871) — — — — Q. — B.fl.VII,51
T. anceps, R. Brown, prodr. 278 (1810) ... ... — — — — Q. — B.fl.VII,51 M.fr.VII,07.
T. muricata, Baker in Journ. Linn. Soc. XV, 363 (1876) ... ... — — — — — Q. — B.fl.VII,52
T. elatior, R. Brown, prodr. 278 (1810) ... ... W.A. S.A. T. V. N.S.W. — — B.fl.VII,52 M.fr.VII,67;XI,20.
T. simplex, R. Brown, prodr. 278 (1810) ... ... — — — N.S.W. Q. — B.fl.VII,52 M.fr.VII,67.
T. humilis, Endlicher in Lehm. pl. Preiss. II, 36 (1846)... ... W.A. — — — — — — B.fl.VII,53 M.fr.VII,67.
STYPANDRA, R. Brown, prodr. flor. Nov. Holl. 278 (1810).
S. glauca, R. Brown, prodr. 278 (1810) ... ... W.A. — — V. N.S.W. Q. — B.fl.VII,53 M.fr.VII,64.
S. caespitosa, R. Brown, prodr. 279 (1810) ... ... — — T. V. N.S.W. Q. — B.fl.VII,54 M.fr.VII,65.
ARTHROPODIUM, R. Brown, prodr. 276 (1810).   (Dichopogon.)
A. paniculatum, R. Brown, prodr. 276 (1810) ... ... — S.A. T. V. N.S.W. — — B.fl.VII,56 M.fr.VII,66.
A. minus, R. Brown, prodr. 276 (1810) ... ... — S.A. T. V. N.S.W. — — B.fl.VII,56 M.fr.VII,67;XI,20.
A. capillipes, Endlicher in Lehm. pl. Preiss. II, 34 (1846) ... W.A. — — — — — — B.fl.VII,57
A. Preissii, Endlicher in Lehm. pl. Preiss. II, 35 (1846) ... W.A. — — — — — — B.fl.VII,57 M.fr.VII,66.
A. dianellaceum, F. v. M., fragm. X, 58 (1876) ... ... — — — — — — Q B.fl.VII,57 M.fr.X,65.
A. strictum, R. Brown, prodr. 276 (1810) ... ... — S.A. T. V. N.S.W. — — B.fl.VII,58 M.fr.VII,66.
A. laxum, Sieber in Roemer & Schultes, syst. VII, 441 (1830)... — S.A. — V. N.S.W. — — B.fl.VII,59
CHLOROPHYTUM, Ker in Bot. Mag. t. 1071 (1808).   (Chlorophyton, Phalangium partly.)
C. laxum, R. Brown, prodr. 277 (1810) ... ... — — — — — N.A. B.fl.VII,60 M.fr.I,63;VII,71.
C. alpinum, Baker in Journ. Linn. Soc. XV, 329 (1876)... ... — T. — — — — — B.fl.VII,60
HERPOLIRION, J. Hooker, Fl. of New Zeal. I, 258 (1853).
H. Novae Zelandiae, J. Hooker, Fl. of New Zeal. I, 258 (1853) ... — T. V. N.S.W. — — B.fl.VII,61 M.fr.VII,71.
SOWERBAEA, Smith in Transact. Linn. Soc. V, 159 (1800).   (Sowerbya.)
S. juncea, Smith in Transact. Linn. Soc. V, 160, t. 6 (1800) ... — — — V. N.S.W. Q. — B.fl.VII,61 M.fr.VI,180.
S. laxiflora, Lindley, Botan. Regist. t. 10 (1841) ... ... W.A. — — — — — — B.fl.VII,62 M.fr.VI,180;VII,87.
S. alliacea, F. v. M., fragm. V, 180 (1868) ... ... — — — — — N.A. B.fl.VII,62 M.fr.VII,87.
ALLANIA, Endlicher, gen. plant. 151 (1837).   (Alania.)
A. Endlicheri, Kunth, enum. IV, 644 (1843) ... ... — — — — N.S.W. — B.fl.VII,63 M.fr.VI,180.
BARTLINGIA, F. v. M., fragm. VII, 88 (1870).   (Laxmannia.)
B. grandiflora, F. v. M., fragm. VII, 88 (1870) ... ... — — — — — — B.fl.VII,64 M.fr.I,159;VII,88.
 App. LVI (1839)
B. squarrosa, F.v.M.; Laxm.,Lindl.Bot.Reg.XXV,App.LVI(1839) W.A. — — — — — — B.fl.VII,64
B. minor, F. v. M.; Laxmannia, R. Brown, prodr. 286 (1810) ... W.A. — — — — — — B.fl.VII,65 M.fr.VII,88,89.

```
B. gracilis, F v. M., fragm. VII, 88 (1870)      ...   ...   ...   —   —   — V.  N.S.W.  Q.   —   B.fl.VII,65   M.fr.I,159.
B. ramosa, F. v. M., fragm. VII, 88 (1870)      ...   ...   ...   W.A.  —   —   —   —   —   —   B.fl.VII,66   M.fr.VI,255.
B. sessiliflora, F. v. M. in pap. R. S. of Tasm. 116 (1877)   ...   W.A.  S.A.  T.  V.  N.S.W.  —   —   B.fl.VII,66
B. brachyphylla; Laxm., F. v. M. in Benth. Fl. VII, 67 (1878)   W.A.   —   —   —   —   —   —   B.fl.VII,66   M.fr.I,158.
B. sessilis, F.v.M.; Laxm.,Lindl.Bot.Reg. XXV,App.LVI (1839)   W.A.   —   —   —   —   —   —   B.fl.VII,67
    STAWELLIA, F. v. M., fragm. phytogr. Austr. VII, 85 (1870).
S. dimorphantha, F. v. M., fragm. VII, 85 (1870)   ...   ...   W.A.   —   —   —   —   —   —   B.fl.VII,67   M.fr.VII,85.
    JOHNSONIA, R. Brown, prodr. fl. Nov. Holl. 287 (1810).
J. lupulina, R. Brown, prodr. 287 (1810)   ...   ...   ...   W.A.   —   —   —   —   —   —   B.fl.VII,68   M.fr.VII,80.
J. pubescens, Lindley, Bot. Regist. XXV, App. LVII (1839)   ...   W.A.   —   —   —   —   —   —   B.fl.VII,68   M.fr.VI,255.
    ARNOCRINUM, Endlicher & Lehmann in pl. Preiss. II, 41 (1846).
A. Drummondii, Endlicher in Lehm. pl. Preiss. II, 41 (1846)   ...   W.A.   —   —   —   —   —   —   B.fl.VII,69   M.fr.VII,73.
A. Preissii, Lehmann, pl. Preiss. II, 42 (1846)   ...   ...   W.A.   —   —   —   —   —   —   B.fl.VII,70
    BORYA, Labillardiere, Nov. Holl. pl. specim. 81, t. 107 (1804).
B. nitida, Labillardiere, Nov. Holl. pl. spec. 81, t. 107 (1804)...   W.A.   —   —   —   —   —   —   B.fl.VII,71   M.fr.VII,87.
B. septentrionalis, F. v. M., fragm. 41 (1850)   ...   ...   —   —   —   —   —   Q.   —   B.fl.VII,71   M.fr.V,41.
    CALECTASIA, R. Brown, prodr. fl. Nov. Holl. 263 (1810).
C. cyanea, R. Brown, prodr. 264 (1810)   ...   ...   ...   W.A.  S.A.  — V.   —   —   —   B.fl.VII,121
    BAXTERIA, R. Brown in Hook. Lond. Journ. of Bot. II, 492, t. 13-15 (1843).
B. australis, R. Brown in Hook. Lond. Journ. II, 492 (1843)   ...   W.A.   —   —   —   —   —   —   B.fl.VII,120
    ACANTHOCARPUS, Lehmann, pl. Preiss. II, 274 (1847).  (Chamaexeros).
A. Preissii, Lehmann, pl. Preiss. II, 274 (1847)   ...   ...   W.A.   —   —   —   —   —   —   B.fl.VII,111
A. Serra, F. v. M.; Xerotes, Endlicher in pl. Preiss. II, 49 (1847)   W.A.   —   —   —   —   —   —   B.fl.VII,111   M.fr.VIII,211.
A. fimbriatus ; Xerotes, F. v. M., fragm. VII, 211 (1874)   ...   W.A.   —   —   —   —   —   —   B.fl.VII,111   M.fr.VIII,211.
    XEROTES, R. Brown, prodr. 257 (1810).  (Lomandra, 1804.)
X. Banksii, R. Brown, prodr. 263 (1810)   ...   ...   ...   —   —   —   —   Q.   —   B.fl.VII,96   M.fr.VIII,213.
X. dura, F. v. M. in Transact. Vict. Inst. 42 (1854)   ...   ...   — S A.   —   —   —   —   B.fl.VII,97   M.fr.VIII,207.
X. longifolia, R. Brown, prodr. 262 (1810)...   ...   ...   S.A.  T.  V.  N.S.W.  Q.   —   B.fl.VII,98   M.fr.VIII,210.
X. rigida, R. Brown, prodr. 262 (1801)   ...   ...   ...   W.A.   —   —   —   —   —   —   B.fl.VII,99   M.fr.VIII,209.
X. Sonderi, F. v. M., fragm. VIII, 200 (1874)   ...   ...   W.A.   —   —   —   —   —   —   B.fl.VII,99   M.fr.
X. odora, Endlicher in Lehm. pl. Preiss. II, 50 (1846)   ...   W.A.   —   —   —   —   —   —   B.fl.VII,99
X. Brownii, F. v. M., fragm. VIII, 206 (1874)   ...   ...   — S.A.   — V.  N.S.W.  Q. N.A.  B.fl.VII,100   M.fr.VIII,206.
X. Ordii, F. v. M., fragm. XI, 23 (1878)   ...   ...   W.A.   —   —   —   —   —   —   M.fr.XI,137.
X. sororia, F. v. M., first gen. Report 15 (1854)...   ...   S.A.   — V.  N.S.W.  Q.   —   B.fl.VII,100
X. Endlicheri, F. v. M., fragm. VIII, 205 (1874)   ...   W.A.   —   —   —   —   —   —   B.fl.VII,101   M.fr.VIII,205.
X. sericea, Endlicher in Lehm. pl. Preiss. II, 51 (1846) ..   W.A.   —   —   —   —   —   —   B.fl.VII,101   M.fr.VIII,207.
X. purpurea, Endlicher in Lehm. pl. Preiss. II, 49 (1846)   W.A.   —   —   —   —   —   —   B.fl.VII,101   M.fr.VIII,213.
X. Preissii, Endlicher in Lehm. pl. Preiss. II, 52 (1846)...   W.A.   —   —   —   —   —   —   B.fl.VII,102
X. effusa, Lindley in Mitch. Three Exped. II, 151 (1839)   W.A. S A.   — V.  N.S.W.   —   —   B.fl.VII,102   M.fr.VIII,209.
X. micrantha, Endl. in Lehm. pl. Preiss. II, 49 (1846)...   W.A. S.A.   — V.  N.S.W.  Q.   —   B.fl.VII,103   M.fr.VIII,212.
X. Thunbergii, F. v. M., fragm. VIII, 208 (1874)   ...   —  S.A.   — V.  N.S.W.  Q.   —   B.fl.VII,103   M.fr.VIII,208.
X. caespitosa, Benth. Fl. Austr. VII, 104 (1878)   ...   W.A.   —   —   —   —   —   —   B.fl.VII,104   M.fr.212.
X. pauciflora, R. Brown, prodr. 261 (1810)...   ...   ...   W.A.   —   —   —   —   —   —   B.fl.VII,104
X. flexifolia, R. Brown, prodr. 260 (1810)   ...   ...   ...   —   —  N.S.W.   —   —   B.fl.VII,105   M.fr.VIII,207.
X. glauca, R. Brown, prodr. 200 (1810)   ...   ...   ...   W.A. S.A. T. V. N.S.W.   —   —   B.fl.VII,105   M.fr.VIII,208.
X. elongata, Bentham, Fl. Austr. VII, 106 (1878)   ...   —  S.A.   —  N.S.W.  Q.   —   B.fl.VII,106
X. rupestris, Endlicher in Lehm. pl. Preiss. II, 50 (1846)   W.A.   —   —   —   —   —   —   B.fl.VII,107
X. collina, R. Brown, prodr. 260 (1810)   ...   ...   ...   W.A.   —   —   —   —   —   —   B.fl.VII,107
X. suaveolens, Endlicher in Lehm. pl. Preiss. II, 50 (1846)   W.A.   —   —   —   —   —   —   B.fl.VII,107   M.fr.VIII,207.
X. turbinata, Endlicher in Lehm. pl. Preiss. II, 51 (1846)   W.A.   —   —   —   —   —   —   B.fl.VII,107   M.fr.VIII,211.
X. spartea, Endlicher in Lehm. pl. Preiss. II, 51 (1846)...   W.A.   —   —   —   —   —   —   B.fl.VII,108   M.fr.VIII,211.
X. juncea, F. v. M. in Transact. Vict. Inst. 135 (1855) ...   — S.A.   —   —   —   —   —   B.fl.VII,108   M.fr.VIII,210.
X. leucocephala, R. Brown, prodr. 260 (1810)   ...   ...   S.A.   — V. N S.W. Q.   —   B.fl.VII,110   M.fr.VIII,210.
X. hastilis, R. Brown, prodr. 203 (1810)   ...   ...   ...   W.A.   —   —   —   —   —   —   B.fl.VII,110   M.frVIII,210;XI,54.
    DASYPOGON, R. Brown, prodr. fl. Nov. Holl. 263 (1810).
D. bromelifolius, R. Brown, prodr. 263 (1810)   ...   ...   W.A.   —   —   —   —   —   —   B.fl.VII,118
D. Hookeri, Drummond in Hook. Lond. Journ. II, 168 (1843)...   W.A.   —   —   —   —   —   —   B.fl.VII,110   M.fr.II,112;IV,173.
    XANTHORRHOEA, Smith in Transact. Linn. Soc. IV, 219 (1798).
X. gracilis, Endlicher in Lehm. pl. Preiss. II, 30 (1846)...   W.A.   —   —   —   —   —   —   B.fl.VII,114   M.fr.IV,112.
X. pumilio, R. Brown, prodr. 288 (1810)   ...   ...   ...   —   —   —   —   Q.   —   B.fl.VII,114
X. macronema, F. v. M., fragm. IV, 112 (1864)   ...   ...   —   —   —  N.S.W. Q.   —   B.fl.VII,113
X. hastilis, R. Brown, prodr. 288 (1810)   ...   ...   ...   —   —   —  N.S.W. Q.   —   B.fl.VII,115   M.fr.IV,113.
X. arborea, R. Brown, prodr. 288 (1810)   ...   ...   ...   —   —   —  N.S.W. Q.   —   B.fl.VII,115   M.fr.IV,113.
X. semiplana, F. v. M., fragm IV, 111 (1864)   ...   ...   — S.A.   —   —   —   —   B.fl.VII,115   M.fr.IV,111.
X. bracteata, R. Brown, prodr. 288 (1810)...   ...   ...   —   —   —  N.S.W.   —   —   B.fl.VII,116
X. minor, R. Brown, prodr. 288 (1810)   ...   ...   ...   — S.A. T. V. N.S.W.   —   —   B.fl.VII,114   M.fr.IV,112.
X. australis,'R. Brown, prodr. 288 (1810)   ...   ...   ...   —   T. V.   —   —   —   B.fl.VII,116   M.fr.IV,111.
X. quadrangulata, F. v. M., fragm. IV, 111 (1864)   ...   — S.A.   —   —   —   —   —   B.fl.VII,117   M.fr.IV,111.
X. Preissii, Endlicher in Lehm. pl. Preiss. II, 30 (1846)...   W.A.   —   —   —   —   —   —   B.fl.VII,117   M.fr.IV,110.
    KINGIA, R. Brown in King's Narrat. of survey of Austr. coasts II, 520 (1827).
K. australis, R. Brown in King's Narr. of surv. II, 535 (1827)...   W.A.   —   —   —   —   —   —   B.fl.VII,110
```

PALMAE.
Ray, method. pl. emend. 135 (1703).

```
CALAMUS, Linné, spec. plant. 325 (1753).
C. australis, Martius, hist. nat. palm. III, 213 (1839)   ...   ...   —   —   —   —   —   Q.   —   B.fl.VII,134   M.fr.V,48.
C. Muelleri, Wendland & Drude in Linnaea XXXIX, 103 (1871)   —   —   —  N.S.W. Q.   —   B.fl.VII,135
C. caryotoides, A. Cunn. in Martius, hist. nat. palm. III, 212 (1839)   —   —   —   —   —   Q.   —   B.fl.VII,133
```

BACULARIA, F. v. M., fragm. VII, 103 (1870). (Linospadix).
B. monostachya, F. v. M., fragm. VII, 103 (1870) ... ... — — — — N.S.W. Q. — B.fl.VII,136 M.fr.VII,82,103;XI,58.
KENTIA, Blume in Bull. Neerl. 64 (1838). (Grisebachia, Hydriastele, Hodyscope, Kentiopsis, Veitchia partly).
K. minor, F. v. M., fragm. VIII, 235 (1874) ... ... — — — — — Q. — B.fl.VII,137 M.fr.IX,195;XI,58,137.
K. Wendlandiana, F. v. M., fragm. VII, 102 (1870) ... ... — — — — — Q. N.A. B.fl.VII,138 M.fr.VII,102.
K. Belmoreana, C. Moore & F. v. M., fragm. VII, 99 (1870) ... — — — — N.S.W. — B.fl.VII,137 M.fr.VIII,234.
K. Forsteriana, C. Moore & F. v. M., fragm. VII, 100 (1870) ... — — — — N.S.W. — —
K. Baueri, Seemann, fl. Vitiens. 269 (1869) ... ... — — — — N.S.W. — — — M.fr.VII,102;VIII,233.
K. Canterburyana, C. Moore & F. v. M., fragm. VII, 101 (1870) — — — — N.S.W. — B.fl.VII,138 M.fr.VIII,234;X,119.
K. acuminata, Wendland & Drude in Linnea XXXIX, 207 (1871) — — — — — N.A. B.fl.VII,138

CLINOSTIGMA, Herm. Wendland in Seemann's Bonplandia X, 196 (1862). (Cyphokentia). [120.
C. Mooreanum, F. v. M., fragm. VIII, 235 (1874) ... ... — — — — N.S.W. — B.fl.VII,139 M.fr.VII,101;IX, 195; X,
PTYCHOSPERMA, Labillardière in Mém. de l'Inst., ann. 1808, I, 251, pl. X (1809). (Seaforthia, Laccospadix, Archontophoenix).
P. Laccospadix, Bentham, fl. Austr. VII, 140 (1878) ... ... — — — — Q. — B.fl.VII,140
P. Beatricae, F. v. M. in Melb. Chemist. Febr. (1882) ... — — — — Q. — —
P. Alexandrae, F. v. M., fragm. V, 47, t. 43 (1865) ... — — — — Q. — B:fl.VII,140 M.fr.V,47,190,213.
P. Cunninghamii, Herm. Wendland in Bot. Zeit. 346 (1858) ... — — — — N.S.W. Q. — B.fl.VII,141 M.fr.VIII,222.
P. elegans, Blume, Rumphia II, 118 (1836) ... — — — — Q. — B.fl.VII,141 M.fr.VIII,222.
P. Normanbyi, F. v. M., fragm. XI, 56 (1878) ... ... — — — — Q. — B.fl.VII,142 M.fr.VIII,235.

ARECA, Linné, spec. plant. 1189 (1753) from Ray (1688).
A. Alicae, F. v. M. in Regel's Gartenflora, 199-201 (1876) ... — — — — Q. — M.fr.XI,88.
COCOS, Linné, spec. plant. 1188 (1753).
C. nucifera, Linné, spec. plant. 1188 (1753) ... — — — — Q. — B.fl.VII,143
CARYOTA, Linné, gen. plant. 355 (1737).
C. Rumphiana, Martius, hist. nat. palm. III, 195 (1839) ... — — — — Q. — B.fl.VII,144
LICUALA, Rumphius, herb. Amboin. I, 44, t. 9 (1741).
L. Muelleri, Wendland & Drude in Linnaea XXXIX, 223 (1871) — — — — Q. — B.fl.VII,145 M.fr.VIII,221;X,120.
LIVISTONA, R. Brown, prodr. fl. Nov. Holl. 267 (1810). (Corypha partly).
L. humilis, R. Brown, prodr. 268 (1810) ... ... — — — — — N.A. B.fl.VII,146 M.fr.VIII,221.
L. inermis, R. Brown, prodr. 268 (1810) ... ... — — — — — N.A. B.fl.VII,146
L. Mariae, F.v.M. in Giles's Geog. Trav. in Cntr. Austr. 222(1875) — S.A. — — — N.A.
L. australis, Martius, hist. nat. palm. III, 241 (1839) ... — — — — — N.A. — M.fr.XI,54.

PANDANEAE.
R. Brown, prodr. Nov. Holl. (1810)
PANDANUS, Rumphius, herb. Amboin. IV, 139, t. 74 (1750) (Athrodactylis, Fiquetia).
P. odoratissimus, Linné, fil. suppl. 424 (1781) ... — — — — N.A. B.fl.VII,148 M.fr.XI,55.
P. pedunculatus, R. Brown, prodr. 341 (1810) ... ... — — — N.S.W. Q. — B.fl.VII,149 M.fr.VIII,220;IX,193.
P. Forsteri, Moore & F. v. M., fragm. VIII, 220 (1874) ... — — — N.S.W. — B.fl.VII,149 M.fr.VIII,220.
P. aquaticus, F. v. M., fragm. V, 40 (1865) ... ... — — — — Q. N.A. B.fl.VII,148 M.fr.VIII,220. [193
P. monticola, F. v. M., fragm. V, 40 (1865) ... ... — — — — Q. — B.fl.VII,150 M.fr.VII,63;VIII,220; IX,
FREYCINETIA, Gaudichaud in Annal. des scienc. natur. III, 509 (1824).
F. Gaudichaudii, Brown & Bennett, pl. Jav. rar. 31, t. 9 (1838) — — — — Q. — B.fl.VII,151
F. excelsa, F. v. M., fragm. V, 39 (1865) ... ... — — — — Q. — B.fl.VII,151 M.fr.V,30.
F. Baueriana, Endlicher, prodr. fl. ins. Norf. 25 (1833) ... — — — — Q. —
NIPA, Wurmb in Verhandl. Batav. Genootsch. I, 349 (1779), from Camelius, Ray & Rumphius.
N. fruticans, Wurmb in Verhandl. Batav. Genootsch. I, 349 (1779) — — — Q. — — M.fr.XI,128.

AROIDEAE.
A. L. de Jussieu, gen. plant. 23 (1789), from B. de Jussieu (1759).
TYPHONIUM, Schott in Wien. Zeitschr. III, 72 (1829). (Arum partly.)
T. liliifolium, F. v. M. in Schott, prodr. syst. aroid. 107 (1860) — — — — N.A. B.fl.VII,153 M.fr.VIII,187.
T. alismifolium, F. v. M., fragm. VIII, 186 (1874) ... ... — — — — Q. — B.fl.VII,153 M.fr.VIII,186.
T. Brownii, Schott, icon. aroid. II, t. 15 (1857) ... — — — N.S.W. Q. — B.fl.VII,154 M.fr.VIII,187.
T. angustilobum, F. v. M., fragm. X, 66 (1876) ... ... — — — — Q. — B.fl.VII,154 M.fr.X,66.
AMORPHOPHALLUS, Blume in Nouv. Ann. du Mus. III, 306 (1834). (Brachyspatha).
A. variabilis, Blume, Rumphia I, 146, t. 35 (1835) ... ... — — — — . N.A. B.fl.VII,155 M.fr.VIII,87.
COLOCASIA, Ray, Method. 157 (1682). (Caladium partly.)
C. antiquorum, Schott, Meletemat. botan. 18 (1832) ... — — — — Q. — B.fl.VII,155 M.fr.VIII,87.
C. macrorrhiza, Schott, Meletemat. bot. 18 (1832) ... — — — N.S.W. Q. — B.fl.VII,155 M.fr.VIII,87.
RHAPHIDOPHORA, Hasskarl in Fl. Regensb. Beibl. II, 11 (1842). (Epipremnum partly.)
R. pertusa, Schott in Seem. Bonplandia V, 45 (1857) ... — — — — Q. — B.fl.VII,156 M.fr.VIII,87.
GYMNOSTACHYS, R. Brown, prodr. fl. Nov. Holl. 337 (1810).
G. anceps, R. Brown, prodr. 337 (1810) ... ... — — — N.S.W. Q. — B.fl.VII,157 M.fr.VIII,87.
POTHOS, Linné, amoen. acad. I, 410 (1747).
P. Loureiri, Hooker & Arnott, Bot. of Beech. Voy. 220 (1841) ... — — — N.S.W. Q. — B.fl.VII,158 M.fr.I,62;VIII,87.

TYPHACEAE.
A. L. de Jussieu, gener. plant. 25 (1789).
TYPHA, Tournefort, inst. 530, t. 301 (1700), from Theophrastos, Dioscorides and Plinius.
T. angustifolia, Linné, spec. plant. 971 (1753) ... ... W.A. S.A. T. V. N.S.W. Q. N.A. B.fl.VII,159 M.fr.VII,116.
SPARGANIUM, Tournefort, inst. 530, t. 302 (1700).
S. angustifolium, R. Brown, prodr. 338 (1810) ... ... — — V. N.S.W. Q. — B.fl.VII,160 M.fr.VII,117.

121

LEMNACEAE.

J. E. Gray, Nat. arrang. of Brit. pl. II, 729 (1821).

LEMNA, Linné, syst. nat. 9 (1735); Linné, Fl. Lappon. 351 (1737). (Telmatophace, Spirodela.)
L. trisulca, Linné, spec. plant. 970 (1753) ... ... — S.A. T. V. N.S.W. Q. — B.fl.VII,162 M.fr.VIII,188.
L. minor, Linné, spec. plant. 970 (1753) ... ... W.A. S.A. T. V. N.S.W. Q. N.A. B.fl.VII,163 M.fr.VIII,188.
L. oligorrhiza, Kurz in Journ. Linn. Soc. IX, 267 (1867) .. ... — S.A. — V. N.S.W. Q. N.A. B.fl.VII,163 M.fr.VIII,81.
L. polyrrhiza, Linné, spec. plant. 970 (1753) ... ... — — — V. N.S.W. — — B.fl.VII,164 M.fr.VIII,188.
L. gibba, Linné, spec. plant. 970 (1753) ... ... W.A. — — — — — — B.fl.VII,163 M.fr.VIII,188.
WOLFFIA, Horkel & Schleiden in Schlecht. Linnaea XIII, 389 (1839). (Wolfia.)
W. Michelii, Schleiden, Beitr. zur Botan. 233 (1844) .. ... — — V. N.S.W. — — B.fl.VII,162 M.fr.VIII,187.

FLUVIALES.

Ventenat, Tabl. du règn. végét. II, 80, (1799).

TRIGLOCHIN, Rivinus in Ruppius, fl. Jenens. 54 et 366 (1718), from Dalechamps (1586). (Cycnogeton.)
T. centrocarpa, Hooker, icon. plant. t. 728 (1848) ... — W.A. S.A. T. V. N.S.W. — — B.fl.VII,167 M.fr.VI,82.
T. striata, Ruiz & Pavon, Fl. Per. et Chil. III, 72 (1802) ... W.A. S.A. T. V. N.S.W. Q. — B.fl.VII,166 M.fr.VI,83.
T. mucronata, R. Brown, prodr. 343 (1810) ... ... W.A. S.A. — V. — — — B.fl.VII,168 M.fr.VI,81.
T. procera, R. Brown, prodr. 343 (1810) ... ... W.A. S.A. T. V. N.S.W. Q. N.A. B.fl.VII,168 M.fr.VI,83.
T. Maundii, F. v. M., fragm. VI, 83 (1867) ... ... — — — — N.S.W. Q. — B.fl.VII,169

APONOGETON, Linné fil., suppl. plant. 32 (1781).
A. monostachyus, Linné fil., suppl. 214 (1781) ... ... — — — — — Q. N.A. B.fl.VII,188 M.fr.VIII,215.
A. elongatus, F. v. M. in Benth. Fl. Austr. VII, 188 (1878) ... — — — — N.S.W. Q. N.A. B.fl.VII,188 M.fr.VIII,216.

POTAMOGETON, Fuchs, histor. stirp. comment. 651 (1542).
P. natans, Linné, spec. plant. 126 (1753) ... ... W.A. S.A. T. V. N.S.W. Q. — B.fl.VII,170 M.fr.VIII,217.
P. javanicus, Hasskarl in Act. Soc. Ind. Neerl. I, 26 (1857) ... — — — — — Q. N.A. B.fl.VII,171 M.fr.VIII,217.
P. Drummondii, Bentham, Fl. Austr. VII, 171 (1878) ... ... W.A. — — — — — — B.fl.VII,171
P. perfoliatus, Linné, spec. plant. 126 (1753) ... ... — T. V. N.S.W. — — B.fl.VII,172 M.fr.VIII,217.
P. crispus, Linné, spec. plant. 126 (1753) ... ... — S.A. — V. N.S.W. Q. N.A. B.fl.VII,172 M.fr.VIII,217.
P. obtusifolius, Mertens & Koch, Deutschl. Fl. I, 855 (1823) ... — S.A. T. V. N.S.W. Q. — B.fl.VII,172 M.fr.VIII,217.
P. acutifolius, Link in Roem. & Schult. syst. veg. III, 513 (1818) — S.A. — V. N.S.W. — — B.fl.VII,173 M.fr.VIII,216.
P. pectinatus, Linné, spec. plant. 127 (1753) ... ... — S.A. T. V. — — — B.fl.VII,173 M.fr.VIII,217.

RUPPIA, Linné, syst. nat. 9 (1735); Linné, gen. plant. 277 (1737).
R. maritima, Linné, spec. plant. 127 (1753) ... ... W.A. S.A. T. V. N.S.W. Q. — B.fl.VII,174 M.fr.VIII,217

POSIDONIA, Koenig, Annals of Botany II, 95, t. 6 (1806). (Caulinia partly, Kernera partly.)
P. australis, J. Hooker, Fl. Tasman. II, 43 (1860) ... ... W.A. S.A. T. V. — — — B.fl.VII,175 M.fr.VIII,218.

ZOSTERA, Linné, Waesgoeta-Resa, 166-168 et fig. (1747).
Z. nana, F. K. Mertens in Roth, enum. pl. phaner. Germ. 8 (1827) — S.A. T. V. N.S.W. — — B.fl.VII,176 M.fr.VIII,218.
Z. Tasmanica, G. v. Mertens in Garcke, Linnaea XXXV, 68 (1867) — S.A. T. V. — — — B.fl.VII,176 M.fr.VIII,218.

CYMODOCEA, C. Koenig, Annals of Botany II, 96, t. 7 (1806). (Amphibolis, Graumuellera, Kernera partly.)
C. antarctica,F.v.M.; Amphibolis,Agardh.,spec.algar.II,477(1822)W.A. S.A. T. V. — — — B.fl.VII,177 M.fr.VIII,218.
C. ciliata, Ehrenberg in Garcke, Linnaea XXXV, 102 (1867) ... — — — — — Q. N.A. B.fl.VII,178 M.fr.VIII,218.
C. rotundata, Ascherson & Schweinfurth in Neumayer, Anleit. zu
     wissensch. Beob. 361 (1875) ... ... — — — — — — N.A. — M.fr.X,162.
C. serrulata, Ascherson & Magnus in Neumayer, Anleit. zu wiss.
     Beob. 362 (1875) ... ... — — — — — Q. — B.fl.VII,178 M.fr.VIII,218.
C. isoetifolia, Ascherson in Garcke, Linnaea XXXV, 163 (1867) W.A. — — — — — Q. — N.A. B.fl.VII,178

HALODULE, Endlicher, gen. plant. 1368 (1840). (Diplanthera, 1806.)
H. tridentata, Endlicher, gen. plant. 1368 (1840) ... ... — — — — — — N.A. — M.fr.VIII,218.

LEPILAENA, Drummond & Harvey in Hook. Kew misc. VII, 57 (1855). (Zannichellia partly, Hexatheca, Althenia subgenus.)
L. australis, J. Drummond in Hook. Kew misc. VII, 58 (1855)... W.A. — — — — — — B.fl.VII,179 M.fr.VIII,217.
L. Preissii, F, v. M., fragm. VIII, 217 (1874) ... ... W.A. S.A. T. V. — — — B.fl.VII,180 M.fr.VIII,217.

NAJAS, Linné, syst. nat. 9 (1735); Linné, gen. plant. 278 (1737). (Caulinia partly.)
N. major, Allioni, Fl. Pedem. II, 221 (1785) ... ... W.A. — — — — — N.A. B.fl.VII,181 M.fr.VIII,218.
N. tenuifolia, R. Brown, prodr. 345 (1810)... ... ... — S.A. — V. N.S.W. Q. — B.fl.VII,181 M.fr.VIII,219.

ALISMACEAE.

Ventenat, Tabl. du règn. végét. II, 157 (1799).

ALISMA, Rivinus in Ruppius, fl. Jen. 54 (1718), from Dioscorides and Plinius. (Caldesia; Echinodorus partly.)
A. Plantago, Linné, spec. plant. 342 (1753) ... ... — — V. N.S.W. — — B.fl.VII,185 M.fr.VIII,215.
A. acanthocarpum, F. v. M., fragm. I, 23 (1858)... ... ... — — — — — Q. N.A. B.fl.VII,185 M.fr.VIII,214.
A. oligococcum, F. v. M., fragm. I, 23 (1858) ... ... — — — — — Q. N.A. B.fl.VII,185 M.fr.VI,169;VIII,214.
A. parnassifolium, Bassi in act. Bonon. (1768) ... ... — — — — — Q. — B.fl.VII,185 M.fr.VIII,214.

DAMASONIUM, Tournefort, inst. 256 (1700). (Actinocarpus.)
D. australe, Salisbury in Transact. hort. Soc. Lond. I, 268 (1812) W.A. S.A. — V. N.S.W. Q. — B.fl.VII,186 M.fr.VIII,215.

TENAGOCHARIS, Hochstetter in Regensb. bot. Zeit. June (1841). (Butomopsis.)
T. Cordofana, Hochstetter in Regensb. bot. Zeit. June (1841) ... — — — — — Q. N.A. B.fl.VII,187 M.fr.X,103.

PONTEDERIACEAE.

Humboldt, Bonpland & Kunth, nov. gen. et spec. plant. I, 265 (1815).

MONOCHORIA, Presl, reliq. Haenk. I, 127 (1827). (Limnostachys.)
M. cyanea, F. v. M., fragm. VIII, 44 (1872) ... ... — — — — — Q. N.A. B.fl.VII,74 M.fr.VIII,44.

Q

PHILYDREAE.

R. Brown in Flinders' voy. II, 57 (1814).

PHILYDRUM, Banks in Gaertner, de fruct. I, 62 t. 16 (1788). (Philhydrum.)
P. lanuginosum, Banks in Gaertn. de fruct. I. 62 t. 16 (1788) ... — — V. N.S.W. Q. N.A. B.fl.VII,73 M.fr.V,203.
PRITZELIA, F. v. M., Papuan plants 13 (1875). (Hetaeria, Philydrella.)
P. pygmaea, F. v. M., Papuan plants 13 (1875) ... ... W.A. — — — — — — B.fl.VII,74 M.fr.V,203.
HELMHOLTZIA, F. v. M., fragm. V, 202 (1866).
H. acorifolia, F. v. M., fragm. V, 203 (1866) ... ... — — — — N.S.W. Q. — B.fl.VII,75 M.fr.V,203.

COMMELINEAE.

R. Brown, prodr. fl. Nov. Holl. 268 (1810).

ZYGOMENES, Salisbury in Transact. hort. Soc. I, 271 (1812). (Cyanotis).
Z. axillaris, Salisbury in Transact. hort. Soc. I, 271 (1812) ... — — — — — Q. N.A. B.fl.VII,82 M.fr.VIII,62.
COMMELINA, Plumier, nov. pl. Amer. gen. 48, t. 38 (1703). (Commelynia).
C. ensifolia, R. Brown, prodr. 269 (1810) .. .. — S.A. — — N.S.W. Q. N.A. B.fl.VII,83 M.fr.VIII,60;IX,191.
C. cyanea, R. Brown, prodr. 269 (1810) ... ... — — — — N.S.W. Q. N.A. B.fl.VII,84 M.fr.VIII,59;IX,191.
C. lanceolata, R. Brown, prodr. 269 (1810) ... ... — — — — — Q. N.A. B.fl.VII,84 M.fr.VIII,59.
C. agrostophylla, F.v. M. in J. Hook., Fl. Tasm. p. XLVII(1860) — — — — — Q. N.A. — M.fr.VIII,59.
ANEILEMA, R. Brown, prodr. fl. Nov. Holl. 270 (1810).
A. acuminatum, R. Brown, prodr. 270 (1810) ... ... — — — — N.S.W. Q. — B.fl.VII,85 M.fr.VIII,61.
A. biflorum, R. Brown, prodr. 270 (1810) ... ... — — — — N.S.W. Q. — B.fl.VII,86
A. sclerocarpum, F. v. M., fragm. VIII, 61 (1873) ... ... — — — — — — B.fl.VII,86 M.fr.VIII,61.
A. siliculosum, R. Brown, prodr. 270 (1810) ... ... — — — — — Q. N.A. B.fl.VII,87 M.fr.VIII,61;IX,191.
A. gramineum, R. Brown, prodr. 270 (1810) ... ... — — — — N.S.W. Q. N.A. B.fl.VII,87 M.fr.VIII,62.
A. gigantoum, R. Brown, prodr. 271 (1810) ... ... — — — — — Q. N.A. B.fl.VII,88 M.fr.VIII,62.
FLORISCOPA, Loureiro, Fl. Cochinch. I, 192 (1790). (Floscopa, Dithyrocarpus.)
F. scandens, Loureiro, fl. Cochinch. I, 193 (1790) ... — — — — — Q. — B.fl.VII,89 M.fr.VIII,63.
POLLIA, Thunberg, nov. gen. plant. I, 11 (1781).
P. cyanococca, F. v. M., fragm. V, 40 (1865) ... ... — — — — N.S.W. Q. — B.fl.VII,90 M.fr.VIII,63.
CARTONEMA, R. Brown, prodr. fl. Nov. Holl. 271 (1810).
C. philydroides, F. v. M., fragm. I, 62 (1858) ... ... W.A. — — — — — — B.fl.VII,91 M.fr.VIII,64.
C. spicatum, R. Brown, prodr. 271 (1810) ... ... — — — — — — N.A. B.fl.VII,91 M.fr.I,62;VIII,64.
C. parviflorum, Hasskarl in Flora Regensb. 305 (1869) ... — — — — — — N.A. B.fl.VII,91
C. trigonospermum, Clarke in De Cand., mon. phaner. III, 24 (1881) — — — — — N.A. — M.fr.XI,139.
C. brachyantherum, Bentham, Fl. Austr. VII, 92 (1878) ·· — — — — Q. — B.fl.VII,92
C. tenue, Bentham in De Cand., monogr. phaner. III, 264 (1881) — — — — — N.A. — M.fr.XI,139.

XYRIDEAE.

Salisbury in Transact. of the Hortic. Soc. I, 326 (1812).

XYRIS, Gronovius in Linné, genera plantar. 11 (1737).
X. complanata, R. Brown, prodr. 256 (1810) ... ... — — — N.S.W. Q. N.A. B.fl.VII,77 M.fr.VIII,205.
X. pauciflora, Willdenow, phytogr. 2, t. 1 (1794)... ... ... — — Q. N.A. B.fl.VII,78 M.fr.VIII,205.
X. lacera, R. Brown, prodr. 257 (1810) ... ... W.A. — — — — — B.fl.VII,78 M.fr.VIII,204.
X. flexifolia, R. Brown, prodr. 256 (1810) ... ... W.A. — — — — — B.fl.VII,78 M.fr.VIII,204.
X. gracilis, R. Brown, prodr. 256 (1810) ... ... — S.A. T. V. N.S.W. Q. B.fl.VII,79 M.fr.VIII,204.
X. operculata, Labillardiere, Nov. Holl. pl. spec. I, 14 t. 10 (1804) — S.A. T. V. N.S.W. Q. B.fl.VII,79 M.fr.VIII,204.
X. lanata, R. Brown, prodr. 257 (1810) ... ... W.A. — — — — — B.fl.VII,80 M.fr.VIII,204.
X. laxiflora, F. v. M., fragm. VIII, 203 (1874) ... ... W.A. — — — — — B.fl.VII,80 M.fr.VIII,203.
X. gracillima, F. v. M., fragm. VIII, 203 (1874) ... ... W.A. — — — — — B.fl.VII,80 M.fr.VIII,203.

JUNCEAE.

R. Brown, prodr. fl. Nov. Holl. 257 (1810).

LUZULA, De Candolle, Fl. franc. III, 159 (1805).
L. campestris, De Candolle, Fl. franc. III, 158 (1805) ... ... W.A. S.A. T. V. N.S.W. Q. — B.fl.VII,123
L. longiflora, Bentham, Fl. Austr. VII, 123 (1878) ... ... — — — N.S.W. — B.fl.VII,123
JUNCUS, Tournefort, inst. 246, t. 127 (1700), from Camerarius, J. & C. Bauhin, Ray & Morison.
J. gracilis, R. Brown, prodr. 259 (1810) ... ... W.A. — — — — B.fl.VII,123
J. planifolius, R. Brown, prodr. 259 (1810) ... ... W.A. S.A. T. V. N.S.W. — B.fl.VII,125
J. caespititius, E. Meyer in Lehm. pl. Preiss. II, 47 (1846) W.A. S.A. T. V. N.S.W. — B fl.VII,126
J. falcatus, E. Meyer, synops. luzular. 34 (1823) ... ... — — T. V. N.S.W. — B.fl.VII,126
J. bufonius, Linné, spec. plant. 328 (1753)... ... ... W.A. S.A. T. V. N.S.W. Q. B.fl.VII,127
J. homalocaulis, F. v. M., first gen. Rep. 19 (1853) ... — S.A. — V. N.S.W. — B.fl.VII,128
J. Brownii, F. v. M., first gen. Rep. 19 (1853) ... ... — T. V. N.S.W. — B.fl.VII,128 M.fr.IX,78.
J. communis, E. Meyer, synops. Juncor. 12 (1822) ... W.A. S.A. T. V. N.S.W. Q. B.fl.VII,128
J. vaginatus, R. Brown, prodr. 258 (1810)... ... ... — — T. V. N.S.W. Q. B.fl.VII,129
J. pauciflorus, R. Brown, prodr. 259 (1810) ... ... — S.A. T. V. N.S.W. Q. B.fl.VII,130
J. pallidus, R. Brown, prodr. 258 (1810) ... ... W.A. S.A. T. V. N.S.W. Q. ·· B.fl.VII,130
J. maritimus, Lamarck, Encycl. meth. 264 (1789) ... W.A. S.A. T. V. N.S.W. Q. B.fl.VII,130
J. prismatocarpus, R. Brown, prodr. 259 (1810) ... ... W.A. S.A. T. V. N.S.W. Q. ·· B.fl.VII,131
J. pusillus, Buchenau in Abhandl. naturw. Vereins zu Bremen VI,
    395 (1879) ... ... — — T. V. N.S.W. — — B.fl.VII,132

ERIOCAULEAE.

Humboldt, Bonpland & Kunth, nov. gen. et spec. plant. I, 251 (1815).

ERIOCAULON, Linné, gen. plant. ed. sec. 35 (1742). (Électrosperma.)

E. setaceum, Linné, spec. plant. 87 (1753) ...	...	...	...	—	—	—	—	Q. N.A. B.fl.VII,191
E. australe, R. Brown, prodr. 254 (1810) ...	...	...	...	—	—	—	N.S.W. Q. — B.fl.VII,192
E. quinquangulare, Linné, spec. plant. 87 (1753) ...	...	...	...	—	—	—	— N.A. B.fl.VII,102
E. Smithii, R. Brown, prodr. 254 (1810) ...	...	...	—	—	— V. N.S.W. Q. — B.fl.VII,102
E. nanum, R. Brown, prodr. 254 (1810) ...	...	...	...	—	—	—	— Q. — B.fl.VII,193
E. cinereum, R. Brown, prodr. 254 (1810) ...	...	...	...	—	—	—	— N.A. B.fl.VII,193 M.fr.I,95.
E. pusillum, R. Brown, prodr. 254 (1810) ...	...	...	...	—	—	—	— Q. — B.fl.VII,194
E. pallidum, R. Brown, prodr. 254 (1810) ...	...	...	...	—	—	—	— Q. — B.fl.VII,194
E. nigricans, R. Brown, prodr. 254 (1810) ...	...	...	...	—	—	—	— Q. N.A. B.fl.VII,194
E. electrospermum; Electrosperma Australasicum, F. v. M. in Trans. phil. Soc. Vict. I, 24 (1854)	...	...	—	—	— V. N.S.W. — B.fl.VII,195
E. lividum, F. v. M., fragm. I, 92 (1858) ...	...	...	...	—	—	—	— N.A. B.fl.VII,195 M.fr.I,02.
E. concretum, F. v. M., fragm. I, 92 (1858)	...	...	...	—	—	—	— N.A. B.fl.VII,195 M.fr.I,02.
E. Schultzii, Bentham, Fl. Austr. VII, 195 (1878)	...	...	—	—	—	— N.A. B.fl.VII,195
E. tortuosum, F. v. M., fragm. I, 94 (1858)	...	...	...	—	—	—	— N.A. B.fl.VII,196 M fr.I,91.
E. monoscapum, F. v. M., fragm. I, 94 (1858)	...	...	...	—	—	—	— N.A. B.fl.VII,196 M.fr I,94.
E. spectabile, F. v. M., fragm. I, 95 (1858)	...	...	...	—	—	—	— N.A. B.fl.VII,196 M.fr.I,95.
E. scariosum, R. Brown, prodr. 255 (1810) ...	...	...	—	—	—	— Q. N.A. B.fl.VII,197
E. depressum, R. Brown, prodr. 255 (1810)	...	...	...	—	—	—	— N.A. B.fl.VII,197 M.fr.I,02.

RESTIACÈAE.

R. Brown, prodr. flor. Nov. Holl. 243 (1810).

TRITHURIA, J. Hooker, Fl. Tasman. II, 79, t. 138 (1860). (Juncella, 1854.)								
T. submersa, J. Hooker, Fl. Tasman. II, 79, t. 138 (1860)	...	W.A. S.A. T. V. N.S.W. — B.fl.VII,109 M.fr.VIII,237.
APHELIA, R. Brown, prodr. fl. Nov. Holl. 251 (1810). (Brizula.)								
A. cyperoides, R. Brown, prodr. 252 (1810)	...	W.A. —	—	—	— B.fl.VII,200 M.fr.VIII,236.
A. nutans, J. Hooker in Benth. Fl. Austr. VII, 200 (1878)	...	W.A. —	—	—	— B.fl.VII,200
A. gracilis, Sonder in Schlecht. Linnaea XXVIII, 227 (1855)	...	S.A. T. V. N.S.W. — B.fl.VII,201 M.fr.VIII,237.
A. Drummondii, Bentham, Fl. Austr. VII, 201 (1878)	...	W.A. —	—	—	— B.fl.VII,201
A. Pumillo, F. v. M., Schlecht. in Linnaea XXVIII, 226 (1855)	— S.A. T. V. N.S.W. — B.fl.VII,201 M.fr.VIII,237.
A. Brizula, F. v. M., fragm. V, 203 (1866) ...	...	W.A. —	—	—	— B.fl.VII,202 M.fr.VIII,237.
CENTROLEPIS, Labillardière, Nov. Holl. pl. specim. I, 7 (1804). (Desvauxia, Alepyrum.)								
C. humillima, F. v. M. in Benth. Fl. Austr. VII, 203 (1878)	W.A. —	—	—	— B.fl.VII,203
C. polygyna, Hieronymus in Abh. naturf. Ges. Halle XII, 96 (1873)	W.A. S.A. T. V. N.S.W. — B.fl.VII,203 M.fr.VIII,237.
C. alepyroides, Hieronymus in Abh. nat. Ges. Halle XII, 96 (1873)	W.A. —	—	—	— B.fl.VII,204 M.fr.VIII,237.
C. mutica, Hieronymus in Abh. nat. Ges. Halle XII, 97 (1873) ...	W.A. —	—	—	— B.fl.VII,204
C. glabra, F. v. M. in Abh. nat. Ges. Halle XII, 95 (1873)	...	W.A. —	—	—	— B.fl.VII,204 M.fr.VIII,237.
C. inuscoides, Hieronymus in Abh. nat. Ges. Halle XII, 95 (1873)	—	— T. —	—	— B.fl.VII,205
C. monogyna, Bentham, Fl. Austr. VII, 205 (1878)	...	—	— T. —	—	— B.fl.VII,205
C. pulvinata, Desvaux in Ann. des sc. nat. XIII, 42, t. 2 (1828)	—	— —	—	—	— Q. — B.fl.VII,205
C. pusilla, Roemer & Schultes, syst. veg. I, 44 (1817) ...	...	W.A. —	—	—	— Q. — B.fl.VII,205
C. aristata, Roemer & Schultes, syst. veg. I, 44 (1817)	...	W.A. S.A. T. V. N.S.W. — B.fl.VII,206 M.fr.VIII,237.
C. Drummondii, Hieronymus in Abh. nat. Ges. Halle XII, 98 (1873) W.A. —	—	—	— B.fl.VII,206 M.fr.VIII,237.
C. Banksii, Roemer & Schultes, syst. veg. I, 44 (1817) ...	...	—	—	—	— Q. N.A. B.fl.VII,207 M.fr.VIII,237.
C. fascicularis, Labillardière, Nov. Holl. pl. spec. I, 7, t. 1 (1804)	—	S.A. T. V. N.S.W. — B.fl.VII,207
C. pilosa, Hieronymus in Abh. nat. Ges. Halle 102 (1873)	W.A. —	—	—	— B.fl.VII,207
C. strigosa, Roemer & Schultes, syst. veg. I, 43 (1817) ..	W.A. S.A. T. V. N.S.W. — B.fl.VII,207
C. exserta, Roemer & Schultes, syst. veg. I, 44 (1817) ...	...	—	— Q. N.A. B.fl.VII,208 M.fr.VIII,237.
LYGINIA, R. Brown, prodr. fl. Nov. Holl. 248 (1810). (Schoenodum partly.)								
L. barbata, R. Brown, prodr. 248 (1810) ...	...	W.A. —	—	—	— B.fl.VII,210 M.fr.VIII,79.
ECDEIOCOLEA, F. v. M., fragm. phytogr. Austr. VIII, 236 (1874).								
E. monostachya, F. v. M., fragm. VIII, 236 (1874)	...	W.A. —	—	—	— B.fl.VII,211 M.fr.VIII,236.
ANARTHRIA, R. Brown, prodr. fl. Nov. Holl. 248 (1810).								
A. scabra, R. Brown, prodr. 249 (1810) ...	...	...	W.A. —	—	—	— B.fl.VII,212 M.fr.VIII,80.
A. laevis, R. Brown, prodr. 249 (1810) ...	...	...	W.A. —	—	—	— B.fl.VII,212 M.fr.VIII,81.
A. gracilis, R. Brown, prodr. 249 (1810) ...	...	...	W.A. —	—	—	— B.fl.VII,213 M.fr.VIII,72,81.
A. prolifera, R. Brown, prodr. 249 (1810) ...	...	...	W.A. —	—	—	— B.fl.VII,213 M.fr.VIII,82.
A. polyphylla, Nees in Lehm. pl. Preiss. II, 63 (1846) ...	W.A. —	—	—	— B.fl.VII,214 M.fr.VIII,83.
LEPYRODIA, R. Brown, prodr. fl. Nov. Holl. 247 (1810).								
L. scariosa, R. Brown, prodr. 248 (1810) ...	...	...	—	— N.S.W. — B.fl.VII,215 M.fr.VIII,72.
L. Muelleri, Bentham, Fl. Austr. VII, 215 (1878) ...	...	—	S.A. T. V. N.S.W. — B.fl.VII,215
L. gracilis, R. Brown, prodr. 247 (1810) ...	...	—	— N.S.W. — B.fl.VII,216
L. Tasmanica, J. Hooker, Fl. Tasman. II, 72, t. 135 (1860)	—	— T. V. —	— B.fl.VII,216 M.fr.VIII,73,75;IX,104.
L. interrupta, F. v. M., fragm. VIII, 74 (1873) ...	...	—	— N.S.W. — B.fl.VII,217 M.fr.VIII,74.
L. hermaphrodita, R. Brown, prodr. 248 (1810) ...	W.A. —	—	—	— B.fl.VII,217 M.fr.VIII,76.
L. monoides, F. v. M., fragm. VIII, 76 (1873) ...	W.A. —	—	—	— B.fl.VII,217 M.fr.VIII,76.
L. Muirii, F. v. M., fragm. VIII, 78 (1873) ...	...	W.A. —	—	—	— B.fl.VII,218 M.fr.VIII,76.
L. stricta, R. Brown, prodr. 248 (1810) ...	...	W.A. —	—	—	— B.fl.VII,218 M.fr.VIII,74.
L. macra, Nees in Lehm. pl. Preiss. II, 60 (1846) ...	...	W.A. —	—	—	— B.fl.VII,218
L. Drummondiana, Steudel, syn. pl. glum. II, 248 (1855)	W.A. —	—	—	— B.fl.VII,219 M.fr.VIII,75.
L. glauca, F. v. M., fragm. VIII, 77 (1873) ...	...	W.A. —	—	—	— B.fl.VII,219 M.fr.VIII,77.
L. anaectocolea, F. v. M., fragm. VIII, 78 (1873) ...	W.A. —	—	—	— B.fl.VII,220 M.fr.VIII,78.

RESTIO, Linné, syst. nat. edit. duo dec. II, 735 (1767). (Megalotheca.)

R. fastigiatus, R. Brown, prodr. 246 (1810)	—	— — — N.S.W. —	—	B.fl.VII,222	M.fr.VIII,67.
R. Megalotheca, F. v. M., fragm. VIII, 99 (1873)	W.A.	— — — —	—	B.fl.VII,222	
R. tropicus, R. Brown, prodr. 246 (1810) ...	—	— — — —	N.A.	B.fl.VII,223	
R. applanatus, Sprengel, syst. I, 185 (1825)	W.A.	— — — —	—	B.fl.VII,223	M.fr.VIII,09.
R. monocephalus, R. Brown, prodr. 245 (1810)	—	— T. — —	—	B.fl.VII,224	M.fr.VIII,68.
R. dimorphus, R. Brown, prodr. 246 (1810)	—	— — — N.S.W. Q.	—	B.fl.VII,224	M.fr.VIII,68.
R. conferto-spicatus, Steudel, syn. glum. II, 256 (1855) ...	W.A.	— — — —	—	B.fl.VII,224	
R. sphacelatus, R. Brown, prodr. 246 (1810)	W.A.	— — — —	—	B.fl.VII,225	
R. deformis, R. Brown, prodr. 245 (1810)	W.A.	— — — —	—	B.fl.VII,225	M.fr.
R. crispatus, R. Brown, prodr. 246 (1810) ...	W.A.	— — — —	—	B.fl.VII,225	
R. nitens, Nees in Lehm. pl. Preiss. II, 59 (1846) ...	W.A.	— — — —	—	B.fl.VII,226	M.fr.VIII,70.
R. gracilior, F. v. M. in Benth. Fl. Austr. VII, 226 (1878)	W.A.	— — — —	—	B.fl.VII,226	
R. chaunocoleus, F. v. M., fragm. VIII, 64 (1873)	W.A.	— — — —	—	B.fl.VII,226	M.fr.
R. australis, R. Brown, prodr. 245 (1810)	—	— T. V. N.S.W. —	—	B.fl.VII,227	M.fr.VIII,69.
R. gracilis, R. Brown, prodr. 245 (1810)	—	— T. V. N.S.W. Q.	—	B.fl.VII,227	M.fr.VIII,69.
R. complanatus, R. Brown, prodr. 245 (1810)	—	S.A. T. V. N.S.W. Q.	—	B.fl.VII,228	M.fr.VIII,67.
R. tetraphyllus, Labillardière, Nov. Holl. pl. spec. II, 77 (1806)	—	S.A. T. V. N.S.W. Q.	—	B.fl.VII,228	M.fr.VIII,66.
R. laxus, R. Brown, prodr. 245 (1810)	W.A.	— — — —	—	B.fl.VII,228	M.fr.VIII,71.
R. ornatus, Steudel, syn. pl. glum. II, 256 (1855)	W.A.	— — — —	—	B.fl.VII,229	M.fr.VIII,70.
R. leptocarpoides, Bentham, Fl. Austr. VII, 229 (1878)	W.A.	— — — —	—	B.fl.VII,229	M.fr.VIII,65.
R. amblycoleus, F. v. M., fragm. VIII, 65 (1873)	W.A.	— — — —	—	B.fl.VII,230	M.fr.VIII,65.
R. tremulus, R. Brown, prodr. 230 (1810)	W.A.	— — — —	—	B.fl.VII,230	M.fr.VIII,70.

LOXOCARYA, R. Brown, prodr. fl. Nov. Holl. 249 (1810). (Desmocladus.)

L. densa, Bentham, Fl. Austr. VII, 241 (1878)	W.A.	— — — —	—	B.fl.VII,241	
L. Benthami, F. v. M.; Hypolaena, Masters in Cand. mon. phan. I, 367 (1878) ...	W.A.	— — — —	—	B.fl.VII,242	
L. pubescens, Bentham, Fl. Austr. VII, 242 (1878)	W.A.	— — — —	—	B.fl.VII,242	M.fr.VIII,100.
L. fasciculata, Bentham, Fl. Austr. VII, 242 (1878)	W.A.	— — — —	—	B.fl.VII,242	M.fr.VIII,08.
L. flexuosa, Bentham, Fl. Austr. VII, 243 (1878) ...	W.A.	— — — —	—	B.fl.VII,243	M.fr.VIII,100.
L. cinerea, R. Brown, prodr. 249 (1810)	W.A.	— — — —	—	B.fl.VII,243	

CALOSTROPHUS, Labillardière, Nov. Holl. pl. spec. II, 78 (1806). (Calorophus, Hypolaena.)

C. elongatus, F. v. M., fragm. VIII, 86 (1873)	—	— T. — —	—	B.fl.VII,238	M.fr.VIII,86;IX,94;X,120
C. lateriflorus, F. v. M., fragm. VIII, 87 (1873)	—	S.A. T. V. N.S.W. Q.	—	B.fl.VII,238	M.fr.VIII,87.
C. gracillimus, F. v. M., fragm. VIII, 88 (1873)	W.A.	— — — —	—	B.fl.VII,230	M.fr.VIII,88.
C. fastigiatus, F. v. M. in pap. R. Soc. Tasm. 117 (1878)	—	S.A. T. V. N.S.W. —	—	B.fl.VII,239	M.fr.VIII,84.
C. exsulcus, F. v. M.; Hypolaena, R. Brown, prodr. 251 (1810)	W.A.	— — — —	—	B.fl.VII,240	M.fr.VIII,86.

LEPTOCARPUS, R. Brown, prodr. 250 (1810). (Schoenodum partly.)

L. scariosus, R. Brown, prodr. 250 (1810) ...	W.A.	— — — —	—	B fl.VII,232	M.fr.VIII,94.
L. tenax, R. Brown, prodr. 250 (1810)	W.A.	S.A. T. V. N.S.W. —	—	B.fl.VII,232	M.fr.VIII,96.
L. Brownii, J. Hooker, Fl. Tasman. II, 73, t. 136 (1860)	—	S.A. T. V. N.S.W. —	—	B.fl.VII,233	M.fr.VIII,91.
L. canus, Nees in Ann. of nat. hist. VI, 50 (1841)	W.A.	— — — —	—	B.fl.VII,234	M.fr.VIII,95.
L. coangustatus, Nees in Lehm. pl. Preiss. II, 65 (1846)	W.A.	— — — —	—	B.fl.VII,234	M.fr.VIII,94.
L. aristatus, R. Brown, prodr. 250 (1810)	W.A.	— — — —	—	B.fl.VII,235	M.fr.VIII,90.
L. erianthus, Bentham, Fl. Austr. VII, 235 (1878)	W.A.	— — — —	—	B.fl.VII,235	
L. ramosus, R. Brown, prodr. 250 (1810) ...	—	— — — Q.	—	B.fl.VII,236	M.fr.VIII,92.
L. elatior, R. Brown, prodr. 250 (1810)	—	— — — Q.	N.A.	B.fl.VII,236	
L. spathaceus, R. Brown, prodr. 250 (1810)	—	— — — N.A.	—	B.fl.VII,236	M.fr.VIII,93.
L. Schultzii, Bentham, Fl. Austr. VII, 237 (1878)	—	— — — N.A.	—	B.fl.VII,237	M.fr.VIII,93.

LEPIDOBOLUS, Nees in Lehm. pl. Preiss. II, 66 (1846).

L. drapetocoleus, F. v. M., fragm. VIII, 84 (1873)	—	S.A. — V.	—	B.fl.VII,244	M.fr.VIII,84.
L. Preissianus, Nees in Lehm. pl. Preiss. II, 66 (1846) ...	W.A.	— — — —	—	B.fl.VII,245	M.fr.VIII,83.
L. chaetocephalus, F. v. M., fragm. VIII, 84 (1873)	W.A.	— — — —	—	B.fl.VII,245	

CHAETANTHUS, R. Brown, prodr. 251 (1810). (Prionosepalum.)

C. leptocarpoides, R. Brown, prodr. 251 (1810)	W.A.	— — — —	—	B.fl.VII,246	M.fr.VIII,97.

ONYCHOSEPALUM, Steudel, syn. pl. glum. II, 249 (1855).

O. laxiflorum, Steudel, syn. pl. glum. II, 249 (1855) ...	W.A.	— — — —	—	B.fl.VII,246	M.fr.IX,51.

ACALYCEAE HYPOGYNEAE.

F. v. M. in Woolls's plants of the neighb. of Sydney 48 (1880).

CYPERACEAE.

Haller, enum. stirp. Helvet. I, 234 (1742).

KYLLINGIA, Rottboell, descr. et icon. rar. et nov. plant. 13, t. 4 (1773). (Killingia.)

K. intermedia, R. Brown, prodr. 219 (1810)	—	S.A. — V. N.S.W. Q.	—	B.fl.VII,251	M.fr.VIII,27.
K. monocephala, Rottboell, descr. et icon. pl. 13, t. 4 (1773)	—	S.A. — N.S.W. Q.	—	B.fl.VII,251	M.fr.VIII,271;IX,54.
K. cylindrica, Nees in Wight's Contr. to Bot. of Ind. 91 (1834)	—	— — — N.S.W. Q.	—	B.fl.VII,252	M.fr.VIII,271.
K. triceps, Rottboell, descr. et icon. pl. 14, t. 4 (1773) ...	—	— — — —	Q.	B.fl.VII,252	M.fr.VIII,271.

CYPERUS, Tournefort, inst. 527, t. 299 (1700), from Hippocr., Theophr. and Plinius. (Pycreus, Mariscus, Anosporum, Sorostachys.)

C. pumilus, Linné, amoen. acad. IV, 302 (1759) ...	—	— — — — Q.	—	B.fl.VII,258	M.fr.VIII,267.
C. Eragrostis, Vahl, enum. plant. II, 322 (1806) ...	—	S.A. — V. N.S.W. Q.	—	B.fl.VII,258	M.fr.VIII,260;IX,52.
C. flavescens, Linné, spec. plant. 68 (1753)	—	— — — N.S.W. Q.	—	B.fl.VII,259	
C. globosus, Allioni, auctuar. ad fl. Pedem. 49 (1789) ...	—	— — V. — Q.	—	B.fl.VII,260	M.fr.VIII,260;IX,52.
C. unioloides, R. Brown, prodr. 216 (1810)	—	— — V. N.S.W. Q.	—	B.fl.VII,260	M.fr.VIII,259;IX,52.

C. polystachyus, Rottboell, descr. et icon. pl. 30, t. 11 (1773) ... — — — N.S.W. Q. N.A. B.fl.VII,261 M.fr.VIII,265,270;IX,53.
C. flavicomus, Cl. Richard in Michaux, fl. bor. Amer. I, 27 (1803) — — — — — N.A. B.fl.VII,261 M.fr.VIII,265;IX,53.
C. pygmaeus, Rottboell, descr. et icon. pl. 20, t. 14 (1773) ... — — V. N.S.W. Q. N.A. B.fl.VII,262 M.fr.VIII,265;IX,53.
C. cephalotes, Vahl, enum. II, 311 (1806) ... ... — — — — Q. — B.fl.VII,263 M.fr.VIII,272;IX,54.
C. laevigatus, Linné, mantissa alt. 170 (1771) ... ... W.A. — — — — B.fl.VII,263 M.fr.VIII,266;IX,53.
C. platystylis, R. Brown, prodr. 241 (1810) ... ... — — — N.S.W. — — B.fl.VII,264
C. alopecuroides, Rottboell, descr. et icon. pl. 38, t. 6 (1773) .. — — — — Q. — B.fl.VII,264 M.fr.VIII,263;IX,52.
C. pulchellus, R. Brown, prodr. 213 (1810) ... ... — — — — Q. N.A. B.fl.VII,265 M.fr.VIII,271.
C. tenellus, Linné, fil. suppl. 103 (1781) ... ... W.A., S.A. — N.S.W. — — B.fl.VII,265 M.fr.VIII,261;IX,53.
C. gracilis, R. Brown, prodr. 213 (1810) ... ... — S.A. — V. N.S.W. Q. — B.fl.VII,265 M.fr.VIII,264;IX,53.
C. enervis, R. Brown, prodr. 213 (1810) ... ... — — — N.S.W. Q. — B.fl.VII,266
C. debilis, R. Brown, prodr. 213 (1810) ... ... — — — N.S.W. — — B.fl.VII,266
C. castaneus, Willdenow, spec. plant, I, 278 (1797) ... ... — — — — Q. — B.fl.VII,267
C. cuspidatus, Humboldt, Bonpland & Kunth, nov. gen. et sp. pl. I, 204 (1815) ... ... — — — — Q. — B.fl.VII,267
C. squarrosus, Linné, amoen. acad. IV, 303 (1759) ... — S.A. — — Q. N.A. B.fl.VII,268 M.fr.VIII,262;IX,53.
C. difformis, Linné, amoen. acad. IV, 302 (1759)... ... W.A. S.A. — V. N.S.W. Q. N.A. B.fl.VII,268 M.fr.VIII,262;IX,52.
C. tetraphyllus, R. Brown, prodr. 214 (1810) ... ... — — — N.S.W. Q. — B.fl.VII,269 M.fr.VIII,264.
C. trinervis, R. Brown, prodr. 213 (1810) ... ... — S.A. — V. N.S.W. Q. — B.fl.VII,269 M.fr.VIII,267;IX,53.
C. Haspan, Linné, spec. plant. 45 (1753) ... ... — — — N.S.W. Q. N.A. B.fl.VII,270 M.fr.VIII,260;IX,52.
C. concinnus, R. Brown, prodr. 214 (1810)... ... ... — — V. N.S.W. Q. — B.fl.VII,271 M.fr.VIII,261.
C. filipes, Bentham, Fl. Austr. VII, 271 (1878) ... — — — N.S.W. Q. — B.fl.VII,271
C. pedunculosus, F. v. M., fragm. VIII, 266 (1874) ... — — — — Q. — B.fl.VII,272 M.fr.
C. vaginatus, R. Brown, prodr. 213 (1810)... ... W.A. S.A. — V. N.S.W. Q. N.A. B.fl.VII,272 M.fr.VIII,261;IX,53.
C. Holoschoenus, R. Brown, prodr. 215 (1810) ... — S.A. — — Q. N.A. B.fl.VII,273 M.fr.I,200; VIII,262.
C. dactylotes, Bentham, Fl. Austr. VII, 273 (1878) ... — — — — Q. N.A. B.fl.VII,273
C. Gilesii, Bentham, Fl. Austr. VII, 274 (1878) ... — S.A. — N.S.W. Q. — B.fl.VII,274
C. fulvus, R. Brown, prodr. 215 (1810) ... ... — S.A. — N.S.W. Q. — B.fl.VII,274 M.fr.VIII,268.
C. carinatus, R. Brown, prodr. 216 (1810)... ... — — — N.S.W. Q. — B.fl.VII,274
C. alterniflorus, R. Brown, prodr. 216 (1810) ... W.A. S.A. — — — — B.fl.VII,275
C. pilosus, Vahl, enum. II, 354 (1806) ... ... — — — N.S.W. Q. — B.fl.VII,275 M.fr.VIII,260;IX,52.
C. ornatus, R. Brown, prodr. 217 (1810) ... ... — — — — Q. — B.fl.VII,276
C. Iria, Linné, spec. plant. 45 (1753) ... ... — S.A. — N.S.W. Q. N.A. B.fl.VII,276 M.fr.VIII,266;IX,53.
C. clusinoides, Wallich in Kunth, enum. II, 39 (1837) ... — — — — Q. N.A. B.fl.VII,277 M.fr.VIII,264;IX,53.
C. distans, Linné, fil. suppl. 103 (1781) ... ... — — — — Q. — B.fl.VII,277 M.fr.VIII,266;IX,53.
C. tegetiformis, Roxburgh, hort. Benghal. (1814) ... — — — — — N.A. B.fl.VII,278 M.fr.VIII,260;IX,52.
C. articulatus, Linné, spec. plant. 44 (1753) ... — — — — — N.A. B.fl.VII,278
C. diphyllus, Retzius, observ. bot. V, 11 (1789) ... — — — — — N.A. B.fl.VII,279
C. rotundus, Linné, spec. plant. 45 (1753) ... ... — S.A. — V. N.S.W. Q. N.A. B.fl.VII,279 M.fr.VIII,269;IX,53.
C. stenostachys, Bentham, Fl. Austr. VII, 280 (1878) ... W.A. — — — — — B.fl.VII,280
C. congestus, Vahl, enum. II, 350 (1806) ... ... W.A. — — — — — B.fl.VII,280 M.fr.VIII,269;IX,53.
C. subulatus, R. Brown, prodr. 217 (1810)... ... — S.A. — N.S.W. — — B.fl.VII,281
C. sporobolus, R. Brown, prodr. 215 (1810) ... ... — — — — — N.A. B.fl.VII,281
C. angustatus, R. Brown, prodr. 214 (1810) ... ... — — — — Q. N.A. B.fl.VII,282 M.fr.IX,54.
C. inornatus, Boeckeler in Fl. Regensb. 86 (1875) ... — — — — Q. — B.fl.VII,282
C. lucidus, R. Brown, prodr. 218 (1810) ... ... — S.A. T. V. N.S.W. Q. N.A. B.fl.VII,283 M.fr.VIII,270;IX,53.
C. pennatus, Lamarck, illustr. des genr. I, 144 (1791) ... — — — — Q. N.A. B.fl.VII,284 M.fr.VIII,203;IX,53.
C. exaltatus, Retzius, observ. bot. V, 11 (1780) ... — S.A. — V. N.S.W. Q. N.A. B.fl.VII,285 M.fr.VIII,203;IX,53.
C. haematodes, Endlicher, prodr. fl. Norf. 22 (1839) ... — — — N.S.W. — — B.fl.VII,285 M.fr.IX,54.
C. auricomus, Sieber in Sprengel, syst. I, 230 (1825) ... — — — — Q. — B.fl.VII,286 M.fr.VIII,263.
C. ferax, Cl. Richard in act. soc. hist. nat. Par. I, 106 (1792) ... — — — — — — B.fl.VII,286
C. Bowmanni, F. v. M. in Benth. Fl. Austr. VII, 287 (1878) ... — — — N.S.W. Q. — B.fl.VII,287
C. trichostachys, Bentham, Fl. Austr. VII, 287 (1878) ... — — — — Q. — B.fl.VII,287
C. leiocaulis, Bentham, Fl. Austr. VII, 287 (1878) ... — — — N.S.W. Q. — B.fl.VII,287
C. scaber, Bentham, Fl. Austr. VII, 288 (1878) ... — — — — Q. N.A. B.fl.VII,288
C. decompositus, F. v. M., fragm. VIII, 267 (1874) ... — — — — Q. N.A. B.fl.VII,288 M.fr.
C. Armstrongii, Bentham, Fl. Austr. VII, 289 (1878) ... — — — — Q. N.A. B.fl.VII,289
C. umbellatus, Bentham, Fl. Hongk. 380 (1861) ... ... — — — — Q. N.A. B.fl.VII,289
C. conicus, Boeckeler in Schlecht. Linnaea XXXVIII, 371 (1874) W.A. — — — Q. N.A. B.fl.VII,290 M.fr.VIII,268;IX,53.

HELEOCHARIS, R. Brown, prodr. 224 (1810). (Eleocharis, Eleochiton, Scirpidium.)
H. sphacelata, R. Brown, prodr. 224 (1810) ... ... — S.A. T. V. N.S.W. Q. N.A. B.fl.VII,292 M.fr.VI,94;VIII,239.
H. spiralis, R. Brown, prodr. 224 (1810) ... ... — — — — Q. N.A. B.fl.VII,292
H. compacta, R. Brown, prodr. 224 (1810) ... ... — — — N.S.W. Q. N.A. B.fl.VII,293 M.fr.VIII,239.
H. fistulosa, J. A. & J. H. Schultes, mantiss. II, 89 (1824) ... — — — — Q. — B.fl.VII,293 M.fr.VIII,93;VIII,239.
H. tetraquetra, Nees in Wight's Contrib. Bot. of Ind. 113 (1834) ... — — — N.S.W. Q. — B.fl.VII,294 M.fr.VIII,239.
H. cylindrostachys, Boeckeler in Regensb. Flora, 109 (1875) ... — — — — Q. — B.fl.VII,294 M.fr.VIII,240.
H. acuta, R. Brown, prodr. 224 (1810) ... ... W.A. S.A. T. V. N.S.W. Q. — B.fl.VII,294 M.fr.VIII,240.
H. multicaulis, Smith, Engl. Flor. II, 64 (1824) ... W.A. S.A. — V. N.S.W. — — B.fl.VII,296
H. atricha, R. Brown, prodr. 225 (1810) ... ... — — — N.S.W. Q. N.A. B.fl.VII,295 M.fr.VII,252;IX,100.
H. capitata, R. Brown, prodr. 225 (1810) ... ... W.A. — — — Q. — B.fl.VII,296 M.fr.V,93;VIII,240.
H. atropurpurea, Kunth, enum. II, 151 (1837) ... ... — — — N.S.W. Q. N.A. B.fl.VII,296 M.fr.VIII,240.
H. acicularis, R. Brown, prodr. 224 (1810). ... ... — — — N.S.W. Q. N.A. B.fl.VII,296 M.fr.VIII,240.

FIMBRISTYLIS, Vahl, enum. plant. II, 285 (1800). (Abildgaardia, Trichelostylis, Oncostylis.)
F. acicularis, R. Brown, prodr. 226 (1810)... ... ... — — — — Q. N.A. B.fl.VII,301 M.fr.IX,54.
F. acuminata, Vahl, enum. II, 285 (1800) ... ... — — — N.S.W. Q. N.A. B.fl.VII,301 M.fr.VIII,274;IX,54.
F. punctata, R. Brown, prodr. 226 (1810) ... ... — — — — — — B.fl.VII,302
F. rhyticarya, F. v. M., fragm. I, 215 (1859) ... ... — — — — Q. — B.fl.VII,302 M.fr.I,215.
F. leucostachya, Boeckeler in Garcke, Linnaea XXXVIII, 385 (1874) — — — — — N.A. B.fl.VII,302 M.fr.IX,54.
F. nutans, Vahl, enum. II, 285 (1800) ... ... — — — N.S.W. Q. N.A. B.fl.VII,303 M.fr.VIII,274;IX,54.
F. pauciflora, R. Brown, prodr. 225 (1810)... ... ... — — — — Q. N.A. B.fl.VII,303

F. cardiocarpa, F. v. M., fragm. I, 104 (1859) ... ... — — — — — — — N.A. B.fl.VII,303 M.fr.I,194.
F. leucocolea, Bentham, Fl. Austr. VII, 304 (1878) ... ... — — — — N.A. B.fl.VII,304
F. polytrichoides, R. Brown, prodr. 226 (1810) ... ... — — — — Q. N.A. B.fl.VII,304
F. androgyna, R. Brown, prodr. 226 (1810) ... ... — — — — — N.A. B.fl.VII,304
F. subbulbosa, Bentham, Fl. Austr. VII, 305 (1878) ... ... — — — — Q. — B.fl.VII,305
F. totragona, R. Brown, prodr. 226 (1810)... ... ... — — — — — N.A. B.fl.VII,305 M.fr.I,194;VIII,274.
F. trigastrocarya, F. v. M., fragm. I, 104 (1859) ... ... — — — — — N.A. B.fl.VII,305 M.fr.I,194.
F. monandra, F. v. M., fragm. I, 195 (1859) ... ... — — — — N.A. B.fl.VII,306 M.fr.I,195.
F. pterygosperma, R. Brown, prodr. 226 (1810) ... ... — — — — — N.A. B.fl.VII,306 M.fr.I,193;IX,55.
F. sphaerocephala, Bentham, Fl. Austr. VII, 306 (1878)... ... ... — — — — Q. N.A. B.fl.VII,306
F. xyridis, R. Brown, prodr. 226 (1810) ... ... — — — — Q. N.A. B.fl.VII,307 M.fr.VIII,274;IX,54.
F. oxystachya, F. v. M., fragm. I, 195 (1859) ... ... — — — — N.A. B.fl.VII,307 M.fr.VIII,272.
F. macrantha, Boeckeler in Garcke in Linnaea XXXVIII, 388 (1874) — — — — ... — N.A. B.fl.VII,307 M.fr.IX,55.
F. squarrulosa, F. v. M., fragm. I, 216 (1859) ... ... — — — — N.A. B.fl.VII,308 M.fr.I,216.
F. monostachya, Hasskarl, pl. Jav. rar. 61-64 (1848) ... ... — — — N.S.W. Q. N.A. B.fl.VII,308 M.fr.VIII,272;IX,54.
F. Brownii, Bentham, Fl. Austr. VII, 308 (1878) ... ... — — — — — N.A. B.fl.VII,308 M.fr.VIII,273.
F. Dallachyi, F. v. M. in Benth. Fl. Austr. VII, 309 (1878) ... ... — — — Q. — B.fl.VII,309 M.fr.VIII,273.
F. velata, R. Brown, prodr. 227 (1810) ... ... — S.A. — V. N.S.W. Q. N.A. B.fl.VII,309 M.fr.IX,11,54.
F. aestivalis, Vahl, enum. II, 288 (1806) ... ... — — V. N.S.W. Q. N.A. B.fl.VII,310 M.fr.IX,11,54.
F. dichotoma, Vahl, enum. II, 287 (1806) ... ... — — — N.S.W. Q. N.A. B.fl.VII,310 M.fr.IX,10,54.
F. depauperata, R. Brown, prodr. 227 (1810) ... ... — — — — — N.A. B.fl.VII,311
F. spirostachya, F. v. M. in Benth. Fl. Austr. VII, 311 (1878)... ... — — — — N.A. B.fl.VII,311
F. communis, Kunth, enum. II, 234 (1837) ... ... — S.A. — V. N.S.W. Q. N.A. B.fl.VII,311 M.fr.I,196;IX,10,54.
F. ferruginea, Vahl, enum. 291 (1806) ... ... W.A. S.A. — N.S.W. Q. N.A. B.fl.VII,312 M.fr.I,107;IX,10,54.
F. denudata, R. Brown, prodr. 227 (1810)... ... ... — — — — Q. N.A. B.fl.VII,313 M.fr.IX,9.
F. olata, R. Brown, prodr. 227 (1810) ... ... — — — — — N.A. B.fl.VII,313
F. caespitosa, R. Brown, prodr. 228 (1810) ... ... — — — — Q. N.A. B.fl.VII,313 M.fr.I,199.
F. spiralis, R. Brown, prodr. 226 (1810) ... ... — — — — — N.A. B.fl.VII,314
F. subaristata, Bentham, Fl. Austr. VII, 314 (1878) ... ... — — — — — N.A. B.fl.VII,314
F. leptoclada, Bentham, Fl. Austr. VII, 314 (1878) ... ... — — — — Q. — B.fl.VII,314
F. debilis, F. v. M., fragm. I, 198 (1859) ... ... — — — — N.A. B.fl.VII,315 M.fr.I,198.
F. corynocarya, F. v. M., fragm. I, 197 (1859) ... ... — — — — N.A. B.fl.VII,315 M.fr.I,187.
F. solidifolia, F. v. M., fragm. I, 198 (1859) ... ... — — — — N.A. B.fl.VII,315 M.fr.I,198.
F. obtusangula, F. v. M., fragm. I, 198 (1859) ... ... — — — — Q. N.A. B.fl.VII,315 M.fr.I,198.
F. miliacea, Vahl, enum. II, 287 (1806) ... ... — — — — Q. N.A. B.fl.VII,316 M.fr.I,199;IX,12,54.
F. rara, R. Brown, prodr. 227 (1810) ... ... — — — — — N.A. B.fl.VII,316
F. microcarya, F. v. M., fragm. I, 200 (1859) ... ... — — — — Q. N.A. B.fl.VII,316 M.fr.I,200.
F. quinquangularis, Kunth, enum. II, 229 (1837) ... ... — — — — N.A. B.fl.VII,317
F. cyperoides, R. Brown, prodr. 228 (1810) ... ... — — — N.S.W. Q. N.A. B.fl.VII,317 M.fr.VIII,273;IX,11,54.
F. furva, R. Brown, prodr. 228 (1810) ... ... — — — — — Q. — B.fl.VII,318
F. cymosa, R. Brown, prodr. 229 (1810) ... ... — — — — — N.A. B.fl.VII,319
F. multifolia, Boeckeler in Garcke, Linnaea XXXVIII, 397 (1874) — — — — — N.A. B.fl.VII,319 M.fr.IX,55.
F. sericea, R. Brown, prodr. 228 (1810) ... ... — — — — — N.A. B.fl.VII,319
F. macrostachya, Boeckeler in Garcke, Linn. XXXVIII, 386 (1874) — — — — — S.A. — N.S.W. — B.fl.VII,319 M.fr.IX,54.
F. Neilsoni, F. v. M., fragm. IX, 79 (1875) ... ... — S.A. — N.S.W. — B.fl.VII,320 M.fr.IX,70.
F. capitata, R. Brown, prodr. 228 (1810) ... ... — — — — Q. N.A. B.fl.VII,320 M.fr.I,196.
F. Schultzii, Boeckeler in Garcke, Linnaea XXXVIII, 391 (1874) — — — — — N.A. B.fl.VII,321 M.fr.IX,55.
F. barbata, Bentham, Fl. Austr. VII, 321 (1878) ... ... — — S.A. — N.S.W. Q. N.A. B.fl.VII,321 M.fr.IX,55.
F. capillaris, A. Gray, Man. of Bot. N.U.S. fifth ed. 567 (1867) W.A. — — — Q. N.A. B.fl.VII,322 M.fr.IX,55.

SCIRPUS, Tournefort, inst. 528, t. 300 (1700), from Terentius. (Isolepis, Malacochaete.)
S. humillimus, Bentham, Fl. Austr. VII, 324 (1878) ... ... — — N.A. B.fl.VII,324 M.fr.IX,7.
S. fluitans, Linné, spec. plant. 48 (1753) ... ... W.A. S.A. T. V. N.S.W. — B.fl.VII,325 M.fr.IX,55.
S. arenarius, Bentham, Fl. Austr. VII, 325 (1878) ... ... W.A. — — — — B.fl.VII,325
S. lenticularis, Sprengel, syst. I, 208 (1825) ... ... — — N.S.W. — B.fl.VII,326 M.fr.IX,55.
S. crassiusculus, J. Hooker in Benth. Fl. Austr. VII, 326 (1878) ... — T. V. N.S.W. — B.fl.VII,326
S. cyperoides, Sprengel, syst. I, 208 (1825) ... ... W.A. — — — — B.fl.VII,326
S. setaceus, Linné, spec. plant. 49 (1753) ... ... W.A. S.A. T. V. N.S.W. Q. — B.fl.VII,327 M.fr.IX,55.
S. riparius, Sprengel, syst. I, 208 (1825) ... ... W.A. S.A. T. V. N.S.W. — B.fl.VII,327
S. cartilagineus, Sprengel, syst. I, 208 (1823) ... ... W.A. S.A. T. V. N.S.W. — B.fl.VII,328 M.fr.IX,55.
S. squarrosus, Linné, mant. plant. alt. 181 (1771)... ... ... — — Q. N.A. B.fl.VII,329
S. inundatus, Sprengel, syst. I, 207 (1825)... ... ... — S.A. T. V. N.S.W. Q. N.A. B.fl.VII,329 M.fr.IX,55.
S. prolifer, Rottboell, descr. et icon. pl. 55 t. 17 (1773) ... — S.A. T. V. N.S.W. — B.fl.VII,330 M.fr.IX,55.
S. articulatus, Linné, spec. plant. 47 (1753) ... ... — — N.A. B.fl.VII,331 M.fr.IX,55.
S. nodosus, Rottboell, descr. et icon. pl. 52, t. 8 (1773) ... W.A. S.A. T. V. N.S.W. — B.fl.VII,331 M.fr.IX,55.
S. supinus, Linné, spec. plant. 49 (1753) ... ... — S.A. — — N.S.W. Q. N.A. B.fl.VII,330 M.fr.IX,55.
S. debilis, Pursch, flor. Amer. Sept. I, 55 (1814) ... ... — — N.A. B.fl.VII,332
S. mucronatus, Linné, spec. plant. 50 (1753) ... ... — S.A. — — N.S.W. Q. — B.fl.VII,332 M.fr.IX,8,56.
S. pungens, Vahl, enum. II, 255 (1806) ... W.A. S.A. T. V. N.S.W. Q. N.A. B.fl.VII,333 M.fr.IX,8,56.
S. lacustris, Linné, spec. plant. 48 (1753) ... W.A. S.A. T. V. N.S.W. Q. N.A. B.fl.VII,333 M.fr.IX,7,56.
S. litoralis, Schrader, Fl. Germ. I, 142, t. 5 (1806) ... ... W.A. S.A. — — N.S.W. — B.fl.VII,334 M.fr.IX,7,56.
S. maritimus, Linné, spec. plant. 51 (1753) ... W.A. S.A. T. V. N.S.W. Q. N.A. B.fl.VII,335 M.fr.IX,8,56.
S. grossus, Linné, fil. suppl. plant. 104 (1781) ... ... — — N.A. B.fl.VII,335
S. polystachyus, F. v. M. in Transact. phil. Soc. Vict. I, 108 (1854) ... — — N.S.W. — B.fl.VII,335 M.fr.IX,9,56.

LIPOCARPHA, R. Brown in Tuckey's narrat. of an Exped. to Congo, 459 (1818). (Hypaelyptum.)
L. argentea, R. Brown in Tuck. Congo, 459 (1818) ... ... — — N.A. B.fl.VII,336 M.fr.VIII,238;IX,57.
L. microcephala, R. Brown in Tuck. Congo, 459 (1818) ... ... — V. N.S.W. Q. N.A. B.fl.VII,337 M.fr.VIII,238;IX,55.

FUIRENA, Rottboell, descr. et icon. rar. pl. illustr. 70 (1773).
F. umbellata, Rottboell, descr. et icon. pl. 70, t. 10 (1773) ... ... — — Q. N.A. B.fl.VII,337 M.fr.VIII,238;IX,57.
F. glomerata, Lamarck, Illustr. des genr. I, 150 (1791) ... ... — — N.S.W. Q. N.A. B.fl.VII,338 M.fr.VIII,238;IX,57.

HYPELYTRUM, L. C. Richard in Persoon, synops. I, 70 (1805).   (Hypolytrum.)
H. latifolium, L. C. Richard in Pers. syn. I, 70 (1805) ... ... -- — — — — — Q. — B.fl.VII,339 M.fr.VIII,238;IX,57.

EXOCARYA, Bentham in J. Hooker, icon. plant. t. 1206 (1877).
E. scleroides, Bentham in Hook. icon. pl. t. 1206 (1877)... ... ... — — — — N.S.W. Q. — B.fl.VII,339 M.fr.IX,12.

MAPANIA, Aublet, Hist. pl. Guian. 47, t. 17 (1775).   (Pandanophyllum.)
M. hypelytroides, F. v. M., fragm. IX, 16 (1875)... ... ... — — — — — Q. — B.fl.VII,341 M.fr.IX,16.

SCIRPODENDRON, Zippelius in Journ. Asiat. Soc. Beng. XXXVIII, 85 (1869).
S. costatum, Kurz in Journ. Asiat. Soc. Beng. XXXVIII, 85 (1869) — — — — — Q. — B.fl.VII,341 M.fr.X,104.

LEPIRONIA, L. C. Richard in Persoon, synops. I, 70 (1805).   (Chondrachne).
L. mucronata, L. C. Richard in Pers. syn. I, 70 (1805) ... ... — — — — N.S.W. Q. — B.fl.VII,342 M.fr.IX,17,58.

CHORIZANDRA, R. Brown, prodr. 221 (1810).   (Chorisandra.)
C. spacrocephala, R. Brown, prodr. 221 (1810) ... ... — — N.S.W. Q. — B.fl.VII,344 M.fr.IX,16,58.
C. enodis, Nees in Lehm. pl. Preiss. II, 73 (1846) .. .. W.A. S.A. T. V. N.S.W. — B.fl.VII,344 M.fr.IX,18.
C. multiarticulata, Nees in Ann. and Mag. of nat. Hist. VI, 48 (1842)W.A. — — — — — B.fl.VII,345 M.fr.IX,18.
C. cymbaria, R. Brown, prodr. 221 (1810) ... ... W.A. — V. N.S.W. Q. — B.fl.VII,345 M.fr.IX,18,58

OREOBOLUS, R. Brown, prodr. fl. Nov. Holl. 235 (1810).
O. Pumilio, R. Brown, prodr. 236 (1810) ... ... — — — T. V. N.S.W. — B.fl.VII,346 M.fr.IX,20,58.

REMIREA, Aublet, Hist. des pl. de la Guian. franc. I, 44, t. 16 (1775).
R. maritima, Aublet, Hist. pl. Guian. I, 45, t. 16 (1775) ... ... — — — — Q. — B.fl.VII,347 M.fr.V,92;IX,20,58.

RHYNCHOSPORA, Vahl, enum. plant. II, 229 (1806).   (Morisia.)
R. aurea, Vahl, enum. II, 229 (1806) ... ... — — — — — Q. — B.fl.VII,348 M.fr.IX,17,57.
R. glauca, Vahl, enum. II, 229 (1806) ... ... — — N.S.W. Q. — B.fl.VII,349 M.fr.IX,17,57.
R. Wallichiana, Kunth, enum. II, 289 (1837) ... ... -- — — — Q. N.A. B.fl.VII,349 M.fr.IX,17,52.
R. longisetis, R. Brown, prodr. 230 (1810).. ... ... — — Q. N.A. B.fl.VII,350 M.fr.IX,17,58.
R. tenuifolia, Bentham, Fl. Austr. VII, 350 (1878) ... ... — — Q. N.A. B.fl.VII,350

CYATHOCHAETE, Nees in Lehm. pl. Preiss. II, 86 (1846).   (Tetralepis.)
C. clandestina, Bentham, Fl. Austr. VII, 351 (1878) ... ... W.A. — — — — — B.fl.VII,351 M.fr.IX,40.
C. avenacea, Bentham, Fl. Austr. VII, 351 (1878) ... ... W.A. — — — — -- — B.fl.VII,351 M.fr.IX,40.
C. diandra, Nees in Lehm. pl. Preiss. II, 86 (1846) ... ... — V. N.S.W. — — B.fl.VII,352 M.fr.IX,39.

CARPHA, Banks & Solander in R. Brown, prodr. 230 (1810).
C. alpina, R. Brown, prodr. 280 (1810) ... ... — T. V. N.S.W. — B.fl.VII,381 M.fr.IX,39.

SCHOENUS, Linné, coroll. gen. 2 (1737).   (Chaetospora, Elynanthus, Tricostularia, Helothrix, Isoschoenus, Gymnochaete, Gymnoschoenus, Mesomelaena, Discopodium.)
S. cruentus, F. v. M., fragm. IX, 36 (1875) ... ... W.A. — — — — — B.fl.VII,357 M.fr.IX,37.
S. Benthami, F.v.M.; S.compressus, Benth. Fl.Austr.VII,357(1878)W.A. — — — — — B.fl.VII,357 M.fr.IX,37.
S. lanatus, Labillard., Nov. Holl. pl. spec. I, 19, t. 20 (1804) ... W.A. — — — — — B.fl.VII,357 M.fr.IX,37.
S. curvifolius, Poiret, Encycl. méth. suppl. II, 251 (1811) ... W.A. — — — — — B.fl.VII,358 M.fr.IX,36.
S. capitatus, F. v. M., fragm. IX, 58 (1875) ... ... W.A. — — — — — B.fl.VII,358 M.fr.IX,37.
S. setifolius, Bentham, Fl. Austr. VII, 359 (1878) ... ... W.A. — — — — — B.fl.VII,359
S. Drummondii, Bentham, fl. Austr. VII, 359, non Steud. (1870) W.A. — — — — — B.fl.VII,359 M.fr.IX,37.
S. turbinatus, Poiret, Encycl. méth. suppl. II, 251 (1811) ... ... — — N.S.W. — B.fl.VII,359 M.fr.IX,33.
S. barbatus, Boeckeler in Garcke, Linnaea XXXVIII, 277 (1874) W.A. — — — — — B.fl.VII,360 M.fr.IX,30.
S. flavus, Boeckeler in Garcke, Linnaea XXXVIII, 278 (1874) W.A. — — — — — B.fl.VII,360 M.fr.IX,30.
S. brevisetis, Poiret, Encycl. méth. suppl. II, 251 (1811) ... ... W.A. — — — — — B.fl.VII,360 M.fr.IX,30.
S. Armeria, Boeckeler in Garcke, Linnaea XXXVIII, 279 (1874) W.A. — — — — — B.fl.VII,361 M.fr.IX,30.
S. aphyllus, Boeckeler in Garcke, Linnaea XXXVIII, 280 (1874) — S.A. — V. N.S.W. — B.fl.VII,361 M.fr.IX,28.
S. imberbis, R. Brown, prodr. 231 (1810) ... ... — V. N.S.W. — B.fl.VII,361 M.fr.IX,28.
S. erictorum, R. Brown, prodr. 231 (1810) ... ... — — N.S.W. Q. — B.fl.VII,362 M.fr.IX,58.
S. nitens, Poiret, Encycl. méth. suppl. II, 251 (1811) ... W.A. S.A. T. V. N.S.W. Q. — B.fl.VII,362 M.fr.IX,35.
S. cygneus, Nees in Lehmann, pl. Preiss. II, 81 (1846) ... ... W.A. — — — — — B.fl.VII,363 M.fr.IX,80.
S. minutulus, F. v. M., fragm. IX, 32 (1875) ... ... W.A. — — — — — B.fl.VII,363 M.fr.IX,32.
S. trachycarpus, F. v. M., fragm. IX, 33 (1875) ... ... W.A. — — — — — B.fl.VII,363 M.fr.IX,33.
S. Tepperi, F. v. M., fragm. XI, 106 (1880) ... ... — S.A. — — — — B.fl.VII,364 M.fr.IX,28.
S. nanus, F. v. M., fragm. IX, 36 (1875) ... ... W.A. — — — — — B.fl.VII,364 M.fr.IX,32.
S. pleiostemoneus, F. v. M., fragm. IX, 52 (1875) ... ... W.A. — — — — — B.fl.VII,364 M.fr.IX,32.
S. breviculmis, Bentham, Fl. Austr. VII, 364 (1878) ... ... W.A. — — — — — B.fl.VII,364
S. deformis, Poiret, Encycl. méth. suppl. II, 251 (1811) ... — S.A. — — — — B.fl.VII,364 M.fr.IX,39.
S. unispiculatus, F. v. M. in Benth. Fl. Austr. VII, 365 (1878) W.A. — — — — — B.fl.VII,365
S. obtusifolius, Boeckeler in Garcke, Linnaea XXXVIII,281 (1874) W.A. — — — — — B.fl.VII,366 M.fr.IX,31.
S. granitophyllus, F. v. M., fragm. IX, 31 (1875) ... ... W.A. — — — — — B.fl.VII,366 M.fr.IX,31.
S. asperocarpus, F. v. M., fragm. IX, 29 (1875) ... ... W.A. — — — — — B.fl.VII,366 M.fr.IX,29.
S. Moorei, Bentham, Fl. Austr. VII, 367 (1878) ... ... — — N.S.W. — B.fl.VII,367 M.fr.IX,28.
S. villosus, R. Brown, prodr. 231 (1810) ... ... — — N.S.W. — B.fl.VII,367 M.fr.IX,33.
S. grandiflorus, F. v. M., fragm. IX, 30 (1875) ... ... — — — — — M.fr.IX,33.
S. distans ; Chaetospora, F. v. M., fragm. IX, 35 (1875) W.A. — — — — — B.fl.VII,368
S. calostachyus, Poiret, Encycl. méth. suppl. II, 251 (1811) ... W.A. — — N.S.W. Q. — B.fl.VII,368
S. scabripes, Bentham, Fl. Austr. VII, 368 (1878) ... ... W.A. — — — — Q. — B.fl.VII,368
S. multiglumis, Bentham, Fl. Austr. VII, 308 (1878) ... W.A. — — — — — B.fl.VII,368
S. efoliatus, F. v. M., fragm. IX, 32 (1875) ... ... W.A. — — — — — B.fl.VII,369 M.fr.IX,32.
S. acuminatus, R. Brown, prodr. 231 (1810) ... ... W.A. — — — — — B.fl.VII,369 M.fr.IX,80.
S. pedicellatus, Poiret, Encycl. méth. suppl. II, 251 (1811) ... W.A. — — — — — B.fl.VII,369
S. fasciculatus, Nees in Ann. and Mag. of nat. Hist. VI, 48 (1842) W.A. — — — — — B.fl.VII,370
S. brevifolius, R. Brown, prodr. 231 (1810) ... ... W.A. S.A. — V. N.S.W. Q. — B.fl.VII,370 M.fr.IX,29,80.
S. melanostachys, R. Brown, prodr. 231 (1810) ... ... — V. N.S.W. Q. — B.fl.VII,370 M.fr.IX,29.
S. spartens, R. Brown, prodr. 231 (1810) ... ... — — Q. N.A. B.fl.VII,371
S. vaginatus, F. v. M. in Benth. Fl. Austr. VII, 371 (1878) ... — N.S.W. Q. — B.fl.VII,371

128

S. falcatus, R. Brown, prodr. 232 (1810) ... ... — — — — — Q. N.A. B.fl.VII,372 M.fr.IX,29.
S. punctatus, R. Brown, prodr. 232 (1810)... ... ... — — — — — N.A. B.fl.VII,372
S. indutus, F. v. M. in Benth. Fl. Austr. VII, 372 (1878) ... W.A. — — — — — — B.fl.VII,372
S. bifidus, Boeckeler in Garcke, Linnaea XXXVIII, 282 (1874) W.A. — — — — — B.fl.VII,373 M.fr.IX,38,80.
S. apogon, Roemer et Schultes, syst. veg. II, 77 (1817) .. ... S.A. T. V. N.S.W. Q. — B.fl.VII,373 M.fr.IX,56;X,120.
S. odontocarpus, F. v. M., fragm. IX, 32 (1875) ... ... W.A. — — — — — B.fl.VII,374 M.fr.IX,32.
S. humilis, Bentham, Fl. Austr. VII, 374 (1878) ... ... W.A. — — — — — B.fl.VII,374
S. sculptus, Boeckeler in Garcke, Linnaea XXXVIII, 236 (1874) W.A. — — — — — B.fl.VII,374 M.fr.IX,30.
S. axillaris, Poiret, encycl. méth. suppl. II, 251 (1811) ... ... W.A. S.A. T. V. N.S.W. — B.fl.VII,375 M.fr.IX,34.
S. tenellus, Bentham, Fl. Austr. VII, 375 (1878)... ... ... W.A. — — — — — B.fl.VII,375
S. natans, F. v. M., fragm. IX, 36 (1875) ... ... W.A. — — — — — B.fl.VII,375 M.fr.IX,38.
S. fluitans, J. Hooker, Fl. Tasman. II, 81, t. 141 (1860)... — T. — — — — R.fl.VII,376 M.fr.IX,28.
S. octandrus, F. v. M., fragm. IX, 31 (1875) ... ... W.A. — — — — — B.fl.VII,377 M.fr.IX,31.
S. capillaris, Chaetospora, F. v. M., fragm. IX, 377 (1875) ... W.A. S.A. T. V. — — B.fl.VII,377 M.fr.IX,34,80.
S. stygius, Poiret, encycl. méth. suppl. II, 251 (1811) ... ... W.A. — — — — — B.fl.VII,378 M.fr.IX,30.
S. deustus, F. v. M.; Carpha deusta, R. Brown, prodr. 230 (1810) — — — — N.S.W. Q. — B.fl.VII,379 M.fr.IX,39.
S. tetragonus, Poiret, Encycl. méth. suppl. II, 251 (1811) ... W.A. — — — — — B.fl.VII,380 M.fr.IX,36.
S. subtetragonus, Poiret, Encycl. méth. suppl. II, 251 (1811)... — T. V. N.S.W. — — B.fl.VII,380 M.fr.IX,33.
S. anceps, Poiret, Encycl. méth. suppl. II, 251 (1811) ... ... W.A. — — — — — B.fl.VII,380 M.fr.IX,30.
S. pauciflorus, Poiret, Encycl. méth. suppl. II, 251 (1811) ... — — — — N.S.W. Q. — B.fl.VII,382 M.fr.IX,35.
S. pauciflorus; Lepidosperma, F. v. M., fragm. IX, 23 (1875) ... — — — V. — — B.fl.VII,383 M.fr.IX,23.
S. Tricostularia, F. v. M.; Tricostularia compressa, Nees in pl.
    Preiss. II, 83 (1846) ... ... W.A. — — — — — B.fl.VII,383 M.fr.IX,37.
S. Neesii, F.v.M.; Tricostularia, Lehmann, pl.Preiss. II,83(1846) W.A. — — — — — B.fl.VII,383 M.fr.IX,37.
S. fimbristyloides; Chaetospora, F. v. M., fragm. IX, 34 (1875) — — — — — N.A. B.fl.VII,384 M.fr.IX,34.

    LEPIDOSPORA, F. v. M., fragm. phytograph. Austral. IX, 34 (1875).
L. tenuissima, F. v. M., fragm. IX, 34 (1875) ... ... — S.A. T. V. N.S.W. — — B.fl.VII,365 M.fr.IX,34.

    LEPIDOSPERMA, Labillardière, Nov. Holl. pl. spec. I, 14 (1804).
L. gladiatum, Labillardière, Nov. Holl. pl. spec. I, 15, t. 12 (1804) W.A. S.A. T. V. N.S.W. — B.fl.VII,387
L. effusum, Bentham, Fl. Austr. VII, 387 (1878)... ... ... W.A. — — — — — B.fl.VII,387
L. rupestre, Bentham, Fl. Austr. VII, 388 (1878)... ... ... W.A. — — — — — B.fl.VII,388
L. elatius, Labillardière, Nov. Holl. pl. spec. I, 14 (1804) — S.A. T. V. — — — B.fl.VII,388 M.fr.IX,23.
L. tetraquetrum, Nees in Lehm. pl. Preiss. II, 90 (1846) W.A. — — — — — B.fl.VII,388 M.fr.IX,24.
L. Oldfieldii, J. Hooker, Fl. Tasman. II, 91, t. 146 (1860) ... — T. — — — — B.fl.VII,389
L. exaltatum, R. Brown, prodr. 234 (1810) ... ... W.A. S.A. — V. N.S.W. Q. — B.fl.VII,389
L. longitudinale, Labillardière, Nov. Holl. pl.spec. I, 16, t. 13(1804) W.A. S.A. T. V. N.S.W. — B.fl.VII,389 M.fr.IX,25.
L. concavum, R. Brown, prodr. 234 (1810) ... ... — S.A. T. V. N.S.W. Q. — B.fl.VII,390 M.fr.IX,24.
L. angustatum, R. Brown, prodr. 235 (1810) ... ... W.A. — — — — — B.fl.VII,391 M.fr.IX,29,80.
L. Drummondii, Bentham, Fl. Austr. VII, 391 (1878) ... W.A. — — — — — B.fl.VII,391
L. Brunonianum, Nees in Lehm. pl. Preiss. II, 92 (1846) W.A. — — — — — B.fl.VII,392
L. tuberculatum, Nees in Lehm. pl. Preiss. II, 90 (1846) W.A. — — — — — B.fl.VII,392 M.fr.IX,20.
L. resinosum, F. v. M. in Benth. Fl. Austr. VII, 392 (1878) W.A. — — — — — B.fl.VII,392
L. viscidum, R. Brown, prodr. 234 (1810) ... ... — S.A. — V. N.S.W. — — B.fl.VII,393
L. costale, Nees in Lehm. pl. Preiss. II, 92 (1846) ... — S.A. — V. N.S.W. Q. — B.fl.VII,393
L. laterale, R. Brown, prodr. 234 (1810) ... ... — S.A. T. V. N.S.W. — — B.fl.VII,393
L. congestum, R. Brown, prodr. 234 (1810) ... ... — S.A. — — — — B.fl.VII,394
L. globosum, Labillardière, Nov. Holl. pl. spec. I, 16, t. 14 (1804) — S.A. T. V. — — — B.fl.VII,394
L. lineare, R. Brown, prodr. 235 (1810) ... ... — S.A. T. V. N.S.W. — — B.fl.VII,395 M.fr.IX,26.
L. aphyllum, R. Brown, prodr. 235 (1810) ... ... W.A. — — — — — B.fl.VII,395
L. gracile, R. Brown, prodr. 235 (1810) ... ... W.A. — — — — — B.fl.VII,395
L. semiteres, F. v. M. in Garcke, Linnaea XXXVIII, 327 (1874) — S.A. — V. — — — B.fl.VII,396 M.fr.IX,24.
L. canescens, Boeckeler in Garcke, Linnaea XXXVIII, 330 (1874) — S.A. — V. N.S.W. — — B.fl.VII,396 M.fr.IX,24.
L. pubisquameum, Steudel, syn. pl. glum. II, 158 (1855) W.A. — — — — — B.fl.VII,397 M.fr.IX,27.
L. scabrum, Nees in Lehm. pl. Preiss. II, 92 (1846) W.A. — — — — — B.fl.VII,397 M.fr.IX,27.
L. tenue, Bentham, Fl. Austr. VII, 397 (1878) ... ... W.A. — — — — — B.fl.VII,397
L. leptostachyum, Bentham, Fl. Austr. VII, 397 (1878)... W.A. — — — — — B.fl.VII,397
L. leptophyllum, Bentham, Fl. Austr. VII, 398 (1878) ... W.A. — — — — — B.fl.VII,398 M.fr.IX,24.
L. tortuosum, F. v. M., general Report 9 (1855) ... ... — — V. — — — B.fl.VII,398 M.fr.IX,24.
L. flexuosum, R. Brown, prodr. 235 (1810) ... ... — — — N.S.W. — — B.fl.VII,398 M.fr.IX,27.
L. filiforme, Labillardière, Nov. Holl. pl. spec. I, 17, t. 15 (1804) — S.A. T. V. N.S.W. — — B.fl.VII,399 M.fr.IX,27.
L. Neesii, Kunth, enum. II, 319 (1837) ... ... — — V. N.S.W. — — B.fl.VII,399 M.fr.IX,20,27.
L. carphoides, F. v. M. in Benth. Fl. Austr. VII, 400 (1878) ... W.A. S.A. — V. — — — B.fl.VII,400

    CLADIUM, P. Browne, Civ. and Nat. Hist. of Jamaic. 114 (1756). (Gahnia, Lampocarya, Melachne, Eplandra, Didymonema, Baumea,
        Chapelliera, Morelotia.)
C. Mariscus, R. Brown, prodr. 236 (1810) ... ... — S.A. — V. N.S.W. Q. N.A. B.fl.VII,402 M.fr.IX,14,57.
C. Insulare, Bentham, Fl. Austr. 403 (1878) ... ... — — — N.S.W. — — B.fl.VII,403
C. articulatum, R. Brown, prodr. 237 (1810) ... ... W.A. S.A. — V. N.S.W. Q. — B.fl.VII,403 M.fr.IX,14,57.
C. arthrophyllum, F. v. M., fragm. IX, 14 (1875) ... W.A. — — — — — B.fl.VII,403 M.fr.IX,14.
C. glomeratum, R. Brown, prodr. 237 (1810) ... ... W.A. S.A. — V. N.S.W. Q. — B.fl.VII,404 M.fr.IX,15,57.
C. Pressii, F. v. M. in Benth. Fl. Austr. VII, 405 (1878) ... W.A. — — — — — B.fl.VII,405 M.fr.IX,13,15,56.
C. laxum, Bentham, Fl. Austr. VII, 405 (1878) ... ... W.A. — — — — — B.fl.VII,405
C. riparium, Bentham, Fl. Austr. VII, 405 (1878) ... ... W.A. — — — — — B.fl.VII,405
C. teretifolium, R. Brown, prodr. 237 (1810) ... ... — — — N.S.W. Q. — B.fl.VII,405 M.fr.IX,15,56.
C. tetraquetrum, J. Hooker, Fl. Tasman. II, 95, t. 149 (1860)... — S.A. T. V. N.S.W. — — B.fl.VII,406
C. schoenoides, R. Brown, prodr. 237 (1810) ... ... W.A. — — — — — B.fl.VII,406
C. Gunnii, J. Hooker, Fl. Tasman. II, 95, t. 148 (1860)... — T. — — — — B.fl.VII,407 M.fr.IX,15,57.
C. junceum, R. Brown, prodr. 237 (1810) ... ... W.A. S.A. — V. N.S.W. Q. N.A. B.fl.VII,407 M.fr.IX,16,57.
C. vaginale, Bentham, Fl. Austr. VII, 409 (1878)... ... W.A. — — — — — B.fl.VII,408
C. elynanthoides, F. v. M., fragm. IX, 31 (1875)... ... W.A. — — — — — B.fl.VII,409 M.fr.IX,31.

C. Filum, R. Brown, prodr. 237 (1810) ... ... — S.A. T. V. — — — B.fl.VII,409 M.fr.IX,14,57.
C. trifidum, F. v. M. in l'ap. R. S. Tasm. 117 (1878) ... ... W.A. S.A. T. V. N.S.W. — B.fl.VII,413
C. decompositum, R. Brown, prodr. 237 (1810) ... ... W.A. — — — — — B.fl.VII,417 M.fr.IX,18.
C. microstachyum, F.v.M.; Gahnia, Benth. Fl. Austr.VII,414(1878) — — — V. N.S.W. — B.fl.VII,414
C. polyphyllum, F.v.M.; Gahnia, Benth. Fl. Austr.VII,415(1878) W.A. — — — — — B.fl.VII,415
C. ancistrophyllum, F. v. M. in Benth. Fl. Austr. VII, 415 (1878) W.A. — — — — — B.fl.VII,415
C. lanigerum, R. Brown, prodr. 237 (1810)... ... ... W.A. S.A. — — — — B.fl.VII,415 M.fr.IX,14,57.
C. aristatum, F. v. M. in Benth. Fl. Austr. 416 (1878) ... W.A. — — — — — B.fl.VII,416
C. deustum, R. Brown, prodr. 237 (1810) ... ... W.A. S.A. — — — — B.fl.VII,416 M.fr.IX,14.
C. Radula, R. Brown, prodr. 237 (1810) ... ... — T. V. — — — B.fl.VII,417 M.fr.IX,13,56.
C. Sieberi, F. v. M., fragm. IX, 14 (1875)... ... — — — N.S.W. — B.fl.VII,414 M.fr.IX,14,57.
C. tetragonocarpum, F. v. M.; Gahnia, Boeckeler in Linnaea
   XXXVIII, 347 (1874) ... ... — — V. — — B.fl.VII,418
C. melanocarpum, F. v. M., fragm. IX, 13 (1875) — T. V. — — — B.fl.VII,414 M.fr.IX,13,56.
C. psittacorum, F. v. M., fragm. IX, 13 (1875) ... — S.A. T. V. N.S.W. Q. — B.fl.VII,418 M.fr.IX,13,56.
C. asperum, F. v. M., fragm. IX, 12 (1875) ... — — — N.S.W. Q. — B.fl.VII,412 M.fr.IX,12,56.
C. xauthocarpum, F. v. M., fragm. IX, 13 (1875)... — — — N.S.W. — B.fl.VII,418 M.fr.IX,13,56.

CAUSTIS, R. Brown, prodr. 239 (1810). (Eurostorrhiza.)
C. pentandra, R. Brown, prodr. 240 (1810)... — S.A. T. V. N.S.W. Q. — B.fl.VII,420 M.fr.IX,19,58.
C. flexuosa, R. Brown, prodr. 239 (1810) ... — — V. N.S.W. Q. N.A. B.fl.VII,421 M.fr.IX,58.
C. recurvata, Sprengel, syst. cur. post. 26 (1827)... — — — N.S.W. — B.fl.VII,421 M.fr.IX,19.
C. restiacea, F. v. M. in Benth. Fl. Austr. VII, 421 (1878) — — V. N.S.W. — B.fl.VII,421 M.fr.IX,19.
C. dioica, R. Brown, prodr. 239 (1810) ... ... W.A. — — — — — B.fl.VII,422 M.fr.IX,19,58.

ARTHROSTYLIS, R. Brown, prodr. 229 (1810). (Arthrostylce.)
A. aphylla, R. Brown, prodr. 229 (1810) ... ... — — — — Q. N.A. B.fl.VII,423 M.fr.IX,9,55.

REEDIA, F. v. M., fragm. I, 240, t. 10 (1859).
R. spathacea, F. v. M., fragm. I, 240, t. 10 (1859) ... W.A. — — — — — B.fl.VII,423 M.fr.IX,15.

EUANDRA, R. Brown, prodr. 230 (1810). (Evandra.)
E. aristata, R. Brown, prodr. 239 (1810) ... ... W.A. — — — — — B.fl.VII,424 M.fr.IX,18.
E. pauciflora, R. Brown, prodr. 239 (1810) ... ... W.A. — — — — — B.fl.VII,425 M.fr.IX,18.

SCLERIA, Bergius in Svensk. Vet. Acad. Handl. 149, t. 4 (1765). (Diplacrum, Hypoporum, Sphaeropus.)
S. caricina, Bentham, Fl. Austr. VII (1878) — — — — Q. — B.fl.VII,426 M.fr.IX,58.
S. pygmaea, R. Brown, prodr. 240 (1810) ... ... — — — — Q. N.A. B.fl.VII,427 M.fr.IX,22,58.
S. rugosa, R. Brown, prodr. 240 (1810) ... ... — — — — Q. N.A. B.fl.VII,428 M.fr.IX,22,58.
S. laxa, R. Brown, prodr. 240 (1810) ... ... — — — — Q. N.A. B.fl.VII,428 M.fr.IX,21.
S. Brownii, Kunth, enum. II, 349 (1837) ... ... — — — — Q. N.A. B.fl.VII,429 M.fr.IX,21,58.
S. lithosperma, Willdenow, spec. plant. IV, 316 (1805)... — — — — Q. N.A. B.fl.VII,429 M.fr.IX,21,58.
S. tessellata, Willdenow, spec. plant. IV, 315 (1805) — — — — Q. — B.fl.VII,430
S. margaritifera, Willdenow, spec. plant. IV, 312 (1805) — — — — Q. — B.fl.VII,431
S. Graeffeana, Boeckeler in Regensb. Flora, 121 (1875) ... — — — — Q. N.A. B.fl.VII,431
S. hebecarpa, Nees in Wight, Contrib. to Bot. of Ind. 117 (1834) — — — N.S.W. Q. N.A. B.fl.VII,431 M.fr.IX,21,58.
S. Chinensis, Kunth, enum. II, 357 (1837)... ... — — — — Q. — B.fl.VII,431 M.fr.IX,20,58.
S. oryzoides, Presl, reliq. Haenk. I, 201 (1830) ... — — — — N.A. B.fl.VII,432
S. sphacelata, F. v. M., fragm. IX, 20 (1875) ... — — — N.S.W. Q. — B.fl.VII,432 M.fr.IX,20.

UNCINIA, Persoon, synope. plant. I, 534 (1807).
U. tenella, R. Brown, prodr. 241 (1810) ... ... — — T. V. — — — B.fl.VII,433 M.fr.VIII,151.
U. compacta, R. Brown, prodr. 241 (1810)... ... — — T. V. — — — B.fl.VII,434 M.fr.VIII,152.
U. riparia, R. Brown, prodr. 241 (1810) ... ... — — T. V. N.S.W. — — B.fl.VII,434 M.fr.VIII,152.
U. debilior, F. v. M., fragm. VIII, 151 (1874) ... — — — N.S.W. — B.fl.VII,435 M.fr.VIII,151.

CAREX, Ruppius, Fl. Jencns. 306 (1718).
C. cephalotos, F. v. M. in Transact. phil. Soc. Vict. I, 110 (1854) — — V. N.S.W. — B.fl.VII,437 M.fr.VIII,251.
C. acicularis, Boott in J. Hook. Fl. N. Zel. I, 280, t. 63 (1853) — — T. V. N.S.W. — B.fl.VII,437 M.fr.VIII,251.
C. capillacea, Boott, Illustr. of the gen. Carex I, 44, t. 110 (1858) — — — N.S.W. — B.fl.VII,438 M.fr.IX,191.
C. inversa, R. Brown, prodr. 242 (1810) ... W.A. S.A. T. V. N.S.W. — B.fl.VII,438 M.fr.VIII,252.
C. canescens, Linné, spec. plant. 974 (1753) ... — — V. N.S.W. — B.fl.VII,439 M.fr.VIII,253.
C. echinata, Murray, prodr. design. stirp. Goett. 76 (1770) — — V. N.S.W. — B.fl.VII,439 M.fr.VIII,253.
C. hypandra, F. v. M., fragm. VIII, 259 (1874) ... — — V. N.S.W. — B.fl.VII,439 M.fr.VIII,259.
C. chlorantha, R. Brown, prodr. 242 (1810) — — S.A. T. V. N.S.W. — B.fl.VII,440 M.fr.VIII,256.
C. paniculata, Linné, amoen. acad. IV, 294 (1759) ... W.A. S.A. T. V. N.S.W. Q. B.fl.VII,440 M.fr.VIII,256.
C. declinata, Boott, Ill. of the gen. Carex IV, 17, t. 590 (1867) — — V. N.S.W. — B.fl.VII,441 M.fr.VIII,256.
C. tereticaulis, F. v. M., fragm. VIII, 236 (1874) W.A. S.A. T. V. N.S.W. — B.fl.VII,441 M.fr.VIII,256.
C. flestilis, Boott, Ill. of the gen. Carex II, 86, t. 245 (1860) ... — — — Q. B.fl.VII,441 M.fr.VIII,249;IX,80.
C. gracilis, R. Brown, prodr. 242 (1810) ... — — — N.S.W. — B.fl.VII,442 M.fr.VIII,250.
C. contracta, F. v. M., fragm. VIII, 258 (1874) ... — — — N.S.W. — B.fl.VII,442 M.fr.VIII,257.
C. Gaudichaudiana, Kunth, enum. plant. II, 417 (1837)... — S.A. V. N.S.W. — B.fl.VII,442 M.fr.VIII,257.
C. acuta, Linné, spec. plant. 978 (1753) ... — — V. N.S.W. — B.fl.VII,443 M.fr.VIII,259.
C. lobolepis, F. v. M., fragm. VIII, 259 (1874) ... — — — N.S.W. — B.fl.VII,443 M.fr.VIII,258.
C. flava, Linné, spec. plant. 973 (1753) ... — — T. — — — B.fl.VII,444 M.fr.VIII,258.
C. Buxbaumii, Wahlenberg in Act. Holmiens. 139 (1803) ... — — V. N.S.W. — B.fl.VII,444 M.fr.VIII,232.
C. pumila, Thunberg, Fl. Japon. 38 (1784)... — S.A. T. V. N.S.W. — B.fl.VII,444 M.fr.VIII,251.
C. breviculmis, R. Brown, prodr. 242 (1810) ... — S.A. T. V. N.S.W. — B.fl.VII,443 M.fr.VIII,255.
C. Preissii, Nees in Lehm. pl. Preiss. II, 94 (1846) ... W.A. — — — — — B.fl.VII,446 M.fr.VIII,231.
C. Gunniana, Boott in Transact. Linn. Soc. XX, 143 (1851) — S.A. T. V. N.S.W. — B.fl.VII,446 M.fr.VIII,231.
C. Bichenoviana, Boott in J. Hook. Fl. Tasm. II, 101 (1860) ... — — T. — — — B.fl.VII,446
C. maculata, Boott in Transact. Linn. Soc. XX, 128 (1851) ... — — V. N.S.W. — B.fl.VII,447 M.fr.VIII,238.
C. Brownii, Tuckerman, enum. meth. Caric. 21 (1849) ... — — V. N.S.W. — B.fl.VII,447 M.fr.IX,250.
C. aleophila, F. v. M., fragm. VIII, 257 (1874) ... — — V. N.S.W. — B.fl.VII,447 M.fr.VIII,257.
C. longifolia. R. Brown, prodr. 242 (1810)... ... — T. V. N.S.W. Q. — B.fl.VII,448 M.fr.VIII,250.
C. Pseudo-Cyperus, Linné, spec. plant. 978 (1753) ... W.A. S.A. T. V. N.S.W. Q. — B.fl.VII,448 M.fr.VIII,249.

R

130

GRAMINEAE.
Haller, enum. stirp. Helv. I, 203 (1742).

ERIOCHLOA, Humboldt, Bonpland & Kunth, nov. gen. et sp. pl. I, 94, t. 30 (1815). (Helopus.)
E. punctata, Hamilton, prodr. pl. Ind. occ. 5 (1825) ... ... — — — V. N.S.W. Q. N.A. B.fl.VII,462 M.fr.VI,83;VIII,126
E. annulata, Kunth, Graminéos I, 30 (1832) ... ... — S.A. — — N.S.W. Q. — B.fl.VII,463 M.fr.VI,83.

PASPALUM, Linné, syst. nat. ed. decim. 855 (1759).
P. scrobiculatum, Linné, mantiss. 29 (1767) ... ... — — — N.S.W. Q. N.A. B.fl.VII,360 M.fr.VIII,156.
P. distichum, Linné, syst. nat. ed. decim. 855 (1759) ... ... W.A. — — — N.S.W. Q. — B.fl.VII,460 M.fr.VIII,156.
P. brevifolium, Fluegge, gram. monogr. 150 (1810) ... ... — — — N.S.W. Q. N.A. B.fl.VII,461 M.fr.VIII,195.
P. minutiflorum, Steudel, sym. glum. I, 17 (1855) ... ... — — — — Q. — B.fl.VII,461

PANICUM, Tournefort, inst. 515, t. 298 (1700), from Plinius. (Isachne, Digitaria, Echinochloa, Coridochloa, Chamaeraphis.)
P. coculcolum, F. v. M. in Transact. Vict. Inst. 45 (1854) ... W.A. S.A. — V. N.S.W. — B.fl.VII,467
P. divaricatissimum, R. Brown, prodr. 192 (1810) ... ... — S.A. — V. N.S.W. Q. — B.fl.VII,487 M.fr.VIII,154;X,76.
P. macractinum, Bentham, Fl. Austr. VII, 468 (1878) ... ... — — — N.S.W. Q. — B.fl.VII,468
P. papposum, R. Brown, prodr. 192 (1810) ... ... — — — — — N.A. B.fl.VII,468 M.fr.VIII,155,196.
P. sanguinale, Linné, spec. plant. 57 (1753) ... ... W.A. — — V. N.S.W. Q. N.A. B.fl.VII,469 M.fr.VIII,153.
P. ctenanthum, F. v. M., fragm. VIII, 153 (1874) ... ... — — — — — N.A. B.fl.VII,469 M.fr.VIII,153.
P. stenostachyum, Bentham, Fl. Austr. VII, 470 (1878) ... — — — — — N.A. B.fl.VII,470
P. tenuissimum, Bentham, Fl. Austr. VII, 470 (1878) ... — — — — N.S.W. Q. — B.fl.VII,470
P. parviflorum, R. Brown, prodr. 192 (1810) ... ... — — — N.S.W. Q. — B.fl.VII,470 M.fr.VIII,190;XI,129.
P. Baileyi, Bentham, Fl. Austr. VII, 470 (1878) ... ... — — — N.S.W. Q. — B.fl.VII,471
P. gibboaum, R. Brown, prodr. 193 (1810) ... ... — — — — — Q. N.A. B.fl.VII,471 M.fr.VIII,135;XI,129.
P. leucophaeum, Humb.,Bonpl.& Kunth,nov.gen.etspec.I,97(1815) — S.A. — V. N.S.W. Q. — B.fl.VII,472 M.fr.VIII,155.
P. semialatum, R. Brown, prodr. 192 (1810) ... ... — — — N.S.W. Q. N.A. B.fl.VII,472 M.fr.VIII,190.
P. rarum, R. Brown, prodr. 189 (1810) ... ... — — — — — N.A. B.fl.VII,473
P. argentsum, R. Brown, prodr. 190 (1810) ... ... — — — — — Q. N.A. B.fl.VII,473
P. holosericeum, R. Brown, prodr. 190 (1810) ... ... — — — — — Q. N.A. B.fl.VII,473 M.fr.VIII,155.
P. flavidum, Retzius, observ. IV, 15 (1787) ... ... — — — N.S.W. Q. N.A. B.fl.VII,474 M.fr.VIII,189.
P. gracile, R. Brown, prodr. 190 (1810) ... ... — W.A. S.A. — V. N.S.W. Q. N.A. B.fl.VII,475 M.fr.VIII,189.
P. prostratum, Lamarck, illustr. I, 171 (1791) ... ... — — — N.S.W. Q. N.A. B.fl.VII,476 M.fr.VIII,191.
P. helopus, Trinius in Spreng. Neue Entd. II, 84 (1821) ... — S.A. — — N.S.W. Q. N.A. B.fl.VII,477
P. Gilesii, Bentham, Fl. Austr. VII, 477 (1878) ... ... — S.A. — — — — N.A. B.fl.VII,477 M.fr.XI,129.
P. piligerum, F. v. M. in Benth. Fl. Austr. VII, 477 (1878) ... — — — — — N.A. B.fl.VII,477
P. polyphyllum, R. Brown, prodr. 190 (1810) ... ... — — — — — N.A. B.fl.VII,477 M.fr.VIII,194.
P. dietschyon, Linné, mantiss. alt. pl. 138 (1771) ... — S.A. — — N.S.W. Q. — B.fl.VII,478 M.fr.VIII,194.
P. reversum, F. v. M., fragm. VIII, 152 (1874) ... ... W.A. S.A. — — N.S.W. Q. — B.fl.VII,478 M.fr.X,76.
P. colonum, Linné, syst. nat. ed. decim. 5 (1758)... ... — — — — — N.A. B.fl.VII,478
P. Crus Galli, Linné, spec. plant. 56 (1753) ... ... W.A. S.A. — V. N.S.W. Q. N.A. B.fl.VII,479 M.fr.VIII,198.
P. myosuroides, R. Brown, prodr. 189 (1810) ... ... — — — — — Q. N.A. B.fl.VII,480
P. indicum, Linné, mantiss. pl. alt. 184 (1771) ... ... — — — N.S.W. Q. — B.fl.VII,480 M.fr.VIII,197;XI,129.
P. Myurus, Lamarck, illustr. I, 172 (1791) ... ... — — — — — — B.fl.VII,480 M.fr.X,114.
P. foliosum, R. Brown, prodr. 191 (1810) ... ... — — — N.S.W. Q. — B.fl.VII,480 M.fr.VIII,194.
P. adspersum, Trinius, spec. gram. t. 189 (1830) ... ... — S.A. — — — — Q. — B.fl.VII,481
P. inaequale, F. v. M., fragm. VIII, 189 (1874) ... ... — — — — — — Q. — B.fl.VII,482 M.fr.XI,129. [129.
P. uncinulatum, R. Brown, prodr. 191 (1810) ... ... — — — N.S.W. Q. — B.fl.VII,482 M.fr.VIII,193;X,76;XI,
P. majusculum, F. v. M. in Benth. Fl. Austr. VII, 482 (1878)... — — — — — — B.fl.VII,482
P. pauciflorum, R. Brown, prodr. 483 (1810) ... ... — S.A. — — — — Q. N.A. B.fl.VII,483 M.fr.VIII,193.
P. semitonsum, F. v. M. in Benth. Fl. Austr. VII, 483 (1878) .. — — — — — N.A. B.fl.VII,483
P. antidotale, Retzius, observ. bot. IV, 17 (1876)... ... — — — — — N.A. B.fl.VII,483
P. repens, Linné, spec. pl. sec. 87 (1762) ... ... — — — V. N.S.W. Q. N.A. B.fl.VII,484
P. Buncei, F. v. M. in Benth. Fl. Austr. VII, 487 (1878) ... — — — — — Q. — B.fl.VII,487
P. capillipes, Bentham, Fl. Austr. VII, 484 (1878) ... ... — — — — — Q. N.A. B.fl.VII,484 M.fr.VIII,193.
P. pygmaeum, R. Brown, prodr. 191 (1810) ... ... — — — N.S.W. Q. — B.fl.VII,484 M.fr.VIII,193.
P. brevifolium, Linné, spec. plant. 59 (1753) ... ... — — — — — Q. N.A. B.fl.VII,485 M.fr.VIII,193;XI,129.
P. hermaphroditum, Steudel, syn. glum. I, 67 (1855) ... ... — — — — — — B.fl.VII,485
P. marginatum, R. Brown, prodr. 190 (1810) ... ... — — — V. N.S.W. Q. — B.fl.VII,485 M.fr.VIII,190.
P. lachnophyllum, Bentham, Fl. Austr. VII, 486 (1878) ... ... — — — — — — B.fl.VII,486
P. obseptum, Trinius, gram. panic. diss. alt. 149 (1826)... ... — — — — — — B.fl.VII,486
P. bicolor, R. Brown, prodr. 191 (1810) ... ... — — — N.S.W. Q. — B.fl.VII,487
P. melananthum, F. v. M. in Transact. Vict. Inst. 47 (1854) ... — — — V. N.S.W. Q. — B.fl.VII,488 M.fr.VIII,192.
P. effusum, R. Brown, prodr. 191 (1810) ... ... — S.A. — — N.S.W. Q. — B.fl.VII,488 M.fr.VIII,191.
P. Mitchelli, Bentham, Fl. Austr. VII, 489 (1878) ... ... — S.A. — — — — N.A. B.fl.VII,489 M.fr.X,76.
P. decompositum, R. Brown, prodr. 191 (1810) ... ... W.A. S.A. — V. N.S.W. Q. N.A. B.fl.VII,490 M.fr.VIII,86;VIII,191.
P. trachyrhachis, Bentham, Fl. Austr. VII, 490 (1878) ... — — — — — N.A. B.fl.VII,490 M.fr.VIII,192;XI,129.
P. prolutum, F. v. M. in Transact. Vict. Inst. 46 (1854) ... — — — — — N.A. B.fl.VII,490 M.fr.VIII,197.
P. spinescens, R. Brown, prodr. 193 (1810) ... ... W.A. S.A. — — N.S.W. Q. — B.fl.VII,499 M.fr.VIII,197;XI,129.
P. paradoxum, R. Brown, prodr. 193 (1810) ... ... — — — V. N.S.W. Q. N.A. B.fl.VII,499 M.fr.VIII,197.
P. Chamaeraphis, Trinius in Mem. acad. Petersb. ser.6,III,217(1835) — — — — — N.A. B.fl.VII,500
P. atro-virens, Trinius in Spreng. neue Entdeckung. II, 89 (1821) — — — V. N.S.W. Q. — B.fl.VII,499 M.fr.VIII,193;XI,129.
P. Myosotis, Steudel, syn. glum. I, 96 (1855) ... ... — — — — — Q. — B.fl.VII,625 M.fr.VIII,193.

OPLISMENUS, Palisot, Flore d'Oware et de Benin. II, 14 (1807). (Orthopogon.)
O. compositus, Palisot, agrostogr. 54 (1812) ... ... — — — V. N.S.W. Q. — B.fl.VII,491 M.fr.VIII,199.

SETARIA, Palisot, Essai d'une nouv. Agrostogr. 51, t. 13 (1812).
S. glauca, Palisot, Agrostogr. 51, t. 13 (1812) ... ... S.A. — V. N.S.W. Q. N.A. B.fl.VII,492 M.fr.VIII,110.
S. macrostachya, Humb., Bonpl. & Kunth, nov. gen. et spec. I, 110 (1815) ... ... S.A. — — V. N.S.W. Q. N.A. B.fl.VII,493 M.fr.VIII,110;XI,129.
S. viridis, Palisot, Agrostogr. 51, (1812) ... ... W.A. S.A. — — — N.A. B.fl.VII,494
S. verticillata, Palisot, Agrostogr. 51 (1812) ... ... W.A. — — — N.A. B.fl.VII,494 M.fr.VIII,110.

PENNISETUM, L. C. Richard in Persoon, synops. plant. I. 72 (1805). (Gymnothrix, Plagiosetum.)
P. compressum, R. Brown, prodr. 195 (1810) ... ... — — — — N.S.W. Q. — B.fl.VII,495 M.fr.VIII,109.
P. Arnhemicum, F. v. M., fragm. VIII, 109 (1874) ... ... — — — — — N.A. B.fl.VII,496 M.fr.VIII,109.
P. refractum, F. v. M., fragm. VIII, 100 (1874) ... ... — S.A. — — — Q. N.A. B.fl.VII,495 M.fr.III,147;IX,194.
CENCHRUS, Linné, coroll. gen. 20 (1737).
C. australis, R. Brown, prodr. 196 (1810) ... ... — — — — N.S.W. Q. — B.fl.VII,497 M.fr.VIII,107;XI,129.
C. inflexus, R. Brown, prodr. 195 (1810) ... ... — — — — — N.A. B.fl.VII,497 M.fr.VIII,107.
C. elymoides, F. v. M., fragm. VIII, 107 (1874) ... ... — — — — Q. N.A. B.fl.VII,497 M.fr.VIII,107.
XEROCHLOA, R. Brown, prodr. fl. Nov. Holl. 196 (1810).
X. imberbis, R. Brown, prodr. 197 (1810) ... ... — — — — — N.A. B.fl.VII,501 M.fr.VIII,117.
X. barbata, R. Brown, prodr. 197 (1810) ... ... — — — — — N.A. B.fl.VII,501 M.fr.VIII,117.
X. laniflora, Bentham, Fl. Austr. VII, 502 (1878) ... ... — — — -- — N.A. B.fl.VII,502
THUAREA, Persoon, synops. plant. I, 110 (1803). (Thouarea.)
T. sarmentosa, Persoon, synops. I, 110 (1805) ... ... — — — — Q. N.A. B.fl.VII,502 M.fr.VI,96.
SPINIFEX, Linné, syst. nat. ed. XII, 757 (1767).
S. hirsutus, Labillard., Nov. Holl. pl. spec. II, 81, t. 230 (1806) W.A. S.A. T. V. N.S.W. Q. — B.fl.VII,503 M.fr.VIII,138.
S. longifolius, R. Brown, prodr. 198 (1810) ... ... W.A. — — — Q. N.A. B.fl.VII,504 M.fr.VIII,130. [IX,195.
S. paradoxus, Bentham in Hook. icon. plant. t. 1243 (1877) ... — S.A. — V. N.S.W. Q. — B.fl.VII,504 M.fr.VI,86;VIII,117,199;
LEPTASPIS, R. Brown, prodr. fl. Nov. Holl. 211 (1810).
L. Banksii, R. Brown, prodr. 211 (1810) ... ... — — — — — Q. — B.fl.VII,548 M.fr.VIII,116.
PEROTIS, Aiton, hort. Kewens. I, 85 (1789).
P. rara, R. Brown, prodr. 172 (1810) ... ... — S.A. — — N.S.W. Q. N.A. B.fl.VII,509 M.fr.VIII,115.
POLYPOGON, Desfontaines, flor. Atlantic. I, 66 (1798).
P. fugax, Nees in Steudel, syn. glum. I, 184 (1855) ... ... W.A. — — — — — — B.fl.VII,547
P. tenellus, R. Brown, prodr. 173 (1810) ... ... W.A. — — — — — — B.fl.VII,547 M.fr.VIII,114.
ARUNDINELLA, Raddi, agrostograph. Brasil. 37 (1823).
A. Nepalensis, Trinius, spec. gram. t. 268 (1833) .. .. — — — N.S.W. Q. N.A. B.fl.VII,545 M.fr.VIII,139.
A. Schultzii, Bentham, Fl. Austr. VII, 545 (1878) ... ... — — — — — — N.A. B.fl.VII,545
TRAGUS, Haller, hist. stirp. Helvet. n. 1413 (1768). (Lappago.)
T. racemosus, Haller, stirp. Helv. n. 1413 (1768)... ... — S.A. — V. N.S.W. Q. N.A. B.fl.VII,507 M.fr.VIII,107.
NEURACHNE, R. Brown, prodr. fl. Nov. Holl. 196 (1810). [129.
N. alopecuroides, R. Brown, prodr. 196 (1810) ... ... W.A. S.A. — V. N.S.W. — B.fl.VII,507 M.fr.VI,86;VIII,200; XI,
N. Mitchelliana, Nees in Hook. Lond. Journ. of Bot. II, 410 (1843) — S.A. — V. N.S.W. — B.fl.VII,508 M.fr.VI,86;VIII,200.
N. Munroi. F. v. M., fragm. VIII, 200 (1874) ... ... — S.A. — V. N.S.W. — B.fl.VII,508 M.fr.V,204;XI,129.
ZOYSIA, Willdenow, Neue Schrift. nat. Freunde zu Berl. III, 440 (1801).
Z. pungens, Willdenow, Neue Schrift. nat. Fr. Berl. III, 440 (1801) — — T. V. N.S.W. Q. — B.fl.VII,506 M.fr.VIII,116.
IMPERATA, Cyrillo, plant. rar. Neap. II, 1. 11 (1792).
I. arundinacea, Cyrillo, plant. rar. Neap. II, t. 11 (1792) ... W.A. S.A. T. V. N.S.W. Q. N.A. B.fl.VII,530 M.fr.VIII,126.
ERIANTHUS, L. C. Richard in Michaux, fl. bor. Amer. I, 54 (1803). (Pollinia, Eulalia, Pogonatherum, Leptatherum, Saccharum partly.)
E. articulatus, F. v. M., fragm. VIII, 118 (1874) ... ... — — — — Q. N.A. B.fl.VII,525 M.fr.X,76.
E. irritans, Kunth, Graminées I, 160 (1832) ... ... — — — — Q. N.A. B.fl.VII,526 M.fr.VIII,118.
E. fulvus, Kunth, Graminées I, 160 (1832) ... ... W.A. S.A. — N.S.W. Q. N.A. B.fl.VII,526 M.fr.VIII,117.
E. Roxburghii, F. v. M., fragm. VIII, 117 (1874) ... ... — — — — Q. — B.fl.VII,527 M.fr.VIII,117.
E. Mackinlayi; Pollinia, F.v.M.in Benth. fl. Austr. VII, 527 (1878) — — — — — N.A. B.fl.VII,527 M.fr.VIII,118.
DIMERIA, R. Brown, prodr. fl. Nov. Holl. 204 (1810).
D. acinaciformis, R. Brown, prodr. 204 (1910) ... ... — — — — Q. — B.fl.VII,523
D. tenera, Trinius in Mém. de l'Acad. de Petersb. six. sér. II,
  223 (1833) ... ... — — — — Q. N.A. B.fl.VII,523 M.fr.VII,104;XI,129.
ARTHRAXON, Palisot, Agrostograph. 111, t. XI, f. 6 (1812). (Batratherum.)
A. ciliare, Palisot, Agrostogr. 111, t. 11 (1812) ... ... — — — — N.S.W. Q. — B.fl.VII,524 M.fr.VIII,119;X,76.
ELIONURUS, Willdenow, spec. plant. IV, 941 (1805).
E. citreus, Munro in Benth. Fl. Austral. VII, 510 (1878) ... — — — — — — B.fl.VII,510 M.fr.XI,129.
LEPTURUS, R. Brown, prodr. fl. Nov. Holl. 207 (1810).
L. incurvatus, Trinius, fundam. agrostogr. 123 (1820) ... — S.A. — V. N.S.W. — B.fl.VII,668 M.fr.VIII,117.
L. cylindricus, Trinius, fundam. agrostogr. 123 (1820) ... W.A. S.A. — V. N.S.W. — B.fl.VII,668
L. repens, R. Brown, prodr. 207 (1810) ... ... — — — — Q. N.A. B.fl.VII,665 M.fr.VI,86.
ROTTBOELLIA, Linné, Gl. gram. gen. 22 (1779). (Coelorhachis.)
R. formosa, R. Brown, prodr. 206 (1810) ... ... — — — — Q. N.A. B.fl.VII,513 M.fr.VI,85;XI,129.
R. exaltata, Linné, fil. suppl. 114 (1781) ... ... — — — — Q. N.A. B.fl.VII,513
R. muricata, Retzius, observ. bot. III, 12 (1786)... ... — — — — Q. — B.fl.VII,514
R. ophiuroides, Bentham, Fl. Austr. VII, 514 (1878) ... — — — — Q. N.A. B.fl.VII,514 M.fr.VII,123.
OPHIUROS, K. F. Gaertner, de fruct. III, 3, t. 181 (1805). (Ophiurus.)
O. corymbosus, K. F. Gaertner, de fruct. III, 3, t. 181 (1805)... — — — — Q. N.A. B.fl.VII,512 M.fr.VI,85.
MANISURIS, Linné, syst. nat. ed. duodec. 762 (1767).
M. granularis, Swartz, Fl. Ind. occid. I, 186 (1797) ... ... — — — — Q. N.A. B.fl.VII,514 M.fr.VII,116;XI,129.
HEMARTHRIA, R. Brown, prodr. fl. Nov. Holl. 207 (1810).
H. compressa, R. Brown, prodr. 207 (1810) ... ... W.A. S.A. T. V. N.S.W. Q. — B.fl.VII,511 M.fr.XI,129.
ISCHAEMUM, Linné, spec. pl. 8 (1735); Linné, gen. pl. ed. sec. 525 (1742). (Holcogamium, Spodiopogon partly.)
I. truncatiglume, F. v. M. in Benth. Fl. Austr. VII, 518 (1878) — — — — — — N.A. B.fl.VII,518
I. arundinaceum, F. v. M. in Benth. Fl. Austr. VII, 519 (1878) — — — — — — N.A. B.fl.VII,519
I. triticeum, R. Brown, prodr. 205 (1810) ... ... — N.S.W. Q. — B.fl.VII,519

132

I. australe, R. Brown, prodr. 205 (1810) ... ... — — — — N.S.W. Q. N.A. B.fl.VII,519
I. muticum, Linné, spec. plant. 1049 (1753) ... ... — — — — — Q. — B.fl.VII,520 M.fr.VIII,120.
I. ciliare, Retzius, observ. bot. VI, 36 (1791) ... ... — — — — N.S.W. — — B.fl.VII,520 M.fr.VIII,119.
I. decumbens, Bentham, Fl. Austr. VII, 521 (1878) ... ... — — — — — — N.A. B.fl.VII,521
I. pectinatum, Trinius in Mem. de l'Acad. de Petersb. six. sér.
    II, 296 (1833)... ... ... — — — — N.S.W. Q. — B.fl.VII,521 M.fr.VIII,118.
I. fragile, R. Brown, prodr. 205 (1810) ... ... — — — — — — B.fl.VII,522 M.fr.VIII,123;XI,130.
I. laxum, R. Brown, prodr. 205 (1810) ... ... — — — — — Q. N.A. B.fl.VII,522 M.fr.VIII,119.

ANDROPOGON, Royen, prodr. exhib. pl. hort. acad. Lugd. 52 (1740). (Sorghum, Chrysopogon, Heteropogon, Schizachyrium, Trachypogon, Holcus partly, Spodiopogon partly.)
A. erianthoides, F. v. M., fragm. X, 75 (1876) ... ... — — N.S.W. Q. — B.fl.VII,529 M.fr. X, 75.
A. sericeus, R. Brown, prodr. 201 (1810) ... ... W.A. S.A. — V. N.S.W. Q. N.A. B.fl.VII,529
A. affinis, R. Brown, prodr. 201 (1810) ... ... — — — N.S.W. Q. — B.fl.VII,530
A. pertusus, Willdenow, spec. plant. IV, 922 (1805) ... — S.A. — V. N.S.W. Q. — B.fl.VII,530 M.fr.VIII,122;X,76.
A. annulatus, Forskael, Fl. Aegypt. Arab. 173 (1775) ... — S.A. — — — Q. N.A. B.fl.VII,531 M.fr.VIII,123.
A. Ischaemum, Linné, spec. plant. 1047 (1753) ... ... W.A. — — — — — — B.fl.VII,531 M.fr.VIII,122.
A. intermedius, R. Brown, prodr. 202 (1810) ... ... — S.A. — V. N.S.W. Q. N.A. B.fl.VII,531
A. procerus, R. Brown, prodr. 202 (1810) ... ... — — — — — — N.A. R.fl.VII,532 M.fr.VIII,124.
A. exaltatus, R. Brown, prodr. 202 (1810) ... ... W.A. S.A. — — — — N.A. B.fl.VII,532 M.fr.VIII,124.
A. bombycinus, R. Brown, prodr. 202 (1810) ... ... W.A. S.A. — V. N.S.W. Q. N.A. B.fl.VII,533 M.fr.XI,20.
A. Schoenanthus, Linné, spec. plant. 1046 (1753)... ... — — — — — — Q. — B.fl.VII,534 M.fr.VIII,124.
A. refractus, R. Brown, prodr. 202 (1810) ... ... — — — V. N.S.W. Q. N.A. B.fl.VII,534 M.fr.VIII,124;X,76.
A. lachnatherus, Bentham, Fl. Austr. VII, 534 (1874) ... — — — — N.S.W. — — B.fl.VII,534
A. fragilis, R. Brown, prodr. 202 (1810) ... ... — — — — — Q. N.A. B.fl.VII,535
A. contortus, Linné, spec. plant. 1045 (1753) ... ... W.A. — — N.S.W. Q. N.A. B.fl.VII,517 M.fr.VIII,120.
A. triticeus, R. Brown, prodr. 201 (1810) ... ... — — — — — Q. N.A. B.fl.VII,537 M.fr.VIII,120.
A. Gryllus, Linné, amoen. acad. IV, 332 (1759) ... ... — S.A. — V. N.S.W. Q. N.A. B.fl.VII,537 M.fr.VIII,121;XI,130.
A. montanus, Roxburgh, Flor. Indic. edit. Carey. I, 271 (1820) — — — V. N.S.W. Q. N.A. B.fl.VII,538 M.fr.VIII,122;X,76;XI,[130.
A. acicularis, Retzius, observ. bot. V, 22 (1789) ... ... — — — — — — Q. N.A. B fl.VII,538 M.fr.VIII,122.
A. elongatus, Sprengel, syst. I, 289 (1825)... ... ... — — — — — — Q. N.A. B.fl.VII,539 M.fr.VIII,121.
A. Halepensis, Sibthorp & Smith, Fl. Graeca, I, 52, t. 68 (1806) W.A. — — N.S.W. — — B.fl.VII,540 M.fr.VIII,119.
A. australis, Sprengel, syst. I, 287 (1825) ... ... — — — V. N.S.W. Q. N.A. B.fl.VII,541
A. tropicus, Sprengel, syst. I, 287 (1825) ... ... — — — — — — Q. N.A. B.fl.VII,541
A. intrans, Sorghum, F. v. M. in Benth. Fl. Austr. VII, 541 (1878) — — — — — — N.A. B.fl.VII,541

ANTHISTIRIA, Linné, fil. gram. gen. 35 (1779). (Iscilema.)
A. ciliata, Linné fil., gram. gen. 35 (1779)... ... ... W.A. S.A. T. V. N.S.W. Q. N.A. B.fl.VII,542 M.fr.V,207;VIII,139.
A. frondosa, R. Brown, prodr. 200 (1810) ... ... — — — — N.S.W. Q. — B.fl.VII,542 M.fr.VIII,139;XI,130.
A. avenacea, F. v. M., fragm. V, 206 (1866) ... ... W.A. S.A. — N.S.W. — — B.fl.VII,543 M.fr.VI,253;VIII,139.
A. membranacea, Lindley in Mitch. Trop. Austr. 88 (1848) — S.A. — N.S.W. Q. N.A. B.fl.VII,543 M.fr.V,207;VIII,139.

APLUDA, Linné, spec. plant. 82 (1753).
A. mutica, Linné, spec. plant. 82 (1753) ... ... — — — — N.S.W. Q. — B.fl.VII,544 M.fr.VIII,140.

CHIONACHNE, R. Brown in Horsfield's pl. Jav. rar. 15 (1838).
C. barbata, R. Brown in Bennett, pl. Javan. rar. 18 (1838) ... — — — — — Q. — B.fl.VII,515
C. cyathopoda, F. v. M. in Benth. Fl. Austr. VII, 516 (1878) ... — — — — — — N.A. B.fl.VII,515 M.fr.VIII,116.

ALOPECURUS, Linné, syst. nat. 9 (1735); Linné, gen. plant. 18 (1737). (Perhaps immigrated.)
A. geniculatus, Linné, spec. plant. 60 (1753) ... ... W.A. S.A. T. V. N.S.W. Q. — B.fl.VII,555 M.fr.VIII,138.

LEERSIA, Solander in Swartz, nov. gen. et spec. plant. 1 et 21 (1788). (Asprella, Asperella.)
L. hexandra, Swartz, nov. gen. et spec. pl. 21 (1788) ... ... — — — N.S.W. Q. — B.fl.VII,549 M.fr.VIII,115.

ORYZA, Tournefort, inst. 513, t. 296 (1700), from Theophrastos & Dioscorides.
O. sativa, Linné, spec. plant. 333 (1753) ... ... — — — — — — N.A. B.fl.VII,550 M.fr.VIII,115;XI,130.

POTAMOPHILA, R. Brown, prodr. fl. Nov. Holl. 211 (1810).
P. parviflora, R. Brown, prodr. 211 (1810) ... ... — — — N.S.W. — — B.fl.VII,550 M.fr.VIII,126.

EHRHARTA, Thunberg in Svensk. Ventsk. Acad. Handl. 216, t. 8 (1779). (Tetrarrhena, Microlaena, Diplax.)
E. distichophylla, Labill., Nov. Holl. pl. spec. I, 90, t. 117 (1804) — T. V. — — — B.fl.VII,554 M.fr.VII,90.
E. juncea, Sprengel, syst. II, 114 (1825) ... ... — T. V. N.S.W. — — B.fl.VII,554 M.fr.VII,90.
E. laevis, Sprengel, syst. II, 115 (1825) ... ... W.A. — — — — — B.fl.VII,554 M.fr.VII,90.
E. acuminata, Sprengel, syst. II, 114 (1825) ... ... — T. V. — — — B.fl.VII,555 M.fr.VII,90;XI,130.
E. stipoides, Labillard., Nov. Holl. pl. spec. I, 91, t. 118 (1804) W.A. S.A. T. V. N.S.W. Q. — B.fl.VII,552 M.fr.VII,90.
E. diarrhena, F. v. M., fragm. VII, 89 (1870) ... ... — T. V. — — — B.fl.VII,553 M.fr.VII,89.

HIEROCHLOE, J. G. Gmelin, Fl. Sibiric. I, 100 (1747). (Hierochloa, Disarrhenum.)
H. redolens, R. Brown, prodr. 209 (1810) ... ... — T. V. N.S.W. — — B.fl.VII,558 M.fr.VIII,137.
H. rariflora, J. Hooker, Fl. Austr. I, 93 (1844) ... ... — T. V. — — — B.fl.VII,558 M.fr.VIII,138.

ARISTIDA, Linné, spec. plant. 82 (1753). (Chaetaria, Arthratherum.)
A. hygrometrica, R. Brown, prodr. 174 (1810) ... ... — — — — — — N.A. B.fl.VII,561
A. stipoides, R. Brown, prodr. 174 (1810) ... ... W.A. S.A. — N.S.W. Q. N.A. B.fl.VII,561 M.fr.VIII,111.
A. arenaria, Gaudichaud in Freyc. voy. Bot. 407 (1826)... W.A. S.A. — N.S.W. Q. N.A. B.fl.VII,561
A. Behriana, F. v. M. in Transact. Vict. Inst. 44 (1854) — S.A. — — — — B.fl.VII,562 M.fr.XI,130.
A. leptopoda, Bentham, Fl. Austr. VII, 562 (1878) ... — S.A. — N.S.W. Q. N.A. B.fl.VII,562 M.fr.XI,130.
A. vagans, Cavanilles, icon. et descr. pl. V, 45, t. 471 (1799) ... — S.A. — N.S.W. Q. N.A. B.fl.VII,562 M.fr.VIII,111.
A. ramosa, R. Brown, prodr. 173 (1810) ... ... — S.A. — — N.S.W. Q. — B.fl.VII,563
A. calycina, R. Brown, prodr. 173 (1810) ... ... — — — N.S.W. Q. N.A. B.fl.VII,563 M.fr.XI,130.
A. depressa, Retzius, observ. bot. IV, 22 (1787) ... ... — — — N.S.W. Q. — B.fl.VII,563

STIPA, Linné, spec. plant. 78 (1737). (Streptachne, Urachne partly.)
S. elegantissima, Labillardière, Nov. Holl. pl. spec. I, 23, t. 29(1804) W.A. S.A. — V. N.S.W. Q. N.A. B.fl.VII,565 M.fr.VIII,103.
S. Tuckeri, F. v. M., fragm. XI, 128 (1881) ... ... — V. N.S.W. — — M.fr.XI,129.
S. verticillata, Nees in Spreng. syst. veg. cur. post. 30 (1827) ... — — — N.S.W. Q. — B.fl.VII,566 M.fr.VIII,105.

S. flavescens, Labillardière, Nov. Holl. pl. spec. I, 24, t. 30 (1804) W.A. S.A. T. V. N.S.W. — — B.fl.VII,506 M.fr.VIII,104.
S. teretifolia, Steudel, syn. glum. I, 128 (1855) ... ... W.A. — T. V. — — — B.fl.VII,567 M.fr.VIII,104.
S. compressa, R. Brown, prodr. 175 (1810)... ... ... W.A. — — — — — — B.fl.VII,567
S. Drummondii, Steudel, syn. glum. I, 128 (1855) ... ... W.A. — — — — — — B.fl.VII,567
S. pycnostachya, Bentham, Fl. Austr. VII, 568 (1878) ... ... W.A. — — — — — — B.fl.VII,568
S. setacea, R. Brown, prodr. 174 (1810) ... ... W.A. S.A. T. V. N.S.W. Q. — B.fl.VII,568
S. semibarbata, R. Brown, prodr. 174 (1810) ... ... W.A. S.A. T. V. N.S.W. Q. — B.fl.VII,568 M.fr.VIII,104;XI,129.
S. hemipogon, Bentham, Fl. Austr. VII, 569 (1878) ... ... W.A. — — — — — — B.fl.VII,569 M.fr.VIII,104.
S. pubescens, R. Brown, prodr. 174 (1810)... ... ... W.A. S.A. T. V. N.S.W. Q. — B.fl.VII,569 M.fr.VIII,104.
S. aristiglumis, F. v. M. in Transact. Vict. Inst. 43 (1854) ... — S.A. — V. N.S.W. Q. — B.fl.VII,570 M.fr.VIII,103.
S. Eriopus, Bentham, Fl. Austr. VII, 570 (1878)... ... ... W.A. — — — — — — B.fl.VII,570
S. trichophylla, Bentham, Fl. Austr. VII, 570 (1878) ... ... W.A. — — — — — — B.fl.VII,570
S. scabra, Lindley in Mitch. Trop. Austr. 31 (1848) ... W.A. S.A. — V. N.S.W. Q. — B.fl.VII,570
S. Streptachne, F. v. M. in journ. R.S. of N.S.W. 237 (1881) ... — — — — — Q. — B.fl.VII,572

DICHELACHNE, Endlicher, prodr. fl. Norfolk. 20 (1833). (Muehlenbergia partly.)
D. crinita, J. Hooker, Fl. N. Zel. I, 293 (1833) ... ... W.A. S.A. T. V. N.S.W. Q. — B.fl.VII,574 M.fr.VIII,105.
D. sciurea, J. Hooker, Fl. N. Zel. I, 294 (1853) ... ... — S.A. T. V. N.S.W. Q. — B.fl.VII,574 M.fr.VIII,105;XI,129.

PENTAPOGON, R. Brown, prodr. fl. Nov. Holl. 173 (1810).
P. Billardieri, R. Brown, prodr. 173 (1810)... ... ... — S.A. T. V. N.S.W. — — B.fl.VII,572 M.fr.VIII,106.

ECHINOPOGON, Palisot, Essai d'une nouv. Agrostogr. 42, t. 9, f. 5 (1812). (Cinna partly.)
E. ovatus, Palisot, Agrostogr. 42, t. 9 (1812) ... W.A. S.A. T. V. N.S.W. Q. — B.fl.VII,599 M.fr.VIII,106.

DIPLOPOGON, R. Brown, prodr. fl. Nov. Holl. 170 (1810). (Dipogonia.)
D. setaceus, R. Brown, prodr. 176 (1810) ... ... W.A. — — — — — — B.fl.VII,573 M.fr.VI,86;VIII,106.

AMPHIPOGON, R. Brown, prodr. fl. 175 (1810). (Aegopogon partly, Pentacraspedion, Gamolythrum.)
A. debilis, R. Brown, prodr. 175 (1810) ... ... W.A. — — — — — — B.fl.VII,597 M.fr.VIII,201.
A. strictus, R. Brown, prodr. 175 (1810) ... ... W.A. S.A. — V. N.S.W. Q. N.A. B.fl.VII,597 M.fr.VIII,201.
A. laguroides, R. Brown, prodr. 175 (1810) ... ... W.A. — — — — — — B.fl.VII,598 M.fr.VI,86;VIII,201.
A. cygnorum, Nees in Lehm. pl. Preiss. II, 100 (1846) ... W.A. — — — — — — B.fl.VII,599
A. turbinatus, R. Brown, prodr. 175 (1810) ... ... W.A. — — — — — — B.fl.VII,599 M.fr.VI,86;VIII,201.

PAPPOPHORUM, Schreber, gen. pl. II, 787 (1791). (Enneapogon.)
P. commune, F. v. M., plants of Gregory's Exp. 10 (1859) ... W.A. S.A. — V. N.S.W. Q. N.A. B.fl.VII,600 M.fr.VIII,200.

SPOROBOLUS, R. Brown, prodr. fl. Nov. Holl. 169 (1810). (Vilfa partly.)
S. Virginicus, Humboldt & Kunth, gram. I, 67 (1832) ... W.A. S.A. — V. N.S.W. Q. N.A. B.fl.VII,621 M.fr.VI,84.
S. Indicus, R. Brown, prodr. 170 (1810) ... ... W.A. S.A. — V. N.S.W. Q. — B.fl.VII,622 M.fr.VI,84;X,76.
S. diander, Palisot, Agrostogr. 26 (1812) ... ... — — — — N.S.W. Q. — B.fl.VII,622 M.fr.VI,84;VIII,140.
S. pulchellus, R. Brown, prodr. 170 (1810)... ... ... — — — — N.S.W. Q. N.A. B.fl.VII,623 M.fr.VIII,140.
S. Lindleyi, Bentham, Fl. Austr. VII, 623 (1878)... ... W.A. S.A. — V. N.S.W. Q. — B.fl.VII,623 M.fr.XI,130. [130.
S. actinocladus, F. v. M., fragm. VIII, 140 (1874) ... — S.A. — V. N.S.W. Q. N.A. B.fl.VII,623 M.fr.V,84;IX,194;XI,

AGROSTIS, Linné, syst. nat. 6 (1735); Linné, gen. plant. 19 (1737). (Deyeuxia, Didymochaeta, Trichodium, Bromidium, Lachnagrostis partly.)
A. Muelleri, Bentham, Fl. Austr. VII, 576 (1878) ... ... — — — V. N.S.W. — — B.fl.VII,576 M.fr.VI,86.
A. scabra, Willdenow, spec. plant. I, 378 (1797) ... ... — S.A. T. V. N.S.W. — — B.fl.VII,576
A. venusta, Trinius in Mém. de l'Acad. de Petersb. VI, 340 (1850) W.A. S.A. T. V. N.S.W. — — B.fl.VII,576
A. aequata, Nees in Hook. Lond. Journ. of Bot. II, 412 (1843)... — — T. — — — — B.fl.VII,579
A. Solandri, F. v. M., Veget. of the Chath. Isl. 60 (1864) ... W.A. S.A. T. V. N.S.W. Q. — B.fl.VII,579 M.fr.XI,27,130.
A. montana, R. Brown, prodr. 171 (1810) ... ... — S.A. T. V. N.S.W. — — B.fl.VII,581
A. quadriseta, R. Brown, prodr. 171 (1810) ... ... W.A. S.A. T. V. N.S.W. — — B.fl.VII,581
A. cylindrica, R. Brown, prodr. 171 (1810)... ... ... W.A. — — — — — — B.fl.VII,582
A. densa, F. v. M.; Deyeuxia, Benth. Fl. Austr. VII, 582 (1878) — S.A. — V. — — — B.fl.VII,582
A. frigida, F. v. M. in Benth. Fl. Austr. VII, 583 (1878) ... — T. V. N.S.W. — — B.fl.VII,583
A. scabra, R. Brown, prodr. 172 (1810) ... ... — T. V. N.S.W. Q. — B.fl.VII,583 M.fr.VI,85.
A. nivalis, F. v. M. in Transact. Vict. Inst. 43 (1854) ... — V. N.S.W. — — B.fl.VII,584
A. Gunniana, F. v. M.; Deyeuxia, Benth. Fl. Austr. VII, 584 (1878) — T. — — — — B.fl.VII,584
A. brevigluma, F.v.M.; Deyeuxia, Benth. Fl. Austr. VII, 584 (1878) — — — N.S.W. — — B.fl.VII,584

COELACHNE, R. Brown, prodr. fl. Nov. Holl. 187 (1810).
C. pulchella, R. Brown, prodr. 187 (1810) ... ... — — — — — Q. — B.fl.VII,626

MICRAIRA, F. v. M., fragm. V, 208 (1866).
M. subulifolia, F. v. M., fragm. V, 208 (1866) ... ... — — — — — Q. — M.fr.V,208.

AIRA, Linné, gen. plant. 335 (1737). (Aera, Deschampsia.)
A. caespitosa, Linné, spec. plant. 64 (1753) ... ... — S.A. T. V. N.S.W. — — B.fl.VII,587 M.fr.VIII,138.

TRISETUM, Persoon, synops. plant. I, 97 (1805).
T. subspicatum, Palisot, Agrostogr. 88 (1812) ... ... — T. V. N.S.W. — — B.fl.VII,588 M.fr.VIII,136.

ERIACHNE, R. Brown, prodr. fl. Nov. Holl. 183 (1810).
E. stipacea, F. v. M., fragm. V, 206 (1866) ... ... — — — — — Q. N.A. B.fl.VII,626 M.fr.VIII,137;X,76.
E. Armitii, F. v. M. in Benth. Fl. Austr. VII, 627 (1878) ... — — — — — Q. N.A. B.fl.VII,627
E. squarrosa, R. Brown, prodr. 183 (1810)... ... ... — — — — — Q. N.A. B.fl.VII,628
E. glauca, R. Brown, prodr. 184 (1810) ... ... — — — — — — N.A. B.fl.VII,628
E. rara, R. Brown, prodr. 183 (1810) ... ... — — — — — Q. — B.fl.VII,628
E. agrostidea, F. v. M., fragm. VII, 82 (1870) ... ... — — — — — N.A. B.fl.VII,628 M.fr.VIII,137.
E. ciliata, R. Brown, prodr. 184 (1810) ... ... — — — — — Q. N.A. B.fl.VII,628
E. setacea, Bentham, Fl. Austr. VII, 629 (1878) ... ... — — — — — — N.A. B.fl.VII,629
E. avenacea, R. Brown, prodr. 184 (1810) ... ... — — — — — — N.A. B.fl.VII,629
E. aristidea, F. v. M., fragm. V, 205 (1866) ... ... W.A. S.A. — N.S.W. Q. — B.fl.VII,629 M.fr.V,205.
E. pallescens, R. Brown, prodr. 184 (1810) ... ... — — — — — Q. — B.fl.VII,630
E. festucacea, F. v. M., fragm. V, 205 (1866) ... ... — — — — — — N.A. B.fl.VII,630 M.fr.V,205.

E. ovata, Nees in Hook. Lond. Journ. II, 418 (1843) ... ... W.A. S.A. — — — — N.A. B.fl.VII,630 M.fr.VIII,137.
E. mellacea, F. v. M., fragm. V, 205 (1866) ... ... — — — — — N.A. B.fl.VII,631 M.fr.V,205.
E. pallida, F. v. M. in Benth. Fl. Austr. VII, 631 (1878) ... — — — — — N.A. B.fl.VII,631
E. scleranthoides, F. v. M., fragm. VIII, 233 (1874) ... — S.A. — — — — B.fl.VII,631
E. mucronata, R. Brown, prodr. 184 (1810) ... ... — — — — Q. B.fl.VII,632
E. obtusa, R. Brown, prodr. 184 (1810) ... ... — — — — N.S.W. Q. N.A. B.fl.VII,632 M.fr.VIII,137;XI,130.
E. capillaris, R. Brown, prodr. 184 (1810)... ... ... — — — — — N.A. B.fl.VII,632

ANISOPOGON, R. Brown, prodr. fl. Nov. Holl. 176 (1810).
A. avenaceus, R. Brown, prodr. 176 (1810) ... ... — — — N.S.W. Q. — B.fl.VII,600 M.fr.VIII,106.

DANTHONIA, De Candolle, Fl. franç. III, 32 (1805). (Amphibromus, Monacather, Plinthacanthus.)
D. paradoxa, R. Brown, prodr. 177 (1810) ... ... — — N.S.W. — B.fl.VII,591
D. bipartita, F. v. M., fragm. I, 160 (1859) ... ... W.A. S.A. — V. N.S.W. — B.fl.VII,592 M.fr.VIII,136.
D. carphoides, F. v. M. in Benth. Fl. Austr. VII, 592 (1878) — S.A. — V. N.S.W. — B.fl.VII,592 M.fr.XI,130.
D. penicillata, F. v. M., fragm. VIII, 135 (1873)... ... W.A. S.A. T. V. N.S.W. Q. — B.fl.VII,592 M.fr.VIII,135.
D. robusta, F. v. M. in Transact. Vict. Inst. 44 (1854) ... — — — V. N.S.W. — B.fl.VII,593 M.fr.VIII,136.
D. pauciflora, R. Brown, prodr. 177 (1810) ... ... — — — V. N.S.W. — B.fl.VII,596
D. nervosa, J. Hooker, Fl. Tasman. II, 121, t. 163 (1860) ... W.A. S.A. T. V. N.S.W. — B.fl.VII,580 M.fr.VIII,135.

ASTREBLA, F. v. M., fragm. phytogr. Austr. X, 76 (1876).
A. pectinata, F. v. M., fragm. X, 76 (1876) ... ... — S.A. — — N.S.W. Q. N.A. B.fl.VII,602 M.fr.VIII,134.
A. triticoides, F. v. M., fragm. X, 76 (1876) ... ... — S.A. — — N.S.W. Q. N.A. B.fl.VII,602 M.fr.VIII,134.

MICROCHLOA, R. Brown, prodr. fl. Nov. Holl. 208 (1810).
M. setacea, R. Brown, prodr. 208 (1810) ... ... — — — — — N.A. B.fl.VII,608

CYNODON, L. C. Richard in Persoon, synops. I, 85 (1805).
C. Dactylon, L. C. Richard in Persoon, synops. I, 85 (1805) ... W.A. S.A. — V. N.S.W. Q. — B.fl.VII,609 M.fr.VIII,113.
C. tenellus, R. Brown, prodr. 187 (1810) ... ... — — — — Q. N.A. B.fl.VII,609 M.fr.VIII,113.
C. convergens, F. v. M., fragm. VIII, 113 (1874)... ... — — — — Q. N.A. B.fl.VII,610 M.fr.X,76.
C. ciliaris, Bentham, Fl. Austr. VII, 610 (1878) ... — S.A. — — — — B.fl.VII,610

CHLORIS, Swartz, nov. gen. et spec. plant. 25 (1788).
C. unispicea, F. v. M., fragm. VII, 118 (1871) ... ... — — — Q. — B.fl.VII,611 M.fr.
C. pumilio, R. Brown, prodr. 186 (1810) ... ... — — — Q. N.A. B.fl.VII,611
C. pectinata, Bentham, Fl. Austr. VII, 612 (1878) ... — S.A. — — Q. N.A. B.fl.VII,612 M.fr.VI,85.
C. divaricata, R. Brown, prodr. 186 (1810) ... ... W.A. S.A. — V. N.S.W. Q. — B.fl.VII,612 M.fr.VI,85.
C. acicularis, Lindley in Mitch. Trop. Austr. 33 (1848) ... — S.A. — V. N.S.W. Q. — B.fl.VII,612 M.fr.VI,85;XI,130.
C. truncata, R. Brown, prodr. 186 (1810) ... ... — — — N.S.W. Q. — B.fl.VII,613 M.fr.VI,85.
C. ventricosa, R. Brown, prodr. 186 (1810) ... ... — — — N.S.W. Q. — B.fl.VII,613 M.fr.VI,85.
C. barbata, Swartz, nov. gen. et spec. plant. 25 (1788) ... — S.A. — — — N.A. B.fl.VII,614 M.fr.XI,130.
C. scariosa, F. v. M., fragm. VI, 85 (1868) ... ... — — — — — N.A. B.fl.VII,614

ELEUSINE, J. Gaertner, de fruct. I, 7 (1788). (Leptochloa, Dactyloctenium, Acrachne).
E. Aegyptiaca, Persoon, syn. I, 82 (1805) ... ... — S.A. — V. N.S.W. Q. N.A. B.fl.VII,615 M.fr.VIII,111.
E. Indica, J. Gaertner, de fruct. I, 7 (1788) ... ... — — — N.S.W. Q. — B.fl.VII,615 M.fr.VIII,112.
E. verticillata, Roxburgh, Fl. Iud. ed. Car. & Wall. I, 346 (1820) — — — — Q. N.A. B.fl.VII,615 M.fr.VIII,112. [130.
E. digitata, Sprengel, syst. cur. post. 36 (1827) ... W.A. S.A. — N.S.W. Q. N.A. B.fl.VII,617 M.fr.I,216;VIII,112;XI,
E. Chinensis, F. v. M.; Poa, Koenig in Roth, nov. pl. spec. 65 (1821) — — — — N.S.W. Q. — B.fl.VII,617 M.fr.VIII,132.
E. digitata, Sprengel, syst. veg. cur. post. 36 (1827) ... — — — — Q. N.A. B.fl.VII,617 M.fr.VIII,113.

CENTOTHECA, Desvaux in Palisot, Agrost. 69, t. 14 (1812). [76.
C. lappacea, Desvaux in Palisot, Agrost. 69, t. 14 (1812) ... — — — — Q. — B.fl.VII,640 M.fr.VI,85;VIII,116;X,

POA, Linné, gen. plant. 20 (1737). (Glyceria partly.)
P. Billardieri, Steudel, syn. glum. I, 262 (1855) ... ... W.A. S.A. T. V. — N.A. B.fl.VII,651
P. homomalla, Nees in Lehm. pl. Preiss. II, 104 (1846)... ... W.A. — — B.fl.VII,651
P. caespitosa, G. Forster, fl. ins. Austr. prodr. 89 (1786) ... W.A. S.A. T. V. N.S.W. Q. B.fl.VII,651 M.fr.VIII,133;XI,13.
P. Maxwelli, Bentham, Fl. Austr. VII, 653 (1878) ... ... W.A. — — B.fl.VII,652
P. nodosa, Nees in Lehm. pl. Preiss. II, 105 (1846) ... W.A. S.A. — V. — B.fl.VII,653 M.fr.VIII,132.
P. saxicola, R. Brown, prodr. 654 (1810) ... ... — T. — B.fl.VII,654
P. lepida, F. v. M., fragm. VIII, 130 (1874) ... ... W.A. S.A. — — B.fl.VII,654 M.fr.XI,130.
P. Fordeana, F. v. M., fragm. VIII, 130 (1874) ... ... — S.A. — V. N.S.W. Q. — B.fl.VII,657 M.fr.VIII,130.
P. fluitans, Scopoli, flor. Carniol. 106 (1772) ... ... W.A. S.A. — V. — B.fl.VII,658 M.fr.VIII,129.
P. latispicea; Festuca, F. v. M., VIII, 127 (1873) ... ... — — — N.S.W. — B.fl.VII,658 M.fr.VIII,128.
P. syrtica, F. v. M. in Transact. Vict. Inst. 45 (1854) ... W.A. S.A. T. V. — — B.fl.VII,658 M.fr.VIII,130. [130.
P. dives; Festuca, F. v. M., fragm. III, 147 (1863) ... ... — — — V. — B.fl.VII,650 M.fr.VI,147;VIII,129,XI,
P. ramigera, F. v. M. in Transact. Vict. Inst. 45 (1854)... ... W.A. S.A. — V. N.S.W. — B.fl.VII,650 M.fr.VIII,131,XI, 130.

FESTUCA, Dillenius, nov. gen. 90, t. 3 (1719). (Glyceria partly, Vulpia partly, Schedonorus, Schenodorus, Brizopyrum.)
F. duriuscula, Linné, spec. plant. 74 (1753) ... ... — S.A. T. V. N.S.W. — B.fl.VII,663 M.fr.VI,65.
F. scirpoides, F. v. M., fragm. VIII, 120 (1874) ... ... W.A. — — B.fl.VII,653 M.fr.XI,130.
F. litoralis, Labillardière, Nov. Holl. spec. plant. I, 22, t. 27 (1804) W.A. S.A. — V. N.S.W. Q. — B.fl.VII,656 M.fr.VIII,129.
F. Hookeriana, F. v. M. in J. Hooker, Fl. Tasm. II, 122, t. 115(1860) ... — T. V. N.S.W. — B.fl.VII,656 M.fr.VIII,131.

DIPLACHNE, Palisot, Essai d'une nouv. Agrostogr. 9, t. 16 (1812). (Uralepis.)
D. loliiformis, F. v. M. in Benth. fl. Austr. VII, 618 (1878) ... — S.A. — V. N.S.W. Q. — B.fl.VII,616 M.fr.VIII,128;XI,130.
D. Muelleri, Bentham, Fl. Austr. VII, 619 (1878) ... ... — S.A. — — — N.A. B.fl.VII,619 M.fr.VIII,127.
D. fusca, Palisot, Agrost. 163 (1812) ... ... W.A. S.A. — V. N.S.W. Q. N.A. B.fl.VII,619 M.fr.VIII,127.
D. parviflora, Bentham, Fl. Austr. VII, 620 (1878) ... ... — — — — — N.A. B.fl.VII,620 M.fr.VIII,129.

TRIODIA, R. Brown, prodr. fl. Nov. Holl. 182 (1810).
T. Mitchelli, Bentham, Fl. Austr. VII, 606 (1878) ... ... — — — — N.S.W. Q. — B.fl.VII,606
T. pungens, R. Brown, prodr. 182 (1810) ... ... — S.A. — — — N.A. B.fl.VII,606 M.fr.VIII,129.
T. Cunninghamii, Bentham, Fl. Austr. VII, 606 (1878) ... ... — — — — — Q. N.A. B.fl.VII,606
T. irritans, R. Brown, prodr. 182 (1810) ... ... W.A. S.A. — V. N.S.W. Q. — B.fl.VII,607 M.fr.VIII,129.
T. procera, R. Brown, prodr. 182 (1810) ... ... — — — — — N.A. B.fl.VII,607
T. microstachya, R. Brown, prodr. 182 (1810) ... ... — — — — — N.A. B.fl.VII,607
T. microdon, F. v. M.; Triraphis, Benth. Fl. Austr. VII, 605 (1878) — — — — N.S.W. — — B.fl.VII,605

DISTICHLIS, Rafinesque in Journ. de Phys. LXXXIX, 104 (1819). (Uniola partly, Brizopyrum partly.)
D. maritima, Rafinesque in Journ. de Phys. LXXXIX, 104 (1819) — S.A. T. V. N.S.W. — — B.fl.VII,637 M.fr.VIII,129.

BROMUS, Dillenius in Linné, syst. nat. 8 (1735); Linné, gen. plant. 15 (1737).
B. arenarius, Labillardière, Nov. Holl. pl. spec. I, 23, t. 28 (1804) W.A. S.A. — V. N.S.W. Q. — B.fl.VII,660 M.fr.VI,126;XI,20.

ERAGROSTIS, Palisot, Essai d'une nouv. Agrostogr. 70, t. 14 (1812). (Poa partly.)
E. tenella, Palisot in Roem. et Schult. syst. veg. II, 576 (1817) — S.A. — V. N.S.W. Q. N.A. B.fl.VII,643 M.fr.VIII,132;XI,130
E. nigra, Nees in Steud. nomencl. bot. I, 563 (1841) ... ... — — — — N.S.W. — B.fl.VII,643
E. imbecilla, Bentham, Fl. Austr. VII, 643 (1878) ... ... — — — — — Q. B.fl.VII,643
E. trichophylla, Bentham, Fl. Austr. VII, 644 (1878) ... ... — S.A. — — — — B.fl.VII,644
E. leptocarpa, Bentham, Fl. Austr. VII, 644 (1878) ... ... — S.A. — — Q. N.A. B.fl.VII,644
E. megalosperma, F. v. M. in Benth. Fl. Austr. VII, 644 (1878) ... — — — N.S.W. Q. — B.fl.VII,644
E. pilosa, Palisot, Agrost. 71 (1812)... ... ... — S.A. — V. N.S.W. Q. — B.fl.VII,645 M.fr.VIII,133;XI,130
E. leptostachya, Steudel, syn. glum. I, 279 (1855) ... ... — — — N.S.W. Q. — B.fl.VII,645
E. Schultzii, Bentham, Fl. Austr. VII, 646 (1878) ... ... — — — — — N.A. B.fl.VII,646
E. diandra, Steudel, syn. glum. I, 279 (1855) ... ... W.A. S.A. — N.S.W. Q. N.A. B.fl.VII,646 M.fr.XI,130.
E. Brownii, Nees in Steud. nomencl. bot. I, 562 (1841) ... W.A. S.A. — V. N.S.W. Q. N.A. B.fl.VII,646
E. concinna, Steudel, syn. glum. I, 279 (1855) ... ... — S.A. — — Q. N.A. B.fl.VII,647
E. speciosa, Steudel, syn. glum. I, 279 (1855) ... ... — S.A. — — Q. N.A. B.fl.VII,648
E. laniflora, Bentham, Fl. Austr. VII, 648 (1878) ... ... — S.A. — N.S.W. Q. — B.fl.VII,648
E. eriopoda, Bentham, Fl. Austr. VII, 648 (1878) ... ... — S.A. — N.S.W. — N.A. B.fl.VII,648
E. setifolia, Nees in Hook. Lond. journ. II, 419 (1843) ... W.A. S.A. — V. N.S.W. Q. N.A. B.fl.VII,648
E. lacunaria, F. v. M., first gen. Rep. 20 (1853) ... ... — S.A. — V. N.S.W. Q. — B.fl.VII,649
E. falcata, Gaudichaud in Voy. Freyc. Bot. 408, t. 25 (1826) ... W.A. S.A. — V. N.S.W. Q. N.A. B.fl.VII,649
E. stenostachya, Steudel, syn. glum. I, 279 (1855) ... ... — — — — Q. N.A. B.fl.VII,650

HETERACHNE, Bentham in J. Hook. icon. plant. XIII, 39 (1877).
H. Brownii, Bentham in J. Hook. icon. XIII, 40 (1877)... ... — — — — Q. N.A. B.fl.VII,635 M.fr.VIII,132.
H. Gulliveri, Bentham in J. Hook. icon. XIII, 39, t. 1250 (1877) — — — — Q. N.A. B.fl.VII,635

ECTROSIA, R. Brown, prodr. fl. Nov. Holl. 185 (1810).
E. Schultzii, Bentham, Fl. Austr. VII, 633 (1878) ... ... — — — — — N.A. B.fl.VII,633
E. leporina, R. Brown, prodr. 186 (1810) ... ... — — — N.S.W. Q. N.A. B.fl.VII,633 M.fr.VIII,109.
E. agrostoides, Bentham, Fl. Austr. VII, 634 (1878) ... ... — — — — — N.A. B.fl.VII,634
E. Gulliveri, F. v. M., fragm. VIII, 201 (1874) ... ... — — — — Q. N.A. B.fl.VII,634 M.fr.VIII,201.

ELYTHROPHORUS, Palisot, Essai d'une nouv. Agrostogr. 67, t. 14 (1812).
E. articulatus, Palisot, Agrost. 67, t. 14 (1812) ... ... — S.A. — V. N.S.W. Q. N.A. B.fl.VII,638 M.fr.VIII,109.

TRIRAPHIS, R. Brown, prodr. fl. Nov. Holl. 185 (1810). (Triraphis.)
T. mollis, R. Brown, prodr. 185 (1810) ... ... W.A. S.A. — V. N.S.W. Q. N.A. B.fl.VII,603 M.fr.VIII,108.
T. pungens, R. Brown, prodr. 185 (1810) ... ... — — — — — N.A. B.fl.VII,604 M.fr.VIII,108,125.
T. bromoides, F. v. M., fragm. VIII, 108 (1874) ... ... W.A. — — — — — B.fl.VII,604 M.fr.VIII,108.
T. danthonioides, F. v. M., fragm. VIII, 125 (1874) ... ... W.A. — — — — — B.fl.VII,604 M.fr.VIII,108.

AGROPYRON, J. Gaertner, nov. comm. Petrop. XIV, 539 (1770). (Agropyrum, Triticum partly, Anthosachne, Vulpia partly.)
A. scabrum, Palisot, Agrost. 102 (1812) ... ... W.A. S.A. T. V. N.S.W. Q. — B.fl.VII,665 M.fr.VI,85;XI,27,130.
A. velutinum, Nees in Hook. Lond. Journ. of Bot. II, 417 (1843) ... — — T. V. N.S.W. — — B.fl.VII,665
A. pectinatum, Palisot, Agrost. 102 (1812)... ... ... — — T. V. N.S.W. — — B.fl.VII,666 M.fr.VI,85.

ARUNDO, Tournefort, inst. 526 (1700), from Varro. (Phragmites.)
A. Phragmites, Dodoens, stirp. hist. pemptad. 602 (1583) ... — S.A. T. V. N.S.W. Q. — B.fl.VII,637
A. Roxburghii, F. v. M.; Phragmites, Kunth in nov. act. acad.
   Caes. XIX, suppl. I, 152 (1843) ... ... — — — — Q. N.A. — M.fr.XI,139.

BAMBUSA, Schreber, gener. plantar. I, 236 (1759).
Species (three, of this or allied genera, not yet found in flower)
   undetermined ... ... — — — — Q. N.A. — M.fr.VI,85;XI,129.

ACOTYLEDONEAE.

A. L. de Jussieu, genera plantarum 1 (1789).

ACOTYLEDONEAE VASCULARES.

Meissner, plantarum vascularium genera, 420 (1843).

RHIZOSPERMAE.

G. Weber, primit. Fl. Holsat. 74 (1780).

AZOLLA, Lamarck, Encycl. méthod. I, 343 (1783).
A. pinnata, R. Brown, prodr. 167 (1810) ... ... — S.A. — V. N.S.W. Q. — B.fl.VII,679 M.fr.V,140.
A. filiculoides, Lamarck, Encycl. méth. I, 343 (1783) ... — S.A. T. V. N.S.W. Q. — B.fl.VII,680 M.fr.V,140;XI,127.

MARSILEA, Linné, syst. nat. 8 (1735); Linné, gen. plant. 326 (1737). (Marsiglia.)
M. quadrifolia, Linné, spec. plant. 1099 (1753) ... ... W.A. S.A. — V. N.S.W. Q. N.A. B.fl.VII,683 M.fr.V,140.

PILULARIA, Vaillant, Bot. Paris. 159, t. 15 (1727).
P. globulifera, Linné, spec. plant. 1100 (1753) ... ... W.A. S.A. T. V. N.S.W. — — B.fl.VII,684 M.fr.V,140.

ISOETES, Linné, skanska resa 420 (1751).
I. Drummondii, A. Braun in Berl. Monatsb. 503 (1863)... ... W.A. — — V. — — — — B.fl.VII,672 M.fr.V,140.
I. elatior, F. v. M. & A. Braun in Linnaea XXV, 722 (1852) ... — — T. — — — — — M.fr.V,140;XI,130
I. humilior, F. v. M. & A. Braun in Linnaea XXV, 722 (1852)... — — T. — — — — — M.fr.V,140;XI,139.
I. Gunnii, A. Braun in Berl. Monatsb. 535 (1863) ... ... — — T. — — — — — M.fr.XI,139.
I. Muelleri, A. Braun in Berl. Monatsb. 535 (1868) ... ... — — — Q. — — M.fr.XI,139.

LYCOPODINEAE.

Swartz, synops. filic. XV (1806).

PSILOTUM, Swartz in Schrader, Journ. fuer die Bot. II, 109 (1800)
P. triquetrum, Swartz, synops. filic. 187 (1806) ... ... — — — N.S.W. Q. — B.fl.VII,681 M.fr.V,112;X,118.
P. complanatum, Swartz, synops. filic. 187 (1806) ... ... — — — — Q. — B.fl.VII,682 M.fr.V,112.

TMESIPTERIS, Bernhardi in Schrader, Journ. fuer die Bot. II, 131 (1800).
T. Tannensis, Bernhardi in Schrad. Journ. II, 131, t. 2 (1800)... — — T. V. N.S.W. Q. — B.fl.VII,680 M.fr.V,112;X,118.

LYCOPODIUM, Ruppius, Fl. Jenens. 32 (1718), from Tabernaemontanus (1588).
L. Selago, Linné, spec. plant. 1102 (1753) ... ... — — T. V. N.S.W. — — B.fl.VII,074 M.fr.V,111;X,118.
L. Phlegmaria, Linné, spec. plant. 1101 (1753) ... ... — — T. V. N.S.W. — Q. — B.fl.VII,674 M.fr.V,111;X,118.
L. clavatum, Linné, spec. plant. 1101 (1753) ... ... — — T. V. N.S.W. — — B.fl.VII,075 M.fr.V,111.
L. Carolinianum, Linné, spec. plant. 1104 (1753)... ... W.A. — T. V. N.S.W. Q — B.fl.VII,075 M.fr.V,111;VI,124;X,118
L. laterale, R. Brown, prodr. 165 (1810) ... ... — S.A. T. V. N.S.W. Q. N.A. B.fl.VII,676 M.fr.V,111;X,118.
L. densum, Labill., Nov. Holl. plant. spec. II, 104, t. 251 (1806) — S.A. T. V. N.S.W. — — B.fl.VII,676 M.fr.V,111;X,118.
L. scariosum, G. Forster, florul. ins. austr. prodr. 87 (1786) ... — — T. V. — — B.fl.VII,676 M.fr.V,112.
L. volubile, G. Forster, florul. ins. austr. prodr. 86 (1786) ... — — T. V. — — B.fl.VII,077 M.fr.V,112.

SELAGINELLA, Palisot, prodr. des fam. de l'Aethéogamie 101 (1805). (Lycopodium partly.)
S. Preissiana, Spring in Mém. de l'Ac. de Brux. 61 (1842) ... W.A. S.A. T. V. N.S.W. Q. — B.fl.VII,677 M.fr.V,112;VIII,275.
S. uliginosa, Spring in Mém. de l'Ac. de Brux. II, 60 (1842) ... — S.A. T. V. N.S.W. Q. — B.fl.VII,678 M.fr.V,112;VIII,275.
S. flabellata, Spring in Mém. de l'Ac. de Brux. 174 (1849) ... — — — — — Q. — B.fl.VII,678 M.fr.VIII,175,275.
S. concinna, Spring in Mém. de l'Ac. de Brux. 109 (1849) ... — — — — — Q. — B.fl.VII,678 M.fr.V,112;VIII,274.
S. Belangeri, Spring in Mém. de l'Ac. de Brux. 242 (1849) ... — — — — — Q. N.A. B.fl.VII,679

PHYLLOGLOSSUM, Kunze in Mohl & Schlechtendal, Bot. Zeitung, 721 (1843).
P. Drummondii, Kunze in der Bot. Zeit. 721 (1843) ... ... W.A. — T. V. N.S.W. — — B.fl.VII,672 M.fr.V,112;X,118.

FILICES.

Linné, gen. plant. 322 (1737).

OPHIOGLOSSUM, Tournefort, inst. rei herb. 548, t. 325 (1700), from Bock (1539). (Ophioderma.)
O. vulgatum, C. Bauhin, pinax theatr. bot. 354 (1623) ... — S.A. T. V. N.S.W. Q. N.A. B.fl.VII,688 M.fr.V,112;VI,124.
O. pendulum, Linné, sp. pl. edit. sec. 1518 (1763) ... ... — — — — N.S.W. Q. — B.fl.VII,689 M.fr.V,113;VI,124.

BOTRYCHIUM, Swartz in Schrader, Journ. fuer die Bot. II, 110 (1800).
B. Lunaria, Swartz in Schrad. Journ. II, 110 (1800) ... — T. V. N.S.W. — — B.fl.VII,690 M.fr.V,113.
B. ternatum, Swartz, synops. filic. 172 (1806) ... ... — T. V. N.S.W. — — B.fl.VII,690 M.fr.V,113.

HELMINTHOSTACHYS, Kaulfuss, enum. filic. Chamiss. 28 (1824).
H. Zeylanica, Kaulfuss, enum. fil. 28 (1824) ... ... — — — — Q. — B.fl.VII,690 M.fr.V,113;IX,190.

LYGODIUM, Swartz in Schrader, Journ. fuer die Bot. II, 7 & 106 (1800). (Hydroglossum, Lygodictyon.)
L. scandens, Swartz in Schrad. Journ. II, 106 (1800) ... ... — — N.S.W. Q. N.A. B.fl.VII,691 M.fr.V,VII,83;VIII,94,X,118.
L. reticulatum, Schkuhr. kryptog. Gewaechse 139, t. 139 (1809) ... — — — — Q. — B.fl.VII,692 M.fr.V,VII,83;VIII,137.
L. Japonicum, Swartz, syn. filic. 154 (1806) ... ... — — — — Q. N.A. B.fl.VII,692 M.fr.V,113;VII,84,156.

SCHIZAEA, Smith in Mem. de l'Acad. Turin. V, 149, t. 19 (1791). (Lopidium, Actinostachys.)
S. rupestris, R. Brown, prodr. 162 (1810) ... ... — — N.S.W. — — B.fl.VII,093 M.fr.V,113.
S. fistulosa, Labillardière, Nov. Holl. pl. spec. II, 103, t. 250 (1806) — S.A. T. V. — — B.fl.VII,694 M.fr.V,113.
S. dichotoma, Smith in Mém. de l'Ac. Tur. V, 149 (1791) ... — — — — Q. — B.fl.VII,694 M.fr.V,113;X,118.
S. Forsteri, Sprengel, Anleit. zur Kenntn. der Gew. III, 157 (1804)— — — — — — M.fr.VIII,275.

ANGIOPTERIS, G. Hoffmann in Comment. Goett. XII, 29 (1796). [118.
A. evecta, G. Hoffmann in Comment. Goett. XII, 29 (1796) ... — — — Q. — B.fl.VII,694 M.fr.V,114;VII,120;X,

MARATTIA, Swartz, nov. gen. et spec. plant. 8 & 128 (1788).
M. fraxinea, Smith, plant. icon. hact. ined. t. 48 (1790)... ... — — — — N.S.W. Q. — B.fl.VII,695 M.fr.V,114;X,118.
TRICHOMANES, Linné, syst. nat. 8 (1835); Linné, hort. Cliffort. 476 (1737).
T. peltatum, Baker in Journ. Linn. Soc. IX, 336, t. 8 (1865) ... ... — Q. — B.fl.VII,701 M.fr.X,117.
T. Vitiense, Baker in Journ. Linn. Soc. IX, 338, t. 8 (1865) ... — — — — N.S.W. Q. — B.fl.VII,701 M.fr.VIII,32;IX,132.
T. Motleyi, Boech, Hymenoph. Javan (1861) ... ... — Q. — —
T. cuspidatum, Willdenow, spec. plant. V, 499 (1810) ... ... — — — — Q. — M.fr.XI,131&139.
T. parvulum, Poiret, Encycl. méth. VIII, 44 (1810) ... ... — — — N.S.W. Q. — B.fl.VII,701 M.fr.VII,155.
T. digitatum, Swartz, synops. filic. 370 & 422 (1806) ... — — — N.S.W. — B.fl.VII,702
T. venosum, R. Brown, prodr. 159 (1810) ... ... — T. V. N.S.W. Q. — B.fl.VII,702 M.fr.V,116;IX,191.
T. Javanicum, Blume, flor. Jav. filic. 224 (1836) ... ... — — — — Q. — B.fl.VII,702
T. rigidum, Swartz, nov. gen. et spec. pl. 137 (1788) ... ... — — N.S.W. Q. — B.fl.VII,702 M.fr.V,115.
T. bipunctatum, Poiret, Encycl. méth. VIII, 44 (1810) ... ... — — — — Q. — B.fl.VII,703 M.fr.X,117;XI,132.
T. humile, G. Forster, florul. ins. Austr. prodr. 84 (1786) ... — — — N.S.W. —
T. caudatum, Brackenridge, Bot. of Wilk. U.S. Exped. fil. 256,
  t. 36 (1854) ... ... — — — N.S.W. Q. — B.fl.VII,703 M.fr.V,116;VII,122.
T. Baucrianum, Endlicher, prodr. fl. Norf. 17 (1833) ... ... — — — N.S.W. Q. — B.fl.VII,703 M.fr.VII,121;XI,132.
T. parviflorum, Poiret, Encycl. méth. VIII, 44 (1810) ... ... — — — — Q. — B.fl.VII,704 M.fr.VII,122.

HYMENOPHYLLUM, Smith in Roemer, Archiv I, 58 (1797).
H. marginatum, Hooker & Greville, icon. filic.I, t. 34 (1829) ... — — — N.S.W. — — B.fl.VII,705
H. nitens, R. Brown, prodr. 159 (1810) ... ... — T. V. N.S.W. Q. — B.fl.VII,705
H. Javanicum, Sprengel, syst. veg. IV, 132 (1827) ... ... — T. V. N.S.W. Q. — B.fl.VII,706 M.fr.V,116.
H. Tunbridgense, Smith in Roemer, Archiv. I, 56 (1797) ... — T. V. N.S.W. Q. — B.fl.VII,706 M.fr.V,116.
H. multifidum, Swartz, synops. filic. 149 & 378 (1806) ... ... — — N.S.W. — B.fl.VII,707

CYATHEA, Smith in Mém. de l'Acad. Turin. V, 416 (1791). (Hemitelia, Amphicosmia.)
C. Lindsayana, Hooker, syn. filic. 25 (1865) ... ... — — — N.S.W. Q. — B.fl.VII,708 M.fr.VIII,177.
C. arachnoidea, Hooker, syn. filic. 24 (1866) ... ... — — — N.S.W. — B.fl.VII,708 M.fr.VII,200.
C. Macarthurii, F. v. M., fragm. VII, 177 (1874) ... ... — N.S.W. — B.fl.VII,708 M.fr.VIII,176;X,120.
C. medullaris, Swartz, synops. filic. 140 (1806) ... ... — T. V. N.S.W. — B.fl.VII,708 M.fr.VII,116;VIII,177.
C. Cunninghami, J. Hooker, icon. plant. 985 (1854) ... ... — T. V. — —
C. brevipinnea, Baker in Bentham fl. Austr. VII, 709 (1878) ... — — N.S.W. — B.fl.VII,709 M.fr.XI,139.
C. Moorei, F. v. M.; Hemitelia, Baker in Gardn. Chron. 252 (1872) — — N.S.W. — B.fl.VII,709 M.fr.VIII,157.

CERATOPTERIS, Brogniart in Bull. de la Soc. philom. 186 (1821). (Parkeria.)
C. thalietroides, Brogniart in Bull. Soc. philom. 186 (1821) ... — — Q. N.A. B.fl.VII,695 M.fr.V,122;VIII,158.

GLEICHENIA, Smith in Mém. de l'Acad. Turin. V, 418 (1791). (Mertensia, Platyzoma, Stromatopteris.) [118.
G. platyzoma, F. v. M., fragm. V, 114 (1866) ... ... — — N.S.W. Q. N.A. B.fl.VII,696 M.fr.V,114;VIII,157;X,
G. circinata, Swartz, syn. filic. 105 et 394 (1806) ... ... — S.A. T. V. N.S.W. Q. N.A. B.fl.VII,697 M.fr.V,115.
G. dicarpa, R. Brown, prodr. 161 (1810) ... ... — T. V. N.S.W. Q. — B.fl.VII,698 M.fr.V,114;VII,115.
G. flabellata, R. Brown, prodr. 161 (1810)... ... ... — T. V. N.S.W. Q. — B.fl.VII,698 M.fr.V,114;VIII,157.
G. Hermanni, R. Brown, prodr. 161 (1810) ... ... — — N.S.W. Q. N.A. B.fl.VII,698 M.fr.V,114;X,118.

OSMUNDA, Tournefort, inst. rei herb. 547, t. 324 (1700), from l'Obel (1576). (Todea, Leptopteris.)
O. barbara, Thunberg, prodr. plant. Capens. 171 (1800)... ... — S.A. T. V. N.S.W. Q. — B.fl.VII,699 M.fr.V,114.
O. Fraseri, F.v.M.; Todea, Hooker & Greville,icon. filic. t. 101 (1829) — — N.S.W. — B.fl.VII,699 M.fr.V,114;VII,156.
O. Moorei, F.v.M.; Todea, Baker in Trimen, Journ. of Bot. 16 (1873) ... — — N.S.W. — B.fl.VII,700 M.fr.VIII,157.

ALSOPHILA, R. Brown, prodr. 158 (1810).
A. Rebeccae, F. v. M., fragm. V, 53 (1865) ... ... — — — Q. — B.fl.VII,710 M.fr.V,117,215;VIII,180.
A. Loddigesii, Kunze in Schlecht. Linnaea XXIII, 221 (1850)... — — — N.S.W. Q. — B.fl.VII,710 [VIII,178.
A. australis, R. Brown, prodr. 158 (1810) ... ... — T. V. N.S.W. Q. — B.fl.VII,710 M.fr.V,52,116;VII,120;
A. excelsa, R. Brown, prodr. 158 (1810) ... ... — — N.S.W. — —
A. Leichhardtiana, F. v. M., fragm. V, 53 (1865) ... ... — — — N.S.W. — B.fl.VII,711 M.fr.V,117;VII,200;VII,
A. Robertsiana, F. v. M., fragm. V, 54 (1865) ... ... — — — N.S.W. — B.fl.VII,712 M.fr.V,117,213. [120.

DICKSONIA, L'Heritier, sert. Angl. 31 (1788). (Cibotium, Patania, Dennstaedtia, Deparia, Balantium.)
D. Billardieri, F. v. M., fragm. VIII, 175 (1874)... ... — S.A. T. V. N.S.W. Q. — B.fl.VII,712 M.fr.V,117;VI,190;VII,
D. Youngiae, C. Moore in Hook. & Bak. syn. filic. 51 (1865) ... — — N.S.W. — B.fl.VII,713 M.fr.VI,200. [107.
D. davallioides, R. Brown, prodr. 158 (1810) ... ... — — N.S.W. — B.fl.VII,713 M.fr.VIII,158. [104.
D. nephrodioides; Deparia, Baker in Gardn. Chron. 253 (1872)... — — N.S.W. — B.fl.VII,713 M.fr.VIII,157;IX,78;X,

DAVALLIA, Smith in Mém. de l'Acad. de Turin. V, 414 (1791). (Microlepia, Humata.)
D. solida, Swartz, synops. filic. 132 & 345 (1806)... ... — — — Q. — B.fl.VII,715
D. elegans, Swartz, synops. filic. 132 & 347 (1806) ... ... — — — — Q. — B.fl.VII,715 M.fr.V,118;VIII,157.
D. pyxidata, Cavanilles, descripcion de las plantas n. 694 (1802) — — V. N.S.W. Q. — B.fl.VII,716 M.fr.V,118.
D. podata, Smith in Mém. de l'Ac. de Turin. V, 414 (1791) ... — — — Q. — B.fl.VII,716 M.fr.V,118.
D. dubia, R. Brown, prodr. 157 (1810) ... ... — T. V. N.S.W. Q. — B.fl.VII,717 M.fr.V,118.
D. flaccida, R. Brown, prodr. 157 (1810) ... ... — — — N.S.W. — B.fl.VII,717 M.fr.V,118;VII,156.
D. tripinnata, F. v. M. in Benth. Fl. Austr. VII, 717 (1878) ... — — — Q. — B.fl.VII,717

VITTARIA, Smith in Mém. de l'Acad. de Turin. V, 414 (1791).
V. elongata, Swartz, synops. filic. 109 & 302 (1806) ... ... — — N.S.W. Q. — B.fl.VII,718 M.fr.V,122;VII,121.

LINDSAYA, Dryander in Mém. de l'Acad. de Turin. V, 413 (1791). (Lindsaea, Isoloma, Synophlebium, Schizoloma.)
L. linearis, Swartz, synops. filic. 118 & 318, t. 3 (1806) ... — W.A. S.A. T. V. N.S.W. Q. — B.fl.VII,719 M.fr.V,119.
L. dimorpha, Bailey, Queensland Ferns 19 (1874) ... ... — — — B.fl.VII,719
L. cultrata, Swartz, synops. filic. 119 (1806) ... ... — — — Q. — B.fl.VII,719
L. flabellulata, Dryander in Transact. Linn. Soc. III, 41, t. 8 (1797) — — — Q. N.A. B.fl.VII,720 M.fr.V,119.
L. lobata, Poiret, Encycl. méth. XI (1813) ... ... — — — Q. — B.fl.VII,720
L. trichomanoides, Dryander in Trans. Linn. Soc. III, 43, t. 11 (1797) — — T. — B.fl.VII,720 M.fr.V,118.
L. microphylla, Swartz, synops. filic. 120 (1806) ... ... — — N.S.W. Q. — B.fl.VII,721 M.fr.V,119.
L. incisa, Prentice in Trim. Journ. of Bot. 295 (1873) ... ... — — — N.S.W. — B.fl.VII,721
L. Fraseri, Hooker, spec. filic. I, 221, t. 70 (1846) ... ... — — — Q. — B.fl.VII,721 M.fr.V,118.

s

138

L. ensifolia, Swartz, synops. filic. 118 (1806) ... ... -- — — — — Q. N.A. B.fl.VII,721 M.fr.V,118.
L. lanuginosa, Wallich, numerical list n. 154 (1828) ... ... — — — — — Q. — B.fl.VII,722 M.fr.V,118.

ADIANTUM, Tournefort, in 543, t. 317 (1700), from Hippocrates, Theophrastos, Dioscorides and Plinius.
A. lunulatum, N. Burmann, Fl. Ind. 235 (1768) ... ... — — — — — Q. N.A. B.fl.VII,723 M.fr.V,119.
A. Capillus veneris, Linné, spec. plant. 1096 (1753) ... ... — — — — — Q. — B.fl.VII,723
A. Aethiopicum, Linné, syst. nat. edit. decim. n. 15 (1759) ... W.A. S.A. T. V, N.S.W. Q. — B.fl.VII,724 M.fr.V,119.
A. formosum, R. Brown, prodr. 155 (1810) ... ... — — — V. N.S.W. Q. — B.fl.VII,724 M.fr.V,120;VIII,157.
A. affine, Willdenow, spec. plant. V, 443 (1810) ... ... — — — — N.S.W. Q. — B.fl.VII,724 M.fr.V,119.
A. diaphanum, Blume, flora Javae, filic. 215 (1836) ... ... — — — V. N.S.W. Q. — B.fl.VII,725 M.fr.VII,121.
A. hispidulum, Swartz, syn. filic. 124 & 321 (1806) ... ... — — — V. N.S.W. Q. — B.fl.VII,725 M.fr.V,120.

CHEILANTHES, Swartz, syn. filic. 126, t. 3 (1806). (Notholaena, Nothochlaena.)
C. Prenticei, Luerssen in Uhlw. & Behr. Bot. Centralblatt IX,
    440 (1882) ... ... — — — — — — N.A. —
C. pumilio, F. v. M.; Notholaena, R. Brown, prodr. 146 (1810) ... -- -- — — Q. — B.fl.VII,773 M fr.VIII,175.
C. fragillima, F. v. M., fragm. V, 123 (1866) ... ... — — — — — N.A. B.fl.VII,774 M.fr.V,123.
C. vellea, F. v. M., fragm. V, 123 (1866) ... ... W.A. S.A. — V. N.S.W. Q. N.A. B.fl.VII,773 M.fr.V,123;VIII,176.
C. distans, A. Braun, index sem. hort. Berol. (1850) ... — — — V. N.S.W. Q. — B.fl.VII,774 M.fr.V,122;VIII,176.
C. tenuifolia, Swartz, syn. filic. 126 (1806) .. .. W.A. S.A. T. V. N.S.W. Q. N.A. B.fl.VII,726 M.fr.V,122.

PTERIS, Linné, syst. nat. 9 (1735); Linné, gen. plant. 322 (1737.) (Pellaea, Cheiloplecton, Platyloma, Litobrochia.)
P. geraniifolia, Raddi, syn. filic. Brasil. 46 (1819) ... ... — — — N.S.W. Q. — B.fl.VII,728 M.fr.V,124.
P. paradoxa, Baker in Benth. Fl. Austr. VII, 729 (1878) ... ... — — — N.S.W. Q. — B.fl.VII,729
P. falcata, R. Brown, prodr. 154 (1810) ... ... — — - T. V. N.S.W. Q. — B.fl.VII,729 M.fr.V,123.
P. rotundifolia, G. Forster, florul. ins. Austr. prodr. 79 (1786) .. — — — — — Q. — B.fl.VII,730
P. longifolia, Linné, spec. plant. 1074 (1753) ... ... — — V. N.S.W. Q. — B.fl.VII,730 M fr.V,126.
P. ensiformis, N. Burmann, Fl. Ind. 230 (1768) ... ... -- — — — — Q. — B.fl.VII,730 M.fr.V,123.
P. umbrosa, R. Brown, prodr. 154 (1810) ... ... — — — V. N.S.W. Q. — B.fl.VII,730 M.fr.V,126.
P. quadriaurita, Retzius, observ. bot. VI, 38 (1791) ... ... — — — — — Q. — B.fl.VII,731 M.fr.V,125.
P. arguta, Aiton, hort. Kew. III, 458 (1789) ... ... — — — T. V. N.S.W. Q. — R.fl.VII,731 M.fr.V,126.
P. aquilina, Linné, spec. plant. 1075 (1753) ... ... W.A S.A. T. V. N.S.W. Q. — B.fl.VII,731 M.fr.V,126.
P. incisa, Thunberg, prodr. pl. Capens. 171 (1800) ... — S.A. T. V. N.S.W. Q. — B.fl.VII,732 M.fr.V,124;VIII,158.
P. marginata, Bory. voy. dans les isles d'Afr. II, 192 (1804) ... — — — — — Q. — B.fl.VII,733 M.fr.V,125.
P. comans, G. Forster, florul. ins. Austr. prodr. 79 (1786) ... — — — T. V. N.S.W. Q. — B.fl.VII,733 M.fr.V,125;VII,121.

LOMARIA, Willdenow in Magaz. der Ges. naturf. Freunde zu Berlin, III, 160 (1809). (Stegania, Plagiogyria.)
L. Patersoni, Sprengel, syst. veg. IV, 62 (1827) ... ... — T. V. N.S.W. Q. — B.fl.VII,734 M.fr.V,122;VII,158.
L. vulcanica, Blume, fl. Javae, filic. 202 (1836) ... ... — — T. — — — — B.fl.VII,733 M.fr.V,121.
L. discolor, Willdenow, spec. plant. V, 293 (1810) ... ... — S.A. T. V. N.S.W. — — B.fl.VII,735 M.fr.V,121.
L. lanceolata, Sprengel, syst. veg. IV, 62 (1827) ... ... — S.A. T. V. N.S.W. — — B.fl.VII,735 M.fr.V,121.
L. attenuata, Willdenow, spec. plant V, 290 (1810) ... ... — — — N.S.W. — — B.fl.VII,736
L. alpina, Sprengel, syst. veg. IV, 62 (1827) ... ... — T. V. N.S.W. — — B.fl.VII,736 M.fr.V,121.
L. fluviatilis, Sprengel, syst. veg. IV, 65 (1827) ... ... — T. V. N.S.W. -- — B.fl.VII,736 M.fr.V,121.
L. Fullageri, F. v. M., fragm. VIII, 157 (1874) ... ... — — N.S.W. — — B.fl.VII,737 M.fr.VIII,157.
L. Capensis, Willdenow, spec. plant. V, 291 (1810) ... ... — 8.A. T. V. N.S.W. Q. — B.fl.VII,737 M.fr.V,121.
L. euphlebia, Kunze in Mohl & Schlecht. Bot. Zeit. VI, 521 (1849) — — — — — Q. — B.fl.VII,738 M.fr.VI,124.

BLECHNUM, Linné, spec. plant. 1077 (1753.)
B. cartilagineum, Swartz, syn. filic. 114 & 312 (1806) ... — — V. N.S.W. Q. — B.fl.VII,738 M.fr.V,120.
B. laevigatum, Cavanilles, descripcion de las plantas 650 (1802) — — — — N.S.W. — — B.fl.VII,739 M.fr.V,120.
B. serrulatum, C. Richard in Annal. du Mus. d'Hist. nat. Par.
    I, 114 (1802) ... ... — — — N.S.W. Q. N.A. B.fl.VII,739 M.fr.V,120.
B. orientale, Linné, syst. nat. ed. decim. (1759) ... ... — — — — — Q. N.A. B.fl.VII,740

MONOGRAMMA, Commerçon in Schkuhr's Kryptog. Gew. 82, t. 87 (1809). (Monogramme, Vaginularia, Pleurogramme, Diclidopteris.)
M. Junghuhnii, Hooker, spec. filic. V, 123, t. 289 (1864) ... ... — — — — — Q. — B.fl.VII,740 M.fr.VII,119.

WOODWARDIA, Smith in mém. de l'Acad. de Turin. V, 411 (1791). (Doodia.)
W. aspera, Mettenius, fil. hort. bot. Lips. 65 (1856) ... ... — — — — — Q. — B.fl.VII,740 M.fr.V,130.
W. caudata, Cavanilles, descripcion de las pl. n. 653 (1802) ... — T. V. N.S.W. Q. N.A. B.fl.VII,742 M.fr.V,129;VIII,158.

ASPLENIUM, Linné, gen. plant. 322 (1737) from C. Bauhin (1651). (Scolopendrium, Allantodea, Diplazium, Callipteris, Anisogonium,
    Thamnopteris, Neottopteris, Darea, Caenopteris, Athyrium, Diplora, Asplenum.)
A. Nidus, Linné, spec. plant. 1079 (1753) ... ... — — V. N.S.W. Q. — B.fl.VII,744 M.fr.V,130.
A. simplicifrons, F. v. M., fragm. V, 74 (1865) ... ... — — — — — Q. — B.fl.VII,744 M.fr.V,74,130;VIII,158.
A. attenuatum, R. Brown, prodr. 150 (1810) ... ... — — — V. N.S.W. Q. — B.fl.VII,745 M.fr.V,130.
A. Trichomanes, Linné, spec. plant. 1080 (1753) ... ... — — T. V. N.S.W. — — B.fl.VII,745 M.fr.V,131.
A. flabellifolium, Cavanilles, descripcion de las pl. n. 636 (1802) W.A. S.A. T. V. N.S.W. Q. — B.fl.VII,745 M.fr.V,131.
A. paleaceum, R. Brown, prodr. 150 (1810) ... ... — — — V. N.S.W. Q. — B.fl.VII,746 M.fr.V,131;VIII,158.
A. falcatum, Lamarck, Encycl. méth. I, 303 (1783) ... ... — — — V. N.S.W. — — B.fl.VII,746 M.fr.V,131;VII,156.
A. Hookerianum, Colenso in Journ. of Inst. Sc. Tasm. II, 169 (1842) — — — V. N.S.W. — — B.fl.VII,747
A. furcatum, Thunberg, prodr. pl. Cap. 172 (1800) ... W.A. S.A. — V. N.S.W. Q. — B.fl.VII,748 M.fr.V,131.
A. laserpitiifolium, Lamarck, Encycl. méth. I, 310 (1783) ... — — — — N.S.W. Q. — B.fl.VII,748 M.fr.V,131.
A. marinum, Linné, spec. plant. 1081 (1733) ... ... — — T. V. N.S.W. — — B.fl.VII,747 M.fr.V,132.
A. bulbiferum, G. Forster, florul. ins. Austr. prodr. 433 (1786)... — S.A. T. V. N.S.W. Q. — B.fl.VII,749 M.fr.V,131.
A. pteridoides, Baker, syn. filic. sec. edit. 488 (1874) ... — — — — N.S.W. — — B.fl.VII,749
A. umbrosum, J. Smith in Hook. Lond. Journ. IV, 174 (1842)... — T. V. N.S.W. Q. — B.fl.VII,749 M.fr.V,132.
A. pallidum, Blume, fl. Javae, filic. 176 (1836) ... ... — — — — — Q. — M.fr.XI,131.
A. sylvaticum, Presl, reliq. Haenk. I, 42 (1830) ... ... — — — — — Q. — B.fl.VII,750 M.fr.V,133.
A. maximum, D. Don, prodr. filic. Nepal. 8 (1825) ... ... — — — — N.S.W. Q. — B.fl.VII,751
A. polypodioides, Mettenius, fil. hort. Lips. 78 (1856) ... ... — — — — — Q. — B.fl.VII,751
A. melanochlamys, Hooker, spec. filic. III, 299 (1860) ... ... — — — N.S.W. — — B.fl.VII,751 M.fr.V,132.
A. decussatum, Swartz, spec. fil. 76 & 260 (1806) ... ... — — — — — Q. — B.fl.VII,751

139

CYSTOPTERIS, Bernhardi in Schrader's neuem Journ. II, 26 (1806).
C. fragilis, Bernhardi in Schrad. Journ. I, 26 (1806) ... ... — T. — — — — B.fl.VII,752 M.fr.V,118.
ASPIDIUM, Swartz in Schrader's Journ. II, 429 (1000). (Nephrodium, Nephrolepis, Polystichum, Lastraea, Sagenia, Oleandra.)
A. cordifolium, Swartz, syn. filic. 45 (1806) ... ... — — — — N.S.W. Q. — B.fl.VII,754 M.fr.V,136.
A. exaltatum, Swartz, syn. filic. 45 (1806)... ... ... — — .. — — Q. N.A. B.fl.VII,754 M.fr.V,136
A. ramosum, Palisot, Fl. d'Oware et de Benin II, 53, t. 91 (1807) — — — N.S.W. Q. — B.fl.VII,754 M.fr.V,135.
A. unitum, Swartz, syn. filic. 47 (1806) ... ... W.A. — — — N.S.W. Q. N.A. B.fl.VII,755 M.fr.V,137;VIII,158.
A. pteroides, Swartz, syn. filic. 49 (1806) ... — — — — Q. — B.fl.VII,755
A. molle, Swartz, syn. filic. 49 (1806) ... ... — S.A. — V. N.S.W. Q. N.A. B.fl.VII,756 M.fr.V,135.
A. truncatum, Gaudichaud in Voy. Freyc. Bot. 332, t. 10 (1826) — — — N.S.W. Q. — B.fl.VII,756 M.fr.V,135.
A. confluens, Mettenius in Schlecht. Linnaea XXXVI, 125 (1863) — — — — Q. N.A. B.fl.VII,757 M.fr.V,133;VIII,158.
A. aculeatum, Swartz in Schrad. Journ. II, 37 (1800) ... — S.A. T. V. N.S.W. Q. — B.fl.VII,757 M.fr.V,134.
A. aristatum, Swartz, syn. filic. 53 & 253 (1806) ... — — — — N.S.W. Q. — B.fl.VII,757 M.fr.V,134.
A. Capense, Willdenow, spec. plant. V, 267 (1810) ... — T. V. N.S.W. — B.fl.VII,758 M.fr.V,134.
A. decompositum, Sprengel, syst. IV, 100 (1827)... ... — S.A. T. V. N.S.W. Q. — B.fl.VII,758 M.fr.V,136.
A. tenerum, Sprengel, syst. IV, 100 (1827) ... ... — — N.S.W. Q. — B.fl.VII,759
A. uliginosum, Kunze in Schlecht. Linnaea XX, 6 (1847) ... — — N.S.W. Q. — B.fl.VII,759 M.fr.V,133.
A. hispidum, Swartz, syn. filic. 56 (1806) ... ... — T. V. — — — B.fl.VII,760 M.fr.V,133.
POLYPODIUM, Tournefort, inst. 540, t. 316 (1700), from Theophrastos, Dioscorides & Plinius. (Niphobolus, Goniophlebium, Goniopteris, Phlebodium, Phegopteris, Phymatodes, Pleopeltis, Drynaria, Dictyopteris, Arthropteris, Xiphopteris, Meniscium partly; from Dodoens, l'Ecluse, Bauhin, Morison, Ray, Plumier and particularly Petiver as Polypodium.)
P. australe, Mettenius, filic. hort. bot. Lips. 36 (1856) ... ... — T. V. N.S.W. Q. — B.fl.VII,762 M.fr.V,127.
P. Hookeri, Brackenridge, Bot. Wilk. Exped. Filic. 4 (1854) ... — — — — N.S.W. Q. — B.fl.VII,763 M.fr.V,127.
P. contiguum, Brackenridge, Bot. Wilk. Exp. Fil. 6 t. (1854)... — — — — N.S.W. — B.fl.VII,763 M.fr.V,127.
P. grammitidis, R. Brown, prodr. 147 (1810) ... ... — T. V. N.S.W. — B.fl.VII,764 M.fr.V,127.
P. tenellum, G. Forster, florul. ins. Austr. prodr. 440 (1786) ... — — — N.S.W. Q. — B.fl.VII,764 M.fr.V,126;VII,121.
P. proliferum, Roxburgh in Wall. num. list 11 (1828) ... — — — N.S.W. Q. N.A. B.fl.VII,765 M.fr.V,128.
P. urophyllum, Wallich, numerical list 11 (1828) ... — — — — Q. — B.fl.VII,765 M.fr.IV,163;V,137; VIII,
P. Hillii, Baker in Hook., syn. filic. sec. edit. 505 (1874) — — — — Q. — B.fl.VII,766 [121.
P. poecilophlebium, Hooker, spec. filic. V, 14 (1863) — — — — Q. — B.fl.VII,766 M.fr.V,127.
P. serpens, G. Forster, florul. insul. austr. prodr. 81 (1786) — — — V. N.S.W. Q. — B.fl.VII,767 M.fr.V,129.
P. confluens, R. Brown, prodr. 146 (1810)... ... — — — N.S.W. Q. — B.fl.VII,767 M.fr.V,129.
P. acrostichoides, G. Forster, florul. insul. austr. prodr. 81 (1786) — — — — — Q. — B.fl.VII,768 M.fr.V,129.
P. attenuatum, R. Brown, prodr. 146 (1810) ... ... — — — N.S.W. Q. — B.fl.VII,768 M.fr.V,128;V,121.
P. simplicissimum, F. v. M. in Hook. & Bak. syn. fil. sec. edit. 513 (1874) ... ... — — — — Q. — B.fl.VII,768 M.fr.VII,120,136.
P. nigrescens, Blume, Fl. Javae, filic. 161, t. 70 (1836) ... — — — — Q. — B.fl.VII,769
P. phymatodes, Linné, mant. pl. alt. 300 (1771) ... ... — — — — Q. N.A. B.fl.VII,769
P. pustulatum, G. Forster, florul. ins. Austr. pr. 81 (1786) — — T. V. N.S.W. Q. — B.fl.VII,769 M.fr.V,127.
P. scandens, G. Forster, florul. ins. Austr. pr. 81 (1786) — — — V. N.S.W. Q. — B.fl.VII,770 M.fr.V,128.
P. verrucosum, Wallich, numerical list 11 (1828)... ... — — — — Q. — B.fl.VII,770
P. subauriculatum, Blume, Fl. Javae, filic. 177, t. 83 (1836) — — — — Q. — B.fl.VII,770 M.fr.V,128.
P. rigidulum, Swartz, synops. filic. 38 & 320 (1806) — — — N.S.W. Q. — B.fl.VII,771
P. quercifolium, Linné, spec. plant. 1087 (1753) — — — — Q. N.A. B.fl.VII,772 M.fr.V,127.
P. irioides, Poiret, Encycl. méth. V, 513 (1804) ... ... — — — — Q. — B.fl.VII,771 M.fr.V,127.
P. punctatum, Thunberg, fl. Japonic. 337 (1784)... ... — S.A. T. V. N.S.W. Q. — B.fl.VII,764 M.fr.V,129.
HYPOLEPIS, Bernhardi in Schrader, neu. Journ. I, 34 (1806).
H. tenuifolia, Bernhardi in Schrad. Journ. II, 34 (1806) ... — — — N.S.W. Q. — B.fl.VII,772 M.fr.V,129.
GRAMMITIS, Swartz in Schrader, Journ. II, 3 & 17 (1800). (Gymnogramme, Selligues).
G. Reynoldsii, F. v. M. in Benth. Fl. Austr. VII, 775 (1878) ... — S.A. — — — — B.fl.VII,775 M.fr.VII,175.
G. Muelleri, Hooker in F. v. M. fragm. V, 139 (1866) ... ... — — — — Q. — B.fl.VII,775 M.fr.V,138; VI,124; VII,
G. rutifolia, R. Brown, prodr. 146 (1810) ... ... W.A. S.A. T. V. N.S.W. Q. — B.fl.VII,776 M.fr.V,137. [121.
G. leptophylla, Swartz, syn. fil. 23 et 218, t. 1 (1806) ... W.A. S.A. T. V. N.S.W. — B.fl.VII,776 M.fr.V,137.
G. pinnata, F. v. M., fragm. VI, 124 (1868) ... ... — — — — Q. — B.fl.VII,777 M.fr.VI,124.
G. ampla, F. v. M., fragm. V, 188 (1866) ... ... — — — — Q. — B.fl.VII,777 M.fr.V,188.
ANTROPHYUM, Kaulfuss, enum. filic. 197 (1824).
A. reticulatum, Kaulfuss, enum. filic. 107 (1824)... ... — — — — Q. — B.fl.VII,777 M.fr.V,138.
ACROSTICHUM, Linné, gen. plant. 322 (1737), indicative. (Elaphoglossum, Stenochlaena, Lomariopsis, Hymenolepis, Gymnopteris, Chrysodium.)
A. conforme, Swartz, syn. filic. 10 & 192, t. 1 (1806) — — — — Q. — B.fl.VII,778 M.fr.V,136.
A. scandens, J. Smith in Hook. spec. fil. V, 249 (1864) — — — — Q. N.A. B.fl.VII,778 M.fr.VI,124.
A. sorbifolium, Linné, spec. plant. 1069 (1753) — — — — Q. — B.fl.VII,779 M.fr.VII,110.
A. neglectum, Bailey in proc. Linn. Soc. of N.S.W., V, 32 (1880) — — — — — M.fr.XI,131.
A. repandum, Blume, Fl. Javae 39, t. 14 & 15 (1836) ... — — — — Q. — B.fl.VII,779 M.fr.V,138.
A. aureum, Linné, spec. plant. 1069 (1753) — — — N.S.W. Q. N.A. B.fl.VII,779 M.fr.V,138;VII,120.
A. spicatum, Linné, fil. suppl. pl. 444 (1781) — — — N.S.W. Q. — B.fl.VII,780 M.fr.V,138.
A. pteroides, R. Brown, prodr. 145 (1810)... ... — — — — Q. N.A. B.fl.VII,780 M.fr.X,76.
PLATYCERIUM, Desvaux in Mém. de la Soc. Linn. de Paris VI, 213 (1827).
P. alcicorne, Desvaux in Mém. Soc. Linn. Par. VI, 213 (1827)... — — — N.S.W. Q. — B.fl.VII,780 M.fr.V,139.
P. grande, J. Smith in Hooker, Lond. Journ. III, 402 (1841) .. — — — N.S.W. Q. — B.fl.VII,781 M.fr.V,139.

ADDITIONAL SPECIES.

Pachynema sphenandrum, F. v. M. & Tate in Transact. R. S. of
  S. Austr. V, 79 (1882) ... ... — — — — — — N.A. —
Polyalthia Holtzeana, F. v. M. in Wing's S. Sc. Record II, 230
  (1882) ... ... — — — — — — N.A. —
Philotheca Hassellii, F. v. M., fragm. XII, inedit. ... ... W.A. — — — — — — — —
Tribulus astrocarpus, F. v. M., fragm. XII, 4 (1882) ... ... W.A. — — — — — — — M.fr.XII,
Euphorbia obliqua, Bauer in Endl. prodr. fl. Norfolk. 85 (1833) — — — — N.S.W. — — —
Phyllanthus Tatei, F. v. M. in Wing's S. Sc. Record II, 55 (1882) — S.A. — — N.S.W. — — —
Ficus Pinkiana, F. v. M. in Wing's S. Sc. Record II, 273 (1882) — — — — — Q. — —
Peperomia Baueriana, Miquel, syst. piperac. 120 (1843) ... — — — — N.S.W. — — —
Nepenthes Bernaysii, Bailey in proceed. I. S. of N.S.W. 185 (1880) — — — — — Q. — —
Ptilotus Polakii, F. v. M., fragm. XII, 274 (1882) W.A. — — — — — — — —
Atriplex Bunburyanum, F. v. M. in Wing's S. Sc. Record II,
  274 (1882) ... ... W.A. — — — — — — — M.fr.XII,14
Kochia prosthecochaeta, F. v. M., fragm. XII, 14 (1882) ... W.A. — — — — — — — M.fr.XII,14
Kochia melanocoma, F. v. M., fragm. XII, 14 (1882) ... ... W.A. — — — — — — — M.fr.XII,12
Bassia astracantha, F. v. M., fragm. XII, 12 (1882) ... .. W.A. — — — — — — — M.fr.XII,12
Bassia tridens, F. v. M., fragm. XII, 12 (1882) ... ... W.A. — — — — — — — M.fr.XII,12
Bassia Forrestiana, F. v. M., fragm. XII, 12 (1882) ... ... W.A. — — — — — — — M.fr.XII,12
Bossiaea Scortechinii, F. v. M. (inedit). ... ... — — — — N.S.W. Q. — —
Bossiaea Webbii, F. v. M. in Melb. Chemist, Dec. (1882) W.A. — — — — — — — —
Isotropis Forrestii, F. v. M. in Wing's S. Sc. Record II, 252 (1882) W.A. — — — — — — — —
Daviesia arborea, F. v. M. & Scortechini in proc. L. S. of N.S.W.
  VII, 221 (1882) ... ... — — — — N.S.W. Q. — —
Crotalaria striata, De Candolle, prodr. II, 131 (1825) ... — — — — — Q. — —
Indigofera tinctoria, Linné, spec. plant. 751 (1753) ... — — — — — Q. — —
Swainsona Oliverii, F. v. M. in Wing's S. Sc. Rec. II, 152 (1882) W.A. S.A. — — — — — —
Callistemon pithyoides, Miquel in Nederl. Kruidk. Arch. IV, 142
  (1859) ... ... — — — V. N.S.W. Q. — —
Melaleuca cylindracea, R. Brown in Benth. Fl. Austral. III, 146
  (1866) ... ... — — S.A. — — — — —
Eucalyptus Foelschoana, F. v. M. in Melb. Chemist, Nov. (1882) — — — — — — N.A. —
Cryptandra Waylii; Trymalium, F. v. M. & Tate in Transact.
  R. S. of S. Austr. V, 80 (1882) ... ... — — S.A. — — — — —
Adenanthos Forrestii, F. v. M. in Wing's S. Sc. Record II, 230
  (1882) ... ... W.A. — — — — — — —
Grevillea deflexa, F. v. M. inedit. .. .. W.A. — — — — — — —
Epaltes Tatei, F. v. M. inedit. ... ... — S.A. — — — — — —
PODOSPERMA, Labillardière, Nov. Holl. plant. specim. II, 35, t. 177 (1806). (Podotheca.)
Podosperma Polakii, F. v. M., fragm. XII, 21 (1882) ... ... W.A. — — — — — — — M.fr.XII,21
Helipterum sterilescens, F. v. M. in Wing's S. Sc. Record II,
  274 (1882) ... ... W.A. — — — — — — — —
Helipterum Forrestii, F. v. M. in Wing's S. Sc. Record II, 273
  (1882) ... ... W.A. — — — — — — — —
Velleya macroplectra, F. v. M., fragm. XII, 22 (1882) ... ... W.A. — — — — — — — M.fr.XII,22
Thunbergia fragrans, J. Koenig in Roxb. pl. of Coromand. I, 44,
  t. 67 (1795) ... ... — — — — — — N.A. —
ISANDRA, F. v. M., fragm. phyt. Austr. XII (inedit.) Near Anthocercis.
Isandra Bancroftii, F. v. M., fr. XII (inedit.) ... ... — — — — — — — —
DICLADANTHERA, F. v. M., fragm. phytogr. Austral. XII, 23 (1882).
Dicladanthera Forrestii, F. v. M., fragm. XII, 23 (1882) ... W.A. — — — — — — — M.fr.XII,23
Prostanthera Campbelli, F. v. M. in Wing's S. Sc. Record II,
  252 (1882) ... ... W.A. — — — — — — — —
Chloanthes lepidota, F. v. M. inedit. ... ... W.A. — — — — — — — —
Myoporum obscurum, Endlicher, prodr. fl. Norfolk. 54 (1833) ... — — — — N.S.W. — — —
Eremophila Pantoni, F. v. M. in Wing's S. Sc. Record II, 251 (1882) W.A. — — — — — — — —
ERIA, Lindley in Edwards, Bot. Regist. t. 904 (1825).
Eria Fitzalani, F. v. M. in Wing's S. Sc. Record II, 252 (1882) — — — — — Q. — —
LUISIA, Gaudichaud in Freycin. voy. Bot. 426, t. 37 (1826).
Luisia teretifolia, Gaudichaud in Freyc. voy. 426, t. 37 (1826) — — — — — Q. — —
Prasophyllum Tepperi, F. v. M. in Tepp. pl. about Ardross., Oct.
  (1880) ... ... — — S.A. — — — — —
Smilax purpurata, G. Forster, florul. Ins. Austr. prodr. 70 (1786) — — — — N.S.W. — — —
Pandanus Moorei, F. v. M. inedit. ... ... — — — — N.S.W. — — —
Carex haematostoma, Nees in Wight, contrib. to the Bot. of
  Ind. 125 (1834) ... ... — — T. — — — — —

ADDITIONAL ANNOTATIONS.

Myosurus from l'Obel (1576).
Myosurus minimus ... ... Q.
Ranunculus from Plinius.
Ranunculus aquatilis ... ... N.S.W.
Hibbertia hirsuta... ... ... S.A.
Doryphora Sassafras ... ... Q.
Endiandra Sieberi ... ... Q.
Tristichocalyx pubescens .. N.A.
Papaver from Plinius.
Cleome viscosa ... ... W.A.
Capparis spinosa .. ... S.A.
Capparis laeiantha ... ... W.A.
Barbarea from l'Obel (1576).
Barbarea vulgaris ... ... S.A.
Capsella pilosula ... ... W.A.
Senebiera = Carara, Medicus, Pflanzen-Gattung. 84 (1792), from Coalpini (1583). (Coronopus partly).
Lepidium strongylophyllum ... ... W.A.
Lepidium leptopetalum ... ... W.A.
Lepidium rotundum ... ... S.A.
Cakile from l'Obel (1576).
Hybanthus enneaspermus ... ... W.A.
Hymenanthera Banksii ... ... S.A.
Pittosporum revolutum ... ... M.fr.XII,4.
Pittosporum ferrugineum ... ... M.fr.XII,4.
Pittosporum melanospermum ... ... Q.
Hymenosporum flavum ... ... M.fr.XII,4.
Marianthus microphyllus ... ... M.fr.XII,2.
Citriobatus pauciflorus ... ... N.S.W.
Billardiera varifolia ... ... M.fr.XII,2.
Cheiranthera linearis ... ... M.fr.XII,4.
Cheiranthera filifolia ... ... M.fr.XII,4.
Bergia perennis ... ... S.A.
Polygala from Dioscorides & Plinius.
Polygala persicarifolia ... ... Q.
Polygala Chinensis ... ... S.A.
Comesperma sylvestre ... ... S.A.,N.S.W.
Comesperma viscidulum... ... ... S.A.
Tribulus Hystrix ... ... W.A.
Hedraianthera porphyropetala ... ... N.S.W.
Tribulus macrocarpus ... ... W.A.
Tribulus platypterus ... ... W.A.
Sida virgata ... ... W.A.
Sida petrophila ... ... W.A.
Sida rhombifolia ... ... S.A.
Sida physocalyx ... ... W.A.
Sida lepida ... ... W.A.,S.A.
Abutilon tubulosum ... ... N.A.
Abutilon cryptopetalum ... ... M.fr.IX,131.
Abutilon geranioides ... ... W.A.
Abutilon Avicennae, Gerard, gen. hist. of pl. 790 (1597).
Urena lobata ... ... N.A.
Hibiscus panduriformis ... ... W.A.,Q.
Hibiscus Normani ... ... M.fr.V,44.
Hibiscus Sturtii ... ... W.A.
Gossypium australe ... ... W.A. M.fr.V,44.
Gossypium flaviflorum ... ... M.fr.V,44.
Waltheria Indica ... ... W.A.
Commersonia Fraseri ... ... Q.
Euphorbia erythrantha ... ... W.A.
Euphorbia pilulifera ... ... N.A.
Monotaxis luteiflora ... ... N.A.
Poranthera microphylla .. ... M.fr.XII,11.
Pseudanthus micranthus... ... M.fr.XII,11.
Beyeria viscosa ... ... M.fr.XII,10.
Ricinocarpus pinifolius ... ... Q. M.fr.XII,10.
Bertya rosmarinifolia ... ... M.fr.XII,11.
Bertya rotundifolia ... ... M.fr.XII,11.
Micranthenm hexandrum ... ... M.fr.XII,11.
Dissilaria tricornis ... ... M.fr.XII,11.
Phyllanthus rhytidospermus ... ... S.A. M.fr.XII,11.
Phyllanthus rigens ... ... M.fr.XII,11.

Phyllanthus Gastroemii ... ... M.fr.XII,11.
Phyllanthus australis ... ... M.fr.XII,11.
Securinega Leucopyrus ... ... S.A., N.A.
Adriana tomentosa ... ... M.fr.XII,11.
Adriana quadripartita ... ... M.fr.XII,11.
Tragia Novae Hollandiae ... ... M.fr.XII,11.
Fontainea Pancheri ... ... M.fr.XII,11.
Omalanthus populifolius... ... ... M.fr.XII,11.
Ficus platypoda ... ... W.A.
Ficus scabra ... ... N.A.
Ficus stenocarpa = Ficus subglabra, F. v. M., fragm. IX, 152 (1875) ... ... N.S.W.,Q.
Pouzolzia quinquenervis ... ... N.A.
Casuarina suberosa ... ... S.A.
Casuarina inophloia ... ... N.S.W.
Peperomia Urvilleana ... ... N.S.W.
Leea aculeata = L. Brunoniana, Clarke in Britten's Journ. of Bot. X, 105 (1881).
Castanospora Alphandi ... ... N.S.W.
Atalaya variifolia ... ... Q.
Dodonaea lanceolata ... ... S.A.
Dodonaea truncatiales = D. calycina, Cunningham in A. Gray, Bot. of Wilk. Expl. Exped. I, 262 (1854).
Dodonaea platyptera ... ... W.A.
Dodonaea microxyga ... ... S.A.
Buesseraceae = Burseraceae.
Celastrus bilocularis ... ... N.S.W.
Stackhousia muricata ... ... S.A.
Plumbago Zeilanica ... ... S.A.
Aegialitis annulata ... ... N.A.
Portulaca from W, Turner (1538).
Sagina apetala ... ... N.S.W.
Saponaria from Bock (1539).
Ptilotus incanus ... ... S.A.
Ptilotus psilotrichoides ... ... N.A.
Ptilotus helipteroides ... ... W.A.
Ptilotus rotundifolius ... ... W.A.
Ptilotus latifolius... ... ... W.A.
Rhagodia Billardieri ... ... N.A. M.fr.XII,16.
Rhagodia parabolica ... ... W.A.,V. M.fr.XII,15.
Rhagodia crassifolia ... ... Q M.fr.XII,15.
Rhagodia Preissii ... ... S.A. M.fr.XII,15.
Rhagodia Gaudichaudii ... ... M.fr.XII,15.
Rhagodia spinescens ... ... M.fr.XII,15.
Rhagodia nutans ... ... W.A.,N.A. M.fr.XII,15.
Rhagodia linifolia ... ... M.fr.XII,15.
Chenopodium atriraceum ... ... M.fr.XII,16.
Chenopodium microphyllum ... ... M.fr.XII,16.
Chenopodium rhadinostachyum (1892).
Chenopodium cristatum ... ... M.fr.XII,17.
Chenopodium atriplicinum ... ... M.fr.XII,17.
Chenopodium carinatum ... ... N.A. M.fr.XII,17.
Dysphania plantaginella... ... ... W.A.
Dysphania litoralis ... ... M.fr.XII,17.
Dysphania myriocephala ... ... M.fr.XII,17.
Atriplex from W. Turner (1538).
Atriplex prostratum includes A. microcarpum and A. Pumilio ... ... M.fr.XII,17.
Atriplex fissivalve ... ... M.fr.XII,16.
Atriplex paludosum ... ... M.fr.XII,16.
Atriplex halimoides ... ... W.A.,N.A. M.fr.XII,17.
Atriplex vesicarium ... ... M.fr.XII,17.
Atriplex holocarpum ... ... M.fr.XII,17.
Atriplex nummularium ... ... M.fr.XII,17.
Atriplex leptocarpum ... ... M.fr.XII,17.
Atriplex crystallinum ... ... M.fr.XII,17.
Kochia sedifolia ... ... W.A. M.fr.XII,15.
Kochia lobiflora ... ... M.fr.XII,15.
Kochia lanosa ... ... M.fr.XII,15.
Kochia triptera ... ... M.fr.XII,15.
Kochia oppositifolia ... ... M.fr.XII,15.
Kochia brevifolia ... ... M.fr.XII,15.

Erechtites mixa ... ... S.A.,V.
Erechtites prenanthoides ... ... S.A.
Candollea pilosa, Labillardière in Annal. du Mus. VI., 453, t. 63 (1805).
Candollea glauca, Labillardière in Annal. du Mus. VI., 454, t. 64 (1805).
Wahlenbergia (Strzeleckia).
Lobelia Benthamii ... ... S.A.
Isotoma petraea ... ... N.A.
Dampiera Brownii ... ... Q.
Dampiera diversifolia ... ... M.fr.I,120.
Dampiera incana ... ... N.A.
Leschenaultia hirsuta ... ... M.fr.XII,23.
Leschenaultia longiloba ... ... M.fr.XII,23.
Catosperma Muelleri ... ... S.A.
Scaevola depauperata ... ... N.A.
Scaevola globuliflora = S. globulifera.
Goodenia Chambersii ... ... N.A.
Goodenia Hassallii = G. Hassellii.
Mitrasacme pilosa ... ... S.A.
Logania floribunda ... ... Q.
Plantago from W. Turner (1538).
Ligustrum from W. Turner (1538).
Notelaea linearis ... ... Q.
Cynanchum floribundum ... ... W.A.
Gymnanthera nitida ... ... W.A,
Tylophora erecta ... ... N.A.
Ipomoea Turpethum ... ... N.A.
Ipomoea Muelleri... ... ... S.A.
Convolvulus from W. Turner (1538).
Polymeria angusta ... ... S.A.
Cuscuta from Bock (1539).
Cuscuta australis ... ... W.A.,S.A.
Solanum viride, Solander in G. Forster, prodr. 89 (1786).
Datura Leichhardtii ... ... W.A.
Lindernia, Allioni in misc. Taurin. III, 178 (1766), from Gagnebin (Pyxidaria).
Peplidium Muelleri ... ... W.A.,S.A.
Buchnera parviflora ... ... W.A.
Coldenia procumbens ... ... S.A.
Pollichia latisepalea ... ... Q.,N.A.
Clerodendrun lanceolatum ... ... W.A.
Spartothamnus teucriiflorus ... ... W.A.
Eremophila oppositifolia ... ... S.A.
Eremophila Mitchelli ... ... W.A.
Eremophila exilifolia ... ... W.A.
Eremophila Freelingii ... ... N.A.
Eremophila Macdonnelli... ... ... N.A.
Eremophila Goodwinii ... ... S.A.
Eremophila platycalyx ... ... W.A.
Eremophila alternifolia ... ... W.A.
Eremophila scoparia ... ... W.A.
Prasophyllum Fimbria - P. Reichenbachii, F. v. M. (inedit.)
Chiloglottis Gunnii ... ... N.S.W.
Astelia (Funkia 1805).
Typhonium Brownii ... ... N.A.

Wolffia Michelii ... ... S.A.
Cyperus Eragrostis ... ... M.fr.XII,25.
Cyperus pulchellus ... ... M.fr.XII,25.
Cyperus gracilis ... ... M.fr.XII,25.
Cyperus tenellus ... ... M.fr.XII,25.
Cyperus squarrosus ... ... M.fr.XII,25.
Cyperus fulvus ... ... W.A.
Cyperus rotundus ... ... M.fr.XII,25.
Cyperus polystachys ... ... M.fr.XII,25.
Cyperus Iris ... ... M.fr.XII,25.
Cyperus decompositus ... ... M.fr.XII,25.
Scirpus grossus ... ... M.fr.XII,24.
Scirpus supinus ... ... M.fr.XII,25.
Scirpus litoralis ... ... M.fr.XII,25.
Cyathochaete clandestina ... ... M.fr.XII,25.
Cyathochaete diandra ... ... M.fr.XII,25.
Schoenus aphyllus ... ... M.fr.XII,26.
Schoenus ericetorum ... ... S.A.
Schoenus sphaerocephalus ... ... M.fr.XII,26.
Lepidospora tenuissima ... ... M.fr.XII,26.
Lepidosperma gladiatum ... ... M.fr.XII,26.
Lepidosperma filiforme ... ... M.fr.XII,25.
Cladium junceum... ... ... M.fr.XII,26.
Cladium schoenoides ... ... M.fr.XII,26.
Cladium trifidum ... ... M.fr.XII,26.
Cladium filum ... ... M.fr.XII,26.
Cladium melanospermum ... ... M.fr.XII,25.
Cladium Mariscus ... ... M.fr.XII,25.
Cladium teretifolium ... ... M.fr.XII,26.
Cladium psittacorum ... ... S.A.
Cladium Radula ... ... M.fr.XII,26.
Caustis pentandra ... ... M.fr.XII,26.
Arthrostylis aphylla ... ... Q. M.fr.XII,26.
Carex Brownii ... ... M.fr.XII,26.
Carex tereticaulis... ... ... M.fr.XII,26.
Carex declinata ... ... M.fr.XII,26.
Carex breviculmis ... ... M.fr.XII,26.
Carex inversa ... ... M.fr.XII,26.
Carex acuta ... ... M.fr.XII,26.
Carex pumila ... ... W.A.
Tragus racemosus... ... ... W.A.
Andropogon Gryllus ... ... W.A.
Andropogon procerus ... ... W.A.
Anthistiria membranacea ... ... Q.
Chionachne cyathopoda ... ... N.A.
Sporobolus Lindleyi ... ...
Eriachne obtusa, W.A., S.A.
Eleusine Aegyptia = E. cruciata, Lamarck, tabl. encycl. et meth. I, 203 t. 48 (1791) W.A.
Setaria glauca ... ... W.A.
Setaria macrostachya ... ... W.A.
Eragrostis pilosa ... ... W.A.
Eragrostis tenella ... ... Q.
Trirhaphis pungens ... ... W.A.
Arundo Phragmites ... ... W.A.
Botrychium ternatum ... ... S.A.

SEQUENCE OF ORDERS ACCORDING TO PREDOMINANCE,
WITH INDICATIONS OF THEIR NUMBERS OF SPECIES.

Total number of Vasculares = 8646.   Dicotyledoneae = 6897.   Monocotyledoneae = 1522.   Acotyledoneae = 227.

Leguminosae	...1058	Tiliaceae ...	52	Rosaceae ...	17	Ericaceae ...	5
Myrtaceae ...	651	Loganiaceae	51	Anonaceae ...	16	Nymphaeaceae	4
Proteaceae...	586	Apocyneae...	48	Salicarieae...	16	Magnoliaceae	4
Compositae	529	Droseraceae	46	Junceae ...	16	Samydaceae	4
Cyperaceae	372	Santalaceae	43	Monimieae...	15	Elatineae ...	4
Gramineae...	346	Pittosporeae	38	Celastrinae	15	Lineae	4
Epacrideae...	273	Laurineae ...	37	Olacinae ...	15	Cupuliferae	4
Orchideae ...	255	Meliaceae ...	36	Ebenaceae ...	15	Plumbagineae	4
Euphorbiaceae	224	Saxifrageae	35	Menispermeae	14	Plantagineae	4
Goodeniaceae	212	Campanulaceae	34	Stackhousieae	13	Dioscorideae	4
Filices	209	Polygaleae...	32	Cycadeae ...	13	Gesneriaceae	3
Rutaceae ...	185	Portulaceae	32	Violaceae ...	12	Pedalinae	3
Liliaceae ...	161	Acanthaceae	30	Myrsineaceae	12	Philydreae...	3
Rubiaceae	124	Coniferae	29	Phytolaccaceae	11	Nepenthaceae	2
Labiatae	124	Fluviales ...	29	Scitamineae	11	Guttiferae ...	2
Sterculiaceae	123	Combretaceae	27	Piperaceae...	10	Malpighiaceae	2
Salsolaceae...	112	Ficoideae ...	27	Aroideae ...	10	Burseraceae	2
Malvaceae...	105	Palmae	26	Pandaneae...	10	Connaraceae	2
Umbelliferae	103	Polygonaceae	25	Anacardiaceae	9	Caprifoliaceae	2
Sapindaceae	100	Loranthaceae	25	Hydrocharideae	9	Aquifoliaceae	2
Caudolleaceae	95	Irideae ...	25	Xyrideae ...	9	Styraceae ...	2
Dilleniaceae	95	Caryophylleae	24	Rhizospermeae	9	Hydrophylleae	2
Amarantaceae	94	Capparideae	23	Flacourtieae	8	Burmanniaceae	2
Restiaceae...	93	Casuarineae	23	Simarubeae	7	Typhaceae...	2
Rhamnaceae	89	Cucurbitaceae	23	Geraniaceae	7	Myristicaceae	1
Amaryllideae	86	Gentianeae...	23	Frankeniaceae	7	Papaveraceae	1
Solanaceae...	80	Lentibularinae	23	Bignoniaceae	7	Hypericinae	1
Scrophularinae	78	Araliaceae ...	21	Crassulaceae	6	Ochnaceae ...	1
Myoporinae	78	Zygophylleae	20	Rhizophoreae	6	Podostemoneae	1
Verbenaceae	77	Jasmineae ...	20	Melastomaceae	6	Balenophoreae	1
Thymeleae...	74	Sapotaceae...	19	Primulaceae	6	Hamamelidae	1
Convolvulaceae	67	Commelineae	19	Lemnaceae...	6	Elaeagneae	1
Urticaceae...	63	Vitifereae ...	18	Alismaceae	6	Cornaceae ...	1
Haloragcae	60	Ericauleae	18	Aristolochieae	5	Orobancheae	1
Asclepiadeae	60	Lycopodineae	18	Nyctagineae	5	Taccaceae ...	1
Cruciferae ...	54	Ranunculaceae	17	Onagreae ...	5	Roxburghiaceae	1
Asperifoliae	53	Tremandreae	17	Passifloreae	5	Pontederiaceae	1

Total number of Genera of Vasculares 1355.

INDEX OF ORDERS AND GENERA.

M'Carron, Bird and Co., Printers, 37 Flinders Lane West, Melbourne.

www.ingramcontent.com/pod-product-compliance
Lightning Source LLC
Chambersburg PA
CBHW021811190326
41518CB00007B/547